# METHODS IN MOLECULAR BIOLOGY

*Series Editor*
**John M. Walker**
**School of Life and Medical Sciences**
**University of Hertfordshire**
**Hatfield, Hertfordshire, UK**

For further volumes:
http://www.springer.com/series/7651

For over 35 years, biological scientists have come to rely on the research protocols and methodologies in the critically acclaimed *Methods in Molecular Biology* series. The series was the first to introduce the step-by-step protocols approach that has become the standard in all biomedical protocol publishing. Each protocol is provided in readily-reproducible step-by step fashion, opening with an introductory overview, a list of the materials and reagents needed to complete the experiment, and followed by a detailed procedure that is supported with a helpful notes section offering tips and tricks of the trade as well as troubleshooting advice. These hallmark features were introduced by series editor Dr. John Walker and constitute the key ingredient in each and every volume of the *Methods in Molecular Biology* series. Tested and trusted, comprehensive and reliable, all protocols from the series are indexed in PubMed.

# Cell Viability Assays

## Methods and Protocols

### Second Edition

Edited by

**Oliver Friedrich and Daniel F. Gilbert**

*Institute of Medical Biotechnology (MBT), Department of Chemical and Biological Engineering (CBI), Friedrich-Alexander-University Erlangen-Nüremberg, Erlangen, Germany*

Humana Press

*Editors*
Oliver Friedrich
Institute of Medical Biotechnology (MBT)
Department of Chemical and Biological
Engineering (CBI)
Friedrich-Alexander-University
Erlangen-Nüremberg
Erlangen, Germany

Daniel F. Gilbert
Institute of Medical Biotechnology (MBT)
Department of Chemical and Biological
Engineering (CBI)
Friedrich-Alexander-University
Erlangen-Nüremberg
Erlangen, Germany

ISSN 1064-3745 ISSN 1940-6029 (electronic)
Methods in Molecular Biology
ISBN 978-1-0716-3051-8 ISBN 978-1-0716-3052-5 (eBook)
https://doi.org/10.1007/978-1-0716-3052-5



This Humana imprint is published by the registered company Springer Science+Business Media, LLC, part of Springer Nature.
The registered company address is: 1 New York Plaza, New York, NY 10004, U.S.A.

# Obituary

I was deeply saddened to learn of the untimely and unexpected death of my co-editor and long-time colleague, Dr Daniel Gilbert. He died at home in his sleep on the 25th of March, 2023, aged only 48 - much too young. I knew Daniel for 15 years. We were not only colleagues, but also friends and shared a love of Australia, where we first met in 2008. Daniel was a dedicated and passionate scientist as well as a smart inventor. He passed away during the final stages of this book, his last scientific work and our last joint-project. Daniel will be sorely missed, but his memory will live on through the pages of this book. Rest in peace, Daniel.

Oliver Friedrich, on 27th March 2023

# Preface

Assessing cell viability as a measure for cell fitness under conditions of physiological and pathophysiological stress as well as challenging conditions to cellular and tissue homeostasis has become a universal bioanalytical task not only in life sciences and biomedical research but more and more also in areas of biotechnology, environmental and clinical toxicology, biomedical engineering, tissue engineering, and oncology, to name a few. As such, standardized protocols and routines to assess various aspects of cellular homeostasis, viability, and function are needed, their applicability toward specific tissues and generalization to other cell entities and their limitations must be clarified, and potential combinations of different readouts need to be discussed. Originally developed for classical in vitro 2D culture conditions of cells interacting with stiff substrates as monolayers, recent developments in tissue engineering or whole tissue applications in situ are calling out for reliable methods to be applied to the next dimension for cells interacting with extracellular matrices in 3D. As such, one of the major challenges in the coming years is to validate cell viability assays regarding their upscaling from 2D to 3D and where limitations thereof have to be acknowledged.

Our first edition of *Cell Viability Assays* for the MiMB series in 2017, with its described variety of assays, proved to be quite a success regarding its impact in the community. When the publisher approached us to consider an updated second edition in 2021, the first edition had already been accessed over 100,000 times.

For this updated edition, we therefore aimed to not only update and include new assays but also to account for the ongoing 2D-3D development and planned for topics and assays that targeted cell viability, mobility, and functionality of tissues and organs, natural or bioartificial, in 3D. We approached renowned experts in the fields of 2D and 3D cell viability assays and were able to secure 30 contributions spanning a wide range of viability and functionality assays, from impedance spectroscopy to chemiluminescence, fluorescence, and label-free optical detection methodologies, providing corresponding "Tips & Tricks" regarding their applications and analyses strategies, a hallmark of the MiMB series. Apart from "classical" cell viability aspects represented by "alive-dead" dichotomy, we also were able to include several chapters that cover various aspects of cell function, e.g., electrical excitability, contraction, and mobility, as well as cell adhesion or cell cycle analysis, among others.

The first two-thirds of the chapters are grouped with a focus on 2D aspects of cell viability assays while the remaining third is dedicated to the third dimension.

We thank all the authors who contributed to making this exciting new second edition of the volume possible, and we are confident that it will provide a valuable resource to the growing community of bioinspired life sciences, biomedical sciences, and biotechnology in providing more standardized protocols to probe cells in various environments regarding their "wellbeing."

*Institute of Medical Biotechnology (MBT), Department of Chemical and Biological Engineering (CBI)*
*Friedrich-Alexander University*
*Erlangen-Nürnberg, Erlangen, Germany*

*Oliver Friedrich*
*Daniel F. Gilbert*

# Preface

# Contents

# Contributors


DAVID J. ADAMS • *Illawarra Health and Medical Research Institute (IHMRI), University of Wollongong, Wollongong, NSW, Australia*

JEFFREY AGYAPONG • *Institute of Parasitology, McGill University, Montreal, QC, Canada*

CHRISTOPH ALEXIOU • *Department of Otorhinolaryngology, Head and Neck Surgery, Section of Experimental Oncology and Nanomedicine (SEON), Else Kröner-Fresenius-Stiftung Professorship, Universitätsklinikum Erlangen, Erlangen, Germany*

GAMZE ATES • *Laboratory of Neuro-Aging & Viro-Immunotherapy, Center for Neurosciences, Faculty of Medicine and Pharmacy, Vrije Universiteit Brussel, Brussels, Belgium*

SANTANU KUMAR BASU • *Institute of Fluid Mechanics, Friedrich-Alexander-Universität Erlangen-Nürnberg (FAU), Erlangen, Germany*

JULIAN BAUER • *Institute of Medical Biotechnology, Department of Chemical and Biological Engineering, Friedrich-Alexander University Erlangen-Nürnberg, Erlangen, Germany*

BRAD BAZOW • *Department of Electrical Engineering and Computer Science, The Catholic University of America, Washington, DC, USA*

ARTHUR BEYDER • *Department of Physiology and Biomedical Engineering, Enteric NeuroScience Program (ENSP), Division of Gastroenterology and Hepatology, Department of Medicine, Mayo Clinic, Rochester, MN, USA*

JÜRGEN BRIEGER • *Department of Otorhinolaryngology, University Medical Center Mainz, Mainz, Germany*

MATTHEW BROVOLD • *Wake Forest Institute for Regenerative Medicine, Winston-Salem, NC, USA*

JASMINE BROWN • *Rapid Assay Development Branch, Biomolecular and Computational Toxicology Division, Center for Computational Toxicology and Exposure, Office of Research and Development, US Environmental Protection Agency, Research Triangle Park, NC, USA; Duke Clinical Research Institute, Durham, NC, USA*

LYDIA J. BYE • *Illawarra Health and Medical Research Institute (IHMRI), University of Wollongong, Wollongong, NSW, Australia*

SELINE S. CHOO • *Oak Ridge Institute for Science and Engineering, Oak Ridge, TN, USA; Rapid Assay Development Branch, Biomolecular and Computational Toxicology Division, Center for Computational Toxicology and Exposure, Office of Research and Development, US Environmental Protection Agency, Research Triangle Park, NC, USA*

BYUNG MIN CHUNG • *Department of Biology, The Catholic University of America, Washington, DC, USA*

ZACHARY CONGRESS • *Wake Forest Institute for Regenerative Medicine, Winston-Salem, NC, USA*

JANA KATHARINA DAHLMANNS • *Independent Researcher, Erlangen, Germany*

MARC DAHLMANNS • *Institute for Physiology and Pathophysiology, Friedrich-Alexander-Universität Erlangen-Nürnberg, Erlangen, Germany*

HAIDER DAKHIL • *Institute of Fluid Mechanics, Friedrich-Alexander-Universität Erlangen-Nürnberg (FAU), Erlangen, Germany*

PENG DU • *Auckland Bioengineering Institute, Department of Engineering Science and Biomedical Engineering, University of Auckland, Auckland, New Zealand*

JONAS ECKRICH • *Department of Otorhinolaryngology, University Medical Center Mainz, Mainz, Germany*
SEBASTIAN EGGERT • *cellasys GmbH, Kronburg, Germany*
NICOLAS FÄRBER • *Experimental Physics I, Institute of Physics, University of Augsburg, Augsburg, Germany; Physiology, Institute of Theoretical Medicine, University of Augsburg, Augsburg, Germany*
YONGXIANG FENG • *State Key Laboratory of Precision Measurement Technology and Instrument, Department of Precision Instrument, Tsinghua University, Beijing, China*
ROCIO K. FINOL-URDANETA • *Illawarra Health and Medical Research Institute (IHMRI), University of Wollongong, Wollongong, NSW, Australia*
LENA FISCHER • *Biophysics Group, Department of Physics, University of Erlangen-Nuremberg, Erlangen, Germany*
VICTORIA FONSECA • *Translational Immunology Section, Office of Science Technology (OST), National Institute of Arthritis and Musculoskeletal and Skin Diseases (NIAMS), National Institutes of Health, Bethesda, MD, USA*
OLIVER FRIEDRICH • *Institute of Medical Biotechnology, Department of Chemical and Biological Engineering (CBI), Friedrich-Alexander-University Erlangen-Nürnberg, Erlangen, Germany*
MASSIMO GADINA • *Translational Immunology Section, Office of Science Technology (OST), National Institute of Arthritis and Musculoskeletal and Skin Diseases (NIAMS), National Institutes of Health, Bethesda, MD, USA*
THOMAS GERVAIS • *Department of Engineering Physics, Polytechnique Montréal, Chemin de Polytechnique, Montréal, QC, Canada; Centre de Recherche du Centre Hospitalier de l'Université de Montréal, Montreal, QC, Canada*
MAHSHID GHASEMI • *Future Industries Institute, University of South Australia, Mawson Lakes, SA, Australia*
DANIEL F. GILBERT • *Institute of Medical Biotechnology, Department of Chemical and Biological Engineering (CBI), Friedrich-Alexander-University Erlangen-Nürnberg, Erlangen, Germany*
VALENTINA GIUDICE • *Department of Medicine, Surgery and Dentistry "Scuola Medica Salernitana", University of Salerno, Baronissi, Salerno, Italy; Cell Biology Section, Hematology Branch, National Heart, Lung, and Blood Institute, National Institutes of Health, Bethesda, MD, USA*
STEWART I. HEAD • *School of Medicine, University of Western Sydney, MacArthur, Sydney, NSW, Australia*
LIANG HUANG • *Anhui Province Key Laboratory of Measuring Theory and Precision Instrument, School of Instrument Science and Optoelectronics Engineering, Hefei University of Technology, Hefei, Anhui, China*
CHRISTINA JANKO • *Department of Otorhinolaryngology, Head and Neck Surgery, Section of Experimental Oncology and Nanomedicine (SEON), Else Kröner-Fresenius-Stiftung Professorship, Universitätsklinikum Erlangen, Erlangen, Germany*
VIKRAM JOSHI • *Department of Physiology and Biomedical Engineering, Enteric NeuroScience Program (ENSP), Division of Gastroenterology and Hepatology, Department of Medicine, Mayo Clinic, Rochester, MN, USA*
CALVIN R. JUSTUS • *Department of Internal Medicine, Brody School of Medicine, East Carolina University, Greenville, NC, USA*
JACKSON Y. KEEVER • *Rapid Assay Development Branch, Biomolecular and Computational Toxicology Division, Center for Computational Toxicology and Exposure, Office of Research*

*and Development, US Environmental Protection Agency, Research Triangle Park, NC, USA; Oak Ridge Associated Universities Student Contractor, Oak Ridge, TN, USA*

IVAN KEMPSON • *Future Industries Institute, University of South Australia, Mawson Lakes, SA, Australia*

ALEKSANDR KHUDIAKOV • *Almazov National Medical Research Centre, Saint Petersburg, Russia; Istituto Auxologico Italiano IRCCS, Center for Cardiac Arrhythmias of Genetic Origin and Laboratory of Cardiovascular Genetics, Milan, Italy*

PETER KOHL • *Institute for Experimental Cardiovascular Medicine, University Heart Center Freiburg-Bad Krozingen, Freiburg im Breisgau, Germany; Faculty of Medicine, University of Freiburg, Freiburg im Breisgau, Germany*

ANNA KOSTAREVA • *Almazov National Medical Research Centre, Saint Petersburg, Russia; Department of Women's and Children's Health and Center for Molecular Medicine, Karolinska Institute, Stockholm, Sweden*

ADRIAN KROLINSKI • *Institute of Medical Biotechnology, Department of Chemical and Biological Engineering, Friedrich-Alexander University (FAU) Erlangen-Nürnberg, Erlangen, Germany*

VAN K. LAM • *Department of Biomedical Engineering, The Catholic University of America, Washington, DC, USA*

PETER LARSSON • *Department of Oncology, Institute of Clinical Sciences, Sahlgrenska Academy, University of Gothenburg, Gothenburg, Sweden; Sahlgrenska Center for Cancer Research, Sahlgrenska Academy, University of Gothenburg, Gothenburg, Sweden*

FEI LIANG • *State Key Laboratory of Precision Measurement Technology and Instrument, Department of Precision Instrument, Tsinghua University, Beijing, China*

SISI LIANG • *Future Industries Institute, University of South Australia, Mawson Lakes, SA, Australia*

QUANG MINH LUU • *Future Industries Institute, University of South Australia, Mawson Lakes, SA, Australia*

ANNA MACIÀ • *Biomedical Research Institute of Lleida (IRBLleida); Experimental Medicine Department, University of Lleida, Lleida, Spain*

MONA A. MARIE • *Department of Internal Medicine, Brody School of Medicine, East Carolina University, Greenville, NC, USA*

MOHANA MARIMUTHU • *Department of Biomedical Engineering, Centre for Research, Dhanalakshmi Srinivasan University, Samayapuram, Trichy, Tamil Nadu, India*

ANNA-LENA MERTEN • *Institute of Medical Biotechnology, Department of Chemical and Biological Engineering, Friedrich-Alexander-Universität Erlangen-Nürnberg, Erlangen, Germany*

GEORGE NEHMETALLAH • *Department of Electrical Engineering and Computer Science, The Catholic University of America, Washington, DC, USA*

SIMON V. NEIDINGER • *Experimental Physics I, Institute of Physics, University of Augsburg, Augsburg, Germany; Physiology, Institute of Theoretical Medicine, University of Augsburg, Augsburg, Germany*

STEFANIE NÜBLER • *Institute of Medical Biotechnology, Department of Chemical and Biological Engineering, Friedrich-Alexander University Erlangen-Nürnberg, Erlangen, Germany*

ANAÏS PANOSA • *Biomedical Research Institute of Lleida (IRBLleida); SCT-Microscopy Unit, University of Lleida, Lleida, Spain*

TOSHIMA Z. PARRIS • *Department of Oncology, Institute of Clinical Sciences, Sahlgrenska Academy, University of Gothenburg, Gothenburg, Sweden; Sahlgrenska Center for Cancer Research, Sahlgrenska Academy, University of Gothenburg, Gothenburg, Sweden*

FELIX PFISTER • *Department of Otorhinolaryngology, Head and Neck Surgery, Section of Experimental Oncology and Nanomedicine (SEON), Else Kröner-Fresenius-Stiftung Professorship, Universitätsklinikum Erlangen, Erlangen, Germany*

THUC PHAN • *Department of Electrical Engineering and Computer Science, The Catholic University of America, Washington, DC, USA*

JORDI PIJUAN • *Biomedical Research Institute of Lleida (IRBLleida); SCT-Microscopy Unit, University of Lleida, Lleida, Spain; Laboratory of Neurogenetics and Molecular Medicine-IPER, Institut de Recerca Sant Joan de Déu, Barcelona, Spain; Centro de Investigación Biomédica en Red de Enfermedades Raras (CIBERER), Barcelona, Spain*

ANNE POLJAK • *Bioanalytical Mass Spectrometry Facility (BMSF), Mark Wainwright Analytical Centre (MWAC), University of New South Wales, Sydney, NSW, Australia*

PATSIE POLLY • *Department of Pathology, School of Medical Sciences, University of New South Wales, Sydney, NSW, Australia*

MARK RAFTERY • *Bioanalytical Mass Spectrometry Facility (BMSF), Mark Wainwright Analytical Centre (MWAC), University of New South Wales, Sydney, NSW, Australia*

CHRISTOPHER B. RAUB • *Department of Biomedical Engineering, The Catholic University of America, Washington, DC, USA*

ROBIM M. RODRIGUES • *Department of In Vitro Toxicology and Dermato-cosmetology, Faculty of Medicine and Pharmacy, Vrije Universiteit Brussel, Brussels, Belgium*

PETRA ROHRBACH • *Institute of Parasitology, McGill University, Montreal, QC, Canada*

EDWARD J. SANDERLIN • *Department of Internal Medicine, Brody School of Medicine, East Carolina University, Greenville, NC, USA*

DOMINIK SCHNEIDEREIT • *Institute of Medical Biotechnology, Department of Chemical and Biological Engineering, Friedrich-Alexander University Erlangen-Nürnberg, Erlangen, Germany*

ULRIKE SCHÖLER • *Institute of Medical Biotechnology, Department of Chemical and Biological Engineering, Friedrich-Alexander-Universität Erlangen-Nürnberg, Erlangen, Germany*

SEBASTIAN SCHÜRMANN • *Institute of Medical Biotechnology, Department of Chemical and Biological Engineering, Friedrich-Alexander-Universität Erlangen-Nürnberg, Erlangen, Germany*

CARMINE SELLERI • *Cell Biology Section, Hematology Branch, National Heart, Lung, and Blood Institute, National Institutes of Health, Bethesda, MD, USA*

TIMOTHY J. SHAFER • *Rapid Assay Development Branch, Biomolecular and Computational Toxicology Division, Center for Computational Toxicology and Exposure, Office of Research and Development, US Environmental Protection Agency, Research Triangle Park, NC, USA*

HANNAH C. SHEEHAN • *Department of Biology, Laboratory for Aging and Infertility Research, Northeastern University, Boston, MA, USA*

NATALIA SMOLINA • *Almazov National Medical Research Centre, Saint Petersburg, Russia*

SHAY SOKER • *Wake Forest Institute for Regenerative Medicine, Winston-Salem, NC, USA*

KAI SOMMER • *Institute of Medical Biotechnology, Department of Chemical and Biological Engineering, Friedrich-Alexander University (FAU) Erlangen-Nürnberg, Erlangen, Germany*

SVITLANA STETSENKO • *cellasys GmbH, Kronburg, Germany*

MARTH STINCKENS • *Department of In Vitro Toxicology and Dermato-cosmetology, Faculty of Medicine and Pharmacy, Vrije Universiteit Brussel, Brussels, Belgium*
JENNA D. STRICKLAND • *Axion Biosystems, Atlanta, GA, USA; LabCorp Drug Development, Madison, WI, USA*
INGO THIEVESSEN • *Biophysics Group, Department of Physics, University of Erlangen-Nuremberg, Erlangen, Germany*
JONATHAN L. TILLY • *Department of Biology, Laboratory for Aging and Infertility Research, Northeastern University, Boston, MA, USA*
TAMARA VANHAECKE • *Department of In Vitro Toxicology and Dermato-cosmetology, Faculty of Medicine and Pharmacy, Vrije Universiteit Brussel, Brussels, Belgium*
MARTIN VIELREICHER • *Institute of Medical Biotechnology, Department of Chemical and Biological Engineering, Friedrich-Alexander University (FAU) Erlangen-Nürnberg, Erlangen, Germany*
WENHUI WANG • *State Key Laboratory of Precision Measurement Technology and Instrument, Department of Precision Instrument, Tsinghua University, Beijing, China*
CHRISTOPH WESTERHAUSEN • *Experimental Physics I, Institute of Physics, University of Augsburg, Augsburg, Germany; Physiology, Institute of Theoretical Medicine, University of Augsburg, Augsburg, Germany; Center for NanoScience (CeNS), Ludwig-Maximilians-University Munich, Munich, Germany*
ANDREAS WIERSCHEM • *Institute of Fluid Mechanics, Friedrich-Alexander-Universität Erlangen-Nürnberg (FAU), Erlangen, Germany*
NADINE WIESMANN • *Department of Otorhinolaryngology, University Medical Center Mainz, Mainz, Germany*
JOHANNA WIESNER • *Institute of Medical Biotechnology, Department of Chemical and Biological Engineering, Friedrich-Alexander University (FAU) Erlangen-Nürnberg, Erlangen, Germany*
JOACHIM WIEST • *cellasys GmbH, Kronburg, Germany; Technical University of Munich, Heinz Nixdorf Chair of Biomedical Electronics, TranslaTUM, Munich, Germany*
DORI C. WOODS • *Department of Biology, Laboratory for Aging and Infertility Research, Northeastern University, Boston, MA, USA*
LI V. YANG • *Department of Internal Medicine, Brody School of Medicine, East Carolina University, Greenville, NC, USA*
CALLUM M. ZGIERSKI-JOHNSTON • *Institute for Experimental Cardiovascular Medicine, University Heart Center Freiburg-Bad Krozingen, Freiburg im Breisgau, Germany; Faculty of Medicine, University of Freiburg, Freiburg im Breisgau, Germany*
PENG ZHAO • *State Key Laboratory of Precision Measurement Technology and Instrument, Department of Precision Instrument, Tsinghua University, Beijing, China*

# Part I

## Single Cell Assays in 2D Environments

# Chapter 1

# Assaying Mitochondrial Respiration as an Indicator of Cellular Metabolism and Fitness

**Natalia Smolina, Aleksandr Khudiakov, and Anna Kostareva**

## Abstract

Mitochondrial respiration is an essential component of cellular metabolism. It is a process of energy conversion through enzymatically mediated reactions, the energy of taken-up substrates transformed to the ATP production. Seahorse equipment allows to measure oxygen consumption in living cells and estimate key parameters of mitochondrial respiration in real-time mode. Four key mitochondrial respiration parameters could be measured: basal respiration, ATP-production coupled respiration, maximal respiration, and proton leak. This approach demands the application of mitochondrial inhibitors—oligomycin to inhibit ATP synthase, FCCP—to uncouple the inner mitochondrial membrane and allow maximum electron flux through the electron transport chain, rotenone, and antimycin A to inhibit complexes I and III, respectively. This chapter describes two protocols of seahorse measurements performed on iPSC-derived cardiomyocytes and TAZ knock-out C2C12 cell line.

**Key words** Cellular respiration, Mitochondrial function, Cell viability, iPSC-derived cardiomyocytes, Knock-out cells

## Abbreviations

| | |
|---|---|
| iPSC-CM | iPSC-derived cardiomyocytes |
| KO | Knock-out cells |
| OCR | Oxygen consumption rate |
| WT | Wild type |

## 1 Introduction

Mitochondria play an essential role in muscle cell performance since they supply most of the energy for the maintenance of cellular physiology and homeostasis [1]. Striated muscle cells rely mainly on the mitochondrial oxidative phosphorylation, a pathway converting the energy of oxidized substrates into the energy of ATP

Oliver Friedrich and Daniel F. Gilbert (eds.), *Cell Viability Assays: Methods and Protocols*, Methods in Molecular Biology, vol. 2644, https://doi.org/10.1007/978-1-0716-3052-5_1,

molecules. This pathway is driven by the respiratory complexes located in the inner mitochondrial membrane [2]. The ability of the mitochondria to make ATP and consume oxygen in response to energy demands serves as a reliable hallmark of its functional state, reflecting cell metabolism and fitness.

On the other hand, metabolic abnormalities and mitochondrial dysfunction often accompany various cardiovascular and neuromuscular diseases, including heart failure, cardiomyopathies, Duchenne muscular dystrophy, and Barth syndrome [3]. During the past decade, the model of iPSC-derived cardiomyocytes (iPSC-CM) became widespread in studies of different aspects of cardiovascular biology and pathology [4–6]. The major limitation of the model is the structural, physiological, and metabolic immaturity of the cells. iPSC-CM metabolically depend mostly on glycolysis rather than oxidative phosphorylation [7]. A widely used metabolic maturation approach could shift the phenotype of iPSC-CM from glycolysis to oxidative phosphorylation and stimulate the consumption of fatty acids as an energy source similar to adult cardiomyocytes [8–10]. To assess the maturation of the obtained iPSC-CM, we measured the mitochondrial respiration parameters and showed that applied cardiogenic differentiation protocol results in the development of cells with high rates of oxygen consumption compared to non-differentiated iPSC (Fig. 1).

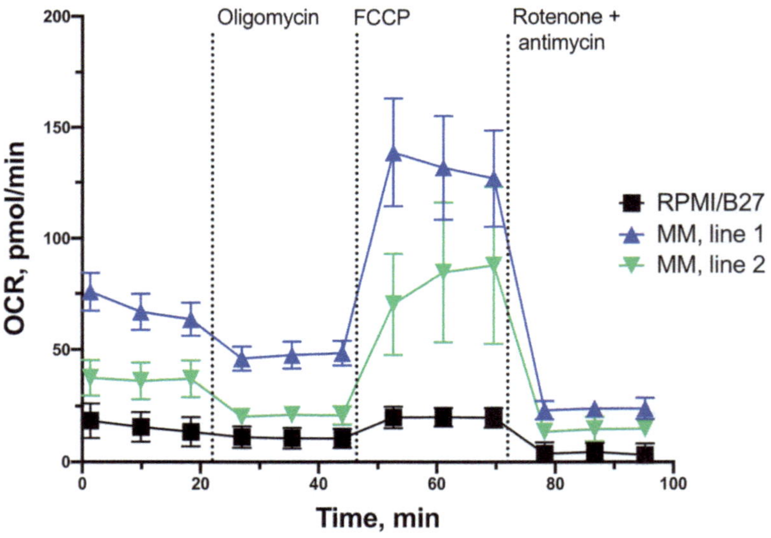


**Fig. 1** Typical traces of OCR obtained for iPSC-CM after metabolic selection in the progress of cardiogenic differentiation (MM, blue and green lines) and control iPSC-CM cultured in iPSC-CM culture medium (RPMI/B27, black line). OCR was significantly higher in cells after applied cardiogenic differentiation protocol. Data are present as mean ± SEM

Knock-out (KO) cell models are widespread in the functional analysis of the genes of interest. In neuromuscular research, KO models are actively used in preclinical drug trials, and the restoration of mitochondrial respiration serves as an optimal measurable endpoint in such experiments [11]. The causing mutations of Barth syndrome are spotted on the X-chromosome on gene TAZ, encoding the tafazzin, a protein participating in cardiolipin remodeling. TAZ KO cells demonstrated the destabilization of respiratory supercomplexes, increased oxidative stress, and reduced mitochondrial membrane stability [3]. To estimate mitochondrial dysfunction in TAZ KO cells, we analyzed respiration performance. Genetic ablation of TAZ induced a significant reduction in basal and maximal respiration compared to cells expressing TAZ (Fig. 2).

The best assay to measure cellular respiratory parameters in situ is a high-throughput real-time mode approach—Agilent Seahorse XF Analyzer. This chapter gives an overview of two protocols we have developed to estimate mitochondrial oxygen consumption rates (OCR) to evaluate the maturation of iPSC-CM in the progress of cardiogenic differentiation and to estimate the mitochondrial dysfunction reflecting the TAZ KO phenotype.

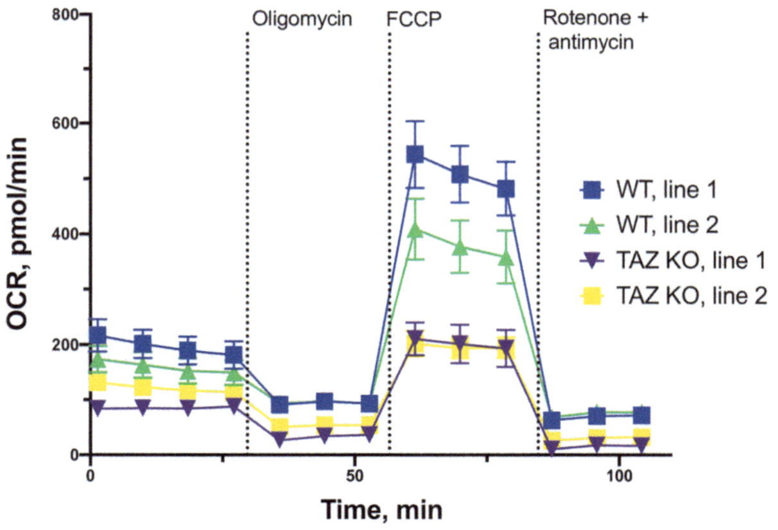


**Fig. 2** Typical traces of OCR obtained for C2C12 TAZ KO (TAZ KO, line 1 and line 2, blue and green lines) and C2C12 WT (WT, line 1 and line 2, purple and yellow lines). KO cells demonstrated greatly reduced OCR compared to WT cells. Data are present as mean ± SEM

## 2 Materials

### *2.1 Cell Culture Reagents and Consumables*

1. A pure population of iPSC-CM generated by standard small molecule-based differentiation protocol [12] with following metabolic purification [13] on 14–16 day of differentiation.
2. C2C12 mouse myoblast cell line.
3. Plating medium for iPSC-CM: cold RPMI 1640, 1% Geltrex LDEV-free reduced growth factor basement membrane matrix.
4. Culture medium for iPSC-CM: RPMI 1640, 2% B27 supplement, 1% Pen-Strep.
5. Maturation medium for iPSC-CM: DMEM, 3 mM glucose, 10 mM L-(+)-Lactic acid, 5 μg/mL (3.7 μM) Vitamin B12, 0.82 μM Biotin, 5 mM Creatine monohydrate, 2 mM Taurine, 2 mM L-carnitine, 0.5 mM Ascorbic acid 2-phosphate, 1× Non-essential Amino Acids, 0.5% (w/v) Albumax, 1x B27 supplement, 1% Knockout Serum Replacement, 1% Pen-Strep (*see* **Note 1**).
6. Culture medium for C2C12: DMEM, 10% Fetal Bovine Serum, 2 mM L-Glu, 1% Pen-Strep.
7. TrypLE Select (Gibco).
8. Y-27632 dihydrochloride (Tocris).
9. PBS.
10. XF24 V7 PS Tissue Culture Microplates.
11. 15 mL tubes.

### *2.2 XF24 Analyzer Assay Reagents and Consumables*

1. Assay medium for iPSC-CM: DMEM (D5030, Sigma), 2 mM L-Glutamine, 1 mM Sodium Pyruvate, 10 mM Glucose, 5 mM HEPES, pH 7.4 adjusted with 1 N NaOH (*see* **Note 2**).
2. Assay medium for C2C12: DMEM (D5030, Sigma), 2 mM L-Glutamine, 1 mM Sodium Pyruvate, 25 mM Glucose, 5 mM HEPES, pH 7.4 adjusted with 1 N NaOH (*see* **Note 2**).
3. Inhibitors of the electron-transport chain (*see* **Note 3**).
   (a) Oligomycin 5 mM.
   (b) FCCP 2 mM.
   (c) Rotenone 5 mM.
   (d) Antimycin A 5 mM.
4. XF calibrant solution.
5. XFe24 sensor cartridge.
6. Sterile filter bottles (0.22 μm filter) and cap.

### 2.3 Equipment Used

1. $CO_2$ incubator (set to 37 °C, 5% $CO_2$).
2. Non-$CO_2$ incubator (set to 37 °C).
3. Water bath (set to 37 °C).
4. pH meter.
5. Stir plate.
6. Cell counter.
7. Seahorse XFe24 Analyzer.

## 3 Methods

### 3.1 iPSC-CM Preparation for the Measurement of $O_2$ Consumption Rates

All procedures are performed under sterile conditions and at room temperature (RT), unless specifically stated.

1. A pure population of iPSC-CM generated by standard small molecule-based differentiation protocol with following metabolic purification is used for plating in Seahorse XF24 plate on 14–16 day of differentiation.
2. Coat XF24 V7 PS Tissue Culture Microplates wells with approximately 30 μL of plating medium. Leave for 30 min in a humidified incubator (set to 37 °C) (*see* **Note 4**).
3. Before plating cells, remove residual plating medium and wash once with PBS.
4. Wash iPSC-CM grown in a 12-well multi-well plate with 1 mL of PBS. Remove PBS, add TrypLE Select 10×, and incubate for 5–20 min at 37 °C until cells become round and detach from the surface.
5. Transfer cell suspension by gentle pipetting to a 15 mL tube and add 5 mL of iPSC-CM culture medium to dilute TrypLE Select.
6. Centrifuge cells at 300 × *g* for 5 min at RT. Discard supernatant.
7. Resuspend iPSC-CM pellet in fresh iPSC-CM culture medium containing 5 μM Y-27632 to get the concentration of 320,000 cells/mL.
8. Plate 80,000 of iPSC-CM (250 μL of cell suspension) per well of Seahorse XF24 plate. Add iPSC-CM culture medium to four wells free of cells for background correction (*see* **Notes 5**, **6**, and 7).
9. Next day, change the iPSC-CM culture medium to the iPSC-CM maturation medium (250 μL).
10. Perform maturation of iPSC-CM monolayers for the next 3 weeks. Change half of the medium volume twice a week.
11. After 3 weeks of differentiation, run measurement.

### 3.2 C2C12 Cell Preparation for the Measurement of $O_2$ Consumption Rates

1. The day before measurement, plate C2C12 cells in XF24 V7 PS Tissue Culture Microplates at 100,000 cells per well in C2C12 culture medium (500 μL of cell suspension); in four wells assigned for background correction, add medium only (*see* **Notes 5**, **6**, **7**, and **8**).

### 3.3 Measurement of Cellular $O_2$ Consumption Rates: Prior to the Day of Assay

1. Hydrate XF24 sensor cartridge with XF Calibrant solution at least 4 h ahead of the assay in a non-$CO_2$ incubator (set to 37 ° C) (*see* **Note 9**).

### 3.4 Measurement of Cellular $O_2$ Consumption Rates: Day of Assay

1. Prepare assay medium and adjust pH to 7.4 using NaOH. Warm assay medium to 37 °C (*see* **Note 10**).
2. Gently remove the culture medium from all wells and add 500 μL of assay medium to all wells (*see* **Note 11**).
3. Incubate cells for 1 h in a non-$CO_2$ incubator (set to 37 °C).
4. During the incubation, prepare intermediate stock solutions of inhibitors in assay medium for loading the sensor cartridge injector ports: oligomycin, FCCP, rotenone, and antimycin A (Tables 1 and 2) (*see* **Note 12**).
5. Load XF24 sensor cartridge placed in hydrated utility plate with intermediate stock solutions of inhibitors. Load 56 μL of oligomycin, 62 μL of FCCP, and 69 μL of a mixture of rotenone and antimycin A to ports A, B, and C, correspondingly (*see* **Note 13**).
6. Set up a protocol in the XFe24 Analyzer software. The protocol represents measurement loops consisting of three times for 3 min measurements, each separated by 2 min wait and 3 min mix for every parameter except for the basal, which is measured four times for better stabilization (Table 3).
7. To calibrate XFe24 analyzer, load XF24 sensor cartridge and utility plate into the instrument tray.
8. After calibration is complete, replace the utility plate with the pre-incubated XF24 microplate with cells.
9. Run the assay.

### 3.5 Data Analysis and Normalization

For data analysis, Seahorse Analytics, the cloud-based tool developed to analyze and interpret data (https://seahorseanalytics.agilent.com/), is recommended. To normalize obtained OCR data across the wells, several approaches could be used, including cell number, genomic DNA, and total protein content normalization. In the present protocol, normalization to the total protein content is recommended as being less time-consuming and more reliable (*see* **Note 14**).

**Table 1**
**Inhibitor preparation for loading sensor cartridge injector ports for measurement of OCR in iPSC-CM**

| Inhibitor | A Stock concentration | B Intermediate stock concentration | C Working concentration | D Volume of stock solution | E Volume of medium | F Volume of injection | G Volume of well |
|---|---|---|---|---|---|---|---|
| Oligomycin (port A) | 5 mM | 25 μM (10×) | 2.5 μM | 7 μL | 1393 μL | 56 μL | 500 μL |
| FCCP (port B) | 2 mM | 20 μM (10×) | 2 μM | 15.5 μL | 1534.5 μL | 62 μL | 560 μL |
| Rotenone/Antimycin A (port C) | 5 mM/5 mM | 25 μM/25 μM (10×/10×) | 2.5 μM/2.5 μM | 8.625 μL/8.625 μL | 1707.75 μL | 69 μL | 690 μL |

To prepare intermediate stock concentration (column B), add the volume of inhibitor stock solution (column D) to the corresponding volume of assay medium (column E)

**Table 2**
**Inhibitor preparation for loading sensor cartridge injector ports for measurement of OCR in C2C12**

| Inhibitor | A<br>Stock concentration | B<br>Intermediate stock concentration | C<br>Working concentration | D<br>Volume of stock solution | E<br>Volume of medium | F<br>Volume of injection | G<br>Volume of well |
|---|---|---|---|---|---|---|---|
| Oligomycin (port A) | 5 mM | 10 μM (10×) | 1 μM | 2.8 μL | 1397.2 μL | 56 μL | 500 μL |
| FCCP (port B) | 2 mM | 20 μM (10×) | 2 μM | 15.5 μL | 1534.5 μL | 62 μL | 560 μL |
| Rotenone/Antimycin A (port C) | 5 mM/5 mM | 10 μM/10 μM (10×/10×) | 1 μM/1 μM | 3.45 μL/3.45 μL | 1718.1 μL | 69 μL | 690 μL |

To prepare intermediate stock concentration (column B), add the volume of inhibitor stock solution (column D) to the corresponding volume of assay medium (column E)

**Table 3**
**Typical mix and measurement cycle times**

| Command | Time (min) | Port | Number of repeats |
|---|---|---|---|
| Calibrate | | | |
| Equilibrate | 12 | | |
| Mix | 3 | | 4 |
| Wait | 2 | | |
| Measure | 3 | | |
| Inject | | A | |
| Mix | 3 | | 3 |
| Wait | 2 | | |
| Measure | 3 | | |
| Inject | | B | |
| Mix | 3 | | 3 |
| Wait | 2 | | |
| Measure | 3 | | |
| Inject | | C | |
| Mix | 3 | | 3 |
| Wait | 2 | | |
| Measure | 3 | | |

## 4 Notes

1. An intermediate stock solution of Vitamin B12 (3 mg/mL, 2.2 mM) in water should be prepared and filtered through 0.2 μm pore-size filter. This stock could be stored at 4 °C for at least 6 months. Sterile-filtered aliquotes of the maturation medium without Albumax could be stored at −20 °C for at least 6 months.
2. Alternatively to commercially available XF Assay Medium, assay medium can be prepared as follows: dissolve DMEM Base (D5030, Sigma) in 800 mL of $dH_2O$, supplement with 10 mL of 200 mM L-glutamine, 10 mL of 100 mM sodium pyruvate, 5 mL of 1 M HEPES, and 15 mg Phenol Red. Add glucose for a final concentration of 10 or 25 mM, depending on the cell type. Stir a medium using a magnetic stirrer until the components completely dissolve. Warm medium to 37 °C and adjust a 7.4 pH using NaOH. pH is temperature-sensitive;

therefore, assay medium needs to be prewarmed ahead of the pH adjustment. Adjust the medium volume up to 1000 mL. Filter the medium through 0.2 μm pore-size filter and store at 4 °C.

3. Alternatively to commercially available, stock compounds could be prepared as follows: dissolve oligomycin A in DMSO to make a stock solution of 5 mM; dissolve FCCP in DMSO to make a stock solution of 2 mM; dissolve rotenone in DMSO to make a stock solution of 5 mM; and dissolve antimycin A in ethanol to make a stock solution of 5 mM. Store all inhibitors at −20 °C for up to 1 month. These agents can be acutely toxic. Wear gloves, protective clothing, face/eye shields, and respiratory protection during preparation. Inhibitors may be light-sensitive; prepare them in dark tubes, and store them in the dark (Tables 1 and 2 describe how to prepare stock compounds to load the cartridge).
4. Plating medium should be prepared at least 1 h ahead of cells plating.
5. At least three technical replicates are desired.
6. Empty wells assigned for background correction should be placed in each plate raw, usually A1, B4, C3, and D6.
7. To define the dynamic range of the instrument and optimal seeding density for a specific cell type, cell density titration assay is performed. Prepare serial dilutions of the cell suspension to get the seeding range from $4 \times 10^4$ to $10 \times 10^4$ cells per well. It will result in 50–90% of cell confluence on the day of the experiment. Run the experiment to measure basal and maximum oxygen consumption rates. Based on the obtained results, choose the optimal seeding density, showing consistent results.
8. When plating, consider that cells of interest might vary in the proliferation rate, for example, KO cells might grow markedly slower compared to wild type (WT) cells.
9. The minimum time of XF24 sensor cartridge calibration, that is, incubation with XF Calibrant, is 4 h.
10. Maximum shelf-life of assay medium is 1 month at 4 °C. However, it is better not to store the medium for longer than 2 weeks due to the instability of sodium pyruvate. pH adjustment is recommended prior to every medium use.
11. Always observe cells under the microscope to ensure cells are healthy, confluent, non-contaminated, and evenly distributed across the wells.
12. Working concentrations of compounds should be titrated for a specific cell line and can range from 0.5 to 2.0 for oligomycin and from 0.125 to 2.0 for FCCP. Seahorse manual

recommends using 1 μM of oligomycin for most cell lines. For a more accurate assessment of concentration working ranges, run the analysis independently for oligomycin and FCCP.

13. XFe24 sensor cartridge loading does not require sterile conditions. Ports corresponding to the empty wells assigned for the background correction should be loaded with the same volume of inhibitors.
14. When performing cell lysis, it is crucial to keep the constant volume of lysis buffer across the wells and ensure complete removing the assay medium since it greatly affects the final sample concentration.

## Acknowledgments

This work was supported by Russian Scientific Foundation grant number 20-15-00271.

### References

1. Hargreaves M, Spriet LL (2020) Skeletal muscle energy metabolism during exercise. Nat Metab 2:817–828. https://doi.org/10.1038/s42255-020-0251-4
2. Hood DA, Memme JM, Oliveira AN, Triolo M (2019) Maintenance of skeletal muscle mitochondria in health, exercise, and aging. Annu Rev Physiol 81:19–41. https://doi.org/10.1146/annurev-physiol-020518-114310
3. Ignatieva E, Smolina N, Kostareva A, Dmitrieva R (2021) Skeletal muscle mitochondria dysfunction in genetic neuromuscular disorders with cardiac phenotype. Int J Mol Sci 22:7349. https://doi.org/10.3390/ijms22147349
4. Giacomelli E, Mummery CL, Bellin M (2017) Human heart disease: lessons from human pluripotent stem cell-derived cardiomyocytes. Cell Mol Life Sci 74:3711–3739. https://doi.org/10.1007/s00018-017-2546-5
5. Brodehl A, Ebbinghaus H, Deutsch MA et al (2019) Human induced pluripotent stem-cell-derived cardiomyocytes as models for genetic cardiomyopathies. Int J Mol Sci 20:4381. https://doi.org/10.3390/ijms20184381
6. Thomas D, Cunningham NJ, Shenoy S, Wu JC (2022) Human-induced pluripotent stem cells in cardiovascular research: current approaches in cardiac differentiation, maturation strategies, and scalable production. Cardiovasc Res 118:20–36. https://doi.org/10.1093/cvr/cvab115
7. Tanosaki S, Tohyama S, Kishino Y et al (2021) Metabolism of human pluripotent stem cells and differentiated cells for regenerative therapy: a focus on Cardiomyocytes. Inflamm Regen 41:5. https://doi.org/10.1186/s41232-021-00156-9
8. Feyen DAM, McKeithan WL, Bruyneel AAN et al (2020) Metabolic maturation media improve physiological function of human IPSC-derived cardiomyocytes. Cell Rep 32:107925. https://doi.org/10.1016/j.celrep.2020.107925
9. Horikoshi Y, Yan Y, Terashvili M, Wells C et al (2019) Fatty acid-treated induced pluripotent stem cell-derived human cardiomyocytes exhibit adult cardiomyocyte-like energy metabolism phenotypes. Cell 8:1095. https://doi.org/10.3390/cells8091095
10. Yang X, Rodriguez ML, Leonard A et al (2019) Fatty acids enhance the maturation of cardiomyocytes derived from human pluripotent stem cells. Stem Cell Rep 13:657–668. https://doi.org/10.1016/j.stemcr.2019.08.013
11. Stocco A, Smolina N, Sabatelli P et al (2021) Treatment with a triazole inhibitor of the mitochondrial permeability transition pore fully corrects the pathology of sapje zebrafish lacking dystrophin. Pharmacol Res 165:105421. https://doi.org/10.1016/j.phrs.2021.105421

12. Lian X, Hsiao C, Wilson G et al (2012) Robust cardiomyocyte differentiation from human pluripotent stem cells via temporal modulation of canonical Wnt signaling. Proc Natl Acad Sci U S A 109. https://doi.org/10.1073/pnas.1200250109

13. Tohyama S, Hattori F, Sano M et al (2013) Distinct metabolic flow enables large-scale purification of mouse and human pluripotent stem cell-derived cardiomyocytes. Cell Stem Cell 12:127–137. https://doi.org/10.1016/j.stem.2012.09.013

# Chapter 2

# The MTT Assay: A Method for Error Minimization and Interpretation in Measuring Cytotoxicity and Estimating Cell Viability

**Mahshid Ghasemi, Sisi Liang, Quang Minh Luu, and Ivan Kempson**

## Abstract

The MTT assay is extensively used, most often to infer a measure of cytotoxicity of treatments to cells. As with any assay though, there are a number of limitations. The method described here is designed with consideration of how the MTT assay fundamentally works to account for, or at least identify, confounding factors in measurements. It also provides a decision-making framework to best interpret and complement the MTT assay to apply it as either a measure of metabolic activity or cell viability.

**Key words** MTT assay, Error minimization, Interpretation, Cell metabolism, Viability assay, Cytotoxicity

## 1 Introduction

The MTT assay is reported extensively in literature based on its simplicity. Unfortunately, over the years since its initial reporting, its interpretation has often been perverted to unequivocally represent cell viability. This is often inaccurate as the assay provides a measure of cell metabolism and often assumes no other interferences in the assay. It is an excellent tool to monitor variations in cell metabolism; however, it needs to be done carefully, and interpretation of data requires some insight into the assay and how it works.

The assay is rapid and easily implemented but should be used with a well thought-out plan to minimize and remove as many potential errors as possible. It is worth asking yourself what the specific purpose is of performing the assay. For example, is it to measure cell viability? Or is it to measure toxicity (or lack thereof) of a treatment on cells? It is more commonly the latter, but interpretation of the assay is often assumed to be a measure of viability. The MTT assay should not be interpreted or reported as a cell viability assay, unless validation and calibration have been conducted.

Oliver Friedrich and Daniel F. Gilbert (eds.), *Cell Viability Assays: Methods and Protocols*, Methods in Molecular Biology, vol. 2644, https://doi.org/10.1007/978-1-0716-3052-5_2,

The assay works by adding a yellow-colored compound (the MTT reagent) to a cell culture. When the reagent is reduced, it turns into formazan (a violet-blue compound). Cells are quite efficient in "metabolizing" the MTT reagent; hence the colorimetric MTT assay provides a measure of cell metabolism, that is, the amount of formazan is increased by a greater cell metabolism, dominated by glycolytic NADH production [1, 2]. This is measured in terms of the optical density, that is, the reduction in the intensity of transmitted light through the sample when illuminating it with a range of wavelengths (bandwidth) that MTT-derived formazan is expected to maximally absorb. For correctness, OD should be reported rather than "absorbance" since other physical processes (e.g., scattering) in addition to absorbance can give rise to an increase in measured OD. The final measured OD is a result of a series of events that can be categorized into two main mechanistic steps; the first step is the reduction of MTT to formazan, and the second step is the optical measurement process that results in a specific OD measured by the spectrophotometer. Different variables can affect the measured OD at each step and thus can have a confounding effect on the data and its interpretation [3]. Being aware of these variables can help minimize measurement errors and data misinterpretations.

In the first step, the concentration of formazan is a result of a multiplex process to which the following factors can contribute:

1. The number of cells capable of taking up the MTT reagent and reducing it to formazan. This is often considered as the number of "viable" cells. However, it is important to remember that "cell viability" is not a simple binary variable, but rather, cells can be in a spectrum of viability statuses. A range of cells within this spectrum can have the capability of MTT uptake and reduction. This also includes cells that are going through the process of cell death yet still maintain membrane integrity. Although these "dying" cells may have lower capability than the cells at the viable end of the spectrum, this difference is not necessarily linear, and therefore, needs to be assessed for each cell line, growth condition, tested treatment, and experimental design.
2. MTT uptake by cells: This can be enhanced or disturbed by other factors such as the cell type, culture media components, tested treatment (e.g., radiation or other factors that increase cell membrane permeability), growth conditions, and the rate of formazan extrusion.
3. The metabolic activity of cells leads to the conversion of MTT to formazan: This can itself be affected by various factors including the cell density (proximity of cells to each other which is also a function of cell number), the MTT reagent itself

(that can disturb cell metabolism), the culture media components, growth conditions, and the tested treatments which can increase or decrease metabolic activity [4–6].

4. Any kind of non-cellular chemical reaction that results in MTT abiotic reduction or production of chemicals that interfere with the measurement by absorbing or scattering light at the illuminating wavelength. In this regard, media components and tested treatment(s) are the most important factors to consider [7, 8].

In the second step, the measured OD can be representative of formazan concentration in the sample if:

1. The MTT-derived formazan has been well solubilized to make a homogeneous solution before the OD measurement. Otherwise, light scattering by non-solubilized formazan precipitations in the solution can confound the OD measurement.
2. The sample is illuminated at the correct wavelength.
3. Absorbance or scattering of light by other components in the sample (i.e., media components, cells, treatments, etc.) do not significantly interfere with the measured OD.

Therefore, when designing the experiment and interpreting the data, it is crucial to keep in mind what the assay is actually measuring and how this measurement can be interpreted regarding your specific research question(s) and consider that the mechanism underlying the MTT assay is a multiplex, and not a univariate, process.

A general approach to the MTT assay has been described in the first edition of this book [9]; however, it is the intention of this chapter to describe an approach to the assay to minimize artifacts and improve interpretation. It also recommends supporting assays to complement the MTT assay data. More in-depth discussion around the confounding factors are also provided in Ghasemi et al. [10].

## 2 Materials

### 2.1 General Equipment

1. Cell counter or hemocytometer.
2. Vortex mixer (optional).
3. Multichannel reagent reservoir (optional).
4. Pipettes and pipettors (multichannel (optional) and single channel).
5. Cell culture flask.
6. Hemocytometer.

7. 96-well flat-bottom tissue culture microtiter plates.
8. 96-well plate spectrophotometer (i.e., plate reader) equipped with 570 ± 10 nm filter.
9. Shaker for microtiter plates.

### 2.2 Reagents

1. Primary cells or cell line(s).
2. Cell culture media appropriate for culturing cells.
3. Serum-free phenol red-free cell culture media.
4. Phosphate-buffered saline (PBS) without $Ca^{2+}$ and $Mg^{2+}$.
5. Milli-Q water.
6. Stock MTT solution in PBS (5 mg/mL): 25 mg of MTT powder dissolved in 5 mL of PBS (*see* **Note 1**).
7. Dimethyl sulfoxide (DMSO), cell culture grade.

## 3 Methods

### 3.1 Making the Right Choice and Planning Ahead

1. In order to assess if the MTT assay is a proper tool to help, at least in part, answer your research question(s), consult the literature, or perform primary or complementary assays to determine potentially confounding variables from your tested treatment(s)/conditions(s) or experimental design. These variables include MTT uptake or formazan extrusion rate (e.g., cell membrane permeability), cell proliferation (e.g. cell senescence), cell viability/cytotoxicity, cell metabolic activity such as mitochondria number/volume/activity (*see* **Note 2**), cell secretome or cell lysis (*See* **Note 3**), abiotic reduction of MTT, and chemical interaction with culture media components or tested treatment(s)/condition(s). In other words, changes in OD level caused by specific treatment/experimental condition(s) (compared to control cells) can result from changes in multiple variables (*see* Table 1) induced by the tested treatment/condition.
2. If multiple of the above variables are affected by your experimental condition(s), it will be more challenging to minimize their confounding effects and thus associate the assay measurements with a single variable such as cell viability or metabolic activity (*see* **Note 4** and Fig. 1). Otherwise, you can consider using the MTT assay as an easy, fast, cheap, and high-throughput screening test of cell metabolic activity or sometimes, if validated, cell viability (*see* **Note 5**).
3. If you find the MTT assay a proper choice for your research aim(s), you now need to have a clear plan for optimizing and performing the assay, taking into account the various confounding factors in order to minimize measurement errors

**Table 1**
**Possible issues when using the MTT assay**

| **Optical density interferences:** | | **References** |
|---|---|---|
| Carbon materials: graphene; SWCNTs; Metal nanoparticles: $TiO_2$; ZnO, AgO, AgNP, $Fe_3O_4$, $CeO_2$ | Nanomaterials giving rise to absorption, scattering, or electronic interactions and changing OD | [11–17] |
| Molecules: Doxorubicin, and so forth | Intrinsic absorbance enhancing OD measurement | [8, 18] |
| **Mitochondria volume/size/membrane integrity change:** | | |
| Ionizing radiation (e.g., X-rays) | Mitochondrial volume and function | [19] |
| Titanium dioxide ($TiO_2$) nanoparticles; Rottlerin | Loss of membrane integrity | [20, 21] |
| Genistein | Increase cell volume and mitochondrial volume | [7] |
| Anticancer drugs modify cell metabolism mitochondria function | | [22] |
| Changes in cell size and mitochondrial activity by f IFNs | | [23] |
| **Redox reactions with MTT:** | | |
| Plant-derived reducing compounds, resveratrol, quercetin, curcumin and naringin polyphenolic compounds: polyphenolic flavonoid genistein | | [24–26] |
| Metal oxide nanoparticles activate or catalyze redox reactions by reactive species (e.g. superoxide, free radicals, hydrogen peroxide) production<br>Porous silica (PSi), $TiO_2$, $CeO_2$, CuO, iron oxide nanomaterials<br>$MoS_2$, $MnO_2$, Au and silver nanomaterials<br>Carbon nanomaterial metal (M) ditellurides (MTe2; M = V, Nb, Ta) react with formazan | | [27–32] |
| Chemical functional groups: thiols, amines, amides, carboxylic acids | | [33] |
| Antioxidants: ascorbic acid, vitamin E and N-acetylcysteine, sulfhydryl-containing compounds | | [34] |

and misinterpretations. As a general guide to planning the assay optimization for adherent cultured cells in 96-well microplates, an exemplary template of a 96-well microplate is provided (*see* Fig. 2). Modifications may still be needed based on your specific experimental design. Consider including all the conditions that you need to test to optimize the assay parameters such as cell seeding number, MTT concentration, and MTT incubation time (*see* **Notes 6** to **11**). Follow your planned templates for optimization and performing the assay which is explained in the next section.

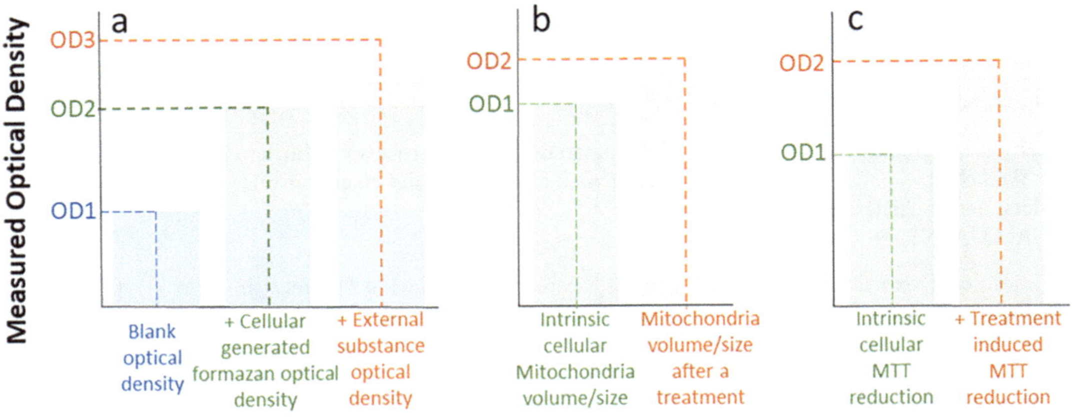


**Fig. 1** (**a**) OD measurements are due to a combination of contributions from the well, the culture media, the formazan produced from the assay, and any treatment added to each well. (**b**) For the same number of cells, a treatment may increase mitochondria activity, leading to greater reduction of the MTT reagent and increase optical density measurements. (**c**) Some chemicals and treatments will lead to reduction of the MTT reagent even when there are no cells present

**Number of seeded cells per well**

| | 0 (no cells) | | | $10^4$ | | | $2\times10^4$ | | | $5\times10^4$ | | |
|---|---|---|---|---|---|---|---|---|---|---|---|---|
| | 1 | 2 | 3 | 4 | 5 | 6 | 7 | 8 | 9 | 10 | 11 | 12 |
| A | 0 | 0 | 0 | 0 | 0 | 0 | 0 | 0 | 0 | 0 | 0 | 0 |
| B | 0.1 | 0.1 | 0.1 | 0.1 | 0.1 | 0.1 | 0.1 | 0.1 | 0.1 | 0.1 | 0.1 | 0.1 |
| C | 0.2 | 0.2 | 0.2 | 0.2 | 0.2 | 0.2 | 0.2 | 0.2 | 0.2 | 0.2 | 0.2 | 0.2 |
| D | 0.3 | 0.3 | 0.3 | 0.3 | 0.3 | 0.3 | 0.3 | 0.3 | 0.3 | 0.3 | 0.3 | 0.3 |
| E | 0.4 | 0.4 | 0.4 | 0.4 | 0.4 | 0.4 | 0.4 | 0.4 | 0.4 | 0.4 | 0.4 | 0.4 |
| F | 0.5 | 0.5 | 0.5 | 0.5 | 0.5 | 0.5 | 0.5 | 0.5 | 0.5 | 0.5 | 0.5 | 0.5 |
| G | W | W | W | | | | | | | | | |
| H | | | | | | | | | | | | |

**Final MTT concentration (mg/ml)**

**Fig. 2** A guide template of a 96-well microplate for designing MTT assay optimization and experiments. MTT concentration (numbers in the wells in mg/mL) and the number of seeded cells (the top row) vary within each plate. This template is replicated for each measurement time point and for each experimental condition (test and control groups) W: water or PBS as blank samples

### 3.2 Optimizing and Performing the Assay

1. Prepare cell suspension from the adherent cells in the source culture vessel (e.g., culture flask), and count the number of cells using a hemocytometer (*see* **Note 12**).
2. According to the cell count of the prepared cell suspension, the cell densities that you aim to test, and the number of wells with each cell density, make a concentration series of cell suspensions with required volumes (each cell concentration in a separate tube).
3. Add 100 μL of the cell suspension with the desired concentration to each well following the planned plate template for your experiment (*see* Fig. 2).
4. For wells of columns 1 to 3, add the same volume of the cell-free culture media of the same type as negative controls (*see* **Note 13**).
5. Allow the cells to grow in the desired culture conditions for the needed time to adhere. This is usually overnight, however, it can differ between cell lines.
6. Remove the media and add 100 μL/well of fresh media containing desired concentrations of tested treatment(s), if any, as planned (*see* **Note 14**). Add the same type and volume of media with no treatment to the wells in the negative control plate (non-treated cells). Incubate the cells in the desired culture condition for the required time for incubation of cells with the tested treatment (*see* **Notes 15** and **16**).
7. In the wells containing cells (Columns 4 to 12 in the template provided in Fig. 2), remove the media and wash the wells gently with serum-free phenol-red free media to completely remove any treatment and previously used culture media remaining in the wells (*see* **Note 17**).
8. Add 90 μL of serum-free phenol red-free fresh media (*see* **Notes 18** to **26**).
9. Add 10 μL of different concentrations of MTT solution (0, 1, 2, 3, 4, and 5 mg/mL) in PBS to each well to make the final MTT concentrations shown in the template (0, 0.1, 0.2, 0.3, 0.4, 0.5 mg/mL). Incubate the plates at the desired culture conditions for the planned durations (2, 3, and 4 h).
10. At the planned time points of OD measurement for each plate (2, 3, and 4 h after MTT addition), add 50 μL of DMSO to each well and mix the wells' contents using a plate shaker until a homogenized solution of formazan is achieved. Incubating at 37 °C can accelerate solubilizing of formazan (*see* **Notes 27–29**).
11. Measure OD levels at 570 nm using a Microplate Reader.

### 3.3 Choosing the Optimum Assay Parameters, Data Analysis, and Interpretation

1. Calculate the mean of the measured OD values of the replicates for each condition.
2. To assess the effect of cell number, plot the mean OD versus cell seeding number. Draw separate plots for each incubation time. Data for each MTT concentration can be represented in each plot by different colors (*see* **Note 30**).
3. To assess the effect of incubation time, plot the mean OD versus incubation time. Draw separate plots for each seeding number or MTT concentration. Data for each respective MTT concentration or seeding number in each plot can be represented by different colors.
4. To assess the effect of MTT concentration, plot the mean OD versus MTT concentration. Draw separate plots for each incubation time. Data for each seeding number can be represented in each plot by different colors.
5. To assess the effect of tested treatment(s)/conditions(s), plot the mean OD versus MTT concentration, mean OD versus cell number, and mean OD versus incubation time in separate plots, indicating different treatments with different colors in each plot.
6. Define the optimum values for cell seeding number, MTT concentration, and MTT incubation time by looking at the plots. The optimum values should maximally reveal the differences between different cell populations in terms of capability in reducing MTT (*see* Fig. 3a) and do not cause cell death/toxicity at a significant level before measuring OD (*see* **Notes 31** and **32**).
7. Assess the confounding effect of culture media on the assay measurements by analyzing the measured OD of cell-free culture media (*see* **Note 18** to **26**).
8. Assess the confounding effect of treatment on the assay measurements by comparing the OD of cell-free culture media containing different concentrations of treatment, including no treatment (columns 1 to 3 in Fig. 2). A significant difference between the OD levels indicates the optical or chemical interference of the treatment on the assay measurements (*see* **Notes 19** to **24**).
9. To analyze the effect of the tested treatment(s)/condition (s) on the OD levels, select the conditions where the optimized assay parameters have been applied.
10. From the mean OD of replicates of each selected condition, subtract the mean OD of replicates of corresponding MTT-free wells (wells A4 to A12 in Fig. 2) which contain the same number of cells and have undergone the same treatment, if

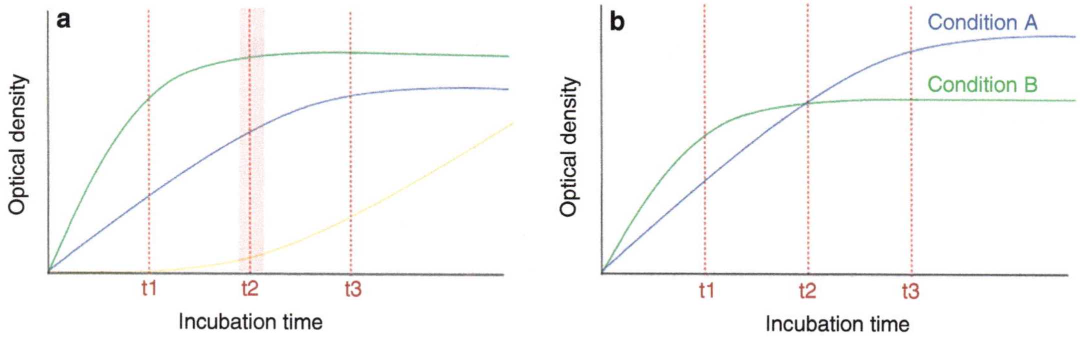


**Fig. 3** Graphs showing OD levels as a function of different MTT incubation times in two hypothetical scenarios. t1, t2, and t3 defined by dotted lines represent different measurement time points. (**a**) Each line color can represent a condition, or assay parameters such as cell number or MTT concentration. For example, the green line can represent a higher MTT concentration, which causes MTT reduction to reach the saturation level (plateau) at an earlier time point (t2) compared to the blue line potentially representing a lower MTT concentration causing a delayed saturation time (t3). The yellow line can represent a condition where the MTT concentration is too low for the number of cells to reach a saturation level in MTT reduction within the time incubated with MTT. The optimum OD measurement time is a time point which optimally reveals the capacity of a cell population to reduce MTT and the differences between various conditions (t2), considering the effect of other assay parameters. For example, assuming each solid line represents a different MTT concentration, the green line can represent the optimum MTT concentration at time t2. (**b**) The optimum OD measurement time also depends on how the OD levels change as a function of incubation time in the tested conditions. Each condition has resulted in a different saturation level and saturation time points. For example, conditions A and B can respectively be representative of a metabolism-inhibiting treatment (causing a delayed saturation of MTT reduction) and a metabolism-enhancing treatment (causing an earlier saturation of MTT reduction). Therefore, changing the measurement time point will result in a different conclusion on the comparison of conditions A and B

any, but have no MTT). This minimizes interference from background including any potential optical interference from cells and any change in cell number caused by the treatment (*see* **Note 33**).

11. Normalize the background-subtracted OD (calculated above in **step 10**) from the treated cells with the following calculation. This gives you the percentage change in the OD level in treated cells compared to non-treated (control) cells:

    (background-subtracted OD for treated cells/background-subtracted OD for non-treated cells) × 100.

12. The result of the above calculation can give you an estimation of how much (in percentage) your tested treatment has increased or decreased the level of MTT reduction by your cells on average, assuming that you have followed all the previous steps to minimize measurement errors (*see* **Note 34**).

13. Preferably repeat the assay at least two more times as technical replicates of the experiment to assure the data are reproducible (*see* **Note 35**).

**Table 2**
**Grouping of assays according to their principle and methods for cell viability/proliferation**

| Principle/methodology | Assay |
|---|---|
| Metabolic activity | MTT; WST; XTT; MTS; WST-1; WST-8; Alamar Blue (resazurin); Calcein AM; Intracellular ATP |
| Membrane integrity | Trypan blue; eosin; propidium iodide; neutral red; lactate dehydrogenase; CFDA-AM; SYTOX; Yo-PRO-1 |
| Protease viability marker activation | Caspase 3/7 |
| Colony formation | Colony counting (manually or gel count systems) |
| DNA synthesis | Isotopically labeled thymidine incorporation; BrdU; EdU incorporation; Ki-67; DAPI |
| Surface membrane composition | Annexin V; Caspase 3/7 |
| Genotoxicity | DNA fragmentation: comet; chromosome disorders; DNA mutation; Tunel |
| Fluorometric assay | Alamar blue; presto blue; natural red, 5-CFDA-AM |
| Colorimetric assay | MTT; XTT; WST-1; WST-8; LDH; CCK8, CVS |
| Redox reaction | MTT; XTT; MTS; WST-1; WST-8; Alamar Blue (resazurin); Calcein AM; Intracellular ATP |

14. The mean calculated percentage from all the technical replicates of the experiment (as explained in **step 11**) can then be interpreted to compare the effect of different treatment(s)/condition(s) with each other or with non-treated (control) cells. However, to make a correct interpretation, you need to consider the type of effect you aim to compare between treatments. These effects may include cell viability/proliferation, treatment toxicity, and/or cellular metabolic activity.

15. To validate the MTT assay as a measure of cell viability or proliferation, a thorough calibration against other assays is necessary and will need to be repeated for any different experimental conditions where confounding effects could be introduced. Various methods, differing in their mechanisms and methodologies, are available for investigating cell viability/proliferation (*see* Table 2 and **Note 36**).

16. To interpret cytotoxicity of a treatment or a chemical from the MTT assay measurements, remember that cytotoxicity can present in a variety of manifestations, and thus, the MTT assay results cannot provide a comprehensive image of a treatment/chemical cytotoxicity. Therefore, multiple measures of integrity of cellular components to assess a treatment toxicity are best practice (*see* **Note 37**), and the MTT assay should be complemented by additional assays, which are not dependent on the same processes (*see* **Note 38**).

17. Ideally, the MTT assay measurements should be validated with complementary assays (*see* Table 2 for examples). This is especially necessary if you observe interferences or effects from experimental conditions (*see* Table 1 for examples). These should be apparent from the various plots produced earlier. The assays in Table 2 have been categorized into similar types of assays either by their methodology or by their fundamental principle, and it is best to complement the MTT assay with assays from different categories. Considering that all methods have their own limitations, it is important to weigh the advantages and disadvantages of each method and make a proper selection of appropriate method(s). Your choice of the complementary assay should depend on the type of variable you aim to measure, such as cell viability, cytotoxicity, or metabolic activity, and so forth, as well as the confounding effects from your experimental conditions. If you identify that the treatment is changing metabolic activity of cells, *see* **Note 39**. If you identify the experimental conditions are altering OD measurements, *see* **Note 40**. If you suspect something other than cells are reducing MTT, *see* **Note 41**.

## 4 Notes

1. For the volumes for preparation of a dilution series from the stock solution, *see* Table 3. The provided volumes will be enough for making 1 mL solutions of each MTT concentration (1, 2, 3, 4, and 5 mg/mL), which, considering the pipetting errors and residues, is sufficient to be used for treating all wells in a 96-well microplate with each concentration of MTT. You can change the suggested volumes accordingly based on the number of wells needed to be treated with each MTT concentration.

**Table 3**
**Volumes to be used for making 1 mL of different concentrations of MTT solutions**

| MTT solution stock concentration (mg/mL) | The volume of the 5 mg/mL stock MTT solution (μL) | PBS volume (μL) |
|---|---|---|
| 0 | 0 | 1000 |
| 1 | 200 | 800 |
| 2 | 400 | 600 |
| 3 | 600 | 400 |
| 4 | 800 | 200 |
| 5 | 1000 | 0 |

2. For example, enhancing metabolic activity at the single-cell level by radiation or chemotherapy may be mistakenly interpreted as increasing cell viability or proliferation.
3. For example, radiation or chemotherapy can induce a specific secretory phenotype, which may cause extracellular reduction of MTT. The effect can be further assessed by measuring the OD of cell secretome or lysate after the addition of MTT as control samples.
4. For example, radiation is known to cause DNA damage and cell death. However, by increasing the permeability of the cell membrane and the number and activity of mitochondria, it results in both enhancing MTT uptake and reduction leading to increased formazan production.
5. You still need to consider the effect of your experimental conditions on any of the mentioned variables in the assay design and the data analysis and interpretation. Hopefully, the experimental methodology described in this chapter would provide a guide on how to minimize errors in this regard. However, choosing how to respond to any artifacts depends on what specific issues arose. Also, it is always assuring and sometimes necessary (depending on the research question) to confirm the assay results using complementary assays that can provide more robust data. As the MTT assay measures a type of metabolic reaction inside the cells and due to the above-mentioned reasons, it is not recommended to use it as a robust quantitative viability test, but rather as a primary qualitative screening test.
6. It is recommended to vary cell number and MTT concentration within each plate, but consider having separate plates, following the same template in Fig. 2, for each measurement time point (time of incubating cells with MTT) so that samples are not disturbed during the desired MTT incubation time.
7. At least three incubation times of 2, 3, and 4 h are recommended to be tested to find the optimum measurement time point. A time point shorter than 2 h is usually too early to measure OD as often cells do not reduce enough MTT in this time frame to enable a comparison of different conditions in terms of MTT reduction. At incubation times beyond 4 h, MTT-induced cytotoxicity often takes place, which increases the risk of erroneous measurements of cell metabolism. However, this can vary slightly for different cell lines and experimental conditions, and you need to decide on the measurement time points to be tested accordingly. You can consult the literature for this purpose or to define the shortest time point to test, you can also monitor when the color change from yellow MTT to purple formazan starts to occur in your wells.
8. There needs to be at least triplicates of each condition.

9. Milli-Q water or PBS will be loaded into three wells as blank samples.
10. Seeding numbers provided here are only exemplary and can vary based on your cell line and experimental conditions. Define the seeding numbers in a way that covers a range of cell densities from 0% to 100% of confluency based on your experimental design. You can also consult the literature to determine this range.
11. All the conditions need to be replicated for each treatment (test) group other than the nontreated (control) group.
12. At this stage, use the type of culture media that is required for cell attachment based on the cell line you are using.
13. Measuring the OD of culture media helps you identify optical or chemical interference by the culture media components.
14. If the solvent of treatment is different from culture media, add the same volume of the solvent to non-treated control cells (as well as wells containing non-treated cell-free media) should the pure effect of treatment be meant to be assessed. For example, if you are adding to your cells 10 μL of a nanoparticle solution in PBS, 10 μL of PBS is needed to be added to your control cells.
15. Considering the treatment incubation time in designing the optimization assay is critical as that can change the extent of the contribution of confounding (optical or chemical) effects of the treatment on the assay measurements.
16. If the tested treatment is not a chemical, such as radiation, you can simply skip **step** 7 and instead treat your cells and incubate them in the desired culture conditions for the planned time after treatment. Then continue from **step 8**.
17. You need to do this step to minimize any possible optical/ chemical interference caused by the presence of the treatment in the supernatant. For example, optical interference from a colored chemical can cause an increase in the OD level when added to cells. If the raw OD levels are used to compare the metabolic activity or viability of the control and treated cells, the observed difference in OD will be incorrectly interpreted as enhancing effect of the tested chemical on the level of cells' metabolic activity, while it is simply a result of the chemical's optical interference with the assay measurements. By washing off the treatment before MTT addition, however, the actual effect of the chemical on the cells' metabolic activity can be revealed more accurately. As the aim of OD measurement in cell-free wells (columns 1 to 3 in the provided template in Fig. 2) is to assess potential interferences from culture media or any added treatment, you do not need to do this step for the wells without cells.

18. The type of culture media depends on the cell line you use as well as your experimental design. With that being said, *see* **Notes 19** to **26** for recommendations on the optimal type of culture media that needs to be used when incubating cells with MTT and how to assess the potential interference from culture media or any chemical/treatment added to the wells in OD measurement.

19. Generally speaking, you need to consider, and preferably test, any optical or chemical interference from the culture media, media additives (such as serum, phenol red, antibiotics, etc.), or any added chemical/treatment if present when incubating cells with MTT.

20. You can use UV-Vis spectroscopy to test the optical interference of well components by measuring the absorbance spectrum of media with and without the tested treatments or media additives. Optical interference is inferred when there is a significant overlap with the assay measurement window.

21. As shown in Fig. 2, you can also compare the OD of cell-free media (± treatment) with the OD of water or PBS as blank controls (row G, Fig. 2). A significant difference in the OD levels between cell-free media (wells A1 to A3, Fig. 2) and PBS is indicative of optical interference by media. A significant difference in the OD levels between cell-free media with different concentrations of treatment (wells A1 to A3, Fig. 2 ± treatment) is indicative of optical interference by the treatment.

22. To assess chemical interference from well components (culture media ± treatment), compare the OD of cell-free culture media with and without different concentrations of MTT (Fig. 2, columns 1 to 3). As MTT itself does not optically interfere with the assay measurement, a significant difference in the OD can be a result of a chemical reaction between MTT and the well components (culture media, added chemicals/treatments, or DMSO).

23. If possible, consider using a type of media that imposes the least optical and chemical interference with the assay measurements. Still, the optical properties and chemical composition of the media (± treatment) may change in the presence of cells. By measuring the OD of cell supernatant and comparing it to the OD of cell-free media, you can rule out this possibility.

24. Subtraction of the corresponding background (well contents with no MTT as explained in Subheading 3.3, **step 10**) from the measured OD of cells can minimize the errors caused by background interferences. However, we generally recommend replacing culture media with fresh media containing no unnecessary additives such as phenol red (*see* **Note 25**), serum (*see* **Note 26**), or tested chemical/treatment(s) before incubating

cells with MTT, as already explained in Subheading 3.2, **step** 7. The more additives present in the media, the more probable and complex the interferences may be.

25. Phenol red absorbance peak (around 560 nm) overlaps with the assay measurement window. However, this can change in the presence of cells and different experimental conditions due to changes in the level of acidity that causes the color change of phenol red. Although these changes may not be consistent between different samples, subtraction of the background OD from the measured values (as explained in Subheading 3.3, **step 10**) should theoretically compensate for such variability. However, as the underlying chemical process of phenol red color change may be affected by addition of MTT and formazan production, we still recommend using phenol red-free media to minimize any potential unknown confounding effect from phenol red.
26. The absence or presence of serum can also affect the metabolic activity of cells and, therefore, influence the assay measurements. However, as the period of MTT incubation only lasts a few hours, this effect may be negligible. This can be tested for each cell line by incubating them with MTT for the required time in serum-free and serum-fed conditions and testing if there is a significant difference between the OD levels. We generally recommend using serum-free media for the duration of MTT incubation. However, if the presence of serum is necessary for your experimental design/research question(s), make sure to use serum from the same batch for all your samples to rule out the effect of serum composition variations on the measurements.
27. You may need to adjust the volume of DMSO, the mixing time on the plate shaker, and/or the incubation time at 37 °C based on your experimental conditions.
28. The required time for incubation with DMSO can be different for different experimental conditions and cell lines and can usually last between 5 and 20 min. You can find the optimum time by macroscopically monitoring the contents of the wells every few minutes until there is no precipitation observed in the wells.
29. It is recommended not to remove or wash the supernatant before adding DMSO as this can cause erroneous measurements by removing intracellularly formed extracellular formazan and loosely attached cells that have contributed to MTT reduction. As MTT itself does not interfere with the measurement, the remaining MTT does not need to be removed. If you have other reasonable justifications for washing or removing media, you need to test and make sure it will not confound your measurements.

30. To do this, you can use Microsoft Excel or GraphPad Prism, or similar software.
31. This could be inferred when there is a drop in the OD level at MTT concentrations or incubation times respectively higher or longer than a specific value.
32. If the pattern of OD difference between tested conditions changes with altering any of the assay parameters (cell seeding number, MTT concentration, and MTT incubation time), the effect of tested conditions needs to be assessed by using more than one optimal value for those parameters. This will provide a more complete perspective on how the tested condition(s)/ treatment(s) affect cell behavior (*see* Fig. 3b).
33. The provided calculation aims to minimize the confounding effects from the background. However, if you suspect chemical interference such as abiotic reduction of MTT by culture media components (*see* **Note 22**), you will need extra consideration in how to interpret these data.
34. Some data have shown that experimental conditions (e.g., pH, cell density, treatments) may also cause interference with the absorbance spectrum of the final formazan-DMSO solution produced by cells. Therefore, to exclude possible interference in the measurements window, measuring the absorbance spectrum of the formazan-DMSO solution from all tested conditions using UV-vis spectroscopy can assist in identifying spectral interference and how significantly it may affect the data.
35. When doing technical replicates, only use the optimized assay parameters in Subheading 3.3, **step 6**, and keep consistency in the experimental setup including, but not limited to, the type of medium, formazan solubilizing agent, serum batch, cell growth phase, and so forth.
36. Methods to determine different aspects of cell proliferation range in complexity for assessing metabolic activity, membrane permeability and integrity, protease activity, congenic cell survival, DNA synthesis, surface membrane composition and genotoxicity. One general approach is to calibrate the OD measurement with cell viability measured by quantifying cell number or DNA content for a specific set of experimental conditions. In this regard, cell counting, stable isotope-labeled thymidine or other DNA measurements can be valid. Dyes incorporated into DNA (such as PI, BrdU) help to provide insight into DNA content changes when evaluating genome instability or changes in proliferation and abundance. A standard measure of clonogenicity by the clonogenic assay is very powerful and, under certain circumstances, can be calibrated

against MTT [35], providing rapid assessment of proliferation capacity without the need to wait many days for colony formation. However, all need to be considered with respect to their own limitations and artifacts (for example [36]).

37. For instance, the International Standard ISO 21115: 2019 (Water quality—Determination of acute toxicity of water samples and chemicals to a fish gill cell line (RTgill-W1)) takes into account different parameters to observe treatment impacts by utilizing three fluorescent indicator stains: 5-CFDA-AM [37] Neutral red [38], and Alamar Blue [39], which elude to cell membrane integrity, lysosomal membrane integrity, and metabolic activity, respectively, to indicate cytotoxicity [40].

38. Examples for such complementary assays are the loss of membrane integrity measured by dye exclusion or lactate dehydrogenase (LDH) release after a cytotoxic insult.

39. For example, some experimental treatments can cause cellular mitochondria to reduce or increase. Consequently, individual cells can reduce less or more MTT to formazan after the respective stimulus impacting OD regardless of changes in viability. Therefore, to complement the MTT assay to measure the effect of such treatments on cell viability, you need to avoid using other assays relying on metabolic function. Instead, consider alternative assays that are mitochondrial-independent to complement the MTT assay (Annexin V staining, Trypan Blue assay, clonogenic assay etc., *see* Table 2).

40. Optical interferences is happening if your data show contribution to the OD signal by components other than formazan. For example, intrinsic absorbance of nanomaterials [11, 41, 42] [or compounds such as doxorubicin (which has an absorption spectrum overlapping formazan [18]) can contribute to the OD measurement. You can test this as explained in Subheading 3.3, **steps** 7 and **8**. In this case, exclude fundamentally similar methods based on colorimetric spectrometry (e.g., XTT, WST-1, WST-8, and LDH). Instead, use different optical methods like fluorometric assays (Alamar blue, presto blue, natural red).

41. Chemical interference is when MTT reduction is happening by abiotic means (something other than cells), for example, when introducing reducing compounds or chemical groups, enzymes, and redox-active nanoparticles. You can test this as explained in Subheading 3.3, **steps** 7 and **8**. The abiotic reduction of MTT causes an overestimation of MTT reduction by cells if raw data are interpreted. In this case, use non-redox-based assays as complementary assays for cell proliferation detection (e.g., Dye exclusion assay, DNA incorporation, clonogenic assay, etc.).

## References


1. Berridge MV, Herst PM, Tan AS (2005) Tetrazolium dyes as tools in cell biology: new insights into their cellular reduction. Biotechnol Annu Rev 11:127–152
2. Berridge MV, Tan AS (1993) Characterization of the cellular reduction of 3-(4,5-dimethylthiazol-2-yl)-2,5-diphenyltetrazolium bromide (MTT): subcellular localization, substrate dependence, and involvement of mitochondrial electron transport in MTT reduction. Arch Biochem Biophys 303(2): 474–482
3. Bernas T, Dobrucki J (2002) Mitochondrial and nonmitochondrial reduction of MTT: interaction of MTT with TMRE, JC-1, and NAO mitochondrial fluorescent probes. Cytometry 47(4):236–242
4. Chakrabarti R et al (2000) Vitamin A as an enzyme that catalyzes the reduction of MTT to formazan by vitamin C. J Cell Biochem 80(1):133–138
5. Collier AC, Pritsos CA (2003) The mitochondrial uncoupler dicumarol disrupts the MTT assay. Biochem Pharmacol 66(2):281–287
6. Pagliacci MC et al (1993) Genistein inhibits tumour cell growth in vitro but enhances mitochondrial reduction of tetrazolium salts: a further pitfall in the use of the MTT assay for evaluating cell growth and survival. Eur J Cancer 29(11):1573–1577
7. Ulukaya E, Colakogullari M, Wood EJ (2004) Interference by anti-cancer chemotherapeutic agents in the MTT-tumor chemosensitivity assay. Chemotherapy 50(1):43–50
8. Stepanenko AA, Dmitrenko VV (2015) Pitfalls of the MTT assay: direct and off-target effects of inhibitors can result in over/underestimation of cell viability. Gene 574(2):193–203
9. Präbst K et al (2017) Basic colorimetric proliferation assays: MTT, WST, and resazurin. In: Methods in molecular biology. pp 1–17
10. Ghasemi M et al (2021) The mtt assay: Utility, limitations, pitfalls, and interpretation in bulk and single-cell analysis. Int J Mol Sci 22(23): 12827
11. Ulukaya E et al (2008) The MTT assay yields a relatively lower result of growth inhibition than the ATP assay depending on the chemotherapeutic drugs tested. Toxicol In Vitro 22(1): 232–239
12. Kong B et al (2011) Experimental considerations on the cytotoxicity of nanoparticles. Nanomedicine 6(5):929–941
13. Kroll A et al (2009) Current in vitro methods in nanoparticle risk assessment: limitations and challenges. Eur J Pharm Biopharm 72(2): 370–377
14. Dhawan A, Sharma V (2010) Toxicity assessment of nanomaterials: methods and challenges. Anal Bioanal Chem 398(2):589–605
15. Schirmer K et al (1998) Ability of 16 priority PAHs to be directly cytotoxic to a cell line from the rainbow trout gill. Toxicology 127(1–3): 129–141
16. Borenfreund E, Puerner JA (1985) Toxicity determined in vitro by morphological alterations and neutral red absorption. Toxicol Lett 24(2):119–124
17. O'Brien J et al (2000) Investigation of the Alamar blue (resazurin) fluorescent dye for the assessment of mammalian cell cytotoxicity. Eur J Biochem 267(17):5421–5426
18. ISO_21115 (2019) Water quality – determination of acute toxicity of water samples and chemicals to a fish gill cell line (RTgill-W1). International Organization for Standardisation, Geneva
19. Kong B et al (2011) Experimental considerations on the cytotoxicity of nanoparticles. Nanomedicine 6:929
20. Detappe A et al (2018) Advancements in nanomedicine for multiple myeloma. Trends Mol Med 24(6):560–574
21. Ahmed KBR et al (2017) Silver nanoparticles: significance of physicochemical properties and assay interference on the interpretation of in vitro cytotoxicity studies. Toxicol In Vitro 38:179–192
22. Seabra AB et al (2014) Nanotoxicity of graphene and graphene oxide. Chem Res Toxicol 27(2):159–168
23. Kroll A et al (2012) Interference of engineered nanoparticles with in vitro toxicity assays. Arch Toxicol 86:1123
24. Henslee EA et al (2016) Accurate quantification of apoptosis progression and toxicity using a dielectrophoretic approach. Analyst 141(23): 6408–6415
25. Rai Y et al (2018) Mitochondrial biogenesis and metabolic hyperactivation limits the application of MTT assay in the estimation of radiation induced growth inhibition. Sci Rep 8(1): 1–15
26. Barkhade T et al (2019) Effect of TiO2 and Fe doped TiO2 nanoparticles on mitochondrial membrane potential in HBL-100 cells. Biointerphases 14(4):041003

27. Maioli E et al (2009) Critical appraisal of the MTT assay in the presence of Rottlerin and uncouplers. Biol Proc Online 11(1):227–240
28. Śliwka L et al (2016) The comparison of MTT and CVS assays for the assessment of anticancer agent interactions. PLoS One 11(5):e0155772
29. Jabbar SAB, Twentyman PR, Watson JV (1989) The MTT assay underestimates the growth inhibitory effects of interferons. Br J Cancer 60(4):523–528
30. Chan SM, Khoo KS, Sit NW (2015) Interactions between plant extracts and cell viability indicators during cytotoxicity testing: implications for ethnopharmacological studies. Trop J Pharm Res 14(11):1991–1998
31. Bruggisser R et al (2002) Interference of plant extracts, phytoestrogens and antioxidants with the MTT tetrazolium assay. Planta Med 68(05):445–448
32. Bernhard D et al (2003) Enhanced MTT-reducing activity under growth inhibition by resveratrol in CEM-C7H2 lymphocytic leukemia cells. Cancer Lett 195(2):193–199
33. Laaksonen T et al (2007) Failure of MTT as a toxicity testing agent for mesoporous silicon microparticles. Chem Res Toxicol 20(12): 1913–1918
34. Kermanizadeh A et al (2015) The role of intracellular redox imbalance in nanomaterial induced cellular damage and genotoxicity: a review. Environ Mol Mutagen 56(2):111–124
35. Liao K-H et al (2011) Cytotoxicity of graphene oxide and graphene in human erythrocytes and skin fibroblasts. ACS Appl Mater Interfaces 3(7):2607–2615
36. Tournebize J et al (2013) Pitfalls of assays devoted to evaluation of oxidative stress induced by inorganic nanoparticles. Talanta 116:753–763
37. Voinov MA et al (2011) Surface-mediated production of hydroxyl radicals as a mechanism of iron oxide nanoparticle biotoxicity. J Am Chem Soc 133(1):35–41
38. Chia HL et al (2018) Cytotoxicity of group 5 transition metal ditellurides (MTe2; M= V, Nb, Ta). Chem Eur J 24(1):206–211
39. Neufeld BH et al (2018) Small molecule interferences in resazurin and MTT-based metabolic assays in the absence of cells. Anal Chem 90(11):6867–6876
40. Bruggisser R et al (2002) Interference of plant extracts, phytoestrogens and antioxidants with the MTT tetrazolium assay. Planta Med 68(5): 445–448
41. Buch K et al (2012) Determination of cell survival after irradiation via clonogenic assay versus multiple MTT assay - a comparative study. Radiat Oncol 7(1):1
42. Hu VW et al (2002) 3H-thymidine is a defective tool with which to measure rates of DNA synthesis. FASEB J 16(11):1456–1457

# Chapter 3

# Assaying Proliferation Characteristics of Cells Cultured Under Static Versus Periodic Conditions

**Daniel F. Gilbert, Oliver Friedrich, and Joachim Wiest**

## Abstract

Two-dimensional in vitro culture models are widely being employed for assessing a vast variety of biological questions in different scientific fields. Common in vitro culture models are typically maintained under static conditions, where the surrounding culture medium is replaced every few days—typically every 48 to 72 h—with the aim to remove metabolites and to replenish nutrients. Although this approach is sufficient for supporting cellular survival and proliferation, static culture conditions do mostly not reflect the in vivo situation where cells are continuously being perfused by extracellular fluid, and thus, create a less-physiological environment. In order to evaluate whether the proliferation characteristics of cells in 2D culture maintained under static conditions differ from cells kept in a dynamic environment, in this chapter, we provide a protocol for differential analysis of cellular growth under static versus pulsed-perfused conditions, mimicking continuous replacement of extracellular fluid in the physiological environment. The protocol involves long-term life-cell high-content time-lapse imaging of fluorescent cells at 37 °C and ambient $CO_2$ concentration using multi-parametric biochips applicable for microphysiological analysis of cellular vitality. We provide instructions and useful information for (i) the culturing of cells in biochips, (ii) setup of cell-laden biochips for culturing cells under static and pulsed-perfused conditions, (iii) long-term life-cell high-content time-lapse imaging of fluorescent cells in biochips, and (iv) quantification of cellular proliferation from image series generated from imaging of differentially cultured cells.

**Key words** Cell growth, Biochip, Microfluidics, HEK293, YFPI152L, Long-term time-lapse microscopy

## 1 Introduction

In vitro culture models have become an indispensable tool for assessing a broad spectrum of biological questions in a vast variety of scientific fields. Conventional cell-based in vitro approaches employed for, for example, drug screening or chemical susceptibility testing, assess the viability or physiology of cells, either maintained in suspension [1], in 2D monolayers [2–8], or in 3D configuration, for example, in spheroids or in multi-cell-configuration such as organ-on-chip approaches [9–13]. These cultures are

Oliver Friedrich and Daniel F. Gilbert (eds.), *Cell Viability Assays: Methods and Protocols*, Methods in Molecular Biology, vol. 2644, https://doi.org/10.1007/978-1-0716-3052-5_3,

almost always established by plating cells in culture media at defined concentration and volume into culture plasticware, including flasks, dishes, multi-titer plates, or lab-on-a-chip devices. Upon plating, the cells are incubated at defined conditions, for example, 37 °C and 5% $CO_2$, for hours, days, or even weeks in accordance with the principles on "Good Cell Culture Practice" (GCCP) [14]. Common in vitro models employed for the aforementioned applications include HEK293 (human embryonic kidney 293) cells. This cell line has also been used for establishing the protocol presented in this chapter.

Cellular proliferation in in vitro cultures is typically assessed using label-free or label-based readout technologies by either continuous time-lapse evaluation or by endpoint observation, where the number of cells is quantified at experiment initiation and at the end of the incubation period. Label-free technologies, such as optical or electrochemical approaches, assess the cell number by measurement of, for example, the cellular contrast in transmission light images, extrusion of $H_3O^+$, cellular consumption of dissolved oxygen, electrophysiological activity, or changes in the dielectricity of cells [6, 13, 15–17]. Label-based methods assess the number or physiology of cells based on luminescence or fluorescence markers indicating, for example, intracellular ATP concentration (e.g., CellTiterGlo® Assay Kit), nuclear DNA content (e.g., Hoechst 33342 stain), activity of intracellular enzymes (e.g., Calcein-AM), or the integrity and function of the cell membrane and membrane proteins such as ion channels (e.g., YFPI152L) [2–5, 18–20].

A commonly employed method for continuous time-lapse evaluation of cellular proliferation uses recombinantly expressed fluorescent proteins such as variants of green fluorescent protein (GFP). Although there is also a large variety of loadable fluorescence indicators or luminescent markers available for label-based monitoring of cellular growth, recombinantly expressed fluorescent proteins are superior to loadable indicators, as proteins are (i) typically very stable with respect to light-induced bleaching, (ii) non-toxic, (ii) cost effective, and (iv) are not diluted by cell division, thereby maintaining strong fluorescence emission and intensity across several generations of cells.

YFPI152L, a genetically engineered variant of yellow fluorescent protein (YFP), fulfills all requirements for fluorescence-based long-term life-cell analysis and has successfully been used in the past for addressing a large variety of biological questions, including continuous time-lapse evaluation of cellular proliferation [6, 21–31]. Implementation of fluorescent indicators, such as YFPI152L, allows quantification of the cellular proliferation rate by analyzing either the number of single cells or the confluence, that is, the percentage of the overall area covered by cells, in images generated from automated fluorescence microscopy using high-content fluorescence imaging infrastructure.

Although many aspects of experimental conditions such as cell culture media composition, culture plastic ware, composition of the gaseous phase, mechanical stimulation, and shear stress have been investigated in detail, the difference in static culture conditions versus periodic exchange of cell culture medium has only marginally been investigated [16, 32–44]. Static culture types are probably the most widely and often used approach, although the culturing conditions may represent a nonphysiological environment. As the culture medium remains unchanged throughout several days, the cells presumably suffer from decreasing availability of nutrients, increasing cellular metabolites, and accumulation of indicators of cell stress as well as altered physiology and fitness, compared to cells cultured under in vivo-like conditions that mimic more closely the native environment of tissue. It is well known that the extracellular microenvironment acidifies in static cultures within short time, creating a nonphysiological situation [45]. In contrast, cells in their native environment in vivo are continuously perfused with extracellular fluid, creating a dynamic extracellular microenvironment of physiological abundance of nutrients and metabolites. To address the issue encountered with conventional in vitro cultures described above, (micro-) fluidics devices, allowing for continuous or periodical perfusion with fresh media, are increasingly being developed and proposed for in vitro models used, for example, in the context of drug susceptibility testing [45–49].

To assess whether the growth characteristics of cells cultured at periodic, that is, dynamically changing, conditions differ from conventional, that is, static, conditions, we have developed a bioassay for observing the growth of differentially cultured cells in cell chips by life-cell time-lapse fluorescence imaging. The assay employs recombinant HEK293-YFP$^{I152L}$ cells, stably expressing a variant of yellow fluorescent protein and quantification of the growth area or confluence in image series [28]. Although being established in our laboratory with recombinant HEK293-YFP$^{I152L}$ cells, the protocol may also be applied with any other fluorescence-labeled adherent cell line. Figure 1a shows the workflow of the assay. Measured time courses of cellular proliferation for cells cultured under static versus periodic conditions are shown in Fig. 1b and indicate differential proliferation characteristics in differentially cultured cells.

## 2 Materials

### 2.1 General Equipment

1. Hemocytometer.
2. Inverted cell culture microscope.
3. 1.5 mL tubes.
4. T75 tissue culture flasks.

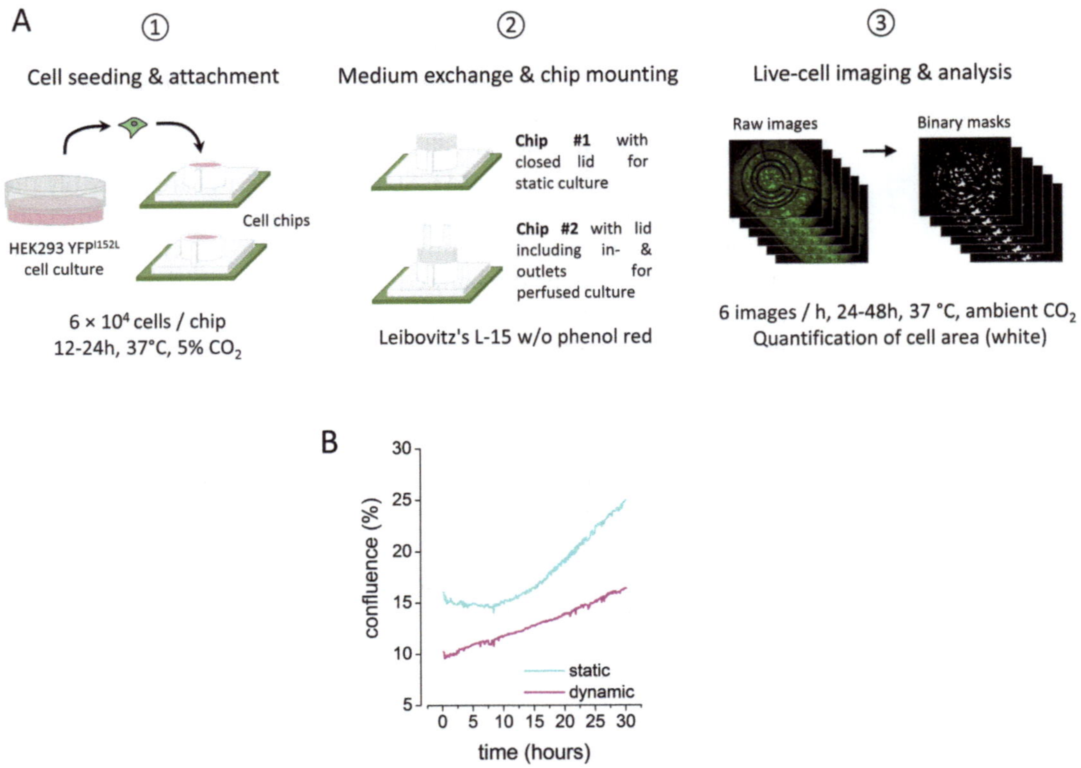


**Fig. 1** (**a**) Experimental workflow for assaying proliferation characteristics of cells cultured under periodic versus static conditions using cell chips and long-term life-cell high-content time-lapse imaging of recombinant HEK293 cells, stably expressing the fluorescent protein YFP$^{I152L}$. Details for ① - ③, *see* in Subheadings 3.1, 3.2, and 3.3. (**b**) Exemplary time courses of the confluence of cells cultured under static (cyan) and dynamic (magenta) culture conditions, respectively. (Modified from Gilbert et al. [28])

5. Parafilm M.
6. EtOH: 70% (v/v).
7. 2× 250 mL bottle for medium and waste.

### *2.2 Biological Material*

1. HEK293-YFP$^{I152L}$ cells [6] (*see* **Note 1**).

### *2.3 Solutions and Reagents*

1. Cell culture medium: 10% fetal bovine serum (FBS), 100 U/mL penicillin, 100 mg/mL streptomycin in Dulbecco's Modified Eagle Medium (DMEM) (*see* **Note 2**).
2. Imaging medium: 10% fetal bovine serum (FBS), 100 U/mL penicillin, 100 mg/mL streptomycin in Leibovitz's L-15 medium without phenol red (*see* **Note 3**).
3. Phosphate buffered saline (PBS).
4. Poly-D-Lysine (PDL): 1× in water.
5. Trypsin-EDTA solution.

### 2.4 Software

1. Image analysis software.
2. Software for data analysis and visualization, for example, SciDAVis, Origin, SigmaPlot, GraphPad Prism, or Microsoft Excel.

### 2.5 Specific Hardware

1. 2× Cell chip with cap (*see* **Note 4**).
2. Automated long-term fluorescence imaging system with motorized stage, cell culture incubator, 10× objective, light source, and YFP excitation/emission filter set (*see* **Note 5**).
3. Peristaltic pump with tubing (*see* **Note 6**).

## 3 Methods

### 3.1 Cell Culture

1. Retrieve frozen HEK293-YFP$^{I152L}$ cells from liquid nitrogen and thaw quickly in a 37 °C water bath.
2. Dilute the cells slowly in 15 mL cell culture medium in a T75 tissue culture flask and incubate at 37 C, 5% $CO_2$, and 95% relative humidity.
3. Exchange the media after 24 h.
4. Passage the cells when approximately 80–90% confluent. Remove the old media, and wash the cells briefly with 5 mL PBS. Add 2 mL of Trypsin-EDTA to detach the cells from the bottom of the flask and incubate at 37 C, 5% $CO_2$ and 95% relative humidity for 3 min. After 3 min trypsinization, add 8 mL media to the flask and pipette up and down to isolate the cells. Seed the cells in a 1:3 dilution into new T75 flasks. Passage the cells every 3–4 days and use for experimentation when approximately 80–90% confluent.

### 3.2 Cell Counting and Cell Chip Preparation

1. To clean and sterilize the two cell chips, fill both chips with approximately 100 μL 70% EtOH each and leave in laminar flow hood at ambient temperature for 10 min.
2. Remove EtOH and wash chips three times with 200 μL PBS.
3. Remove PBS and add 200 μL 1× poly-D-lysine (*see* **Note 7**).
4. Incubate for 10 min at room temperature in a laminar flow hood.
5. Remove poly-D-lysine and leave in the laminar flow hood for approx. 10 min for drying.
6. Trypsinize a confluent T75 flask with 2 mL Trypsin-EDTA as explained in Subheading 3.1. Dilute the cell suspension in a 1: 10 ratio in cell culture medium. Separate the cells by pipetting up and down. Mix 100 μL cell suspension with 900 μL cell culture medium in a 1.5 mL tube in order to count the cells with a hemocytometer (e.g., Neubauer counting chamber).

Pipette 10 μL cell suspension from 1.5 mL tube into each compartment of the counting chamber. Count the number of cells in each of the eight grids using an inverted cell culture microscope. Calculate the average number of cells per milliliter, using the following formula:

$$\frac{\sum \text{cells}}{8} \times 10^5 = \frac{\text{cells}}{\text{mL}}$$

7. Seed a total of $6 \times 10^4$ cells in a volume of 300 μL cell culture medium into each PDL coated chip.
8. Transfer cell-laden chips to incubator and incubate for 12–24 h at 37 C, 5% $CO_2$ and 95% relative humidity to allow cells to settle down and attach to the chip surface (*see* **Note 8**).

### 3.3 Long-Term Culture and Life-Cell Time-Lapse Fluorescence Imaging

1. Approximately 30 min prior to long-term imaging experiments, remove cell culture medium from chips and replace with 300 μL imaging medium.
2. To prevent evaporation of the culture medium during long-term experimentation, close the chip to be used in static culture mode with a cap and seal with Parafilm M.
3. Close the second chip to be used in periodic or pulsed-perfused culture mode with a different lid, equipped with in- and outlets to allow for medium exchange during experimentation.
4. Mount both chips into a microscopy slide carrier (*see* **Note 9**) and transfer to the motorized stage of a high-content long-term imaging system equipped with a temperature-controlled housing and capable of maintaining a constant temperature during the time-lapse experiment.
5. Connect tubing to the second chip and the peristaltic pump and insert the outlet (waste) tube into an empty 250 mL bottle. Insert inlet (imaging medium) tube into a 250 mL bottle filled with imaging solution.
6. Place imaging medium bottle into the temperature-controlled housing of the microscope to ensure equal temperature within both chips during long-term imaging experiment.
7. Set the peristaltic pump to repeatedly perfuse cells within the second chip at a rate of 60 μL per minute for a period of 5 min and in an interval of 10 min.
8. Set the software of the high-content imaging system to image cells in both chips using a 10× objective (*see* **Note 10**) with light passing through a filter suitable to excite and transmit YFP fluorescence signal every 6 min for a total of 24–48 h (*see* **Note 11**).

9. Upon completion of the long-term imaging experiment, quantify the cell area or confluence (in % of total image area) in each recorded image using image analysis software (*see* **Note 12**).
10. Visualize the proliferation characteristics of cells cultured under periodic versus static conditions by plotting the confluence (in %, *y*-axis) against the duration of the experiment (in hours, *x*-axis) for both chips as depicted in Fig. 1b using data analysis software.

## 4 Notes

1. Any other adherent fluorescence-labelled cell line, either stably expressing a fluorescent protein or equipped with a loadable indicator, may be used.
2. Ideally, all reagents should be free of animal components such as FBS or Trypsin [50].
3. We used medium without phenol red to prevent generation of background signal in fluorescence images impairing cell recognition by automated image analysis software for quantitative analysis of cell proliferation
4. We used BioChip-D chips (cellasys GmbH, Germany), but any other chip type suitable for perfusion of cells during culture and fluorescence microscopy may be used. Caps, used to prevent evaporation and for perfusion, should be equipped with in- and outlets, respectively, for connecting tubes compatible with the used peristaltic pump (*see* **Note 5**).
5. The filter set for excitation and detection of cellular fluorescence signal depends on the employed fluorescence label and needs to be adapted according to the recommended label-specific spectral properties.
6. We used an Ismatec Reglo ICC 4CH (Cole-Parmer, USA) peristaltic pump with matching tubing suitable to perfusion at a rate of 60 μL per minute, but any other programmable pump allowing for pulsed or continuous perfusion at a rate of 20–100 μL (depending on chip volume) per minute may be used.
7. Poly-D-lysine promotes adherence of cells onto the growth surface of the employed cell chips. Poly-D-lysine may be exchanged for other reagents promoting adherence of cells.
8. Incubation time may be shortened to a few hours (e.g., 2–4 h) if cells attach fast to shorten preparatory lead time.
9. We have created a custom-designed chip carrier using 3D printing for fixation of chips on a motorized stage, but a conventional slide carrier is also suitable for mounting chips onto a stage.

10. Depending on the size of the cells and the chip type, other objectives, such as a 5× or 20× objective, may also be used.
11. The imaging interval of 6 min reveals ten images per minute (14,400 images per chip and day) and a reasonable temporal resolution for quantitative analysis and comparison of cellular proliferation characteristics. If disk space for storage of captured images is limited or cells are intended to be observed over several days or even in more than a total of two chips per experiment, decreasing the image acquisition rate may be required.
12. Quantification of the cell area (confluence) in cell images is typically conducted by image segmentation, followed by creation of a binary image mask, with the image regions covered by cells displayed in white and the background shown in black (*see* Fig. 1a, ③). We commonly use DetecTiff© software for quantification of the confluence in recorded images [51]. Recommended and free software for quantification of the confluence in image series from fluorescence imaging include, for example, CellProfiler™ [52] or FIJI [53]. Both tools are well documented and provide a large variety of examples facilitating and supporting its use also by unexperienced users.

## References


1. Ritter P, Bye LJ, Finol-Urdaneta RK, Lesko C, Adams DJ, Friedrich O, Gilbert DF (2020) A method for high-content functional imaging of intracellular calcium responses in gelatin-immobilized non-adherent cells. Exp Cell Res 395(2):112210. https://doi.org/10.1016/j.yexcr.2020.112210
2. Gilbert DF, Boutros M (2016) A protocol for a high-throughput multiplex cell viability assay. Methods Mol Biol 1470:75–84. https://doi.org/10.1007/978-1-4939-6337-9_6
3. Gilbert DF, Erdmann G, Zhang X, Fritzsche A, Demir K, Jaedicke A, Muehlenberg K, Wanker EE, Boutros M (2011) A novel multiplex cell viability assay for high-throughput RNAi screening. PLoS One 6(12):e28338. https://doi.org/10.1371/journal.pone.0028338
4. Gilbert DF, Friedrich O (2017) Cell viability assays. Springer New York, New York
5. Kuenzel K, Mofrad SA, Gilbert DF (2017) Phenotyping cellular viability by functional analysis of ion channels: GlyR-targeted screening in NT2-N cells. In: Gilbert DF, Friedrich O (eds) Cell viability assays: methods and protocols. Springer New York, New York, pp 205–214. https://doi.org/10.1007/978-1-4939-6960-9_16
6. Walzik MP, Vollmar V, Lachnit T, Dietz H, Haug S, Bachmann H, Fath M, Aschenbrenner D, Abolpour Mofrad S, Friedrich O, Gilbert DF (2015) A portable low-cost long-term live-cell imaging platform for biomedical research and education. Biosens Bioelectron 64:639–649. https://doi.org/10.1016/j.bios.2014.09.061
7. Gu MB, Mitchell RJ, Kim BC (2004) Whole-cell-based biosensors for environmental biomonitoring and application. Adv Biochem Eng Biotechnol 87:269–305
8. Wiest J, Brischwein M, Ressler J, Otto AM, Grothe H, Wolf B (2005) Cellular assays with multiparametric bioelectronic sensor chips. CHIMIA Int J Chem 59(5):243–246. https://doi.org/10.2533/000942905777676623
9. Alexander F Jr, Eggert S, Wiest J (2017) A novel lab-on-a-chip platform for spheroid metabolism monitoring. Cytotechnology. https://doi.org/10.1007/s10616-017-0152-x
10. Bhise NS, Ribas J, Manoharan V, Zhang YS, Polini A, Massa S, Dokmeci MR, Khademhosseini A (2014) Organ-on-a-chip platforms for studying drug delivery systems. J Control

Release 190:82–93. https://doi.org/10.1016/j.jconrel.2014.05.004
11. Cho S, Yoon JY (2017) Organ-on-a-chip for assessing environmental toxicants. Curr Opin Biotechnol 45:34–42. https://doi.org/10.1016/j.copbio.2016.11.019
12. Schmidt C, Markus J, Kandarova H, Wiest J (2020) Tissue-on-a-chip: microphysiometry with human 3D models on transwell inserts. Front Bioeng Biotechnol 8. https://doi.org/10.3389/fbioe.2020.00760
13. Wiest J (2022) Systems engineering of microphysiometry. Organs-on-a-Chip 4:100016. https://doi.org/10.1016/j.ooc.2022.100016
14. Pamies D, Leist M, Coecke S, Bowe G, Allen DG, Gstraunthaler G, Bal-Price A, Pistollato F, de Vries RBM, Hogberg HT, Hartung T, Stacey G Guidance document on good cell and tissue culture practice 2.0 (GCCP 2.0). https://doi.org/10.14573/altex.2111011
15. Fang Y (2007) Non-invasive optical biosensor for probing cell signaling. Sensors (Basel, Switzerland) 7(10):2316–2329
16. Liu L, Cash TP, Jones RG, Keith B, Thompson CB, Simon MC (2006) Hypoxia-induced energy stress regulates mRNA translation and cell growth. Mol Cell 21(4):521–531. https://doi.org/10.1016/j.molcel.2006.01.010
17. Weiss D, Brischwein M, Grothe H, Wolf B, Wiest J (2013) Label-free monitoring of whole cell vitality. Conf Proc IEEE Eng Med Biol Soc 2013:1607–1610. https://doi.org/10.1109/embc.2013.6609823
18. Kuenzel K, Friedrich O, Gilbert DF (2016) A recombinant human pluripotent stem cell line stably expressing halide-sensitive YFP-I152L for GABAAR and GlyR-targeted high-throughput drug screening and toxicity testing. Front Mol Neurosci 9. https://doi.org/10.3389/fnmol.2016.00051
19. Menzner AK, Abolpour Mofrad S, Friedrich O, Gilbert DF (2015) Towards in vitro DT/DNT testing: assaying chemical susceptibility in early differentiating NT2 cells. Toxicology 338:69–76. https://doi.org/10.1016/j.tox.2015.10.007
20. Menzner A-K, Gilbert DF (2017) A protocol for in vitro high-throughput chemical susceptibility screening in differentiating NT2 stem cells. In: Gilbert DF, Friedrich O (eds) Cell viability assays: methods and protocols. Springer New York, New York, pp 61–70. https://doi.org/10.1007/978-1-4939-6960-9_5
21. Balansa W, Islam R, Fontaine F, Piggott AM, Zhang H, Webb TI, Gilbert DF, Lynch JW, Capon RJ (2010) Ircinialactams: subunit-selective glycine receptor modulators from Australian sponges of the family Irciniidae. Bioorg Med Chem 18(8):2912–2919. https://doi.org/10.1016/j.bmc.2010.03.002
22. Balansa W, Islam R, Fontaine F, Piggott AM, Zhang H, Xiao X, Webb TI, Gilbert DF, Lynch JW, Capon RJ (2013) Sesterterpene glycinyl-lactams: a new class of glycine receptor modulator from Australian marine sponges of the genus Psammocinia. Org Biomol Chem 11(28):4695–4701. https://doi.org/10.1039/c3ob40861b
23. Balansa W, Islam R, Gilbert DF, Fontaine F, Xiao X, Zhang H, Piggott AM, Lynch JW, Capon RJ (2013) Australian marine sponge alkaloids as a new class of glycine-gated chloride channel receptor modulator. Bioorg Med Chem 21(14):4420–4425. https://doi.org/10.1016/j.bmc.2013.04.061
24. Chung SK, Vanbellinghen JF, Mullins JG, Robinson A, Hantke J, Hammond CL, Gilbert DF, Freilinger M, Ryan M, Kruer MC, Masri A, Gurses C, Ferrie C, Harvey K, Shiang R, Christodoulou J, Andermann F, Andermann E, Thomas RH, Harvey RJ, Lynch JW, Rees MI (2010) Pathophysiological mechanisms of dominant and recessive GLRA1 mutations in hyperekplexia. J Neurosci 30(28):9612–9620. https://doi.org/10.1523/jneurosci.1763-10.2010
25. Gebhardt FM, Mitrovic AD, Gilbert DF, Vandenberg RJ, Lynch JW, Dodd PR (2010) Exon-skipping splice variants of excitatory amino acid transporter-2 (EAAT2) form heteromeric complexes with full-length EAAT2. J Biol Chem 285(41):31313–31324. https://doi.org/10.1074/jbc.M110.153494
26. Gilbert D, Esmaeili A, Lynch JW (2009) Optimizing the expression of recombinant alphabetagamma GABAA receptors in HEK293 cells for high-throughput screening. J Biomol Screen 14(1):86–91. https://doi.org/10.1177/1087057108328017
27. Gilbert DF, Islam R, Lynagh T, Lynch JW, Webb TI (2009) High throughput techniques for discovering new glycine receptor modulators and their binding sites. Front Mol Neurosci 2:17. https://doi.org/10.3389/neuro.02.017.2009
28. Gilbert DF, Mofrad SA, Friedrich O, Wiest J (2018) Proliferation characteristics of cells cultured under periodic versus static conditions. Cytotechnology. https://doi.org/10.1007/s10616-018-0263-z
29. Gilbert DF, Wilson JC, Nink V, Lynch JW, Osborne GW (2009) Multiplexed labeling of viable cells for high-throughput analysis of glycine receptor function using flow cytometry.

Cytometry A 75(5):440–449. https://doi.org/10.1002/cyto.a.20703

30. Kahl M, Gertig M, Hoyer P, Friedrich O, Gilbert DF (2019) Ultra-low-cost 3D bioprinting: modification and application of an off-the-shelf desktop 3D-printer for biofabrication. Front Bioeng Biotechnol 7(184). https://doi.org/10.3389/fbioe.2019.00184
31. Talwar S, Lynch JW, Gilbert DF (2013) Fluorescence-based high-throughput functional profiling of ligand-gated ion channels at the level of single cells. PLoS One 8(3): e58479. https://doi.org/10.1371/journal.pone.0058479
32. Abolpour Mofrad S, Kuenzel K, Friedrich O, Gilbert DF (2016) Optimizing neuronal differentiation of human pluripotent NT2 stem cells in monolayer cultures. Develop Growth Differ 58(8):664–676. https://doi.org/10.1111/dgd.12323
33. Dakhil H, Gilbert DF, Malhotra D, Limmer A, Engelhardt H, Amtmann A, Hansmann J, Hübner H, Buchholz R, Friedrich O, Wierschem A (2016) Measuring average rheological quantities of cell monolayers in the linear viscoelastic regime. Rheol Acta 55(7): 527–536. https://doi.org/10.1007/s00397-016-0936-5
34. Demmel F, Brischwein M, Wolf P, Huber F, Pfister C, Wolf B (2015) Nutrient depletion and metabolic profiles in breast carcinoma cell lines measured with a label-free platform. Physiol Meas 36(7):1367–1381. https://doi.org/10.1088/0967-3334/36/7/1367
35. Huh D, Matthews BD, Mammoto A, Montoya-Zavala M, Hsin HY, Ingber DE (2010) Reconstituting organ-level lung functions on a chip. Science 328(5986): 1662–1668. https://doi.org/10.1126/science.1188302
36. Inamdar NK, Borenstein JT (2011) Microfluidic cell culture models for tissue engineering. Curr Opin Biotechnol 22(5):681–689. https://doi.org/10.1016/j.copbio.2011.05.512
37. Khademhosseini A, Langer R (2016) A decade of progress in tissue engineering. Nat Protoc 11(10):1775–1781. https://doi.org/10.1038/nprot.2016.123
38. Liu Q, Wu C, Cai H, Hu N, Zhou J, Wang P (2014) Cell-based biosensors and their application in biomedicine. Chem Rev 114(12): 6423–6461. https://doi.org/10.1021/cr2003129
39. Mahto SK, Yoon TH, Rhee SW (2010) A new perspective on in vitro assessment method for evaluating quantum dot toxicity by using microfluidics technology. Biomicrofluidics 4(3). https://doi.org/10.1063/1.3486610
40. McGillicuddy N, Floris P, Albrecht S, Bones J (2017) Examining the sources of variability in cell culture media used for biopharmaceutical production. Biotechnol Lett. https://doi.org/10.1007/s10529-017-2437-8
41. Pfister C, Bozsak C, Wolf P, Demmel F, Brischwein M (2015) Cell shape-dependent shear stress on adherent cells in a micro-physiologic system as revealed by FEM. Physiol Meas 36(5):955–966. https://doi.org/10.1088/0967-3334/36/5/955
42. van der Valk J, Bieback K, Buta C, Cochrane B, Dirks WG, Fu J, Hickman JJ, Hohensee C, Kolar R, Liebsch M, Pistollato F, Schulz M, Thieme D, Weber T, Wiest J, Winkler S, Gstraunthaler G (2017) Fetal Bovine Serum (FBS): past - present - future. ALTEX. https://doi.org/10.14573/altex.1705101
43. van Midwoud PM, Janse A, Merema MT, Groothuis GM, Verpoorte E (2012) Comparison of biocompatibility and adsorption properties of different plastics for advanced microfluidic cell and tissue culture models. Anal Chem 84(9):3938–3944. https://doi.org/10.1021/ac300771z
44. Yao T, Asayama Y (2017) Animal-cell culture media: history, characteristics, and current issues. Reprod Med Biol 16(2):99–117. https://doi.org/10.1002/rmb2.12024
45. McConnell HM, Owicki JC, Parce JW, Miller DL, Baxter GT, Wada HG, Pitchford S (1992) The cytosensor microphysiometer: biological applications of silicon technology. Science 257(5078):1906–1912
46. Eklund SE, Taylor D, Kozlov E, Prokop A, Cliffel DE (2004) A microphysiometer for simultaneous measurement of changes in extracellular glucose, lactate, oxygen, and acidification rate. Anal Chem 76(3):519–527. https://doi.org/10.1021/ac034641z
47. Marx U, Andersson TB, Bahinski A, Beilmann M, Beken S, Cassee FR, Cirit M, Daneshian M, Fitzpatrick S, Frey O, Gaertner C, Giese C, Griffith L, Hartung T, Heringa MB, Hoeng J, de Jong WH, Kojima H, Kuehnl J, Leist M, Luch A, Maschmeyer I, Sakharov D, Sips AJ, Steger-Hartmann T, Tagle DA, Tonevitsky A, Tralau T, Tsyb S, van de Stolpe A, Vandebriel R, Vulto P, Wang J, Wiest J, Rodenburg M, Roth A (2016) Biology-inspired microphysiological system approaches to solve the prediction dilemma of substance testing. ALTEX 33(3):272–321. https://doi.org/10.14573/altex.1603161

48. Weltin A, Slotwinski K, Kieninger J, Moser I, Jobst G, Wego M, Ehret R, Urban GA (2014) Cell culture monitoring for drug screening and cancer research: a transparent, microfluidic, multi-sensor microsystem. Lab Chip 14(1): 138–146. https://doi.org/10.1039/c3lc50759a
49. Wolf B, Brischwein M, Baumann W, Ehret R, Kraus M (1998) Monitoring of cellular signalling and metabolism with modular sensor-technique: the PhysioControl-Microsystem (PCM). Biosens Bioelectron 13(5):501–509
50. Weber T, Wiest J, Oredsson S, Bieback K (2022) Case studies exemplifying the transition to animal component-free cell culture. Altern Lab Anim:02611929221117999. https://doi.org/10.1177/02611929221117999
51. Gilbert DF, Meinhof T, Pepperkok R, Runz H (2009) DetecTiff: a novel image analysis routine for high-content screening microscopy. J Biomol Screen 14(8):944–955. https://doi.org/10.1177/1087057109339523
52. Carpenter AE, Jones TR, Lamprecht MR, Clarke C, Kang IH, Friman O, Guertin DA, Chang JH, Lindquist RA, Moffat J, Golland P, Sabatini DM (2006) CellProfiler: image analysis software for identifying and quantifying cell phenotypes. Genome Biol 7(10):R100. https://doi.org/10.1186/gb-2006-7-10-r100
53. Schindelin J, Arganda-Carreras I, Frise E, Kaynig V, Longair M, Pietzsch T, Preibisch S, Rueden C, Saalfeld S, Schmid B, Tinevez J-Y, White DJ, Hartenstein V, Eliceiri K, Tomancak P, Cardona A (2012) Fiji: an open-source platform for biological-image analysis. Nat Methods 9(7):676–682. https://doi.org/10.1038/nmeth.2019

# Chapter 4

# Network Reconstruction as a Novel High-Level Marker of Functional Neuronal Viability


## Jana Katharina Dahlmanns and Marc Dahlmanns



## Abstract

Neuronal viability is essential for the maintenance of neuronal networks. Already slight noxious modifications, for example, the selective interruption of interneurons' function, which enhances the excitatory drive inside a network, may already be harmful for the overall network. To monitor neuronal viability on the network level, we implemented a network reconstruction approach that infers the effective connectivity of cultured neurons from live-cell fluorescence microscopy recordings. Neuronal spiking is reported by the fast calcium sensor Fluo8-AM using a relatively high sampling rate (27.33 Hz) to detect fast events such as action potential-evoked rises in intracellular calcium. Spiking records are then subjected to a machine learning-based set of algorithms that reconstruct the neuronal network. Then, the topology of the neuronal network can be analyzed via various parameters, such as the modularity, the centrality, or the characteristic path length. In summary, these parameters describe the network and how it is influenced by experimental modulations, for example, hypoxia, nutrient deficiency, co-culture models, or application of drugs and other factors.



**Key words** Hippocampal culture, Neuronal networks, Algorithm, Connectivity, Fluorescence microscopy, Neuronal viability


## 1 Introduction

Mature neurons are complex cells, whose viability cannot be judged only by means of metabolic parameters, but also by their activity [1]. A healthy neuron also displays healthy activity: it computes various synaptic inputs to generate action potentials and to transmit the information to other neurons inside a neuronal network. On all these stages, neuronal activity can be measured, for example by recording action potentials in whole-cell current clamp mode or by recording inhibitory or excitatory synaptic inputs in terms of postsynaptic potentials (IPSPs or EPSPs). Though isolated parameters can be obtained from these different experimental


Oliver Friedrich and Daniel F. Gilbert (eds.), *Cell Viability Assays: Methods and Protocols*, Methods in Molecular Biology, vol. 2644, https://doi.org/10.1007/978-1-0716-3052-5_4,

approaches, physiological neuronal activity heavily relies on communication inside the network [2], which is not directly addressed by these methods. Recording of large neuronal networks, built up by submodules communicating with each other [3], represents a novel possibility to investigate the neuronal viability. Examination of the network levels allows the experimenter to detect already subtle changes in neuronal viability that still have effects on the entire network.

To extend the capabilities of how neuronal cultures can be examined based on fluorescence microscopy, we have recently implemented a network reconstruction algorithm into an in vitro fluorescence microscopy framework [4]. In this workflow, calcium fluorescence traces from regular recordings are analyzed for their spiking pattern. The spiking data for all cells inside the network are then subjected to algorithms – including transfer entropy [5], mutual information [6], joint entropy [7], and others – and a machine learning approach, to eventually reconstruct the underlying network. This allows the experimenter to assess how a certain drug, a genetic modification, or any other modulation (e.g., hypoxia, nutrient deficiency, or co-culture models), interfere with neuronal function on the network level. This is of particular importance, since it is possible to detect the general output of a network in response to any modifications made on the parts that build up that network. In other words, network analysis includes a new high-level dimension to determine viability.

This protocol is applicable to various cells that can be cultured and to various modifications of neuronal viability. Recently, we demonstrated the suitability of this method by showing how the neurotrophic and neuroprotective transforming growth factor beta member activin A modulates neuronal networks derived from mouse hippocampus [8], and we applied this method to investigate how common antidepressant agents such as serotonin-reuptake inhibitors influence neuronal networks from rat hippocampus [9].

In the first part of this protocol, we briefly describe how to yield viable primary hippocampus cultures, and we explain the way the fluorescence microscopy setup should optimally be designed to enable calcium fluorescence recordings for the scope of network reconstruction. In the second, and main part, we discuss the computational aspects of network reconstruction, which is MATLAB-based and publicly available on GitHub (https://github.com/janawrosch/effective_connectivity). This protocol provides the user with a ready-to-use MATLAB script environment, which requires only little user input to specify imaging data (e.g., objective size).

## 2 Materials

### 2.1 Preparation of Primary Cultures

Here, we describe the preparation of primary hippocampal cultures. Alternatively, especially in case of other sample material such as cerebral cortex or co-cultures, use other protocols known to you and collect all necessary material mentioned in those protocols.

#### 2.1.1 Equipment

1. Animals (mouse or rat).
2. Scissor and Micro-Scissor.
3. Ice/cooling block.
4. Petri dishes (two as basis plus one per each animal).
5. Syringes.
6. Filter for sterile filtering of media.
7. Centrifuge.
8. Microscopes, stereoscopic, and light.
9. Falcons (20 mL, 50 mL).
10. Water bath (or comparable).
11. Cell counter (manual or automatic).
12. Cell culture plates (12 well).
13. Working bench, sterile.
14. Sterile water-soaked paper.
15. Cling film.

#### 2.1.2 Solutions and Media

After preparation, sterile filter all media and store them for the next preparations (at 4 °C, recipe is for around 500 mL).

1. Hank's Balanced Salt Solution (HBBS, Thermo Fisher Scientific).
2. Phosphate buffered Saline (PBS).
3. 2× BES-buffered saline (BBS) solution: in 500 mL $H_2O$, 50 mM BES, 280 mM NaCl, 1.5 mM $Na_2HPO_4$, pH = 7.05 (with 1 M NaOH and 1 M HCl).
4. Eagle's Minimum Essential Medium (MEM).
5. Coating solution: Matrigel (2%), MEM.
6. Culture medium: MEM supplemented with 1% Gibco™ B-27™ (Thermo Fisher Scientific).
7. Basic medium: 500 mL MEM, 2.5 g Glucose, 0.1 g $NaHCO_3$, 0.05 g Transferrin.
8. Starting medium: 500 mL basic medium (**step 6**), 50 mL 10% fetal calf serum (FCS), 5 mL L-glutamine (0.2 M), 1 mL insulin (12.5 mg/mL).

9. Growth medium: 500 mL basic medium, 25 mL 10% FCS, 1.25 mL 0.2 M L-glutamine solution, 10 mL B27™--supplement, 6 μM cytarabine, 535 μL penicillin/streptomycin (100×).
10. Dissociation solution: 100 mL Hank's solution, 12 mM $MgSO_4$* 6 $H_2O$.
11. Digestion solution: 100 mL aqua dest., 137 mM NaCl, 5 mM KCl, 7 mM $Na_2HPO_4$, 25 mM HEPES, pH = 7.2.
12. Imaging buffer: 140 mM NaCl, 10 mM Glucose, 2.4 mM KCl, 10 mM HEPES (4-(2-hydroxyethyl)-1-piperazineethanesulfonic acid), 2 mM $MgCl_2$, and 2 mM $CaCl_2$ (*see* **Note 1**). Use 1 M NaOH and 1 M HCl to set pH = 7.5. Add mannitol to set osmolarity to 320 ± 10 mOsm (*see* **Note 2**) and check that with an osmometer.
13. Staining solution: 1 μM Fluo8-AM (or a comparable fast calcium dye (for the use of genetical calcium sensors, *see* **Note 3**) diluted in imaging buffer.
14. Optional: Drugs to apply to the cultures according to your experimental hypothesis. Add in the appropriate concentration either directly into the imaging buffer (Subheading 2.1.1, **item 11**) or incubate cultures before recording (*see* **Note 4**).
15. Trypsin solution: Trypsin (5 mg/mL), dissociation solution.
16. DNase solution: DNase (375,000 units; 0.05 mg/μL), dissociation solution.

### *2.2 Fluorescence Live-Cell Imaging*

1. Inverted epi-fluorescence microscope.
2. Objective (4×, 0.45 NA; *see* **Note 5**).
3. Dichroic mirror, cut off wavelength of 488 nm in case of Fluo8-AM (*see* **Note 6**).
4. Cover slips.
5. Imaging chamber.
6. Option 1 – Detection of neuronal cells using electrical stimulation equipment: a programmable stimulator and a stimulus isolator, connected to the imaging chamber that contains parallel field electrodes (*see* **Note 7**). If this option does not apply to you, please refer to option 2 described in **step 7**.
7. Option 2 – Detection of neuronal cells with perfusion equipment: a programmable bath application, a suction pump, and potassium chloride (*see* **Note 8**).

### *2.3 Conversion of Raw Camera Images*

1. Software to convert camera raw data to fluorescence images (tagged image file format .tiff).

2. Optional: Driver software of the camera (e.g., Andor Solis water-cooled EM-CCD camera iXon Ultra 897), which enables fast recording from 25 to 30 Hz (*see* **Note 9**).
3. MATLAB (Mathworks Inc., Natick, Massachussetts, USA), only standard toolboxes required.

### 2.4 Files for Network Reconstruction

1. Download all files from https://github.com/janawrosch/effective_connectivity.
2. Save all files in the same directory. Set your MATLAB path in this directory. You do not have to open all these files; however, files termed A–E (used by the experimenter to provide file paths, etc.) require access to all other downloaded files.
3. Download and save into the same directory the following toolboxes, that the connectivity scripts access: https://github.com/hhentschke/measures-of-effect-size-toolbox, https://github.com/piermorel/gramm.

## 3 Methods

### 3.1 Preparation of Primary Hippocampus Cultures

For preparation, at least three animals are recommended, to yield enough cells to plate on one to two multi-well plates, dependent on how dense the cultures are supposed to be. Cover slip coating is performed before preparation, finishing **step** 7 marks the end of day in vitro 0 (DIV 0), and medium is changed on DIV 1 (**step 8**).

1. Before the actual preparation (sometime between 1 day – 1 min before), coat the cover slips with Matrigel (stored at 4 °C) by adding 500 μL (per single 12-well plate) and place it into one well. Shake the well, then take up Matrigel, and add it into the next well, and so on. Incubate the plate (37 °C for 1 h), then take off the excess of the Matrigel.
2. Decapitate the animal (P2–P4) rapidly with a scissor and fixate the head with forceps placed into the eye caveats. After cutting the skull with micro-scissors, scoop out the brain into ice-cooled Hank's solution containing FCS. Remove the cerebellum. Separate both hemispheres and place remaining tissue upside-down under the stereoscopic microscope. Remove basal ganglia until you the hippocampus can be seen right below the cortex. Cut the hippocampus out and extract it from the remaining cortical tissue. Repeat for the other animals.
3. Collect all hippocampus tissue in a fresh petri dish filled with Hank's solution without FCS. Clean the hippocampi from other structures. Cut the hippocampi into smaller parts so that they fit into a pipette, and transfer the tissue into a small falcon (20 mL), filled with Hank's solution without FCS).

Remove the supernatant after tissue has been settled. Again, add Hank's solution (5 mL), wait for settling and remove the supernatant.

4. Add Trypsin solution and DNAse (10 μL) into a syringe, and filter sterile onto the hippocampus pieces. Transfer the mixture into a small petri dish and incubate the mixture (37 °C, 15 min). Transfer it into a falcon containing Hank's solution with FCS. Wash thrice with Hank's solution.
5. After removing Hank's supernatant, add 1 mL dissociation solution and DNase (10 μL, sterile filtered). Separate the cells adherent to each other by pipetting up and down repeatedly. Add Hank's solution with 20% FCS to stop the dissociation reaction.
6. Centrifugate the tissue (10 min, 5000 × *g*). Take off the supernatant and add 2.4 mL starting medium. Calculate the cell number.
7. Use 50 μL cell suspension per well (or whichever volume you calculated for your targeted culture density). Distribute the suspension inside the well by gently shaking. Incubate cells (37 °C for 1 h), then add 950 μL warmed starting medium (Subheading 2.1.1, **item** 7) per well to increase the total volume up to 1 mL. If coverslips float up, press down with a pipette.
8. On the following day, remove 500 μL medium, and add 1 mL fresh pre-warmed growth medium. Press down floating cover slips with a pipette if necessary. Cover all cell culture plates with a sterile water-soaked paper and wrap in cling film. Incubate them until further use (37 °C, 95% $O_2$, 5% $CO_2$).

### 3.2 Fluorescence Live-Cell Imaging of Neuronal Network Activity

1. If applicable, incubate/ treat cell culture wells with any drugs and so forth, then perform experiments between DIV 9 and 14 (*see* **Note 10**).
2. Prepare imaging buffer (store in the incubator or bath at 37 ° C); if required, add any drugs for experimental purposes directly to the imaging buffer; however, we advise to use the incubation model (*see* **Note 11**).
3. Move cover slip with cells into an imaging chamber filled with imaging buffer and place it onto the microscope. Make sure the equipment, which is required for either electrical stimulation (listed in Subheading 2.2, **item 6**) or superfusion with depolarizing agents (listed in Subheading 2.2, **item** 7), is ready.
4. Choose your field of view. Record from a region that has a high number of cells, ideally separated on its border from other cells. Make sure that cells inside your field of view do not overlap

(e.g., no aggregations), since this makes it more difficult to detect the spiking of single cells. If necessary, adjust the number of cells during cell culture preparation (*see* **Note 12**).

5. Set illumination parameters to a brief illumination delivered at high intensity to enable a high sampling rate between 25 and 30 Hz (*see* **Note 13**).
6. Record for 25–30 min and apply electrical or chemical depolarizing stimuli at the end of the recording (*see* **Notes 14, 15**).
7. After the experiment, ensure data are properly stored and data format is set as TIFF (*see* **Note 16**).
8. Take off the imaging chamber, and repeat the experiment with the next cover slips (*see* **Note 17**).
9. Transfer all data (e.g., via hard drives, network, or cloud) to a personal computer with sufficient computational capacity to ensure sufficiently rapid data analysis.

### 3.3 Extraction of Fluorescence Traces and Spike Estimation

1. Run the analysis script A ("Fluorescence trace extraction") that you have downloaded from https://github.com/janawrosch/effective_connectivity, which represents the file required for fluorescence trace extraction.
2. In the graphical user interface (GUI) of file A (Fig. 1), provide the input path where the multilayered TIF files of your experiment are stored (**step 1**, *see* **Note 18**). Next, provide the total number of images for your recording and when you applied chemical or electrical stimulation to excite the neurons (stimstapel), your desired path for saving, and the name of the recording that must end in a running number (**step 2**). Adjust the cell radius and the intensity for the cell detection algorithm (**step 3**), and ensure that these parameters do not overestimate or underestimate your cell dimensions (Fig. 2). To batch process multiple recordings, put them on "hold" after entering inputs and "run trace extraction" later.
3. By running part A, fluorescence traces for each of the detected regions of interest (ROI) are generated as output (Fig. 3).
4. Run part B ("Spike estimation") and enter the file path where results from part A have been stored. Enter the filename stem of your recording files and the running numbers of recordings you want to batch analyze. Ensure to enable the checkbox for control figures if you want to review the correct analysis process (*see* **Note 19**). Start the analysis to perform the spike detection in your traces as illustrated by the displayed control figures (Fig. 4).

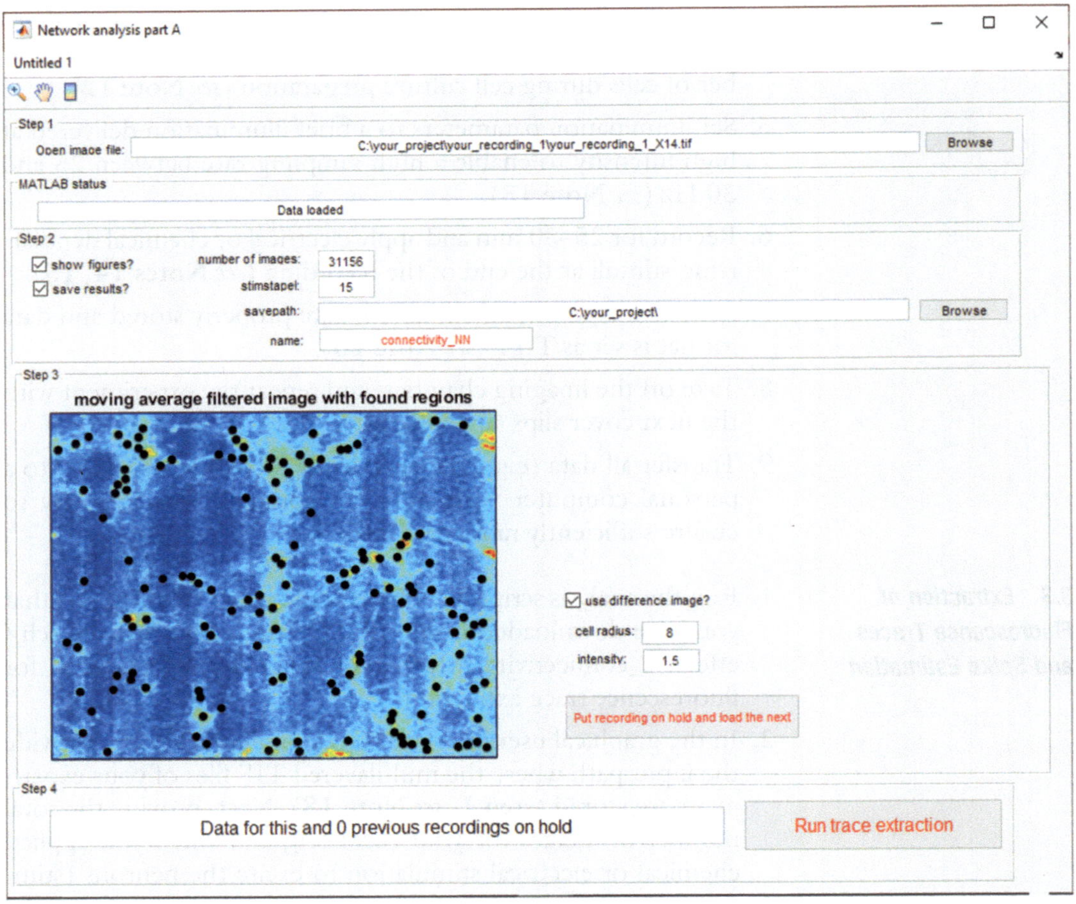


**Fig. 1** GUI from network analysis part. A cell radius of 8 and an intensity of 1.5 were chosen to properly detect all cells in the network

### 3.4 Network Reconstruction

1. Run file C ("Network reconstruction") and enter parameters such as the file path and the running numbers of recordings to process, as already done for part B (Subheading 2.2, **item 2**). Your spiking tables are subjected to a network reconstruction approach (*see* **Note 20**) and yield adjacency matrices for each recording.

### 3.5 Network Topology Analysis and Network Activity Analysis

1. Analyze adjacency matrices according to your specific needs or run file D1 ("Network topology analysis") to calculate several network topology parameters.
2. In the GUI of file D1, enter the file path and the file name. Specify your aims of the analysis by setting "all" parameters (or the type you are interested in). Enter the size of the objective you used (to convert pixels into μm), and set the percentage of how many of the strongest connections you want to analyze to "100" (use less only for very specific research questions). Proceed by clicking "Start" (*see* **Note 21**).

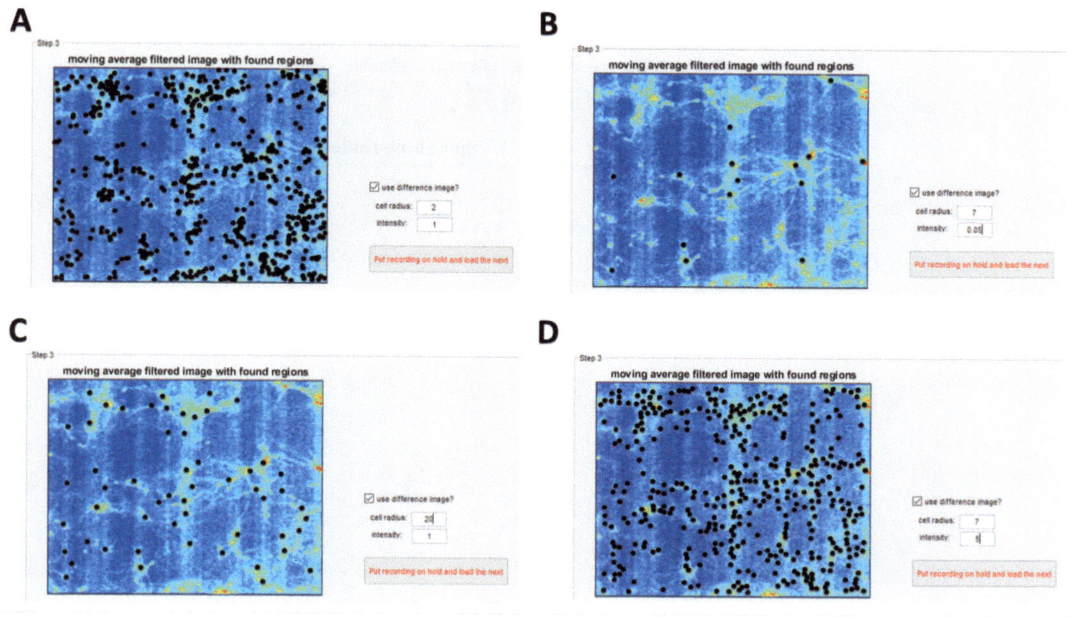


**Fig. 2** Influence of size and intensity parameters on cell detection. (**a**) Chosen cell radius was too low, resulting in too many cells being detected. (**b**) Chosen intensity was too low, so that only the brightest cells were detected. (**c**) Cell radius was chosen too high so that not all cells were detected. (**d**) Intensity was set too high so that too many cells were detected

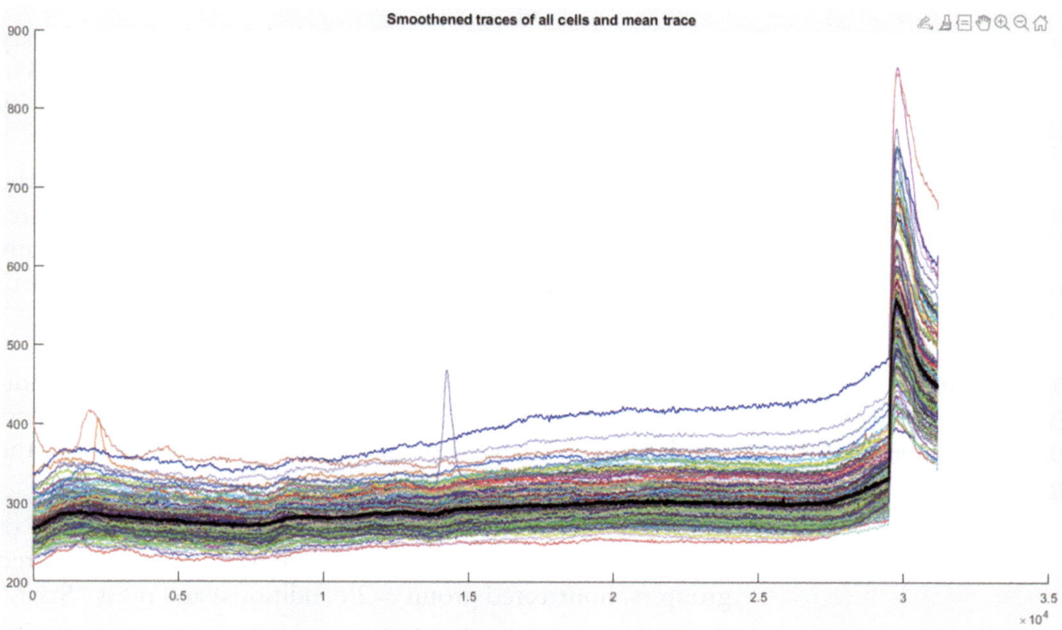


**Fig. 3** Smoothened fluorescence traces. After cell detection, fluorescence traces for all cells are displayed (*y*-axis: arbitrary intensity units: *x*-axis: ms). At the end of the recording, electrical stimulation was applied via field electrodes to excite neuronal cells

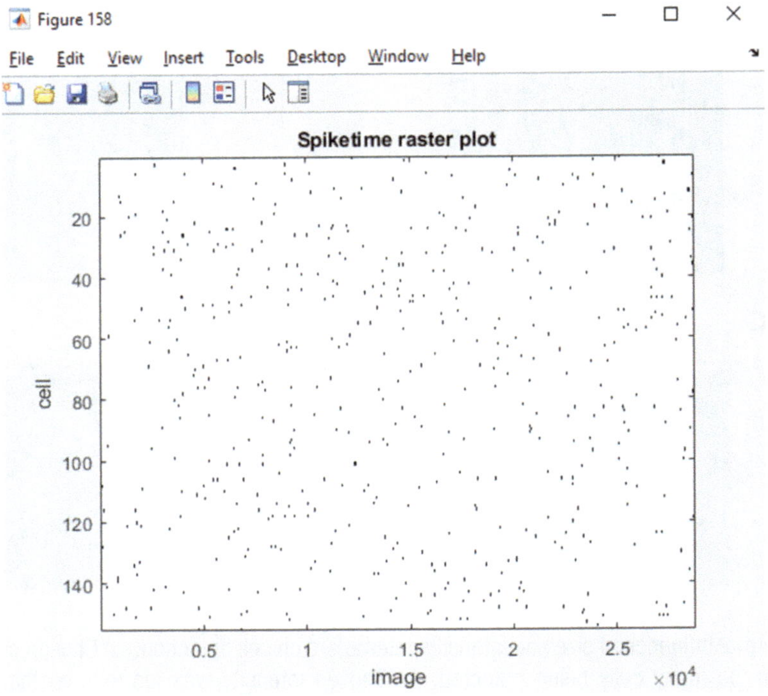


**Fig. 4** Spiketime raster plot. For each cell, the detected spikes were depicted at the corresponding *x*-position (in ms)

3. Run file D2 ("Network activity analysis"), which uses binary spike time traces for each ROI (generated from file B, **step 3** in Subheading 3.3) as input to generate network activity parameters.
4. In the GUI of file D2, enter file path and file name and the recording numbers to be analyzed. Enter your sampling frequency and (if you want to bin neuronal activity) how many seconds one bin should contain and proceed by clicking "Start" (Fig. 5).

### 3.6 Statistical Comparisons Between Different Experimental Groups

1. Run file E ("Compare results of different experimental conditions") to perform statistical comparisons of the parameters that have been analyzed so far for a quick first overview of the results.
2. Enter the file path, the file name, and the total number of conditions in your experiments (e.g., treated group vs. nontreated group = 2 conditions) and press "Start."
3. Enter the names for the different conditions, and specify the running numbers of the recordings that belong to each condition (e.g., 8, 9, 11, and 13). By clicking "Next," repeat this procedure for all conditions in your experiment.

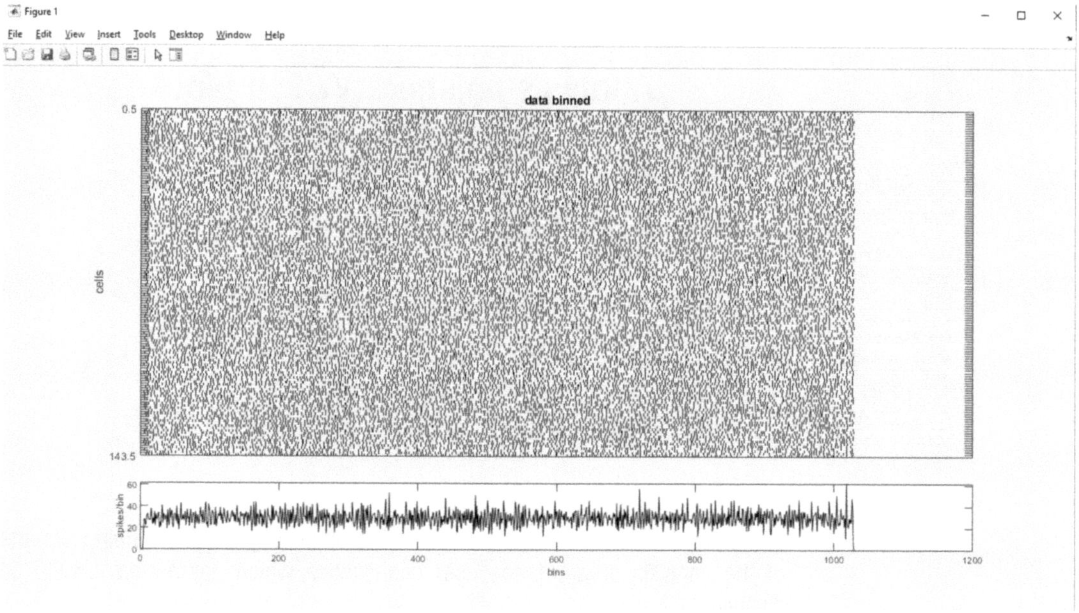


**Fig. 5** Spiking table after binning. For each cell, spiking is displayed per 1 s (chosen bin size here; upper part). For each bin, spikes are summed up to indicate the occurrence of population spiking (bottom part)

4. Now that your analysis is finished, choose among the different parameters which one of them you would like to display (Fig. 6) (*see* **Note 22**).
5. Review your results in the form of figures (as exemplified for the in-degree in Fig. 7 and for the network strength in Fig. 8) and by examining the statistics (illustrated for the in-degree in Fig. 9) (*see* **Note 23**).

## 4 Notes

1. Other ion concentrations may be used as well. To facilitate synaptic transmission, for example, 2.5 mM $CaCl_2$ and 1.5 mM $MgCl_2$ may be used.
2. If you opt for other cultures than primary hippocampus cultures such as cerebral cortex, you may find other imaging buffers more useful. Generally, the recipe for extracellular solution presented in this protocol should work for cortical cultures as well.
3. Genetic calcium sensors bear the risk of systemic errors since the transfection process might not result in a 100% success rate. As a result, not all cells in the network would be detected in the analysis; therefore, the analysis of network interplay will not be reliable. In contrast, the good membrane-permeation rate of the Fluo8-AM dye makes it much more probable that all cells in the culture dish will be stained.

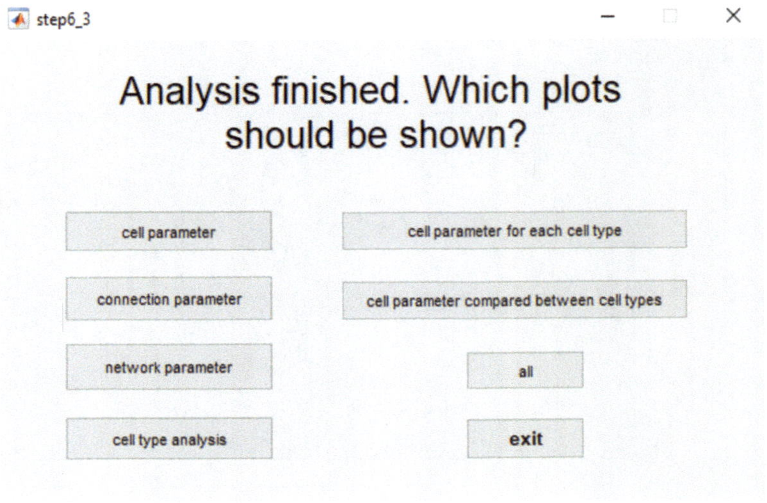


**Fig. 6** Window to choose between results to be displayed. After running all parts of the analysis program, the user can choose which parameters should be displayed

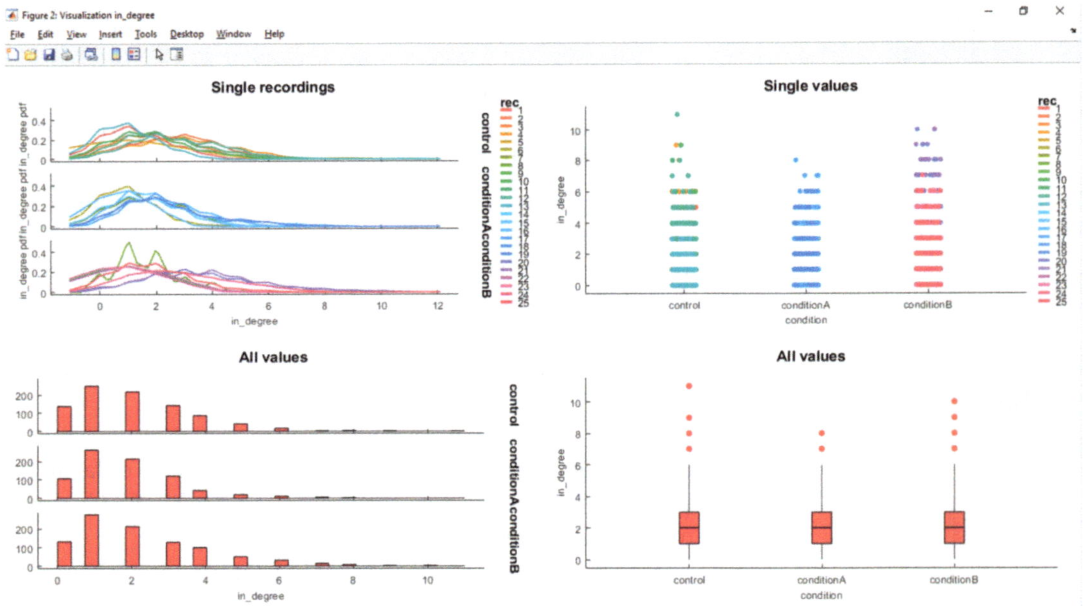


**Fig. 7** Final visualization of the in-degree. Exemplified here is the in-degree parameter for a group of recordings. The user can use this for a first exploration of the results, the value distribution for each parameter for example, to detect outliers and, eventually, to judge the quality of the data set

4. If lipophilic drugs that must be dissolved in dimethyl sulfoxide (DMSO) are used, DMSO concentration should be kept as low as possible, preferably less than 0.01%. If this is not possible, you should add DMSO to your control group experiments.

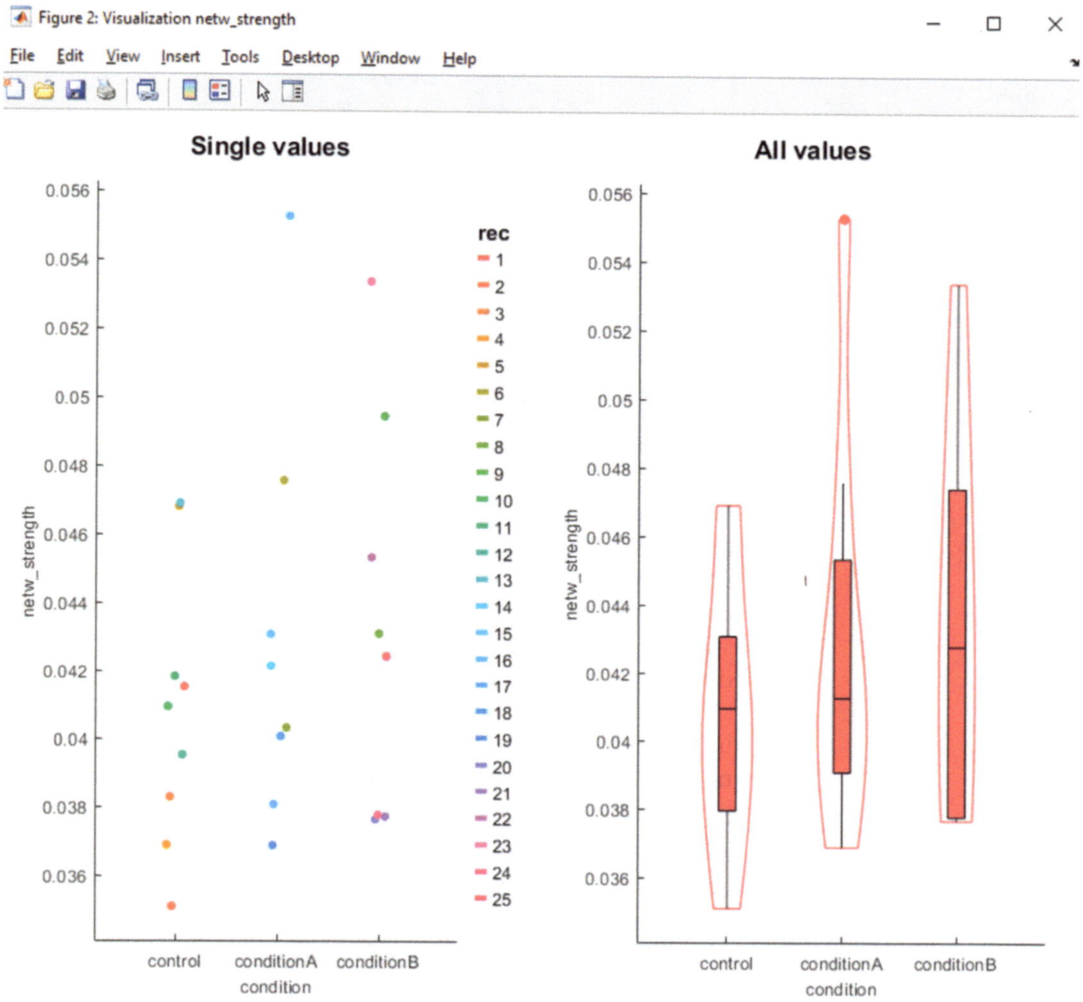


**Fig. 8** Exemplified group comparison of the network strength. Visualization of the network parameter "Network strength" in form of single values (left) and all values (right) to examine data distribution. Please note, that by default no statistical analysis is included in this figure since statistical comparisons must be performed separately

5. A fourfold magnification represents a good trade-off between the number of cells in the field of view and the optical resolution. A good resolution is required to detect spikes inside the noise-containing fluorescence traces. Before reducing magnification to fourfold, it is reasonable to perform preliminary experiments to examine how well spikes are resolved in your setup.
6. If using other fluorescent proteins with a different excitation and emission wavelength, use mirrors, filters, and so forth that fit your dye's requirements.

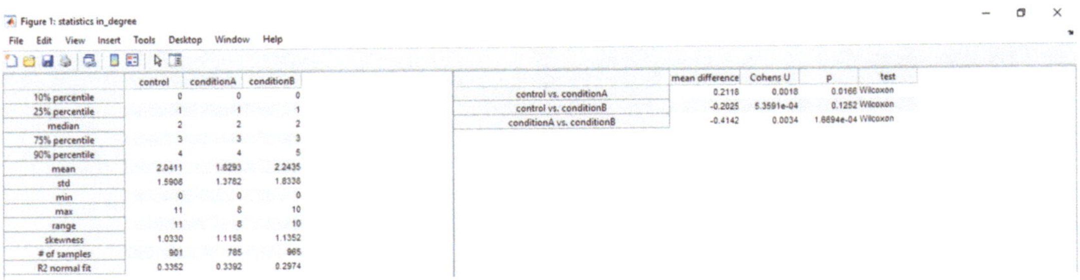
Figure 1: statistics in_degree

File Edit View Insert Tools Desktop Window Help

| | control | conditionA | conditionB |
|---|---|---|---|
| 10% percentile | 0 | 0 | 0 |
| 25% percentile | 1 | 1 | 1 |
| median | 2 | 2 | 2 |
| 75% percentile | 3 | 3 | 3 |
| 90% percentile | 4 | 4 | 5 |
| mean | 2.0411 | 1.8293 | 2.2435 |
| std | 1.5906 | 1.3782 | 1.8338 |
| min | 0 | 0 | 0 |
| max | 11 | 8 | 10 |
| range | 11 | 8 | 10 |
| skewness | 1.0330 | 1.1158 | 1.1352 |
| # of samples | 901 | 785 | 965 |
| R2 normal fit | 0.3352 | 0.3392 | 0.2974 |

| | mean difference | Cohens U | p | test |
|---|---|---|---|---|
| control vs. conditionA | 0.2118 | 0.0018 | 0.0166 | Wilcoxon |
| control vs. conditionB | -0.2025 | 5.3591e-04 | 0.1252 | Wilcoxon |
| conditionA vs. conditionB | -0.4142 | 0.0034 | 1.6694e-04 | Wilcoxon |

**Fig. 9** Exemplified statistics of the in-degree. Statistical analysis in this program includes descriptive and comparative statistics. The data distribution is shown for each condition in form of percentiles, mean, standard deviation etc. All experimental conditions are tested against each other (T-test for normally distributed parameters and Wilcoxon-test otherwise). This is not the correct process to generate correct statistical comparisons for more than two conditions and provides a first overview of results only

7. Electrical stimulation may also be provided by multielectrode arrays. When using field electrodes, alternating current prevents hydrolysis. In our setup, we generally use a constant current amplitude of 50 mA because we found this sufficiently stimulates all neurons in the field of view. As this may be different in your recording setup, preliminary experiments should ensure electrical stimulation to stimulate every neuron in the field of view, because otherwise, some neurons would not be included in the analysis, although they are important parts of the network.
8. A high concentration of potassium chloride (30 mM) will depolarize neuronal membranes, but not glial membranes, which is due to the lack of voltage-gated ion channels that contribute to membrane depolarization.
9. Dependent on your setup, it is possible that frame rates between 25 and 30 Hz cannot be recorded from your usual imaging software, which was the case in our setup. Using plain driver software of the camera, although less user friendly, resolved this problem.
10. After around DIV 14, chances are high that neuronal activity becomes highly synchronized. This impairs proper detection of neuronal connectivity, because the later described detection of neuronal connections is based on reoccurring, asynchronous spiking events.
11. Be aware that the time between placing cells into the imaging chamber and recording network activity needs to be as short as possible: drug effects may occur already during the period before recording. In general, this procedure is not advised because connections are counted during the recording, and experimental changes in the mid of the recording might skew the result. Hence, it is better to choose the incubation method.

If applicable, incubate the cultures for the right time before recording, for example, 2 days before recording (Wrosch et al. 2017; Dahlmanns et al. 2022; Trepl et al. 2022).

12. In Wrosch et al. 2017, the network reconstruction method was validated across of broad range of cell numbers, and it was found that too much and too less cells are decreasing the quality of the reconstruction.
13. Because at a sampling rate of 27.33 Hz the illumination time is very short, illumination still needs to evoke a fluorescence signal large enough to clearly detect cells and their calcium signals separated from background noise. However, there should not be any saturation of the camera by too intense light (perform preliminary experiments to adjust illumination parameters on your setup).
14. In Wrosch et al. 2017, it has been shown that this length is suitable to collect enough spiking events to reconstruct the networks. In case of option 1 – detecting neurons with stimulation equipment – a large electrical stimulus (e.g., 1200 APs, 40 Hz) via field electrodes is required to stimulate neuronal cells selectively and sufficiently. In case of option 2detecting neurons with superfusion equipment—superfuse constantly with imaging buffer and later switch to high-concentration potassium chloride (30 mM) to depolarize neuronal membranes.
15. For superfusion equipment: carefully place superfusion pipette, water level control, and the suction pump inlet so the field of view remains undisturbed by any equipment. Superfusion pipette and suction pump inlet are best positioned opposite each other because this alignment enables a linear flow across the neurons. Stripping of cells from the cover slip is best prevented by setting a 45° angle for the perfusion pipette and a constant flow of 2–2.5 mL/min.
16. If the interplay of your hardware and software does not support save storage of images at recording frequency of 25–30 Hz, it may be advantageous to directly access images from the camera, temporarily store them, and transfer them to long-term storage afterwards.
17. Spread your experimental conditions across neurons prepared at different days to minimize the influence of the preparation itself. Also, try to reach a constant cell number during each preparation.
18. Make sure that in the file step1.m, line 117, the right number of letters is removed from your individual filenames to leave only the filename stem without the running number attached. Examples: first_recording_stack1.tif, first_recording_strack2.

tif, . . ., second_recording_stack1.tif, second_recording_stack2. tif – remove 4 letters (‚x.tif'); first_recording_stack1.tiff, first_-recording_stack2.tiff, . . ., second_recording_stack1.tiff, secon-d_recording_stack2.tiff – remove 5 letters (‚x.tiff'); first_recording_stack_1.tif, first_recording_stack_2.tif, . . ., second_recording_stack_1.tif, second_recording_stack_2.tif – remove 4 letters (‚x.tif').

19. For the control figures of part B, make sure to zoom into these plots, as the number of recording frames usually exceeds the screen resolution, and you will not get everything displayed on your screen.

20. Within part C, the following is analyzed: cross correlation, mutual information, joint entropy, transfer entropy, and generalized transfer entropy.

21. Within part D, following is analyzed: degrees, propagation, path length, clustering, structure, and cell type. Resulting graphs for the in-degree and out-degree are illustrated in Fig. 10.

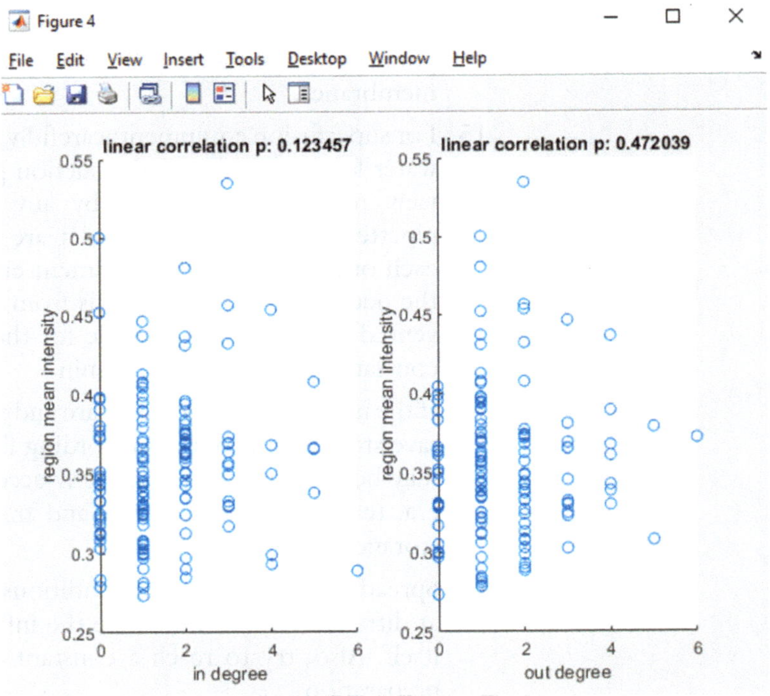


**Fig. 10** Preliminary display of the in-degree and out-degree. The correlation of the in-degree and out-degree with the region mean intensity are shown, indicating that there are no apparent correlations between brightness in the image and the calculated degrees. If you see a correlation here, adjust your imaging and/ or analysis parameters

22. Results can be found in MATLAB workspace in the structures "celldata," "conndata," and "netwdata." Results are displayed using the Gramm toolbox in Matlab, which is implemented in this workflow by default when downloaded from GitHub. Data can easily be transferred to other software by copy-and-paste to meet your illustrative needs.
23. The comparative statistics displayed in the right-hand table serve only the purpose of a first overview and must be correctly calculated independently of the displayed table – especially the included T – and Wilcoxon tests might not be the appropriate tests for your specific experimental design (e.g., more than two conditions).

## References


1. Segal M (2010) Dendritic spines, synaptic plasticity and neuronal survival: activity shapes dendritic spines to enhance neuronal viability. Eur J Neurosci 31(12):2178–2184. https://doi.org/10.1111/j.1460-9568.2010.07270.x
2. Hahn G, Ponce-Alvarez A, Deco G et al (2019) Portraits of communication in neuronal networks. Nat Rev Neurosci 20(2):117–127. https://doi.org/10.1038/s41583-018-0094-0
3. Sporns O (2018) Graph theory methods: applications in brain networks. Dialogues Clin Neurosci 20(2):111–121. https://doi.org/10.31887/DCNS.2018.20.2/osporns
4. Wrosch JK, Einem VV, Breininger K et al (2017) Rewiring of neuronal networks during synaptic silencing. Sci Rep 7(1):11724. https://doi.org/10.1038/s41598-017-11729-5
5. Schreiber T (2000) Measuring information transfer. Phys Rev Lett 85(2):461–464. https://doi.org/10.1103/PhysRevLett.85.461
6. Xu J, Liu Z-r, Liu R et al (1997) Information transmission in human cerebral cortex. Physica D Nonlinear Phenomena 106:363–374. https://doi.org/10.1016/s0167-2789(97)00042-0
7. Lungarella M, Pitti A, Kuniyoshi Y (2007) Information transfer at multiple scales. Phys Rev E Stat Nonlinear Soft Matter Phys 76(5 Pt 2):056117. https://doi.org/10.1103/PhysRevE.76.056117
8. Dahlmanns M, Dahlmanns JK, Schmidt CC et al (2022) Environmental enrichment recruits activin a to recalibrate neural activity in mouse hippocampus. Cereb Cortex. https://doi.org/10.1093/cercor/bhac092
9. Trepl J, Dahlmanns M, Kornhuber J et al (2022) Common network effect-patterns after monoamine reuptake inhibition in dissociated hippocampus cultures. J Neural Transm (Vienna) 129(3):261–275. https://doi.org/10.1007/s00702-022-02477-6

# Chapter 5

# Assaying Mitochondrial Function by Multiparametric Flow Cytometry

**Hannah C. Sheehan, Jonathan L. Tilly, and Dori C. Woods**

## Abstract

Flow cytometry has been a vital tool in cell biology for decades based on its versatile ability to detect and quantifiably measure both physical and chemical attributes of individual cells within a larger population. More recently, advances in flow cytometry have enabled nanoparticle detection. This is particularly applicable to mitochondria, which, as intracellular organelles have distinct subpopulations that can be evaluated based on differences in functional, physical, and chemical attributes, in a manner analogous to cells. This includes distinctions based on size, mitochondrial membrane potential ($\Delta\Psi_m$), chemical properties, and protein expression on the outer mitochondrial membrane in intact, functional organelles and internally in fixed samples. This method allows for multiparametric analysis of subpopulations of mitochondria, as well as collection for downstream analysis down to the level of a single organelle. The present protocol describes a framework for analysis and sorting mitochondria by flow cytometry, termed fluorescence activated mitochondrial sorting (FAMS), based on the separation of individual mitochondria belonging to subpopulations of interest using fluorescent dyes and antibody labeling.

**Key words** Flow cytometry, Organelles, Mitochondria, Mitochondrial heterogeneity, Analytical tools

## 1 Introduction

Although mitochondria are ubiquitous to eukaryotic cells and well-characterized for their role in bioenergetics, these organelles are known to display a high level of heterogeneity in many categories including mitochondrial DNA content, respiration levels, proteomic landscape, morphology, cristae formation and density, subcellular location, membrane potential ($\Delta\Psi_m$), and size [1–4]. Early proteomic studies on bulk-isolated mitochondria demonstrated that only 57% of the identified mitochondrial proteins were consistently expressed across samples taken from the brain, liver, kidney, and heart [5]. These fundamental differences in mitochondrial form and function exist between and within tissues, as well as within individual cells [1, 6]. The variance observed between mitochondrial populations makes these organelles particularly amenable for

Oliver Friedrich and Daniel F. Gilbert (eds.), *Cell Viability Assays: Methods and Protocols*, Methods in Molecular Biology, vol. 2644, https://doi.org/10.1007/978-1-0716-3052-5_5,

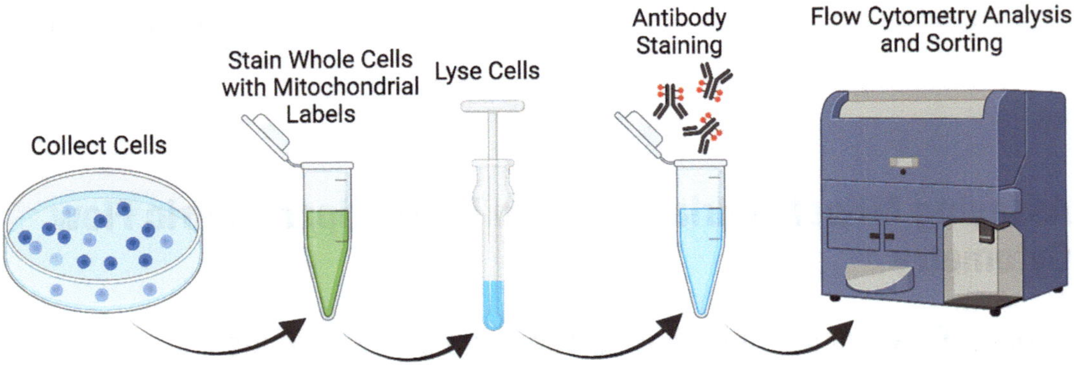


**Fig. 1** Schematic representation of the sample preparation for fluorescence-activated mitochondria sorting. Begin by collecting cells and staining with your desired mitochondrial labels. Next, lyse the cells using a dounce homogenizer, followed by labeling with any required antibodies. Finally, analyze samples via flow cytometry and sort the desired mitochondrial populations

analysis and isolation via flow cytometry. Accordingly, new methods in flow cytometry that include nanoparticle sorting have been developed. We have previously published a validated strategy for multiparametric analysis and single mitochondrion isolation we have termed fluorescence-activated mitochondrial sorting (FAMS) [7]. As an overview, the methodology sample preparation prior to FAMS consists of fluorescent labeling of mitochondria, a cell lysis step optimized for preserving intact mitochondria, and antibody labeling for protein analysis (illustrated in Fig. 1). A fluorescent DNA-tag can also be incorporated for identification of mtDNA. While also applicable for analysis and sorting of mitochondria derived from tissue samples, which require optimized tissue-specific lysis procedures, the step-by-step protocol for FAMS using cultured cells is detailed for this book chapter.

Nanoscale flow cytometry performed with careful size calibration enables detection and sorting of mitochondrial subpopulations based on variations including size, $\Delta\Psi_m$, and antibody-based protein expression, similar to whole-cell flow cytometric applications. In whole-cell cytometry using standard voltages, size resolution is improved using a forward scatter (FSC)-photodiode, which is accurate to approximately 2 μm. To improve performance for nanoparticle analysis and sorting, a FSC-photomultiplier tube (PMT) combined with optimized voltages enables enhanced resolution, including the small size range of mitochondria (0.22–2 μm). Applying these cytometer settings and size gating from approximately 0.22 to 2.0 μm based on SSC versus FSC-PMT in combination with fluorescent labeling, mitochondria can be reliably detected and isolated based on a number of properties. Herein, we provide protocols for mitochondrial labeling and detection using MitoTracker Green FM, the cationic carbocyanine dye JC-1 (5,5′,6,6-′-tetrachloro-1,1′,3,3′-tetraethylbenzimidazolocarbo-cyanine iodide) for the detection of $\Delta\Psi_m$, and antibody-based labeling. These methods can be further combined to evaluate mitochondrial subpopulations.

## 2 Materials

### 2.1 General Equipment

1. BD FACS Aria III Flow Cytometer, equipped with a FSC-PMT in addition to a standard FSC-Diode, as well as 355-, 404-, 488-, 561-, and 633-nm lasers, is used for FAMS in this chapter. Alternatively, other similarly equipped flow cytometers can be used. Prior to sample preparation, the instrument must be calibrated according to manufacturer guidelines specific to the device. On the BD FACS Aria III, cytometer setup and tracking (CS&T) is completed manually, using rainbow fluorescent particles. Before sorting of mitochondria, the BD "Accudrop" program is run (*see* manufacturer's protocol) to ensure successful voltage-aided collection. Additionally, the "Test Sort" method must be completed using the desired sort collection device prior to isolation of mitochondria by FAMS. To enable detection and sorting of submicron events, the instrument should be set to a threshold value of side scatter (SSC) 200, and voltages need to be optimized to detect individual populations of size calibration beads. For size calibration, a series of size calibration beads should be run individually and overlaid to generate a size reference scale (Fig. 2). In this

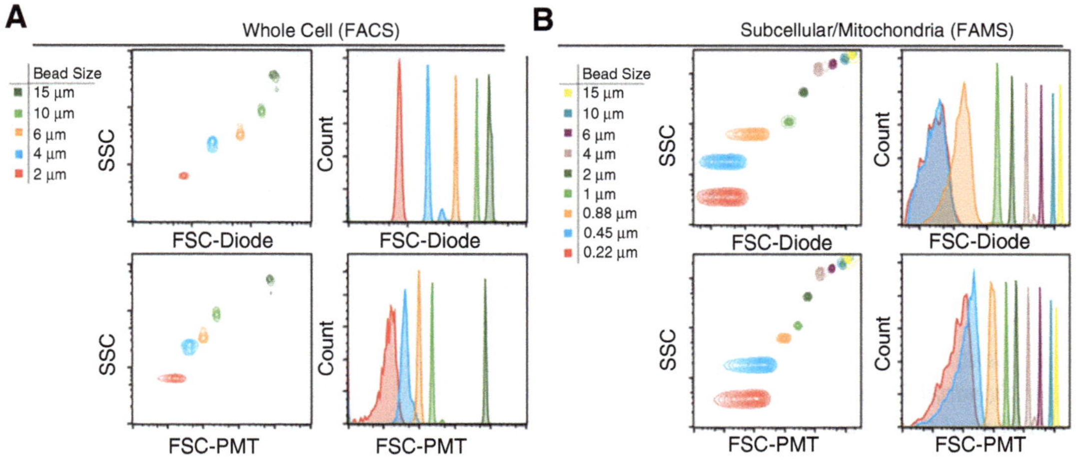


**Fig. 2** Size calibration of nanoparticle beads to optimize cytometer range for isolation of mitochondria. (**a**) A custom FACS Aria II using a standard FSC-diode detector and voltages optimized for whole cell sorting distinguishes size particles down to 2 µm. (**b**) With the use of a FSC-PMT detector and voltages optimized for subcellular particles, the same instrument distinguishes nanoparticles from instrument noise down to 0.22 µm. This figure and legend were first presented in MacDonald et al. 2019 [7]

chapter, plots are generated with FACSDiva Software (version 8.0.2), and data are subsequently analyzed in FlowJo (version 10.8.1).

2. Analytical balance.
3. Refrigerated centrifuge for 15 and 50 mL conical tubes.
4. Refrigerated microcentrifuge.
5. 1 mL dounce tissue grinder/homogenizer.
6. 37 °C incubator.
7. Vortex.
8. Polystyrene FACS tubes.
9. 1.5 mL microcentrifuge tubes.
10. 15 and 50 mL conical tubes.
11. 70 μm cell strainer.
12. Crushed ice or lab armor beads pre-chilled to 4 °C.
13. Pipette aid and 5-, 10-, and 25-mL serological pipettes
14. P10, P20, P200, and P1000 micropipettes and compatible pipette tips

### 2.2 Reagents

1. Cell line or tissue (experiment-specific).
2. 0.85–0.9% blood bank saline.

### 2.3 Buffers

1. PEB buffer: Phosphate-buffered saline (PBS) pH 7.2, 500 mM EDTA pH 8.0, Bovine serum albumin (BSA). Dissolve 0.25 g BSA in 49.80 mL PBS, add 200 μL of 500 mM EDTA, mix and store at 4 °C until use and prepare fresh daily (*see* **Note 1**).
2. Hypo-osmotic lysis buffer: sodium chloride, magnesium chloride, 1 M Tris–HCL pH 7.6, Halt Protease Inhibitor Cocktail EDTA-Free. Dissolve 29.22 mg NaCl and 7.14 mg of $MgCl_2$ into 49.50 mL of MilliQ water. Add 500 μL of 1 M Tris–HCl pH 7.6, vortex to mix. Add Halt Protease Inhibitor Cocktail to a final concentration of 1× only to the volume of RSB Hypo Buffer you will need for your samples. Mix and maintain at 4 °C until use (*see* **Note 1**).

### 2.4 Sample Mitochondrial Stains and Control Reagents

1. MitoTracker Green FM (other MitoTracker dyes can be substituted): Prepare a 1 mM MitoTracker stock solution according to the manufacturer's instructions. This 1 mM stock can be aliquoted, frozen at −20 °C, and used for several months. Prepare MitoTracker working stock (10 μM) by pipetting 10 μL of the 1 mM stock into 1000 μL of PEB. Mix and maintain at room temperature in the dark until needed. Prepare fresh daily from 1 mM aliquots.

2. JC-1 $\Delta\Psi_m$ Dye (*see* **Note 2**): Reconstitute 5 mg JC-1 in 5 mL of DMSO (1 mg/mL). Store aliquots at −20 °C for up to 3 months. Remove one aliquot and equilibrate to room temperature, protected from light. Prepare a 250 μM working stock by mixing 8.15 μL of 1 mg/mL stock in 41.85 μL of DMSO. Mix well and prepare final 100 μM working stock by pipetting 20 μL of 250 μM solution into 30 μL PEB buffer. Mix and maintain at room temperature in the dark until needed. Prepare fresh daily (*see* **Note 3**).
3. When using JC-1, a control sample should be treated with FCCP for a brief incubation in order to depolarize the mitochondrial membrane resulting in a decrease in $\Delta\Psi_m$.
4. Carbonyl cyanide 4-(trifluoromethoxy)phenylhydrazone (FCCP).
5. DAPI (4′,6-diamidino-2-phenylindole): Reconstitute 2 mg DAPI in 1 mL of MilliQ water. This solution can be stored at 4 °C for up to 3 months. Immediately before use, prepare a working stock of DAPI (1:10 dilution) in 0.85–0.9% blood bank saline. Samples will then be stained for 5 min at room temperature with 3 μM DAPI.

### 2.5 Cell Culture Reagents (Cell Line-Specific)

1. Cell culture media.
2. Phosphate-buffered saline (PBS; sterile).
3. Cell detachment solution (we frequently use 0.25% Trypsin-EDTA).

### 2.6 Antibody Labeling Reagents

1. Blocking Buffer: Fetal bovine serum (FBS), Bovine serum albumin (BSA), Phosphate-buffered saline (PBS). Dissolve 100 mg of BSA into 49 mL of PBS, add 1 mL of FBS, and mix and store at room temperature until use (*see* **Note 1**).
2. Mitochondrial Fixation and Permeabilization Buffer: Phosphate-buffered saline (PBS), 16% Paraformaldehyde (PFA), Triton X-100. Add 1.25 mL of 16% PFA and 10 μL of Triton X-100 to 8.74 mL of PBS. Mix and store at room temperature until use (*see* **Note 1**). This buffer is only for use when detection of IMM protein is required, in which case some steps of this protocol will require alternatives (*see* Sect. 3.4).
3. Primary antibodies, as needed for analysis: Direct antibody conjugation kits, as needed for analysis. This protocol utilizes direct antibody conjugation to detect antigens associated with the outer mitochondrial membrane (*see* **Note 4**). If protein expression is to be assessed using antibodies, these antibodies should be conjugated before beginning the experimental procedure to prepare samples for FAMS analysis (*see* **Note 3**). We routinely use Abcam antibody conjugation kits for FAMS experiments but have used other commercially available conjugation kits with success (*see* **Note 5**).

### 2.7 Reagents Used in Instrument Calibration

1. Sphero Rainbow Fluorescent Particles, 3.0–3.4.
2. Flow Cytometry Size Calibration Kit.
3. Nano Fluorescent Size Standard Kit.
4. BDFACS Accudrop Beads.
5. Sort collection device.

## 3 Methods

### 3.1 Mitochondrial Staining

*Timing:* 30 min

Here we use a combination of MitoTracker Green FM (MitoTracker GREEN), which localizes to mitochondria regardless of $\Delta\Psi_m$, and JC-1, which emits a red-orange fluorescence upon conversion from monomeric to aggregate form only in mitochondria exhibiting high-$\Delta\Psi_m$ (*see* **Note 6**). Alternatively, the use of expression vectors that encode photoactivatable proteins with mitochondrial localization signals (MLS) can be used to facilitate mitochondrial identification. Mitochondrial labeling is performed prior to whole-cell lysis.

1. Collect cultured cells by centrifugation following proper trypsinization according to the cell line being used. For many cell lines, 2 mL of pre-warmed (37 °C) 0.25% Trypsin-EDTA is sufficient to detach cells from a 10 cm culture dish in several minutes. Following detachment, deactivate trypsin with 8 mL of pre-warmed cell media containing FBS prior to centrifugation. For consistent results, a recommended sample size is $1 \times 10^6$ cells per sample. In this example, six samples are to be used, equating to $6 \times 10^6$ cells in total.
2. After aspirating or carefully decanting the supernatant, resuspend the cell pellet in 1 mL of PEB buffer per sample (*see* **Note 7**).
3. MitoTracker Green and JC-1 staining. If no OMM proteins are to be antibody-labeled, any antibody-related sample tubes can be omitted. If analysis will be performed on fixed and permeabilized samples for IMM protein analysis, skip **steps 3–10** and proceed to **step 11**.
4. For this analysis, 6 mL of cell suspension will be used. Divide the 6 mL cell suspension between six 1.5 mL microcentrifuge tubes (1 mL of the total volume in each tube) labeled "Unstained," "MitoTracker Green only," "JC-1 only," "Antibody Only," "MitoTracker Green and JC-1," and "MitoTracker Green, JC-1, and Antibody."
5. To the "MitoTracker Green only" tube, add 10 μL of MitoTracker Green working solution.
6. To the "JC-1 only" tube, add 10 μL of JC-1 working solution.

7. To the "MitoTracker Green and JC-1" and the "MitoTracker Green, JC-1, and Antibody" tubes, add 10 μL of MitoTracker Green working solution and 10 μL of JC-1 working solution. The "Unstained" and "Antibody Only" tubes serve as controls and will not require MitoTracker Green or JC-1 stain (*see* **Note 8**).
8. MitoTracker Red labeling (for when mtDNA or IMM protein detection are required). If planning to use DNA labels but not perform antibody-based protein detection, any antibody-related sample tubes can be omitted.
9. For this analysis, 4 mL of cell suspension will be used. Divide the 4 mL of cell suspension between four 1.5 mL microcentrifuge tubes (1 mL of the total volume in each tube) labeled "Unstained," "MitoTracker Red only," "Antibody Stained," and "MitoTracker Red and Antibody."
10. To the "MitoTracker Red only" and "MitoTracker Red and Antibody" tubes, add 10 μL of the MitoTracker RED working solution. The "Unstained" and "Antibody Stained" tubes serve as various controls and will not require MitoTracker Red (*see* **Note 8**).
11. Mix by gently pipetting and incubate for 15 min at 37 °C, protected from light. From the unstained sample, pipet 20 μL of cell suspension into a separate microcentrifuge tube, and do not centrifuge this tube during **step 11**. This will be used in Cell Lysis **step 4** to confirm cell lysis.
12. Centrifuge sample tubes at 500 *g* for 5 min at 4 °C to pellet cells (*see* **Note 9**).

### 3.2 Cell Lysis

*Timing*: 20 min

During this step, cells are incubated in cold hypo-osmotic cell lysis buffer for 10 min prior to gentle dounce homogenization or vortexing to disrupt the cell membrane and release intracellular contents.

1. Resuspend each cell pellet in 1 mL of hypo-osmotic lysis buffer that has been pre-chilled to 4 °C, pipetting up and down 10 times to mix thoroughly.
2. Incubate on crushed ice (or 4 °C Lab Armor beads) for 10 min protected from light to allow for swelling of cells (*see* **Note 10**).
3. Complete cell lysis by vortexing each sample for up to 1 min or dounce homogenizing up to 50 strokes (*see* **Note 11**).
4. Visually confirm cell lysis by microscopy. There should be >90% cell lysis compared to original cell suspension (reserved in Mitochondrial Staining **step 11**).
5. Centrifuge all tubes at 12,000 *g* for 5 min at 4 °C to pellet lysate.

6. Wash by resuspending pellet in 1 mL of PEB buffer and centrifuging tubes at 12,000 *g* for 5 min at 4 °C.
7. Carefully aspirate supernatant. If antibody labeling is not part of the experiment, resuspend pellet in 1 mL of 0.85–0.9% blood bank saline and transfer through a 70 μm cell strainer into a polystyrene FACS tube. Maintain on cold metal beads prior to analysis and sorting (*see* **Note 12**).

### 3.3 Antibody Labeling for OMM (Optional)

*Timing*: 1 h

In this step, cell lysate can be probed using direct antibody labeling of proteins associated with the OMM. These can be detected on a per-organelle basis to further characterize mitochondrial subpopulations.

1. Resuspend pellets in 1 mL of blocking buffer and incubate for 20 min at room temperature protected from light.
2. During this time, dilute conjugated antibody in blocking buffer to desired concentration, preparing 1 mL for each sample to be stained (*see* **Note 13**).
3. Centrifuge sample tubes at 12,000 *g* for 5 min at 4 °C to pellet.
4. Carefully aspirate supernatant and resuspend "Antibody Only" and "MitoTracker Green, JC-1, and Antibody" samples in 1 mL of antibody staining solution prepared in **step 2**, or in blocking buffer for all other samples.
5. Incubate for 20 min at room temperature, protected from light.
6. Centrifuge sample tubes at 12,000 *g* for 5 min at 4 °C to pellet.
7. Carefully aspirate the supernatant.
8. Resuspend samples in 1 mL of PEB buffer to wash. Centrifuge at 12,000 *g* for 5 min at 4 °C to pellet. Carefully aspirate the supernatant.
9. Resuspend pellets in 1 mL of 0.85–0.9% blood bank saline and transfer through a 70 μm cell strainer into a polystyrene FACS tube. Maintain on cold metal beads or crushed ice prior to analysis and sorting.

### 3.4 Antibody Labeling for Fixed Mitochondria (Optional)

*Timing*: 1 h and 40 min

In this step, cell lysate can be probed using indirect antibody labeling of proteins in fixed mitochondria, including those internal to mitochondria, such as those associated with mtDNA, IMM, and cristae. These can be detected on a per-organelle basis to further characterize mitochondrial subpopulations using fixation and permeabilization of cell lysates prior to antibody blocking (*see* **Note 14**). For experiments using fixed mitochondria for protein detection, the basic list of samples that the cell pellet will be divided into in Mitochondrial Staining **step 4** above will be as follows: "Unstained," "MitoTracker Red Only," "Antibody Stained," and "MitoTracker Red + Antibody."

1. Resuspend pellets in 1 mL of fixation and permeabilization solution. Incubate for 5 min at room temperature. Centrifuge at 12,000 *g* for 5 min. Remove supernatant.
2. Resuspend pellets in 1 mL of blocking buffer and incubate for 20 min at room temperature protected from light.
3. During this time, prepare the antibody staining solution by diluting the fluorescent-conjugated primary antibody in blocking buffer to your desired concentration, preparing 200 μL for each sample to be stained (*see* **Note 13**).
4. Centrifuge sample tubes at 12,000 *g* for 5 min at 4 °C to pellet.
5. Carefully aspirate supernatant and resuspend "Antibody stained" samples in 1 mL of antibody staining solution prepared in **step 3** or in blocking buffer for all other samples.
6. Incubate for 20 min at room temperature, protected from light.
7. Centrifuge sample tubes at 12,000 *g* for 5 min at 4 °C to pellet.
8. Carefully aspirate the supernatant.
9. Resuspend samples in 1 mL of PEB buffer to wash. Centrifuge at 12,000 *g* for 5 min at 4 °C to pellet. Carefully aspirate the supernatant.
10. Resuspend pellets in 1 mL of 0.85–0.9% blood bank saline and transfer through a 70 μm cell strainer into a polystyrene FACS tube. Maintain on cold metal beads or crushed ice prior to analysis and sorting.

### 3.5 Mitochondrial Analysis by Flow Cytometry

*Timing*: 30 min for analysis.

1. After running size calibration beads during Equipment Setup (*see* Subheading 3.2), using the same voltages and thresholds, analyze samples.
2. Size gate from approximately 0.2–2 μm to exclude debris and large events. Assign fluorescent gates based on unstained samples (*see* **Note 15,** and Fig. 3).
3. For MitoTracker Green-labeled samples, gate for FITC-positive events by comparing the "Unstained" sample to the "MitoTracker Green Only" sample.
4. Next, if JC-1 is used for detection of $\Delta\Psi_m$, use the FITC-positive events observed in the "MitoTracker Green Only" sample to gate for $FITC^+/PE^-$ and $FITC^+/PE^+$events. $FITC^+/PE^-$ events represent mitochondria with low- to no $\Delta\Psi_m$, while $FITC^+/PE^+$ events represent mitochondria with high-$\Delta\Psi_m$ (*see* **Note 16**).

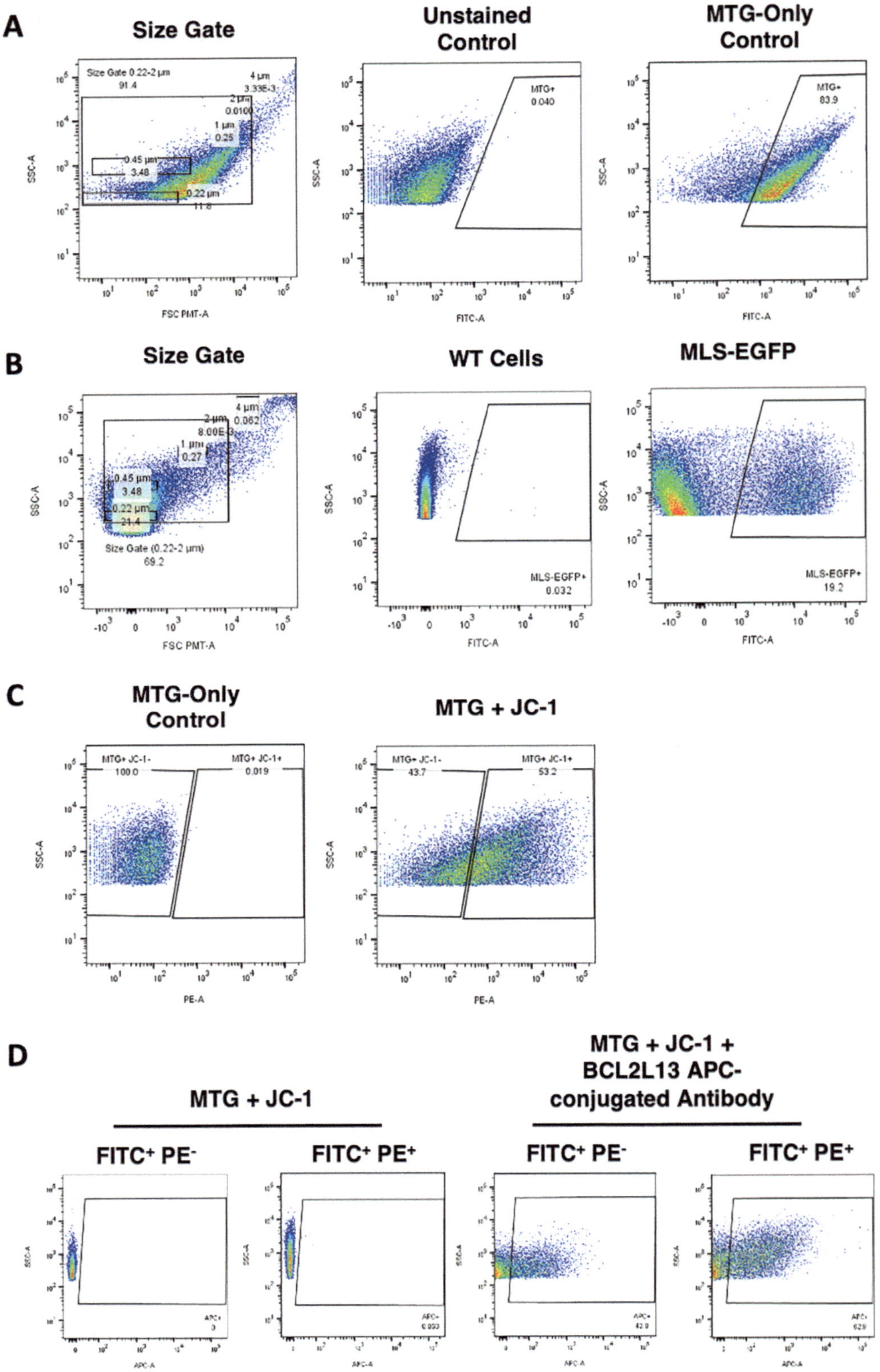


**Fig. 3** Example gating strategies for identification of mitochondria, as well as characterizing $\Delta\Psi_m$ using JC-1 dye, and protein expression on the OMM using a directly conjugated antibody. (**a**) Positive staining for

5. If using MitoTracker Red staining in fixed samples, gate for PE-Texas Red-positive events by comparing the "Unstained" sample to the "MitoTracker Red Only" sample.
6. If DAPI (or another DNA dye) will be used on fixed and permeabilized cell lysate, or dye/antibody-labeled samples, record desired events before adding the DNA dye to the sample tubes. Add the DNA label and then record the sample a second time 5 min after the DNA label has been added, being sure not to overwrite the data from the pre-label recordings.
7. For OMM protein detection: Gate the $FITC^+$ events (MitoTracker $Green^+$) as total mitochondria population. Of the $FITC^+$ events, events positive for the fluorophore conjugated to your antibody represent mitochondria possessing your chosen antigen. Here, APC is used as the example fluorophore: Use the FITC-positive events observed in the "MitoTracker Green Only" sample to gate for $FITC^+/APC^-$ and $FITC^+/APC^+$ events. $FITC^+/APC^-$ events represent mitochondria without the protein present while $FITC^+/APC^+$ events represent mitochondria with the antibody bound.

### 3.6 Sorting of Mitochondria (Optional)

*Timing:* Sorting time will depend on the required number of mitochondria for downstream analysis and sample concentration. We routinely sort $4 \times 10^6$ events per 15 mL collection tube for 40 min.

1. Sort desired events into 15 mL conical tubes containing sheath buffer (*see* **Note 17**). For routine sorting of mitochondria, it is recommended that a 100-μm 20-psi sort nozzle be utilized, but a 70-μm 70-psi nozzle may be used if an extremely high

**Fig. 3** (continued) MitoTracker Green FM (MTG) dye conclusively identifies events as mitochondria in HepG2 liver carcinoma cells after first gating all events from 0.22 to 2 μm (located using sizing beads for reference in FSC-PMT-A vs. SSC-A) to encompass the average size range of mitochondria. Using an MTG stained sample, an event is considered positive for MTG if it exhibits higher fluorescence in the FITC-A channel than the unstained control sample. (**b**) Alternatively, mitochondria can be identified using MLS-EGFP, a photoactivatable protein with a mitochondria localization signal. Using a sample consisting of mitochondria isolated from cells expressing MLS-EGFP, an event is considered to be a mitochondrion if it exhibits higher fluorescence in the FITC-A channel than the wild-type control cell sample. (**c**) $\Delta\Psi_m$ can be measured using an MTG + JC-1-stained sample (derived from HepG2 cells in this example) as compared to an MTG-Only stained sample. A mitochondrion is considered to have high-$\Delta\Psi_m$ if, following size gating and positive fluorescence in the FITC-A channel for MTG dye, it is found to exhibit higher fluorescence in the PE-A channel compared to the MTG-Only stained sample. (**d**) As an example of assessing protein expression within mitochondrial subpopulations, high- and low-$\Delta\Psi_m$ mitochondria [identified using the gating structure detailed in (**a**) and (**b**)] were evaluated for the presence of BCl2L13. For probing of HepG2 cell lysate, the BCL2L13 antibody was directly conjugated to APC. Using a sample stained only with MTG and JC-1, a gate was drawn in the APC channel. Mitochondria were determined to be positive for BCL2L13 if fluorescence in the APC channel was higher in the antibody-stained sample than in the MTG + JC-1 sample

number of mitochondria are needed for downstream analysis (*see* **Note 17**). For sorting of pooled mitochondrial samples, we routinely use the "4-way purity" mask within the BD FACS Diva software. However, for sorting single mitochondria, we use the "single cell" purity mask to increase sample purity. In cases where pooled samples require increased purity, such as mass spectrometry, use of the "single cell" mask should be considered, with the caveat that the increased purity associated with this mask leads to decreased yield of total sorted mitochondria (*see* **Note 18**).

2. Often times, greater than 1.5 mL of blood bank saline containing mitochondria will have been collected, and the total mitochondrial sample will need to be concentrated within a single microcentrifuge tube. To do this, 1.5 mL of the sorted mitochondria (contained in blood bank saline) should be centrifuged at 12,000 *g* for 5 min at 4 °C.
3. Following centrifugation, carefully aspirate the supernatant and add the next 1.5 mL of the sorted mitochondrial sample to the same microcentrifuge tube.
4. Again, repeat the centrifugation step (12,000 *g* for 5 min at 4 °C) followed by careful aspiration of spent blood bank saline. Repeat this process until the entire sorted mitochondria sample has been pelleted. Aspirate supernatant and process mitochondrial pellet for desired downstream applications.
5. Alternatively, single events may be sorted into individual wells of a 96-well plate for downstream analysis by diluting sample to a threshold rate of ~300 events/second and using a high-purity precision mask, such as "Single cell." Sorted samples may be stored at −80 °C prior to downstream analysis where maintenance of respiratory capacity is not required.

## 4 Notes

1. Each of the buffers (prepared in Subheadings 2.3 and 2.6) should be prepared fresh daily and maintained at the proper temperature prior to use.
2. MitoTracker Green FM dye and JC-1 dye are not retained after fixation. When using fixed mitochondria for mtDNA labeling or IMM protein detection, other MitoTracker dyes can be used. Primarily, we use MitoTracker Red CMXRos for this purpose. For MitoTracker Red CMXRos, the stock solution should be made by adding 94.1 μL of DMSO to the lyophilized tube, and then prepare the 10 μM working solution in the same way that the MitoTracker GREEN working solution was prepared in Subheading 2.4, **item 1**. If $\Delta\Psi_m$ and mtDNA

presence are to be analyzed using the same samples, it is possible to first stain using the standard procedure (MitoTracker GREEN + JC-1), sort MitoTracker GREEN+ JC-1+ and MitoTracker GREEN+ JC-1- events to separate high- and low-$\Delta\Psi_m$ mitochondria. After collection, proceed with fixation and permeabilization of the sorted mitochondria, stain with the DNA-tag, and analyze the samples again using a flow cytometer.

3. The concentrations of MitoTracker GREEN, JC-1 and any antibodies used should be determined empirically for each cell line used. The concentrations used here have been tested for HepG2s, MEFs, HEK293s, IMR-90 s, KGNs, MCF7s, and several other commercially available immortalized cell lines.
4. If using antibodies for protein detection, basic application of this protocol requires that antigens of interest are located externally to the mitochondrial membranes and that live cells are used under experimental conditions. However, additional "optional" steps and alternatives have been incorporated to allow for modification of the protocol for use in mtDNA identification within fixed mitochondria and IMM protein identification.
5. Rather than directly conjugating primary antibodies to fluorophores, it is also possible to use fluorescently labeled secondary antibodies. However, we will not detail the use of indirectly conjugated antibodies in this protocol. Additionally, as an alternative to the FACS tubes listed, other polystyrene FACS tubes may be used. In our experience, polypropylene FACS tubes possess particles that can be gated out as noise in cell sorting applications but that cloud analyses in the sub-micron size range needed for analysis of mitochondria (Fig. 4).
6. In intact cells, JC-1 associates with, but does not accumulate within, the outer membrane of non-respiring mitochondria (low-$\Delta\Psi_m$) and emits a green fluorescent spectrum. Accumulation within the inner membrane (high-$\Delta\Psi_m$) causes emission spectra to shift from green to red. This spectral shift makes JC-1 a valuable dye for the ratio metric analysis of low- and high-$\Delta\Psi_m$ within intact cells. In cell lysates, only the red-orange spectra accumulated within mitochondria can be reliably observed due to a lack of association of the green, non-aggregated, form. Therefore, in order to identify and isolate both low- and high-$\Delta\Psi_m$ mitochondrial subpopulations, here we will include MitoTracker GREEN in addition to JC-1. As a control for JC-1, we suggest incorporation of an electron chain uncoupler, such as FCCP.
7. At this step, 1 mL of PEB buffer equates to one sample. If additional replicates, experimental conditions, or controls are

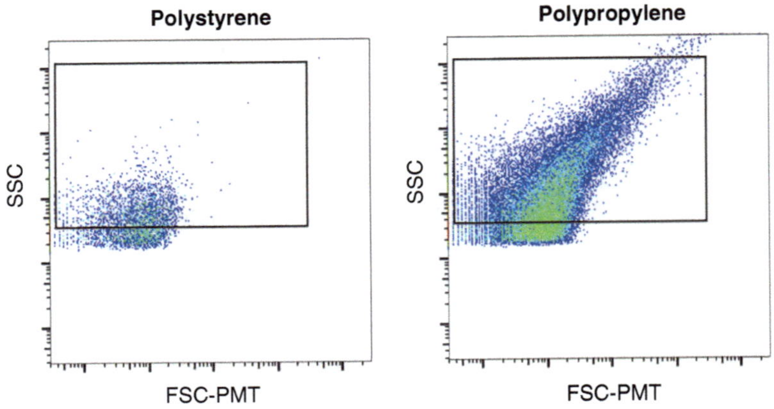


**Fig. 4** MilliQ water was added to clean unused FACS tubes. Events, which in this case were classified as "noise," were recorded for 1 min for each tube type and compared. A gate has been drawn to highlight the approximate size range of mitochondria. During the 1-min time frame, 6951 events were detected in the polystyrene tube, and 47,683 events were detected in the polypropylene tube. The events detected in the polypropylene tube were also seen to be larger in size

desired, alter the volume of PEB buffer used at this step to allow for additional 1 mL samples. However, it is recommended to maintain the $1 \times 10^6$ cells/mL sample concentration.

8. While these are the standard fluorescence controls used in basic experiments, vehicle controls, isotype controls, and other controls are used when appropriate.
9. Use a centrifuge with spinning bucket rotors to minimize cell death.
10. It is important to maintain samples on dry, crushed ice—rather than in the melted ice water—to maintain sample integrity. Alternatively, we routinely utilize Lab Armor beads that have been cooled to 4 °C to keep our samples chilled while working with them.
11. Optimal lysis method must be determined for each cell line. Cell lines with a low cytoplasm: nucleus ratio will require dounce homogenization rather than vortexing to complete lysis. In our hands, many cell lines require only 35 strokes of dounce homogenization rather than 50. Over homogenization is to be avoided, as it compromises mitochondrial integrity.
12. Use of polystyrene FACS tubes is necessary. Using polypropylene tubes is associated with increased debris in the mitochondrial size range, which can skew quantitative results (Fig. 4).
13. Proper antibody concentration must be determined by titration for each antibody and can be both cell type- and antibody lot-specific. For directly conjugated antibodies being used for

OMM protein detection, we suggest titration using five antibody concentrations ranging from 0.25 to 2 μg/uL, followed by analysis of protein expression in mitochondria (events that are between 0.22 and 2 μm and are positive for MitoTracker GREEN fluorescence and antibody-based fluorescence). This allows for the highest signal: noise ratio and prevents excessive antibody use. For primary and secondary antibodies that will be used without direct conjugation, concentrations should be determined by titration, with reference to manufacturer-recommended concentration ranges for flow cytometry.

14. FAMS-isolated mitochondria can also be used for analysis of IMM proteins following fixation and permeabilization, using the same steps listed in this section.
15. Incomplete cell lysis could be visually observed during the lysis step, or appear as a high number of events appearing in a larger size range than where the mitochondria are (typically the 6 μm size range is visible on FAMS plots). If this is observed during analysis, prepare fresh hypo-osmotic lysis buffer, and ensure that it is kept cold. Continue dounce homogenization or vortexing until cells are visibly lysed. Return samples to FACS tubes and re-analyze.
16. After many hours of sorting, fluorescence signal may shift. This could indicate the presence of residual lysis buffer. Dilute samples 1:5 in 0.85–0.9% Blood Bank Saline while waiting to analyze by FAMS.
17. Use of a 70-μm 70-psi nozzle instead of a 100-μm 20-psi could lead to rupture of mitochondrial membranes upon impact with the plastic collection tube due to increased pressure. This nozzle would be fine for certain downstream applications (i.e., mtDNA copy number analysis by PCR) but would not be desirable for others where intact mitochondrial membranes are crucial (i.e., ATP bioluminescence assays).
18. Mitochondria exist as a dense network in physical contact with certain other organelles and intracellular structures—one of the primary examples of this being mitochondrial-ER contact sites [8]. Therefore, we would expect mitochondrial protein samples to be positive for markers of other organelles associated with mitochondria, unless action is taken to disrupt those connections. Increasing sample purity using the "single cell" purity mask on the BD FACS Diva software and maintaining a low threshold event rate (<300 events per second) during sorting decrease the likelihood of contaminating proteins in mass spectrometry or other sensitive applications, but it is not reasonable to expect there to be zero protein remnants in downstream analyses, due to the physical connectivity of

mitochondria to the cellular framework. In other words, this is a biological issue, not a technical issue. In MacDonald et al. 2019 [7], purification of mitochondria in sorted samples for downstream analysis of mtDNA was reported as a comparison of mtDNA to nuclear DNA.

## Acknowledgments

This work was supported by a grant from the National Science Foundation, Award number CAREER 1750996 to D.C.W.

### References

1. Woods DC (2017) Mitochondrial heterogeneity: evaluating mitochondrial subpopulation dynamics in stem cells. Stem Cells Int 2017: 7068567
2. Kuznetsov AV, Margreiter R (2009) Heterogeneity of mitochondria and mitochondrial function within cells as another level of mitochondrial complexity. Int J Mol Sci 10(4): 1911–1929
3. Collins TJ, Berridge MJ, Lipp P, Bootman MD (2002) Mitochondria are morphologically and functionally heterogeneous within cells. EMBO J 21(7):1616–1627
4. Wikstrom JD, Twig G, Shirihai OS (2009) What can mitochondrial heterogeneity tell us about mitochondrial dynamics and autophagy? Int J Biochem Cell Biol 41(10):1914–1927
5. Mootha VK, Bunkenborg J, Olsen JV, Hjerrild M, Wisniewski JR, Stahl E, Bolouri MS, Ray HN, Sihag S, Kamal M, Patterson N, Lander ES, Mann M (2003) Integrated analysis of protein composition, tissue diversity, and gene regulation in mouse mitochondria. Cell 115(5):629–640
6. Wai T, Langer T (2016) Mitochondrial dynamics and metabolic regulation. Trends Endocrinol Metab 27(2):105–117
7. MacDonald JA, Bothun AM, Annis SN, Sheehan H, Ray S, Gao Y, Ivanov AR, Khrapko K, Tilly JL, Woods DC (2019) A nanoscale, multi-parametric flow cytometry-based platform to study mitochondrial heterogeneity and mitochondrial DNA dynamics. Commun Biol 2(1):258
8. Filippin L, Magalhães PJ, Di Benedetto G, Colella M, Pozzan T (2003) Stable interactions between mitochondria and endoplasmic reticulum allow rapid accumulation of calcium in a subpopulation of mitochondria. J Biol Chem 278(40):39224–39234

# Chapter 6

# High-Efficiency Single-Cell Electrical Impedance Spectroscopy

Yongxiang Feng, Liang Huang, Peng Zhao, Fei Liang, and Wenhui Wang

## Abstract

Single-cell impedance measurement is label free and noninvasive in characterizing the electrical properties of single cells. At present, though widely used for impedance measurement, electrical impedance flow cytometry (IFC) and electrical impedance spectroscopy (EIS) are used alone for most microfluidic chips. Here, we describe high-efficiency single-cell electrical impedance spectroscopy, which combines in one chip the IFC and EIS techniques for high-efficiency single-cell electrical property measurement. We envision that the strategy of combining IFC and EIS provides a new thought in the efforts to enhance the efficiency of electrical property measurement for single cells.

**Key words** Single cell analysis, Microfluidics, Electric impedance spectroscopy, Impedance flow cytometry, Electrical property

## 1 Introduction

Biological cells show considerable heterogeneity, which brings single-cell analysis into the recent research focus in understanding the biological processes of cells [1]. Apart from proteomics, genomics, and metabolomics analysis, single-cell analysis also covers measurement of biophysical parameters [2, 3]. Biophysical properties such as mechanical and electrical properties of cells can serve as label-free biomarkers to reveal the physiological status of cells [4]. In particular, cell electrical properties reflect the characteristics of membrane morphology, ion channel status, nuclear size, and cytoplasm [5]. So far, electrical properties are used to characterize physiological states and distinguish different cell populations in applications including but not limited to counting, sorting, and drug screening, and hence, have gained momentum in single-cell research [6, 7].

**Supplementary Information** The online version contains supplementary material available at https://doi.org/10.1007/978-1-0716-3052-5_6.

Oliver Friedrich and Daniel F. Gilbert (eds.), *Cell Viability Assays: Methods and Protocols*, Methods in Molecular Biology, vol. 2644, https://doi.org/10.1007/978-1-0716-3052-5_6,

There are many methods for cell electrical property measurement. The classic Nobel-prize-winning patch clamping method relies on the electrode-embedded micropipette tip sealing with the cell membrane. Though it is widely used in biology, the operation is complicated and time consuming with limited throughput [8]. Recent advances in microelectro-mechanical system (MEMS) and microfluidic techniques allow fabricating large-scale microelectrodes and microfluidic channels, which give rise to several attractive microfluidics-based methods, including electrorotation, electric impedance spectroscopy (EIS), electric cell-substrate impedance spectroscopy (ECIS), and impedance flow cytometry (IFC) [9, 10]. Among these techniques, ECIS is suitable for adherent cells [11], while the other three are for suspended cells, which are of interest in this step-by-step guide.

The cell electrical property measurement methods work properly for particular purposes. Electrorotation, driven by dielectrophoresis (DEP) torque [12], can evaluate electrical properties accurately using the rotation spectrum. But it is time consuming, for example, about 5 min per cell, and troublesome with complex measurement processes involving handling low-conductivity DEP buffer [13]. IFC and EIS obtain the electrical properties of single cells through impedance measurement, which is nondestructive and sensitive. Developed from Coulter counter principle [14], IFC yields a limited number of frequency-dependent instantaneous impedance data points for a cell when it passes through rather than remaining still on coplanar electrodes [15] or parallel-facing electrodes [16]. Thanks to the high-throughput nature, IFC can provide the statistical distribution of a cell population at the single-cell level. EIS yields the impedance spectrum by applying a frequency-sweeping signal excitation to a trapped cell. Ample single-cell trapping techniques can be incorporated into the microfluidic chip, such as dielectrophoresis (DEP) traps [17], hydrodynamic traps [18, 19], and optical traps [20], to trap a single cell.

Here, we provide a step-by-step guide for using a novel microfluidic device combining the IFC and EIS techniques, aiming to screen the electrical properties of single cells more efficiently [21]. IFC and EIS are both impedance-based measurement for the electrical properties of suspended cells using the same cell electrical model, which constitutes the basis for their combination. EIS can obtain the full impedance spectrum, which contains abundant characteristic information of the cell. Its low throughput may be complemented by IFC, which can promptly produce several representative frequency-dependent impedance data points. The device is configured with hydrodynamic trapping sites and coplanar electrodes to allow IFC for passing cells and EIS for trapped cells. With the fabricated device, the individual functionality of IFC and EIS can be verified using three types of cancer cells, and their complementarity can be demonstrated, which holds for a wide range of flow rate. We envision that the strategy of combining IFC and EIS provides a new thought in the efforts to enhancing the efficiency of electrical property measurement for single cells.

## 2 Materials

### 2.1 Analyte Preparation

1. Cell line: Cancer cell lines HeLa, A549, and HepG2 or any other suitable cell line can be used here for experiments described below, and the culturing conditions and medium mentioned below are specific for these cell lines.
2. Cell culture medium: High-glucose Dulbecco's Modified Eagle's Medium (DMEM), supplemented with 10% fetal bovine serum (FBS) and 1% penicillin-streptomycin.
3. 0.05% trypsin.
4. Bovine serum albumin (BSA).
5. Phosphate-buffered saline (PBS).
6. Polystyrene microbeads.

### 2.2 Device Fabrication

1. Soft lithography: Negative photoresist SU-82025 and corresponding developer.
2. Negative photoresist BN303.
3. Polydimethylsiloxane (PDMS) and corresponding curing agent.
4. Hole puncher.
5. Glass substrate.
6. Negative photoresist ROL-7133 and corresponding developer.
7. Deionized water, acetone and absolute ethanol.
8. Gold target.
9. Plasma cleaner.
10. Printed circuit board (PCB).

### 2.3 Impedance Measurement System

1. Signal generator: Direct digital frequency synthesizer AD9958.
2. Transimpedance amplifier (TIA): Operational amplifier OPA657 and 100 kΩ resistor.
3. Lock-in amplifier (LIA): Operational amplifier OPA227, multiplier AD835, resistor, and capacitor.
4. Data acquisition: Data acquisition board PCI-6289.

### 2.4 Software and Analysis

1. Data acquisition: LabVIEW, for example, custom-made code "DataAcqurie-Labview2018.vi" provided as Electronic Supplemental Material, available on link.springer.com.

2. Data analysis and processing: MATLAB, for example, custom-made code "DataProcessing-matlab-R2016b.m," provided as Electronic Supplemental Material, available on link.springer.com.

#### 2.5 Clean Room Equipment

1. Spin-coating system.
2. Hot plates.
3. Electron-beam lithography system.
4. Magnetron Sputtering Machine.
5. Plasma cleaner.
6. Vacuum pump.

#### 2.6 Experimental Equipment

1. CCD camera.
2. Inverted microscope with C-port.
3. Microfluidic pump.

## 3 Methods

#### 3.1 Cell Preparation

1. Culture the cancer cells HeLa, HepG2, and A549 using an incubator at 37 °C in 5% $CO_2$.
2. When the adherent cells are grown to 70%, remove the culture medium from the culture flask, and rinse the flask with 2 mL of PBS gently to remove culture medium residue.
3. Remove the PBS and add 2 mL 0.05% trypsin to detach the adherent cells. Incubate the flask (37 °C) for 3 min when most of the cells are detached and floating.
4. Add 3 mL fresh culture medium to the culture flask, and rinse the flask to wash all cells off from the flask.
5. Centrifuge at 300 g for 5 min to remove the supernatant, and resuspend the collected cells in PBS supplemented with 1% BSA to avoid adhesion and shake gently to obtain the uniform cell suspension.
6. Use cell concentration of $10^6$/mL in experiments for conducting the electrical impedance measurement.

#### 3.2 Microfluidic Chip Design

1. To enable EIS and IFC simultaneously, two types of microchannels are required on one chip for cells to pass or stay. One for IFC should allow the cell to pass, and the other for EIS allow the cell to stay. To this end, the device is designed with upper and lower layers (Fig. 1).
2. The lower layer is configured with an array of coplanar electrodes for impedance sensing in IFC/EIS, while the upper layer is deployed with the two types of microchannels.

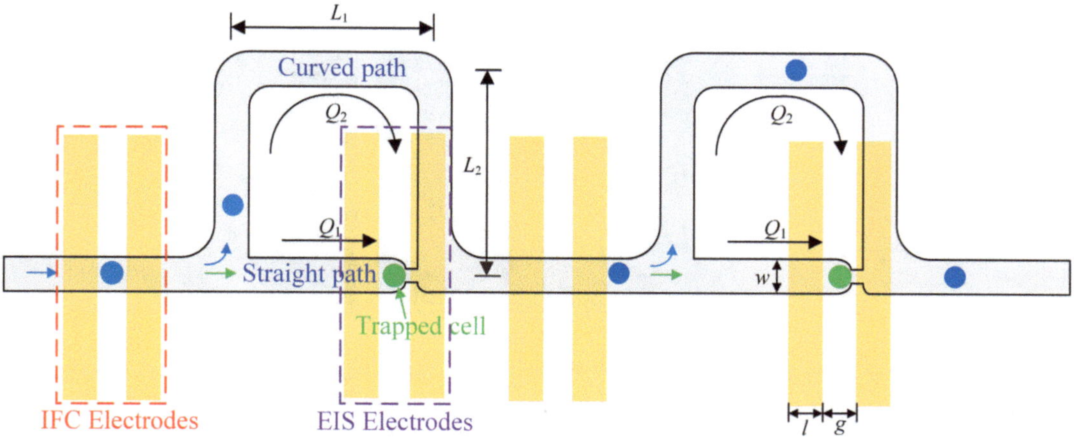


**Fig. 1** The schematic diagram of the microchip structure depicting the cell passing/trapping for IFC/EIS impedance sensing. The channels are designed following the least flow resistance path principle. IFC electrodes are underneath the main channel, and EIS electrodes are underneath the trapping sites. The first incoming cell would pass the main channel for IFC and then be trapped on the trapping site for EIS

3. The upper layer is designed as a passive hydrodynamic microfluidic structure with an array of repeating cell pass-and-stay units (*see* **Note 1**); each unit has a main channel bifurcating to two paths. The curved (bypassing) path allows for cells to pass through and the straight (trapping) path for one cell to be trapped in the narrow constriction (trapping site). Electrodes are placed properly underneath the main channel and the trapping site. The microchannel geometry of the paths ($L_1 = 125$ μm, $L_2 = 125$ μm) is designed such that the volumetric flow rate ratio of $Q_1/Q_2 > 1$ (*see* **Note 2**) to follow the least flow resistance path principle.
4. Under such condition, the flow resistance of the bypassing path is greater than that of the trapping path, and the first cell passing through the main channel would be trapped in the next trapping site. Therefore, the same cell can be measured via IFC and then EIS.
5. To achieve highly sensitive impedance sensing and avoid cell blockage and current leakage, the width and height of the microchannels are both 20 μm to allow easy passing of the cells with diameter of 10–16 μm. The constriction width is 6 μm to stop the cell fully on site for EIS.
6. In case that the trapping site is taken up by a cell, the trapping path is presumably blocked, and its flow resistance will increase drastically so that the next pass-and-stay unit is in action. Considering factors of measurement sensitivity (*see* **Note 3**), space efficiency, fabrication, and cell size, here we set gap $g = 30$ μm, length $l = 30$ μm and width $w = 20$ μm for the electrodes.

**Fig. 2** Microfluidic chip fabrication procedure and fabricated device. (**a**) The fabrication procedure of the microfluidic chip using the lift-off and soft-lithography technique. (**b**) The assembled microfluidic device composed of a PDMS top layer and a glass bottom layer patterned with electrodes, which are wired out through a customized printed circuit board. (**c**) Optical micrograph of the chip

### 3.3 Soft Lithography for Fabricating Microchannel

Figure 2a illustrates the fabrication procedure of the device. For fabricating the upper layer of cell capture, soft lithography is used.

1. Rinse glass substrate with acetone, absolute ethanol, and deionized water for 1 min and blow-dry.
2. Spin-coat the cleaned and dried glass with negative photoresist BN-303 (*see* **Note 4**) at 1000 rpm for 35 s, then bake at 65 °C for 2 min and at 95 °C for 5 min.
3. UV expose (6.9 mJ/cm$^2$) the spin-coated glass slides fully for 10 s utilizing an electron-beam lithography system.
4. Post-bake the exposed slides at 65 °C for 2 min and at 95 °C for 5 min.
5. Spin-coat the glass with negative photoresist SU8 2025 at 500 rpm for 15 s and at 4000 rpm for 35 s, resulting in a photoresist film with a thickness of ~20 μm.
6. Soft-bake at 95 °C for 5 min, UV expose with mask (6.9 mJ/cm$^2$) for 32.6 s and post-bake at 65 °C for 1 min and at 95 °C for 5 min (*see* **Note 5**).
7. Develop the slides at room temperature about 5 min (*see* **Note 6**) and hard-bake at 120 °C for 1 h to obtain the master mold.
8. Prepare a mixture of PDMS base and curing agent (10:1 ratio by weight), degas using vacuum pump, and then pour onto the master mold (*see* **Note** 7).
9. After curing at 70 °C for at least 2 h, peel off the solidified PDMS from the mold and cut into pieces.
10. Punch holes as an inlet and outlet for the cell capture part to work.

### 3.4 Lift-Off Procedure for Fabricating Sensing Electrodes

Lift-off is used for fabricating the lower layer of impedance sensing electrodes.

1. Rinse glass substrate with acetone, absolute ethanol, and deionized water for 1 min and blow-dry it.
2. Spin-coat the negative photoresist ROL-7133 on the clean glass at 500 rpm for 15 s and at 2500 rpm for 60 s and then soft-bake it at 108 °C for 90 s.
3. UV expose (6.9 mJ/cm$^2$) the spin-coated glass slides with mask for 11 s utilizing a lithography machine and post-bake at 108 °C for 90 s.
4. Develop the slides at room temperature about 1 min, and soak them in deionized water for 30 s.
5. Sputter 10 nm Cr (adhesive layer) and 100 nm Au subsequently on the patterned glass utilizing magnetron sputtering machine.
6. Remove the photoresist by soaking in acetone for 10 min and leave the patterned electrodes.

### 3.5 Microchip Bonding

1. Treat the two layers with oxygen plasma for 60 s utilizing plasma cleaner and align-bond them together (*see* **Note 8**).
2. Bake the assembled microfluidic chip at 120 °C for 1 h to further improve bonding strength ultimately.
3. Connect a matching PCB to the electrodes for electrical signal excitation and recording. The overview of the fabricated device and corresponding microscopic image are shown in Fig. 2b and Fig. 2c, respectively.

### 3.6 Impedance Measurement System Configuration

Figure 3 illustrates the schematic diagram of a home-made portable measuring circuit based on the LIA principle of impedance measurement.

1. DDS AD9958 controlled by Microcontroller Unit (MCU) provide the AC signal. The excitation signal is filtered by an elliptic reconstruction filter and then delivered to microfluidic chip.
2. Implement current-to-voltage conversion by using high gain bandwidth operational amplifiers OPA657. Considering the impedance variation range (10–10$^3$ kΩ) and characteristics of the amplifier, select the feedback resistor of the TIA as 100 kΩ.
3. Modulate the voltage signal converted from current by the TIA with the quadrature signal, using the wideband four quadrant multiplier AD835.
4. Demodulate the modulation voltage signal subsequently utilizing the high-precision operational amplifiers OPA227, which

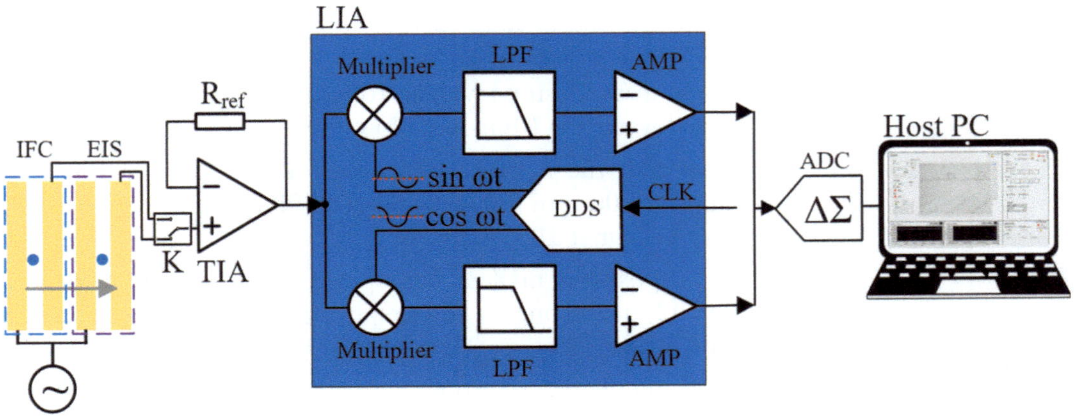


**Fig. 3** The impedance measuring system using the principle of lock-in amplifier. The switch between IFC and EIS is controlled by a relay

serve as low-pass filters (LPF). Select the cut-off frequency of the LPF as 10 Hz to filter out the up-converting signal with $10^3$ Hz as the minimum sweeping frequency (*see* **Note 9**).

5. Amplify the filtered DC voltage 4× to improve the signal-to-noise ratio and to be read by the computer through the data acquisition system PCI-6289.

### *3.7 Experimental Setup and Data Analysis*

1. Place the microfluidic chip on an inverted microscope with enough working space for electrical and fluidic circuits.
2. Connect the home-made impedance measurement system to the electrodes for excitation and signal read-out.
3. Before sample loading, pretreat the microchannel with 1% BSA (in 1 × PBS) to prevent adhesion of samples from PDMS sidewalls.
4. Load samples subsequently by pulling the syringe pump into the microchannel.
5. Apply the input of the IFC and the EIS sensing electrodes with 1 $V_{p\text{-}p}$, 1 MHz and 1 $V_{p\text{-}p}$ sweep (1–$10^3$ kHz) signals, respectively, and connect the output of the sensing electrodes to the impedance measuring system via relays (*see* **Note 10**).
6. Acquire and analyze the electrical impedance data generated by the measuring circuit utilizing home-made LabVIEW and MATLAB program.
7. During the measurement process, record the cell movement at 100 Hz utilizing a CCD camera mounted on the microscope C-port.

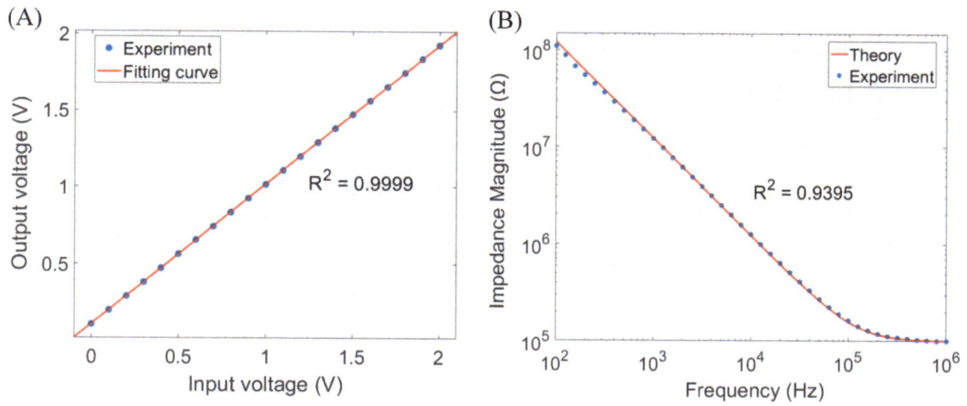


**Fig. 4** Impedance measuring system calibration and verification results. (**a**) The linear relationship between the input voltage and output voltage ($f = 1$ MHz, $R^2 = 0.9999$). (**b**) The measured and theoretical impedance magnitude of the RC network are in excellent match ($R^2 = 0.9395$) for the example settings evaluated here

### 3.8 Measurement System Calibration and Verification

1. Calibrate the measuring system using a standard resistor whose resistance is equal to the feedback resistance at a fixed frequency (*see* **Note 11**).
2. Feed a series of input voltage into the input of the home-made portable measuring circuit, and obtain corresponding output voltage values and confirm the significant linear relationship between them (e.g., $V_{out} = 0.9249\, V_{in} + 0.1488$, $R^2 = 0.9999$, Fig. 4a) (*see* **Note 12**).
3. Verify the feasibility of the impedance measurement system. Connect an *RC* network to the measuring system to test the frequency response. Apply the *RC* network consisting of a 100 kΩ resistor and a 13 pF capacitor with a sweep signal from $10^2$ to $10^6$ Hz, and calculate the impedance magnitude with the calibrated system (*see* **Note 13**).
4. Verify that the impedance analysis system is accurate and, therefore, can be used for the following single-cell impedance measurement. This can be done by looking into the consistent match of impedance spectroscopy between theoretical and experimental results (e.g., $R^2 = 0.9395$, Fig. 4b).

### 3.9 IFC Validation and Cell Classification

1. First, verify that the IFC module worked for the passing cells. For proof, pull cells into the channel at 10 nL/min (*see* **Note 14**).
2. Obtain the image frames for passing cells and the corresponding electrical impedance signal of each cell transit event. For each cell passing through the IFC electrode, the voltage signals exhibit a Gaussian shape, whose peak appears when the cell is midway between the electrodes (Fig. 5a).

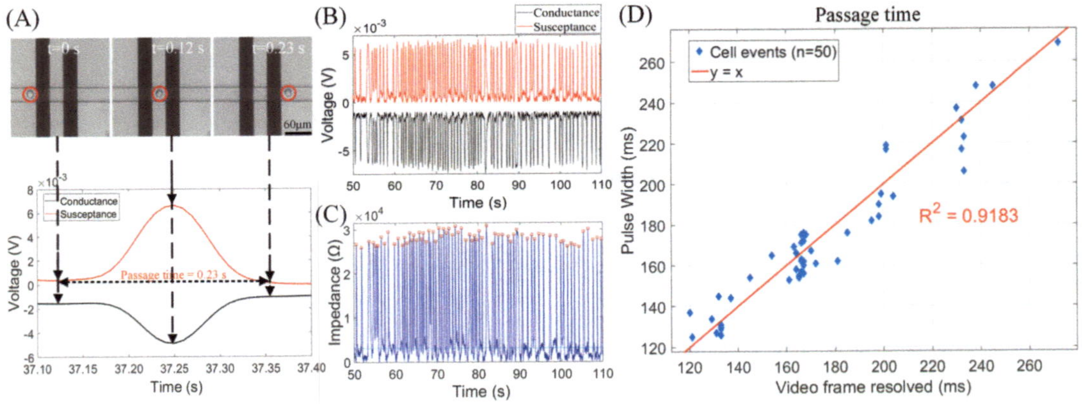


**Fig. 5** IFC functionality verification. ($f = 1$ MHz, $V_{p\text{-}p} = 1$ V). (**a**) The time-lapsed microscopic image snapshots of a HeLa cell flowing through the IFC electrodes. The corresponding electrical impedance signal of this cell transit event is illustrated below. (**b**) Instantaneous time-varying pulse signals measured for continuous cell transit events. (**c**) The corresponding impedance magnitude signal for the cell transit events in (**b**). The blue line represents the baseline for peak detection, and the red points highlight the pulse peaks representing a cell event. (**d**) The passage time measured via the microscopic video clip versus the electrical pulse width. Data points can be fitted linearly ($R^2 = 0.9183$)

3. Analyze the recording of the IFC signals (Fig. 5b) and the resolved impedance magnitude (Fig. 5c) data. In continuous flowing, each peak corresponds to a passing cell.
4. Extract the passage time for each cell event from the video clip, and compare it to the pulse width extracted from the continuous IFC signals. The high correlation (e.g., $R^2 = 0.9183$, Fig. 5d) indicates that the electrical measurement is accurate and reliable.
5. After validating the IFC module, measure different types of cancer cells (e.g., HepG2, A549 and HeLa) independently. Use the microbeads as a reference group because they are of rather uniform material and size distribution.
6. Carry out the impedance measurement at 1 MHz frequency where the overall impedance of the single cell is dominated by size and electrical property (e.g., Fig. 6a).
7. Separate the different cell populations with the difference of IFC-related impedance magnitude (Fig. 6b), which is associated with the size and electrical properties such as cytoplasm conductivity and membrane permittivity.

### 3.10 EIS Validation and Electrical Property Extraction

1. Feed cells into the channel at 10 nL/min, and trap single cells at the trapping sites to perform EIS measurement.
2. Apply a sweep signal excitation to the EIS electrodes, and obtain the impedance spectrum (Figs. 7a, b) of the single cells to resolve the electrical properties via the single-cell electrical model [22] (*see* **Note 15**).

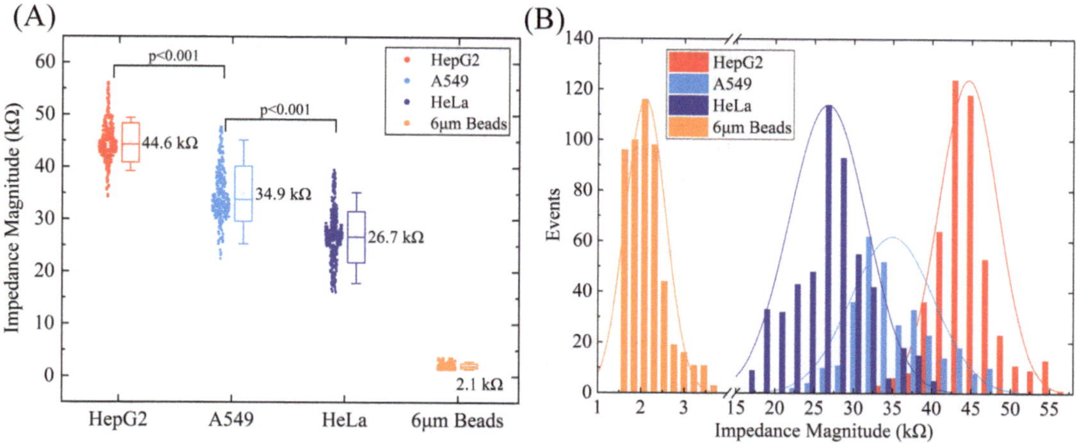


**Fig. 6** Impedance magnitude distribution for three types of cancer cells and 6 μm beads. (**a**) The average impedance magnitudes of HepG2, A549, HeLa cells, and beads, showing significant difference ( $p < 0.001$). (**b**) The impedance magnitudes of cells and beads exhibit Gaussian distribution in population

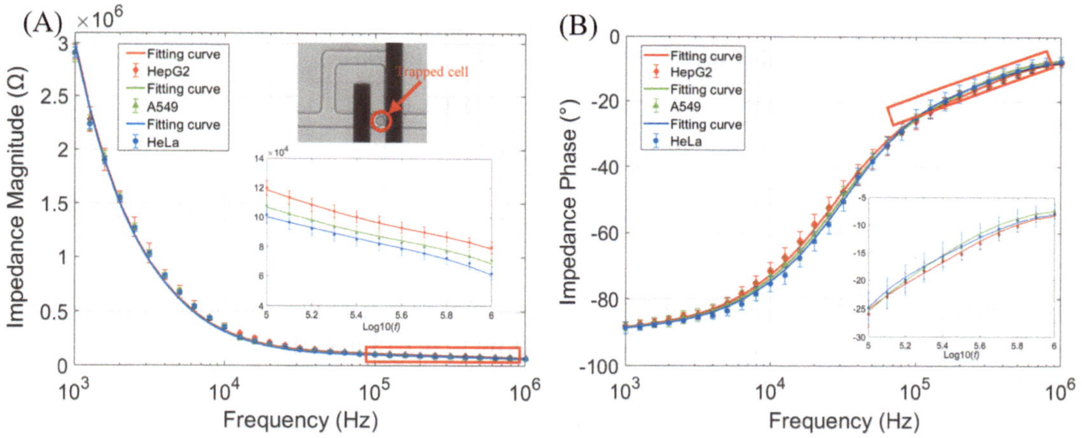


**Fig. 7** EIS impedance measurement results for three types of cancer cells. (**a**) Impedance magnitude and (**b**) phase spectrum. Discrete data points with standard deviation ($N = 10$) were obtained via EIS for the trapped cell and fitted by the curve with least error according to single-cell electrical model. Insets zoom in the high frequency range data points

3. Set the values of medium permittivity $\varepsilon_{\text{med}}$, cytoplasm permittivity $\varepsilon_i$ and cell membrane thickness $d$ in accordance with their representative value and the values of double layer capacitance $C_{\text{DL}}$, stray capacitance $C_{\text{s}}$, medium conductivity $\sigma_{\text{med}}$, and the radius $R$ of the cell from measurement (*see* **Note 16**) as shown in Table 1 and extract the electrical parameters (cytoplasm conductivity $\sigma_i$ and specific membrane capacitance $C_{\text{mem}}$) shown in Table 2 utilizing model fitting method (e.g., least squares model fitting) (*see* **Note 17**).

**Table 1**
**Electrical model parameters used for EIS-based electrical property extraction**

| Model parameters | Values |
|---|---|
| Double-layer capacitance $C_{DL}$ | 90 pF |
| Stray capacitance $C_s$ | 0.5 pF |
| Medium conductivity $\sigma_{med}$ | 1.515 S/m |
| Medium relative permittivity $\varepsilon_{med}$ | 78 |
| cytoplasm relative permittivity $\varepsilon_i$ | 60 |
| Membrane thickness $d$ | 10 nm |
| Cell radius $R$ | 7.5 μm |

**Table 2**
**EIS-extracted electrical properties for three types of cancer cells**

| Cell type | $C_{mem}$ (mF/m$^2$) | $\sigma_i$ (S/m) |
|---|---|---|
| HeLa | 18.06 ± 1.12 | 0.38 ± 0.09 |
| A549 | 24.31 ± 1.43 | 0.36 ± 0.12 |
| HepG2 | 21.78 ± 0.96 | 0.32 ± 0.07 |

### 3.11 Complementarity of EIS and IFC

After verifying the feasibility of both IFC and EIS modules, conduct experiments to show that EIS and IFC can complement each other.

4. Conduct a cell passing through the main channel with IFC first. Due to the limitation of customized circuit, that is, only one single-frequency impedance of the flowing single cell can be obtained in IFC (*see* **Note 18**), pull slowly the same cell back and forth around the IFC electrodes a few (e.g., 6) times. Each time, apply an increasing frequency for excitation, and thus, obtain 6 IFC-based impedance data points.
5. Pull forward the same cell to be trapped on the trapping site for EIS measurement. For EIS measurement, conduct several test runs with an increasing flow rate of 10 nL/min (or 0.185 mm/s), 100 nL/min (1.85 mm/s), and 1000 nL/min (18.5 mm/s), which would generate increasing fluidic pressure on the trapped cell under EIS. In each run, keep the flow rate constant.

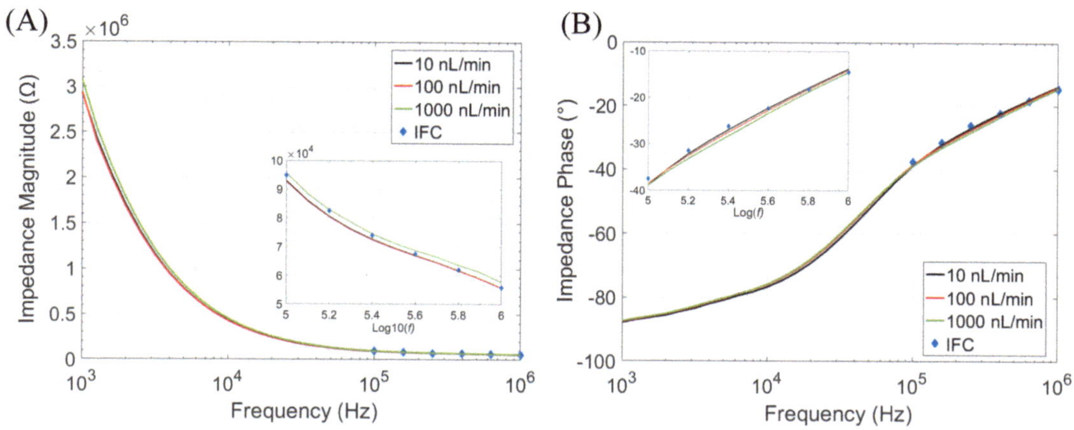


**Fig. 8** Comparison of the IFC results and EIS results for the same cell under the effect of flow rate. (**a**) Impedance magnitude and (**b**) phase spectrum at three levels of flow rate. Note the discrete data points are IFC results, and the curves are fitting EIS results. Insets zoom in the high frequency range data points

6. Sweep the frequency in the range of $10^3$–$10^6$ Hz and obtain the impedance magnitude (Fig. 8a) and phase spectrum (Fig. 8b).
7. Compared to the IFC results, the deviation shall be less than 5% for both magnitude and phase.

## 4 Notes

1. The number of trapping units in this chapter is set to eight for demonstration purposes but can be expanded easily by tens of times through increasing the number of signal channels of the measuring system.
2. The *Q* ratio is an important indicator of the trapping efficiency of the current design. Literature [23] indicates that its suggested values should be higher than 1. With this goal in mind, the design variables of the channels can be fine-tuned for trapping efficiency optimization.
3. In principle, the size and gap of the electrodes affect the sensitivity of measurement. The measurement signal includes three components: cell impedance, medium impedance, and double-layer impedance. Minimization of the latter two components would increase the measurement sensitivity. The medium impedance is proportional to the gap of the electrodes, while the double-layer impedance is inversely proportional to the size of electrodes in direct contact with the sample solution. Thus, in practice, it is desirable to maximize the size and minimize the gap.

4. The negative photoresist BN-303 serves as an adhesive layer to prevent the SU-8 structural layer from falling off. Similar photoresist can also play this role.
5. The fabrication parameters should be adjusted with different thickness of the photoresist film.
6. In this step, insufficient development or excessive development can happen easily. Therefore, master the development time and the amount of the developer. It is better to develop and observe utilizing the microscope multiple times.
7. Slightly press the SU-8 mold, and slowly squeeze out the air bubbles underneath; otherwise, it will easily cause the chip to be uneven.
8. When there is no professional alignment bonding device, put a drop of deionized water on the oxygenated slide, and align the two layers under microscope.
9. The cut-off frequency needs to be optimized carefully by considering the applied signal frequency and the output Gaussian signal frequency (i.e., the flow velocity of cells).
10. Note that the voltage is chosen such that the AC-induced transmembrane potential is far below the threshold value (i.e., 1 V) for cell electroporation.
11. Under this setting, the ideal system response is supposed to be the measured voltage equal to the input voltage ($V_{out} = V_{in}$).
12. The calibration curves for other frequencies can also obtained with the same method and high $R$-square values to facilitate measurement correction in the following experiments.
13. Two DC values $D_c$ and $D_s$ can be obtained from the output of the LIA-based impedance measurement system, given by

$$D_c = \frac{A^2K}{2}\cos\varphi,\quad D_s = -\frac{A^2K}{2}\sin\varphi. \tag{1}$$

where $A$ is the input voltage amplitude, $K = \frac{R_0}{|Z^*_{exp}|}$ is attenuation coefficient with the known reference feedback resistance $R_0$ used in TIA and $|Z^*_{exp}|$ the impedance amplitude, and $\varphi$ is phase shift coefficient. Thus, we can get

$$K = \frac{2}{A^2}\sqrt{{D_c}^2 + {D_s}^2},\quad \varphi = -\arctan\frac{D_s}{D_c}. \tag{2}$$

Substituting the definition of $K$ into Eq. 2 and rearranging the terms yields the impedance amplitude obtained via experiment

$$|Z^*_{exp}| = \frac{R_0A^2}{2\sqrt{{D_c}^2 + {D_s}^2}}. \tag{3}$$

14. Cells may aggregate in the medium, causing some unwanted effects, for example, channel blockage or non-single cell transit events in impedance measurement. Take three measures to consider and solve the problem. (i) Filter cell medium out by the cell strainer (40 μm) before an experiment. (ii) Configure the device with an array of pillars with 30 μm gap as on-chip filter in the upstream of microchannels. (iii) Judge the cell transit events as non-single cell events if the signal amplitude exceeded the average value by 50%.
15. Typically, the electrical properties of the cell are represented by four parameters, namely $\varepsilon_i$, $\sigma_i$, $\varepsilon_{\text{mem}}$, and $\sigma_{\text{mem}}$. They can be extracted by minimizing the squares between the measured impedance $Z^*_{\text{exp}}$ and its model-estimated value $Z^*_{\text{theory}}$ via

$$\min \sum_i \left[ Z^*_{\text{exp}}(\omega_i) - Z^*_{\text{theory}}(\omega_i) \right]^2, \tag{4}$$

where $i$ indexes the experimental data points. Note that $\varepsilon_i$ and $\sigma_{\text{mem}}$ are negligible because they make little change to the impedance spectrum in the medium frequency range. In practice, they can be assigned a reasonable value to facilitate the calculation of the other two parameters.
16. The value of medium conductivity $\sigma_{\text{med}}$ is measured by using the conductivity meter, and the radius $R$ of the cell is measured from image frames. The value of $C_s$ is calculated through the impedance measurement when the electrodes are dry (i.e., there's no solution in the microchannel). The value of $C_{\text{DL}}$ is calculated through the impedance measurement in the low-frequency range when the microchannel is filled with the buffer solution (i.e., PBS in this work).
17. Model fitting is the process of constructing a model, or mathematical function, that has the best fit to a series of data points, possibly subject to constraints. There are many model fitting methods such as least squares [21] and neural network [24]. One can apply these methods in the Matlab Toolbox (e.g., curve fitting toolbox and neural net fitting toolbox).
18. It is important to point out that the home-made impedance measuring system is currently only able to obtain one single-frequency data point, so the same cell is driven back and forth "manually" over the electrodes six times to obtain multifrequency (six single-frequency) data points in IFC. This manual back-and-forth operation could be eliminated if multifrequency data points can be obtained simultaneously via more sophisticated instruments such as HF2IS, FPGA, or MCU [24].

## Acknowledgments

This work is supported by the NSFC (no. 62174096, 21727813, and 52105572) and the One Thousand Young Talent Program of China.

### References

1. Klein AM, Mazutis L, Akartuna I, Tallapragada N, Veres A, Li V, Peshkin L, Weitz DA, Kirschner MW (2015) Droplet barcoding for single-cell transcriptomics applied to embryonic stem cells. Cell 161(5): 1187–1201. https://doi.org/10.1016/j.cell.2015.04.044
2. Galler K, Brautigam K, Grosse C, Popp J, Neugebauer U (2014) Making a big thing of a small cell--recent advances in single cell analysis. Analyst 139(6):1237–1273. https://doi.org/10.1039/c3an01939j
3. Huang L, Feng Y, Liang F, Zhao P, Wang W (2021) Dual-fiber microfluidic chip for multimodal manipulation of single cells. Biomicrofluidics 15(1):014106. https://doi.org/10.1063/5.0039087
4. Mansor MA, Ahmad MR (2015) Single cell electrical characterization techniques. Int J Mol Sci 16(6):12686–12712. https://doi.org/10.3390/ijms160612686
5. Honrado C, Bisegna P, Swami NS, Caselli F (2021) Single-cell microfluidic impedance cytometry: from raw signals to cell phenotypes using data analytics. Lab Chip 21(1):22–54. https://doi.org/10.1039/d0lc00840k
6. Vembadi A, Menachery A, Qasaimeh MA (2019) Cell cytometry: review and perspective on biotechnological advances. Front Bioeng Biotechnol 7:147. https://doi.org/10.3389/fbioe.2019.00147
7. Shrirao AB, Fritz Z, Novik EM, Yarmush GM, Schloss RS, Zahn JD, Yarmush ML (2018) Microfluidic flow cytometry: the role of microfabrication methodologies, performance and functional specification. Technology 6(1): 1–23. https://doi.org/10.1142/S2339547818300019
8. Fertig N, Blick RH, Behrends JC (2002) Whole cell patch clamp recording performed on a planar glass chip. Biophys J 82(6): 3056–3062. https://doi.org/10.1016/S0006-3495(02)75646-4
9. Xu Y, Xie X, Duan Y, Wang L, Cheng Z, Cheng J (2016) A review of impedance measurements of whole cells. Biosens Bioelectron 77:824–836. https://doi.org/10.1016/j.bios.2015.10.027
10. Zheng Y, Nguyen J, Wei Y, Sun Y (2013) Recent advances in microfluidic techniques for single-cell biophysical characterization. Lab Chip 13(13):2464–2483. https://doi.org/10.1039/c3lc50355k
11. Nguyen TA, Yin TI, Reyes D, Urban GA (2013) Microfluidic chip with integrated electrical cell-impedance sensing for monitoring single cancer cell migration in three-dimensional matrixes. Anal Chem 85(22): 11068–11076. https://doi.org/10.1021/ac402761s
12. Huang L, Zhao P, Wang W (2018) 3D cell electrorotation and imaging for measuring multiple cellular biophysical properties. Lab Chip 18(16):2359–2368. https://doi.org/10.1039/c8lc00407b
13. Hughes MP (1998) Computer-aided analysis of conditions for optimizing practical electrorotation. Phys Med Biol 43(12):3639. https://doi.org/10.1088/0031-9155/43/12/019
14. Bull BS, Schneiderman MA, Brecher G (1965) Platelet counts with the Coulter counter. Am J Clin Pathol 44(6):678. https://doi.org/10.1093/ajcp/44.6.678
15. Rodriguez-Trujillo R, Castillo-Fernandez O, Garrido M, Arundell M, Valencia A, Gomila G (2008) High-speed particle detection in a micro-Coulter counter with two-dimensional adjustable aperture. Biosens Bioelectron 24(2):290–296. https://doi.org/10.1016/j.bios.2008.04.005
16. Holmes D, Pettigrew D, Reccius CH, Gwyer JD, van Berkel C, Holloway J, Davies DE, Morgan H (2009) Leukocyte analysis and differentiation using high speed microfluidic single cell impedance cytometry. Lab Chip 9(20): 2881–2889. https://doi.org/10.1039/b910053a
17. Mansoorifar A, Koklu A, Beskok A (2019) Quantification of cell death using an impedance-based microfluidic device. Anal Chem 91(6):4140–4148. https://doi.org/10.1021/acs.analchem.8b05890
18. Mi L, Huang L, Li J, Xu G, Wu Q, Wang W (2016) A fluidic circuit based, high-efficiency and large-scale single cell trap. Lab Chip

16(23):4507–4511. https://doi.org/10.1039/c6lc01120a

19. Chai H, Feng Y, Liang F, Wang W (2021) A microfluidic device enabling deterministic single cell trapping and release. Lab Chip 21(13): 2486–2494. https://doi.org/10.1039/d1lc00302j
20. Lai C-W, Hsiung S-K, Yeh C-L, Chiou A, Lee G-B (2008) A cell delivery and pre-positioning system utilizing microfluidic devices for dual-beam optical trap-and-stretch. Sens Actuators B-Chem 135(1):388–397. https://doi.org/10.1016/j.snb.2008.08.041
21. Feng Y, Huang L, Zhao P, Liang F, Wang W (2019) A microfluidic device integrating impedance flow cytometry and electric impedance spectroscopy for high-efficiency single-cell electrical property measurement. Anal Chem 91(23):15204–15212. https://doi.org/10.1021/acs.analchem.9b04083
22. Morgan H, Sun T, Holmes D, Gawad S, Green NG (2007) Single cell dielectric spectroscopy. J Phys D-Appl Phys 40(1):61–70. https://doi.org/10.1088/0022-3727/40/1/s10
23. Jin D, Deng B, Li JX, Cai W, Tu L, Chen J, Wu Q, Wang WH (2015) A microfluidic device enabling high-efficiency single cell trapping. Biomicrofluidics 9(1):014101. https://doi.org/10.1063/1.4905428
24. Feng Y, Cheng Z, Chai H, He W, Huang L, Wang W (2022) Neural network-enhanced real-time impedance flow cytometry for single-cell intrinsic characterization. Lab Chip 22:240–249. https://doi.org/10.1039/d1lc00755f

# Chapter 7

# Cell Viability Multiplexing: Quantification of Cellular Viability by Barcode Flow Cytometry and Computational Analysis

**Valentina Giudice, Victoria Fonseca, Carmine Selleri, and Massimo Gadina**


## Abstract

Fluorescent cell barcoding (FCB) is a useful flow cytometric technique for high-throughput multiplexed analyses and can minimize technical variations after preliminary optimization and validation of protocols. To date, FCB is widely used for measurement of phosphorylation status of certain proteins, while it can be also employed for cellular viability assessment. In this chapter, we describe the protocol to perform FCB combined with viability assessment on lymphocytes and monocytes using manual and computational analysis. We also provide recommendations for FCB protocol optimization and validation for clinical sample analysis.


**Key words** Flow cytometry, Fluorescent cell barcoding, Viability, Multiplexing, Fixable dyes

## 1 Introduction

Fluorescent cell barcoding (FCB), a high-throughput flow cytometry (FCM) technique for multiplexed assays, is currently employed for phospho-specific FCM (phosphoflow), drug screening and signaling profiling [1–8], intracellular cytokine detection [9, 10], and comprehensive immunophenotyping [1–12]. FCB displays high reproducibility and reliability also in clinical studies showing very low intra- and inter-operator variability by conventional, manual, and semi- or automated, machine learning gating strategy methods [7, 12]. Indeed, the inclusion of a bridge control placed in the same position in a matrix across FCB assays performed in different days and by various operators significantly lowers intra- and inter-assay variability caused by biological and technical differences [7].

Novel technologies investigating modifications in gene and protein expression at the single-cell level (e.g., single cell RNA

Oliver Friedrich and Daniel F. Gilbert (eds.), *Cell Viability Assays: Methods and Protocols*, Methods in Molecular Biology, vol. 2644, https://doi.org/10.1007/978-1-0716-3052-5_7,

sequencing) have markedly widened our knowledge of biological processes; however, researchers are now dealing with massive amounts of data obtained by these single cell assays, such as FCB. Moreover, by combining those techniques, data content even exponentially increases [13, 14]. Manual analysis of high-content data is challenging, time consuming, and operator-dependent, including flow cytometry data where different marker combinations and gate boundaries can be chosen by the researcher as nonstandardized international guidelines are accepted, especially for novel methodologies, like FCB [7, 15]. Computational analysis by unsupervised or semiautomated algorithms is more objective, as data are examined by reproducible mathematical methods [7, 16–19]. Cytobank, Matlab, and Bioconductor offer several bioinformatics tools for all steps of FCM data analysis, including data preprocessing, transformation, visualization, deconvolution, and statistical analysis [16–19]. For example, the software SPADE is largely employed for hierarchical data clustering, and viSNE for t-distributed stochastic neighbor embedding dimensionality reduction [16, 18], while flowCore, flowClust, and DebarcodeR, all Bioconductor's packages, can be integrated in a fully or semiautomated sequential workflow for FCB data analysis [7, 12, 19].

Apoptotic and dead cells can nonspecifically bind monoclonal antibodies and create artifacts, particularly when cell populations with low frequencies are studied [20]. To exclude dead cells from analysis, DNA intercalating agents and intracytoplasmic amine-reactive dyes have been largely used, as membrane-damaged dead cells expose both surface and intracellular epitopes acquiring a higher fluorescence intensity compared to live cells with intact membranes that expose only surface epitopes [20–22]. Because of its broad applications, viability dye stainings have been also combined with FCB staining [7, 12, 20–24].

In this chapter, we describe an FCB procedure for cellular viability by barcode flow cytometry and computational analysis on cryopreserved peripheral blood mononuclear cells (PBMCs). In this protocol, FCB was performed using amine reactive DyLight 350 and DyLight 800 with succinimidyl ester (NHS) dyes for barcoding nine samples stained with Aqua (LIVE/DEAD Fixable Aqua for 405 nm excitation) viability dye. Standardization of FCM analysis is described for both manual and computational gating strategies, and notes for technical recommendations are also reported.

## 2 Materials

### 2.1 FCB Buffers and Dyes

1. Barcoding dyes: DyLight 350–NHS, and DyLight 800–NHS (Thermo Fisher Scientific, Waltham, MA, USA) (*see* **Note 1**).

2. Barcoding buffers: BD Perm Buffer II, BD Perm Buffer III, 5× BD Phosflow Lyse/Fix Buffer, 4× BD Phosflow Barcoding Wash Buffer (BD Biosciences; San Jose, CA, USA).
3. Viability dye: LIVE/DEAD Fixable Aqua (a viability dye) for 405 nm excitation (Thermo Fisher Scientific). Thaw reagents, and first add 50 μL of Component B to 1 vial of Component A of supplied reagents. Mix well and store at <−80 °C protected from light.
4. Ficoll-Paque gradient centrifugation (MP Biomedicals, LLC, Santa Ana, CA, USA), according to manufacturer's instructions for PBMC isolation.
5. Ack Lysing buffer (Quality Biological, Gaithersburg, MD, USA) for red blood cell lysis.
6. Phosphate buffer saline (PBS).
7. Dymethyl-sulfoxide (DMSO) Sigma-Aldrich (St. Louis, MO, USA).
8. 1× BD Perm Buffer II: BD Perm Buffer II, cold PBS. Add 20 mL of BD Perm Buffer II at room temperature (RT) and 20 mL of cold PBS to a 50 mL Falcon Conical tube (Fisher Scientific, Loughborough, UK). Mix by gentle inversion to avoid bubble formation. Store at +4 °C.
9. 1× BD Phosflow Lyse/Fix Buffer: 5× BD Phosflow Lyse/Fix Buffer, PBS. Add 10 mL of 5× BD Phosflow Lyse/Fix Buffer and 40 mL of PBS at RT to a 50 mL Falcon Conical tube. Mix by inversion and store at RT.
10. 1× Barcoding Wash Buffer: 4× Barcoding Wash Buffer, PBS. Add 5 mL of cold 4× Barcoding Wash Buffer and 35 mL of PBS at RT to a 50 mL Falcon Conical tube. Gently mix by inversion to avoid bubble formation, and store at RT. Mixed solution should be used within 1–2 days.
11. FCB dye stock concentration: DyLight 350–NHS, DyLight 800–NHS, DMSO. Dyes are supplied as dry powders. DMSO should be used for reconstitution at 500 μg/mL concentration (stock concentration) to be further diluted (work concentrations). For 1 mg of dry powder, add 1 mL of DMSO to the powder and gently pipet, then transfer the solution to a 15 mL Falcon Conical tube (Fisher Scientific, Loughborough, UK). Subsequently, rinse the tube where dry powder was contained with 1 mL of DMSO and transfer the solution to the 15 mL Falcon Conical tube with reconstituted dye. Gently mix well by pipetting. Aliquot 1 mL of reconstituted dye in two 1.5 mL Eppendorf tubes. Reconstituted dyes should be stored at <−80 °C protected from light (*see* **Note 2**).

### 2.2 Samples

1. Peripheral blood mononuclear cells (PBMCs): 6–7.5 × $10^5$ cells/30 μL/well, 6–7.5 × $10^5$ cells/30 μL for DyLight 350 single-color control, 6–7.5 × $10^5$ cells/30 μL for DyLight 800 single-color control, 6–7.5 × $10^5$ cells/30 μL for unstained control, 6–7.5 × $10^5$ cells/30 μL for stained non-barcoded control, 6–7.5 × $10^5$ cells/30 μL for unstained barcode control. Isolate PBMCs from whole heparinized or EDTA obtained from patients or healthy subjects by Ficoll-Paque density gradient centrifugation (MP Biomedicals, LLC, Santa Ana, CA) according to the manufacturer's instructions and as described in Giudice et al. [11]. Cells can be stored until use or employed right after isolation.

### 2.3 Instruments and Software

1. Flow cytometer: for sample acquisition, the cytometer should be equipped with appropriate lasers: ultraviolet, 355 nm for DyLight 350 excitation; violet, 405 nm for Aqua viability dye excitation; and red laser, 628/640 nm for DyLight 800 excitation (e.g., LSR Fortessa; BD Biosciences).
2. Acquisition analysis software: BD FACSDiva software (BD Biosciences), Navios Software, or Navios EX Software (Beckman Coulter).
3. Post-acquisition analysis software: FlowJo software (BD Biosciences) or Kaluza Analysis Flow Cytometry software (Beckman Coulter).
4. Computational analysis software: RStudio (v.2022.02.2), R software (v.4.2.0) (RStudio, Inc., Boston, MA, USA), BiocManager (Bioconductor, v.3.15).
5. R packages: flowCore, flowClust, flowViz, flowStats, flowWorkspace, ggcyto, ggplot2, flowType, and openxlsx.

## 3 Methods

### 3.1 Sample Preparation

1. Thaw frozen cells prepared in Subheading 2.2 in a water bath at +37 °C for 1–2 min. If starting from freshly isolated samples, go to **step 5** (Fig. 1).
2. Wash with 1 mL of PBS by centrifuging at 400 *g* for 5 min.
3. Discard supernatant and vortex to loosen cell pellets.
4. Add 1 mL of PBS and resuspend cells.
5. Transfer cells to a 5 mL FACS tube.
6. Add 2 mL of PBS and wash by centrifuging at 400 *g* for 5 min.
7. Discard supernatant and vortex briefly to loosen cell pellets.
8. Resuspend cells in an appropriate volume of PBS to obtain a concentration of 1 × $10^6$ cells/100 μL.

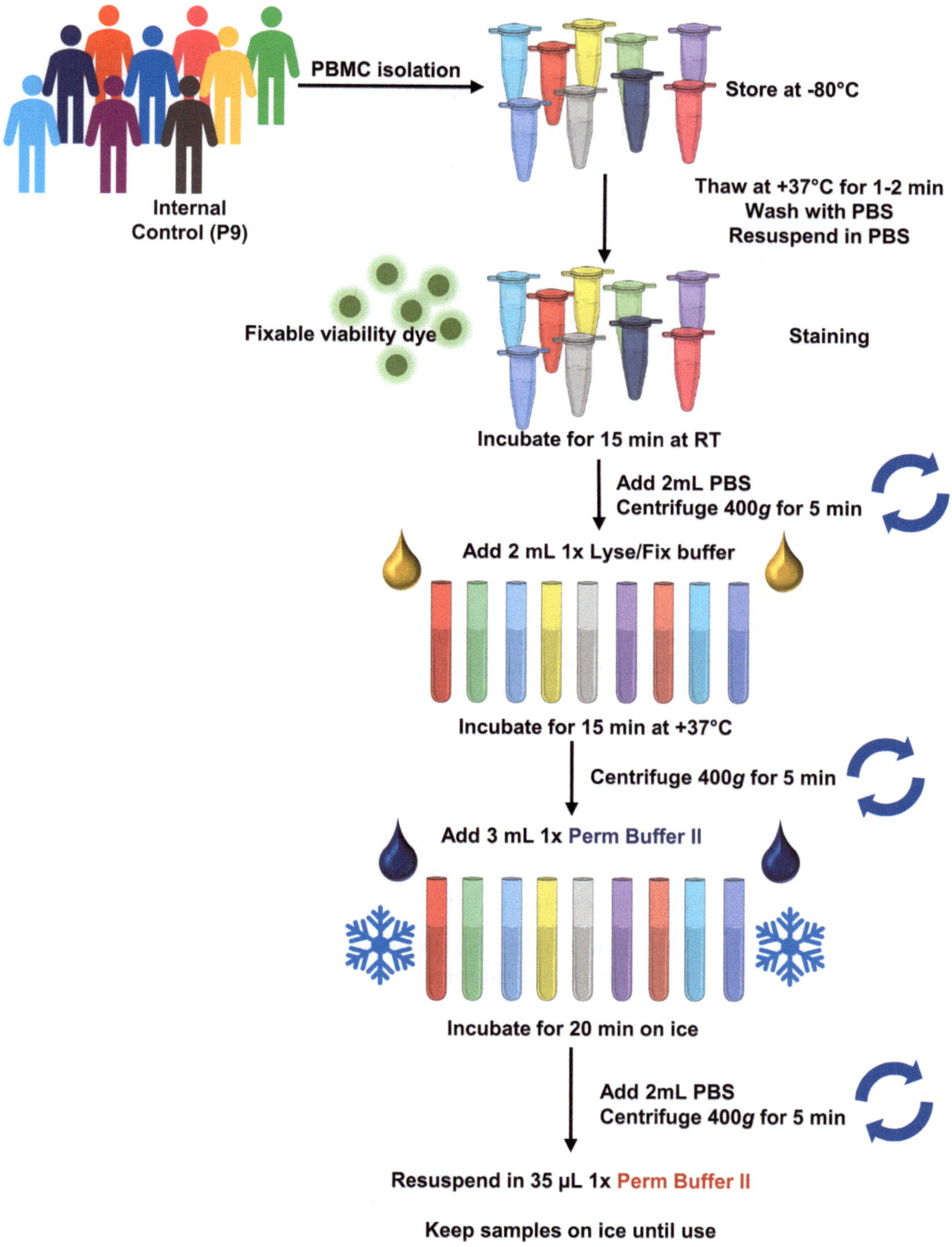


**Fig. 1** Illustration of sample preparation. Peripheral blood mononuclear cells (PBMCs) are isolated from healthy subjects or patients (P from 1 to 8) and from a healthy individual who will be used as internal bridge control in every matrix in position 9 (P9). Cells can be stored at −80 °C until use, then thawed at +37 °C for 1–2 min and washed with phosphate buffer saline (PBS). After washed, a total of $1 \times 10^6$ cells/100 μL is stained with 3 μL of diluted viability dye to each 100 μL of sample, incubated for 15 min at room temperature (RT), washed with

### 3.2 Viability Dye Staining

1. Dilute 3 μL of viability dye in 47 μL of PBS.
2. Add 3 μL of diluted viability dye to each 100 μL of sample.
3. Incubate for 15 min at RT in the dark.
4. Add 2 mL of PBS and centrifuge at 400 *g* for 5 min.
5. Discard supernatant and vortex to loosen cell pellets.

### 3.3 Sample Fixation and Permeabilization

1. Add 2 mL of 1× Lyse/Fix Buffer to each well and gently mix by pipetting.
2. Incubate for 15 min at 37 °C.
3. Centrifuge at 400 *g* for 5 min. Discard supernatant and vortex to loosen the cell pellets.
4. Add 3 mL of cold Phosflow Perm Buffer II.
5. Incubate on ice for 20 min (*see* **Note 3**).
6. Add 2 mL of PBS and centrifuge at 400 *g* for 5 min. Discard supernatant and vortex to loosen cell pellets.
7. Resuspend cells in 35 μL of cold 1× Phosflow Perm Buffer II.
8. Keep cells on ice until use.

### 3.4 Control Preparations

For protocol optimization, an internal bridge control should be added to each matrix in the same position (*see* **Note 4**) to measure intra- and inter-assay variability, as percentage of positive cell populations and median fluorescence intensity (MFI) values should not significantly differ within a matrix and across experiments. The internal bridge control should be processed as other samples, and controls labeled "No FCB" with fixation/permeabilization without barcoding should be added to exclude the presence of differences in percentage of positive cells and MFIs caused by increased autofluorescence of barcoded samples.

1. Thaw frozen cells from one healthy subject used as bridge control in a water bath at +37 °C for 1–2 min.
2. Wash with 1 mL of PBS by centrifuging at 400 *g* for 5 min.
3. Discard supernatant and vortex briefly to loosen cell pellets.
4. Add 1 mL of PBS and resuspend cells.
5. Transfer cells to a 5 mL FACS tube.
6. Add 2 mL of PBS and wash by centrifuging at 400 *g* for 5 min.

**Fig. 1** (continued) 2 mL of PBS, and centrifuged at 400 *g* for 5 min. After resuspension, 2 mL of 1× Lyse/Fix Buffer are added to samples, cells are incubated for 15 min at 37 °C, and subsequently centrifuged at 400 *g* for 5 min. Then, 3 mL of Phosflow Perm Buffer II are added to each sample and, after incubation on ice for 20 min, 2 mL of 1× PBS are added for washing. Cells are centrifuged at 400 *g* for 5 min, supernatants are discarded, cell pellets loosen, and samples are resuspended in 35 μL of cold 1× Phosflow Perm Buffer II until use

7. Discard supernatant and vortex briefly to loosen cell pellets.
8. Resuspend cells in an appropriate volume of PBS to obtain a concentration of $1 \times 10^6$ cells/100 μL.
9. Calculate the number of controls as follows:
   (a) An internal control (viability dye + fixation and permeabilization + FCB) should be included in each matrix.
   (b) One "No FCB" control (viability dye + fixation and permeabilization).
   (c) One "Viab alone" control (viability dye alone without fixation, permeabilization, and FCB).
10. Dilute 3 μL of viability dye in 47 μL of PBS.
11. Add 3 μL of diluted viability dye to each 100 μL of sample.
12. Incubate for 15 min at RT in the dark.
13. Add 2 mL of PBS and centrifuge at 400 *g* for 5 min.
14. Discard supernatant and vortex briefly to loosen cell pellets.
15. Resuspend the "Viab alone" control in 300 μL of PBS and keep at +4 °C until acquisition.
16. Add 2 mL of 1× Lyse/Fix Buffer to internal controls and "No FCB" sample and gently mix by pipetting.
17. Incubate for 15 min at 37 °C.
18. Centrifuge at 500 *g* for 5 min. Discard supernatant and vortex briefly to loosen the cell pellets.
19. Add 3 mL of cold Phosflow Perm Buffer II.
20. Incubate on ice for 20 min (*see* **Note 3**).
21. Add 2 mL of PBS and centrifuge at 500 *g* for 5 min. Discard supernatant and vortex to loosen cell pellets.
22. Resuspend internal controls in 35 μL of cold 1× Phosflow Perm Buffer II and move to Subheading 3.5, while resuspend the "No FCB" control in 300 μL of PBS and keep at +4 °C until acquisition.

### 3.5 Nine-Sample Matrix Preparation (3 × 3)

Prepare working concentrations using DMSO from stock solutions before starting experiments or during sample permeabilization, as DMSO tends to precipitate (Fig. 2).

1. Before starting the experiment, thaw stock solution (500 μg/mL) at RT protected from light. Dyes typically thaw in 10–20 min.
2. Vortex vigorously right before use. Do not spin otherwise DMSO precipitates.
3. Prepare dye working concentrations with DMSO in round bottom 96-well plate, and label wells as "DyL350-0",

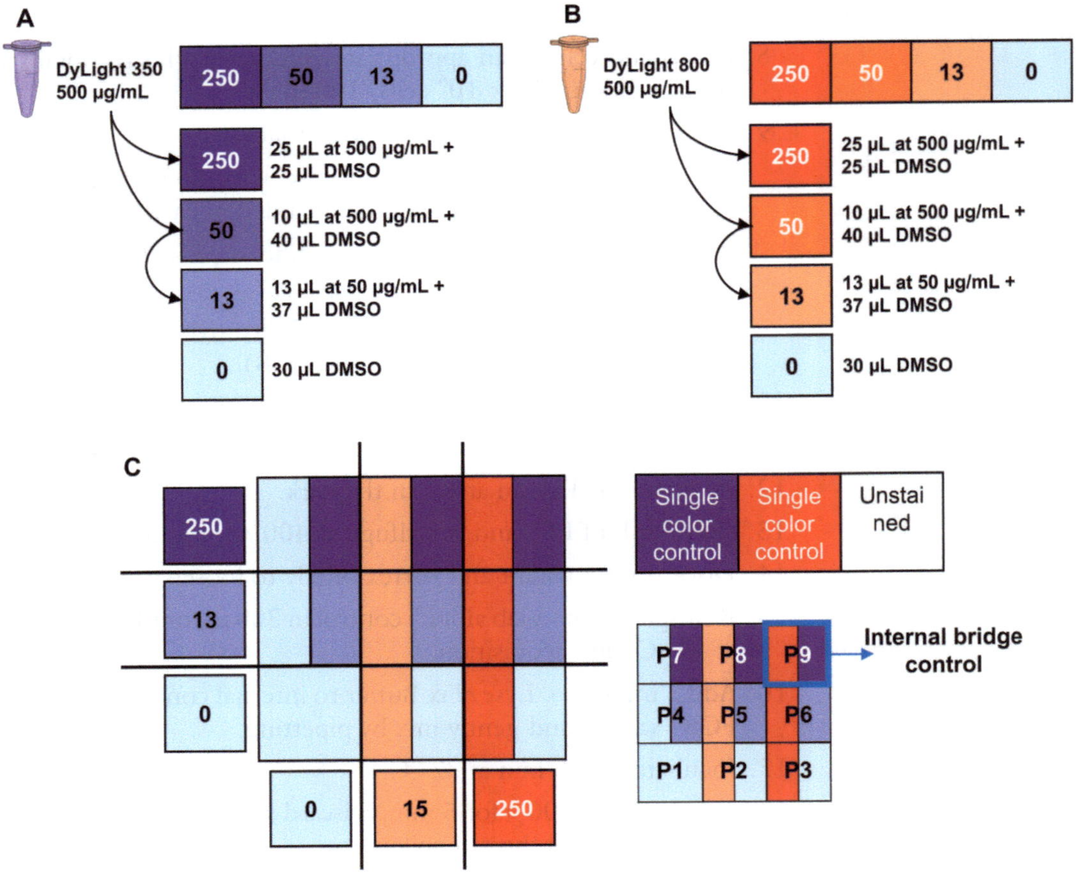


**Fig. 2** Illustration of 3 × 3 matrix preparation. Working concentrations of dyes are prepared using DMSO and starting with stock concentrations (500 µg/mL). (**a**) For DyLight 350, 250- and 50-dilution concentrations are prepared using stock solution and the 13-dilution using the 50-diltuion. (**b**) For DyLight 800, 250- and 50-dilution concentrations are prepared using stock solution, and the 13-dilution is prepared employing the 50-dilution. (**c**) In each well of the matrix, a total of 5 µL of diluted dyes or DMSO are added to appropriate well to reach 10 µL of total dyes/DMSO. In addition, 10 µL of DyLight 350 or DyLight 800 at 250 µg/mL are added to single well for preparation of single-color controls for cytometer setting together with unstained sample (10 µL of DMSO + barcoded sample). Then, a total of 30 µL of each sample are added to appropriate wells according to their position in the correct matrix. After incubation with dyes, samples from the same matrix are combined in one tube for further processing

"DyL350-13", "DyL350-250", and "DyL800-0", "DyL800-13," and "DyL800-250" (*see* **Note 4**).

For each dye, prepare the following dilutions:

250-dilution = 25 µL of 500 µg/mL dilution + 25 µL of DMSO

50-dilution = 10 µL of 250-dilution +40 µL of DMSO

13-dilution = 13 µL of 50-dilution +37 µL of DMSO

0-dilution = 30 µL of DMSO

The 50-dilution is used only for preparing the 13-dilution and not for FCB matrix preparation.

4. In a round bottom 96-well plate, add 5 μL of diluted dyes or DMSO to appropriate well to have 10 μL of total dyes/DMSO.
5. Keep plates at RT protected from light.

### 3.6 Fluorescent Cell Barcoding

All steps are performed at RT unless otherwise stated.

1. After gently mixed by pipetting, add 30 μL of each sample prepared in Subheading 3.3 and internal controls prepared in Subheading 3.4 into appropriate wells (Fig. 3).
2. Mix well by gently pipetting.
3. For single-color control preparation for PMT voltage setting, add:
   (a) 30 μL of specimen into a well with 10 μL of DyLight 350 (single-color control for DyLight 350)
   (b) 30 μL of sample into a well with 10 μL of DyLight 800 (single-color control for DyLight 800)
   (c) 30 μL of sample into a well with 10 μL of DMSO (unstained sample)
4. Seal or cover the plate with aluminum foil.
5. Incubate for 20 min on ice. Longer incubation time increases sample autofluorescence.
6. Label four 5 mL FACS tubes as follows "Matrix," "DyL350," "DyL800," and "Unstained."
7. Pre-fill tubes with 3 mL of 1× Barcoding Wash Buffer (*see* **Note 5**).
8. After incubation with dyes, transfer:
   (a) All barcoded samples into the "Matrix" tube
   (b) Single-color-dye controls in separate "DyL350" or "DyL800" tubes
   (c) Unstained sample in corresponding "Unstained" tube
9. Centrifuge at 400 *g* for 5 min, and discard supernatant. Vortex briefly to loosen cell pellets.
10. Wash again with 3 mL of 1× Barcoding Wash Buffer.
11. Centrifuge at 400 *g* for 5 min, and discard supernatant. Vortex briefly to loosen cell pellets.
12. Resuspend samples in 300 μL of 1× Barcoding Wash Buffer and keep on ice until cytometer setting.

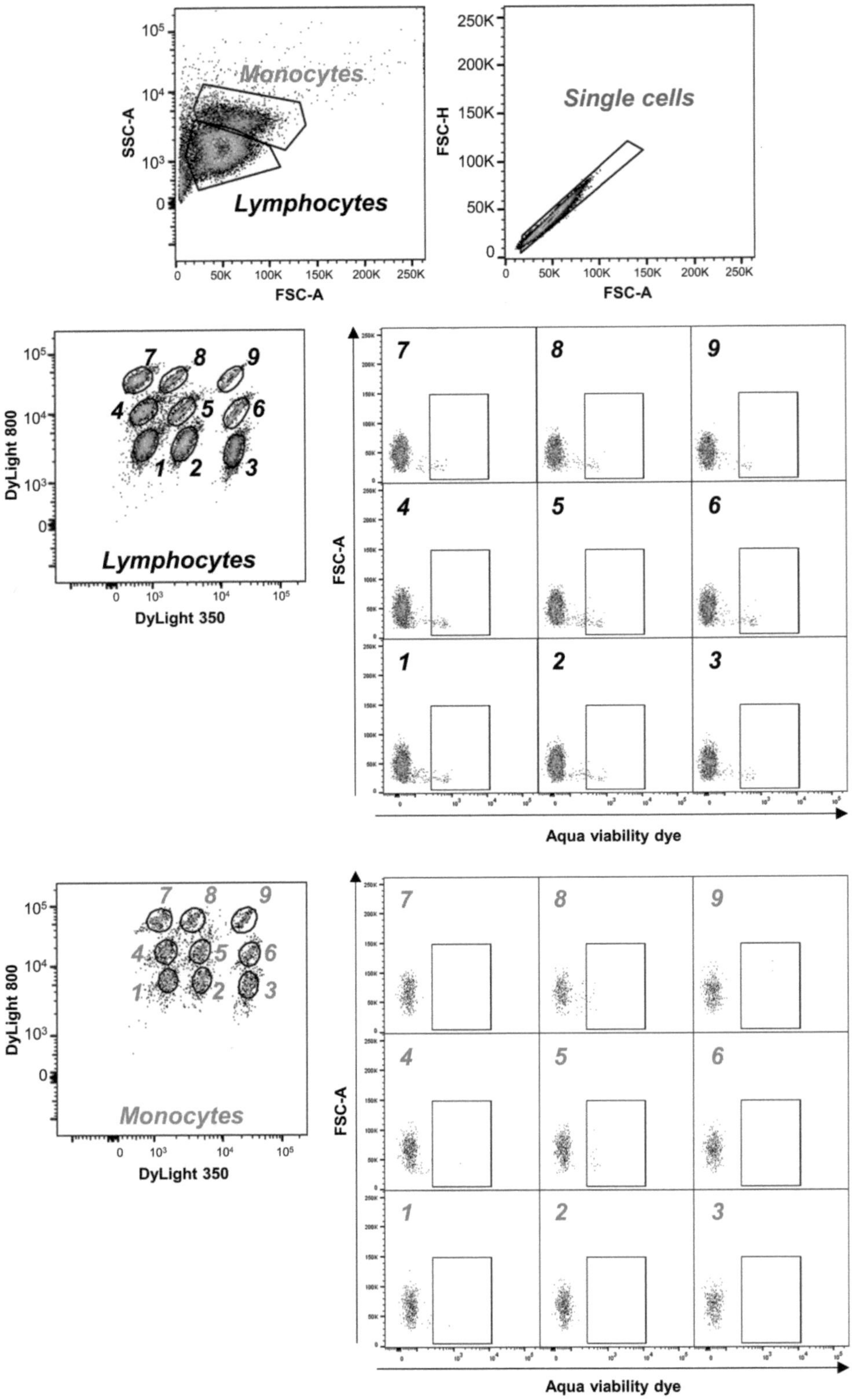


**Fig. 3** Manual gating strategy. Lymphocytes and monocytes are identified based on linear parameters, forward scatter area (FSC-A) and side scatter-area (SSC-A), and doublets are excluded based on FSC-A versus forward scatter height (FSC-H). On lymphocytes (upper) and monocytes (bottom), nine barcoded populations are gated based on FCB dyes (DyLight 350 vs DyLight 800), and on each population of dead cells (positive for Aqua viability dye) are identified

### 3.7 Acquisition and Compensation

1. Calculate pre-acquisition compensation using:
   (a) Barcoded cells with the highest concentration of DyLight 350 and DyLight 800 as prepared in Subheading 3.3.
   (b) Viability dye control as prepared in Subheading 3.4.
   (c) Unstained sample prepared in Subheading 3.3 as negative control for setting PMT voltages.
2. Record a minimum of 5000 single cells or 3000 total lymphocytes for each control to carry out a reliable pre-acquisition compensation.
3. Record a minimum of 300,000 lymphocytes for real samples at medium speed.

### 3.8 Manual Gating Strategy

1. First, check post-acquisition compensation matrix. Check for any spillover of barcoding dyes into other channels (*see* **Note 6**).
2. Apply post-acquisition compensation matrix to samples.
3. Identify lymphocytes and monocytes based on linear parameters, FSC-A and SSC-A (Fig. 3).
4. Exclude doublets based on forward scatter area (FSC-A) versus forward scatter height (FSC-H) or weight (FSC-W) (Fig. 3).
5. On lymphocytes, perform deconvolution by plotting the FCB dye channels (DyLight 350 vs. DyLight 800), and name barcoded populations from one (bottom left) to nine (upper right).
6. On each barcoded lymphocyte population, plot FSC-A versus Aqua viability dye, and gate dead cells (positive for Aqua dye). Create a table with percentage of Aqua$^+$ dead cells for each barcoded lymphocyte population.
7. On monocytes, identify barcoded populations based on FCB dyes (DyLight 350 vs. DyLight 800), and name each population from one (bottom left) to nine (upper right).
8. On barcoded monocyte population, plot FSC-A versus Aqua viability dye, and gate dead cells (positive for Aqua dye). Create a table with percentage of Aqua$^+$ dead monocytes for each barcoded lymphocyte population.
9. Apply this gating strategy to all samples in the experiment and export tables.

### 3.9 Computational Gating Strategy

1. Load FCS files, and define parameters as follows: BUV.396.A as DyLight 350 (FCB dye), APC.Cy7.A as DyLight 800 (FCB dye), and V500.A as Aqua dye (viability dye) (Fig. 4).
2. Read FCS files by read.FCS function of flowCore.

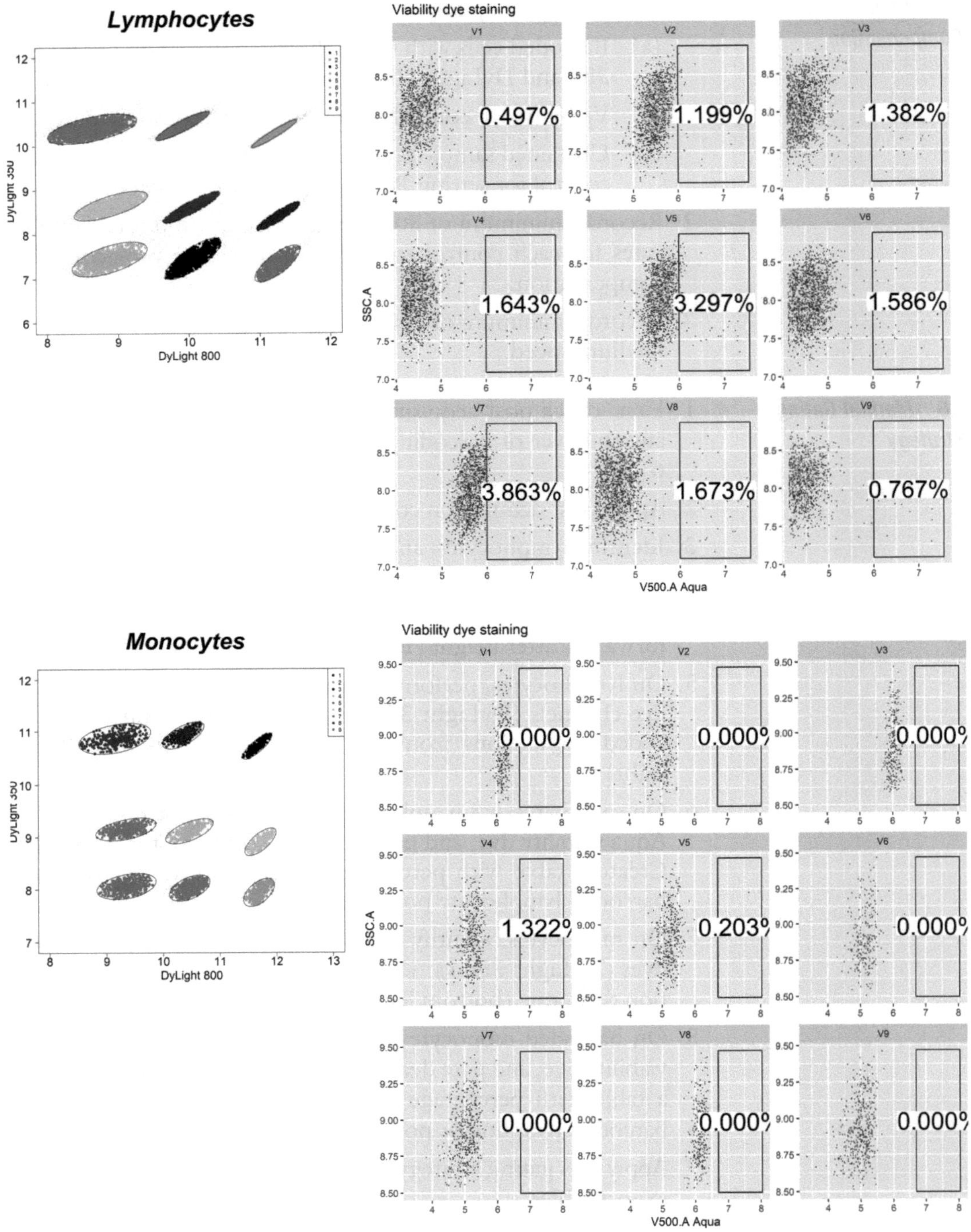


**Fig. 4** Computational gating strategy. Lymphocytes and monocytes are identified based on FSC-A and SSC-A, and debris removed (FSC-A < 20k). After deconvolution by plotting FCB dye channels (DyLight 350 vs. DyLight 800), on each barcoded population, dead cells (positive for Aqua viability dye) are gated

```
BCD_stain <- read.FCS(paste(filenames[1], ".fcs", sep=""),
transformation=FALSE, alter.names = T)
```

3. Remove debris by filtering with rectangleGate function and FCS-A and SSC-A parameters.

```
nodebris <- rectangleGate(filterId="no Debris", "FSC.A"=c
(20,000,1.2e5), "SSC.A"=c(0,Inf))
```

4. Remove events with FSC values lower than 20,000 using Subset method.

```
BCD_stain_nodebris <- Subset(BCD_stain, filter(BCD_stain,no-
debris))
```

5. Automatically cluster lymphocytes and monocytes (two clusters, $K = 2$) using linear parameters (FSC.A and SSC.A) and flowClust function.

```
Lympho_mono <- flowClust(BCD_stain_nodebris, varNames = c
("FSC.A", "SSC.A"), K=2, B=100)
```

6. Plot the results.

```
plot(Lympho_mono, data=BCD_stain_nodebris, level=0.8, z.cut-
off=0, xlim=c(2e4,1.2e5), ylim=c(0,1e4), pch=c(16,16), col = c
("red", "blue"), cex=1, cex.axis=2, xlab="FSC-A", ylab="SSC-
A", cex.lab=2, las=1)
legend("topright", pch = c(16,16), col = c("red", "blue"),
legend = Lympho_mono@prior$order)
```

7. Extract and automatically identify lymphocytes and monocytes based on SSC values, as lymphocytes have lower SSC than monocytes.

```
FCB_pop1 <- split(BCD_stain_nodebris, Lympho_mono, popula-
tion=list(p1=1, p2=2))
G1 <- FCB_pop1$p1
G2 <- FCB_pop1$p2
g1 <- summary(G1)
g2 <- summary(G2)
g1a <- g1[6,4]
g2a <- g2[6,4]
Max_SSC <- c(g1a, g2a)
get_order1 <- Max_SSC[order(Max_SSC, decreasing = TRUE)]
Maximum_SSC <- head(get_order1)[1]
Minimum_SSC <- head(get_order1)[2]
if (g1a == Maximum_SSC) {
```

```
(Mono <- G1)
} else {
(Mono <- G2)
}
if (g1a == Minimum_SSC) {
(Lympho <- G1)
} else {
(Lympho <- G2)
}
```

8. Transform filtered data using biexponential function for both lymphocyte and monocyte populations.

```
biexp <- biexponentialTransform("myTransform",w=0)
after.2 <- transform(Lympho, transformList(c("FSC.A", "FSC.
H", "SSC.A", "V500.A", "BUV.396.A", "APC.Cy7.A", "Time"),
biexp))
tf <- transformList(from=colnames(Lympho), tfun=biexp)
BCD_stain_trans<-tf %on% Lympho
after.3 <- transform(Mono, transformList(c("FSC.A", "FSC.H",
"SSC.A", "V500.A", "BUV.396.A", "APC.Cy7.A", "Time"), biexp))
tf2 <- transformList(from=colnames(Mono), tfun=biexp)
BCD_mono_trans <-tf2 %on% Mono
```

9. First, analyze lymphocyte population, and filter data for FCB dye channels (BUV.396.A and APC.Cy7.A).

```
BCD_gate <- rectangleGate(filterId = "FluoRegion1","APC.Cy7.
A"=c(5,12), "BUV.396.A"=c(0,12))
BCD_filter = Subset(BCD_stain_trans, filter(BCD_stain_trans,
BCD_gate))flowClust function is employed to identify the nine
barcoded populations.
```

10. Filter data for Aqua viability dye channel (V500.A) for removing residual debris (events between the minimum value and the first quartile in the Aqua dye channel).

```
data_L <- as.matrix(summary(BCD_filter))
Filter_gate <- rectangleGate("SSC.A"=c(data_L["Min.", "SSC.
A"], data_L["Max.", "SSC.A"]), "V500.A"=c(data_L["1st Qu.",
"V500.A"], data_L["Max.", "V500.A"]))
FCB_filter = Subset(BCD_filter, filter(BCD_filter, Filter_-
gate))
```

11. Deconvolute FCB data by identifying all nine barcoded populations by setting varNames parameter on FCB dye channels and K on 9, that is, the number of expected populations.

```
FCB_clust <- flowClust(FCB_filter, varNames = c("APC.Cy7.A",
"BUV.396.A"), K=9, B=100)
```

12. Plot results.

```
plot(FCB_clust, data=FCB_filter, level=0.8, z.cutoff=0,
xlim=c(8,12), ylim=c(6,12), pch=c
(16,16,16,16,16,16,16,16,16), col = c("red", "green", "blue",
"cyan", "darkorange1", "gold", "hotpink", "deeppink2", "deeps-
kyblue2"), cex=1, cex.axis=2, xlab="DyLight 800", ylab="Dy-
Light 350", cex.lab=2, las=1)
legend("topright", pch = c(16,16,16,16,16,16,16,16,16), col =
c("red", "green", "blue", "cyan", "darkorange1", "gold", "hot-
pink", "deeppink2", "deepskyblue2"), legend = FCB_clust@prior
$order)
```

13. After clustering, barcoded populations are identified individually, and then create a flowset.

```
FCB_pop <- split(FCB_filter, FCB_clust, population = list
(p1=1, p2=2, p3=3, p4=4, p5=5, p6=6, p7=7, p8=8, p9=9))
pop1 <- FCB_pop$p1
pop2 <- FCB_pop$p2
pop3 <- FCB_pop$p3
pop4 <- FCB_pop$p4
pop5 <- FCB_pop$p5
pop6 <- FCB_pop$p6
pop7 <- FCB_pop$p7
pop8 <- FCB_pop$p8
pop9 <- FCB_pop$p9
FCB_fs <- c(pop1, pop2, pop3, pop4, pop5, pop6, pop7, pop8,
pop9)
fs <- as(FCB_fs, "flowSet")
```

14. To automatically define the limits of each parameter in rectangleGate function for viable cell identification, first create a data frame with the V500.A values of each barcoded population.

```
data1 <- summary(fs)
write.csv(data1, file =paste(filenames,"_for_gate.csv", sep =
""))
data1 <- read.csv(file= "folder/filename_for_gate.csv',
stringsAsFactors = TRUE, sep = ',')
data_Aqua <- cbind(data1$V1.V500.A, data1$V2.V500.A, data1
$V3.V500.A, data1$V4.V500.A, data1$V5.V500.A, data1$V6.V500.
A, data1$V7.V500.A, data1$V8.V500.A, data1$V9.V500.A)
```

15. Identify the maximum value across barcoded population for the third quartile and the maximum of V500.A channel (*see* **Note 8**).

```
third_Aqua <- max(data_Aqua[5,])
third_Aqua <- round(third_Aqua, digits = 0)
max_Aqua <- max(data_Aqua[6,])
```

16. Similarly, identify the maximum value across barcoded population for the minimum and the maximum of SSC.A channel.

```
data_SSC <- cbind(data1$V1.SSC.A, data1$V2.SSC.A, data1$V3.
SSC.A, data1$V4.SSC.A, data1$V5.SSC.A, data1$V6.SSC.A, data1
$V7.SSC.A, data1$V8.SSC.A, data1$V9.SSC.A)
min_SSC <- min(data_SSC[1,])
max_SSC <- max(data_SSC[6,])
```

17. For dead cell discrimination, create a rectangleGate using the minimum and maximum of SSC.A, and the third quartile and the maximum of Aqua viability dye.

```
Aqua_gate <- rectangleGate("SSC.A"=c(min_SSC, max_SSC),
"V500.A"=c(third_Aqua, max_Aqua))
```

18. Plot the entire flowset with the gate identifying dead cells and relative percentages.

```
plot2 <- ggcyto(fs, aes(x = "V500.A", y = "SSC.A")) + geom_hex
(bins = 128) + geom_gate(Aqua_gate) + geom_stats(size = 8)
plot2 <- plot2 + ggtitle("Viability dye staining")
```

19. Create a table with results.

```
live.fs_L <- filter(fs, Aqua_gate)
results <- summary(live.fs_L)
V1 <- results$V1@p*100
V2 <- results$V2@p*100
V3 <- results$V3@p*100
V4 <- results$V4@p*100
V5 <- results$V5@p*100
V6 <- results$V6@p*100
V7 <- results$V7@p*100
V8 <- results$V8@p*100
V9 <- results$V9@p*100
data_Lymp <- cbind(V1, V2, V3, V4, V5, V6, V7, V8, V9)
write.csv(data_Lymp, file =paste(filenames,"_data_live_dea-
d_lymp.csv", sep = ""))
```

20. Next, analyze monocyte populations, and filter and subset data for FCB dye channels (BUV.396.A and APC.Cy7.A).

```
BCD_gate <- rectangleGate(filterId = "FluoRegion1","APC.Cy7.
A"=c(5,14), "BUV.396.A"=c(0,12))
BCD2_filter = Subset(BCD_mono_trans, filter(BCD_mono_trans,
BCD_gate))
```

21. Filter data for Aqua viability dye channel (V500.A) for removing residual debris (events between the minimum value and the first quartile in the Aqua dye channel).

```
data_M <- as.matrix(summary(BCD2_filter))
Filter_gate2 <- rectangleGate("SSC.A"=c(data_M["Min.", "SSC.
A"], data_M["Max.", "SSC.A"]), "V500.A"=c(data_M["1st Qu.",
"V500.A"], data_M["Max.", "V500.A"]))
FCB_filter2 = Subset(BCD2_filter, filter(BCD2_filter,Filter_-
gate2))
```

22. Deconvolute FCB data by identifying all nine barcoded populations by setting varNames parameter on FCB dye channels and K on 9, that is, the number of expected populations.

```
FCB_clust4 <- flowClust(BCD2_filter, varNames = c("APC.Cy7.
A","BUV.396.A"), K=9, B=100)
```

23. Plot results.

```
plot(FCB_clust4, data=BCD2_filter, level=0.8, z.cutoff=0,
xlim=c(8,13), ylim=c(7,12), pch=c
(16,16,16,16,16,16,16,16,16), col = c("red", "green", "blue",
"cyan", "darkorange1", "gold", "hotpink", "deeppink2", "deeps-
kyblue2"), cex=1, cex.axis=2, xlab="DyLight 800", ylab="Dy-
Light 350", cex.lab=2, las=1)
legend("topright", pch = c(16,16,16,16,16,16,16,16,16), col =
c("red", "green", "blue", "cyan", "darkorange1", "gold", "hot-
pink", "deeppink2", "deepskyblue2"), legend = FCB_clust4@prior
$order)
```

24. After clustering, barcoded populations are identified individually, and then create a flowset.

```
FCB_pop4 <- split(BCD2_filter, FCB_clust4, population=list
(m1=1, m2=2, m3=3, m4=4, m5=5, m6=6, m7=7, m8=8, m9=9))
M1 <- FCB_pop4$m1
M2 <- FCB_pop4$m2
M3 <- FCB_pop4$m3
M4 <- FCB_pop4$m4
```

```
M5 <- FCB_pop4$m5
M6 <- FCB_pop4$m6
M7 <- FCB_pop4$m7
M8 <- FCB_pop4$m8
M9 <- FCB_pop4$m9
FCB_mono_fs <- c(M1, M2, M3, M4, M5, M6, M7, M8, M9)
ms <- as(FCB_mono_fs, "flowSet")
```

25. To automatically define the limits of each parameter in rectangleGate function for viable cell identification, first create a data frame with the V500.A values of each barcoded population. Identify the mean across barcoded population for the third quartile and the maximum of V500.A channel. Similarly, identify the maximum value across barcoded population for the minimum and the maximum of SSC.A channel (*see* **Note 7**).

```
data2 <- summary(ms)
write.csv(data2, file =paste(filenames,"_for_gate_mono.csv",
sep = ""))
data2 <- read.csv(file= "folder/filename_010_for_gate_mono.
csv", stringsAsFactors = TRUE, sep = ',')
data_Aqua_M <- cbind(data2$V1.V500.A, data2$V2.V500.A, data2
$V3.V500.A, data2$V4.V500.A, data2$V5.V500.A, data2$V6.V500.
A, data2$V7.V500.A, data2$V8.V500.A, data2$V9.V500.A)
third_Aqua2 <- mean(data_Aqua_M[6,])
third_Aqua2 <- round(third_Aqua2, digits = 1)
max_Aqua2 <- max(data_Aqua_M[6,])
data_SSC2 <- cbind(data2$V1.SSC.A, data2$V2.SSC.A, data2$V3.
SSC.A, data2$V4.SSC.A, data2$V5.SSC.A, data2$V6.SSC.A, data2
$V7.SSC.A, data2$V8.SSC.A, data2$V9.SSC.A)
min_SSC2 <- min(data_SSC2[1,])
max_SSC2 <- max(data_SSC2[6,])
```

26. For dead cell discrimination, create a rectangleGate using the minimum and maximum of SSC.A, and the mean and maximum of Aqua viability dye. Plot data.

```
Aqua_gate2 <- rectangleGate("SSC.A"=c(min_SSC2, max_SSC2),
"V500.A"=c(third_Aqua2, max_Aqua2))
plot4 <- ggcyto(ms, aes(x = "V500.A", y = "SSC.A")) + geom_hex
(bins = 128) + geom_gate(Aqua_gate2) + geom_stats(size = 8)
plot4 <- plot4 + ggtitle("Viability dye staining")
```

27. Create a table with the results.

```
live.fs_M <- filter(ms, Aqua_gate2)
resultsM <- summary(live.fs_M)
```

```
Z1 <- resultsM$V1@p*100
Z2 <- resultsM$V2@p*100
Z3 <- resultsM$V3@p*100
Z4 <- resultsM$V4@p*100
Z5 <- resultsM$V5@p*100
Z6 <- resultsM$V6@p*100
Z7 <- resultsM$V7@p*100
Z8 <- resultsM$V8@p*100
Z9 <- resultsM$V9@p*100
data_Mono <- cbind(Z1, Z2, Z3, Z4, Z5, Z6, Z7, Z8, Z9)
 write.csv(data_Mono, file =paste(filenames,"_data_live_-
dead_mono.csv", sep = ""))
```

### 3.10 Validation of FCB Data: Intra-Assay Variability

FCB requires optimization and standardization, especially for acquisition and analysis of samples from clinical studies [7]. For this reason, an internal bridge control is required for intra- and inter-assay variability assessment and a fixed and permeabilized control without barcoding to exclude any effects of FCB dyes on the results, especially when data are reported as MFI values (*see* **Note 8**) [11, 12].

Intra-assay variability assessment can identify outliers within a matrix, frequently due to compensation issues in a particular matrix position. This is important for samples barcoded with the highest dye concentrations or increased autofluorescence causing a shift in fluorescence of positive cells (Fig. 5).

For intra-assay variability:

1. Calculate percent of dead cells (Aqua viability dye$^+$ cells) in all samples using FlowJo or other software, and export data as .csv or .xls file.
2. Calculate means for each cell population.
3. Define the range of variability as < mean -2SD or > mean +2SD.
4. For each cell population, compare percent of positive cells to calculated means. Values should be within mean +/− 2SD.

### 3.11 Validation of FCB Data: Inter-Assay Variability

Inter-assay variability identifies variations between matrices caused by technical issues probably related to FCB matrix preparation.

1. Calculate percent of dead cells (Aqua viability dye$^+$ cells) using data from samples in the same position across matrices (the bridge control) by FlowJo software, and export data as .csv or . xls file.
2. Calculate means for each cell population.
3. Define the range of variability as < mean -2SD or > mean +2SD.
4. For each cell population, compare percent of positive cells to calculated means. Values should be within mean +/− 2SD.

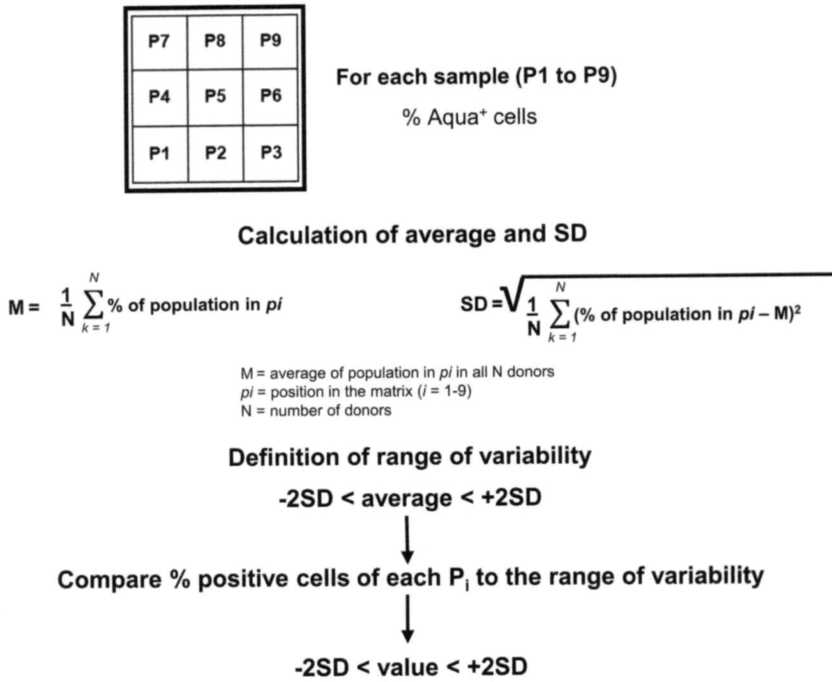


**Fig. 5** Intra- and inter-assay variability assessment. For each sample (from P1 to P9), the percentage of Aqua$^+$ dead cells is calculated, and average and standard deviation (SD) are defined as shown in the range of variability

### *3.12 Validation of FCB Data: Internal Bridge Control*

Intra- and inter-assay variabilities can be assessed only during protocol optimization, while validation of FCB data using the internal bridge control should be performed for every experiment (Fig. 6).

1. Calculate the percentage of dead cells (Aqua viability dye$^+$ cells) of the internal bridge control with and without barcoding by FlowJo software, and export data as .csv or .xls file.
2. Compare the percentage of positive cells across matrices by unpaired t-test or analysis of variance (ANOVA, for multiple group comparison, e.g., if several matrices are analyzed together). Variations should be not statistically significant.
3. Correlate paired stimulated versus unstimulated values of the percentage of positive cells by Pearson correlation analysis. Correlations should have an $r > 0.85$ and a $P$ value $<0.05$.
4. Correlate percentage of positive cells obtained from barcoded internal controls to matched not-barcoded samples ("No FCB"). Correlations should have an $r > 0.9$ and a $P$ value $<0.05$.
5. Calculate means for each cell population obtained from "No FCB" controls (average of controls).

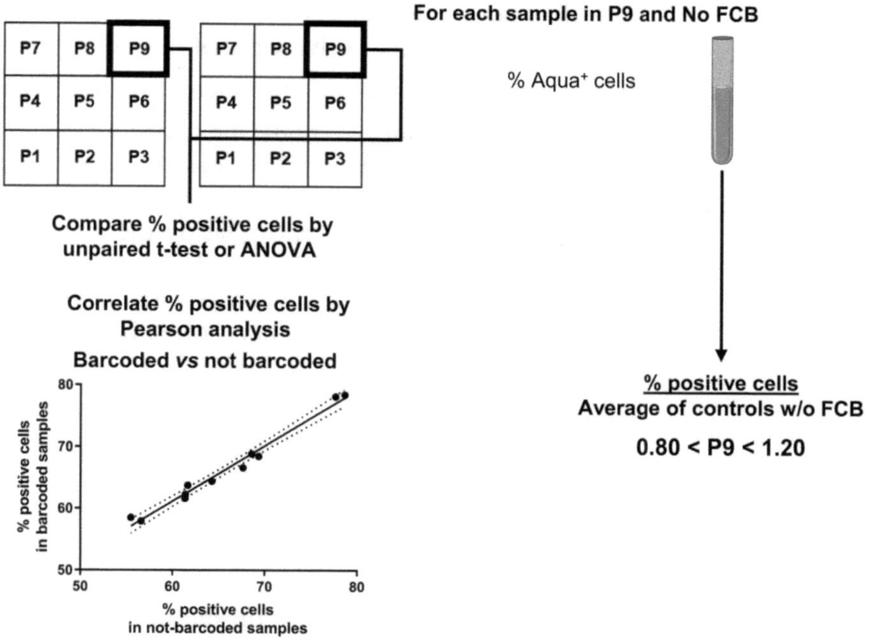


**Fig. 6** Validation of FCB data. Percentage of dead cells from the internal bridge control (P9) is obtained and compared between matrices by unpaired t-test (for two group comparison) or ANOVA (multiple group comparison). Pearson correlation analysis can be additionally employed to show high correlation between data obtained from matched barcoded and not-barcoded internal bridge controls. In addition, percentage of positive cells of barcoded control (P9) should be normalized for the average of percentage of positive cells obtained from controls without barcoding (w/o FCB). Ratios should be within 0.80–1.20

6. Divide each percentage of positive cells obtained from barcoded internal controls to the average of controls. Ratios should be within 0.8 and 1.2 (*see* **Note 9**).

## 4 Notes

1. Dyes should be chosen based on available lasers and detectors on each flow cytometer, and also, the best dye combination should be selected to minimize spillover of dye fluorescence in surface/intracellular marker antibody channels (e.g., avoid DyLight 800 if using PE-Cy7- or BV750/BV785-conjugated antibodies on a LSR Fortessa cytometer).
2. Stock concentrations are usually stable for long period of time at −80 °C; however, for aliquots of 500 μL of stock solution (500 μg/mL) repeated thawing cycles should be avoided to increase dye stability. Preparation of smaller aliquots is not suggested because solutions could be less homogeneous and/or stable compared to larger volume aliquots. Dry

powders can be also reconstituted at 1 mg/mL concentration (1 mL of DMSO to dissolve 1 mg of powder).

3. Longer incubation times do not increase FCB efficiency.
4. Internal bridge controls should be added always in the same position across matrices and experiments to avoid biases caused by technical variations related to different dye concentrations at each experiment. Moreover, by placing the internal control in position 9, samples will be barcoded with the highest concentrations of both dyes. The absence of variations across matrices is also important to exclude effects of FCB dyes on results.
5. Pre-filling tubes with 3 mL of 1× Barcoding Wash Buffer allows a quick dilution in a large volume of residual unbound dyes upon transfer reducing nonspecific binding and minimizing technical variabilities caused by different transfer speeds that may be operator dependent.
6. Post-acquisition compensation matrices should be carefully checked to exclude spillovers of FCB dyes in other channels. For example, particular attention should be paid when Pacific Orange is combined with Brilliant Violet-conjugated antibodies.
7. The mean or the maximum across barcoded population for the third quartile and the maximum of V500.A channel can be used in rectangleGate function for automatic viable cell identification.
8. Higher concentrations of FCB dyes might change MFI values of same fluorochromes because of increased autofluorescence of barcoded samples. For this reason, it is preferable to use higher dye concentrations or 12-sample matrices concentrations for surface marker staining when MFI values are not required (e.g., immunophenotyping or T cell subset characterization).
9. The ratio of variability between results from barcoded internal controls and not-barcoded matched controls should ideally be 1 (no effects of FCB dyes on staining and procedures); however, a 20% of biological variation should be allowed for (between 0.80 and 1.20) [12].

## References


1. Krutzik PO, Nolan GP (2006) Fluorescent cell barcoding in flow cytometry allows high-throughput drug screening and signaling profiling. Nat Methods 3:361–368
2. Krutzik PO, Nolan GP (2003) Intracellular phospho-protein staining techniques for flow cytometry: monitoring single cell signaling events. Cytometry A 55:61–70
3. Krutzik PO, Crane JM, Clutter MR, Nolan GP (2008) High-content single-cell drug screening with phosphospecific flow cytometry. Nat Chem Biol 4:132–142
4. Kalland ME, Oberprieler NG, Vang T, Taskén K, Torgersen KM (2011) T cell-signaling network analysis reveals distinct differences between CD28 and CD2

costimulation responses in various subsets and in the MAPK pathway between resting and activated regulatory T cells. J Immunol 187: 5233–5245

5. Spurgeon BE, Aburima A, Oberprieler NG, Taskén K, Naseem KM (2014) Multiplexed phosphospecific flow cytometry enables large-scale signaling profiling and drug screening in blood platelets. J Thromb Haemost 12:1733–1743
6. Davies R, Vogelsang P, Jonsson R, Appel S (2016) An optimized multiplex flow cytometry protocol for the analysis of intracellular signaling in peripheral blood mononuclear cells. J Immunol Methods 436:58–63
7. Tsai WL, Vian L, Giudice V, Kieltyka J, Liu C, Fonseca V, Gazaniga N, Gao S, Kajigaya S, Young NS, Biancotto A, Gadina M (2020) High throughput pSTAT signaling profiling by fluorescent cell barcoding and computational analysis. J Immunol Methods 477: 112667
8. Manohar S, Shah P, Biswas S, Mukadam A, Joshi M, Viswanathan G (2019) Combining fluorescent cell barcoding and flow cytometry-based phospho-ERK1/2 detection at short time scales in adherent cells. Cytometry A 95(2):192–200
9. Stam J, Abdulahad W, Huitema MG et al (2011) Fluorescent cell barcoding as a tool to assess the age-related development of intracellular cytokine production in small amounts of blood from infants. PLoS One 6(10):e25690
10. Prussin C, Metcalfe DD (1995) Detection of intracytoplasmic cytokine using flow cytometry and directly conjugated anti-cytokine antibodies. J Immunol Methods 188:117–128
11. Giudice V, Feng X, Kajigaya S, Young NS, Biancotto A (2017) Optimization and standardization of fluorescent cell barcoding for multiplexed flow cytometric phenotyping. Cytometry A 91(7):694–703
12. Giudice V, Fantoni G, Biancotto A (2019) Fluorescent cell barcoding for Immunophenotyping. Methods Mol Biol 2032:53–68
13. Gao S, Wu Z, Arnold B, Diamond C, Batchu S, Giudice V, Alemu L, Raffo DQ, Feng X, Kajigaya S, Barrett J, Ito S, Young NS (2022) Single-cell RNA sequencing coupled to TCR profiling of large granular lymphocyte leukemia T cells. Nat Commun 13(1):1982
14. Lu Y, Biancotto A, Cheung F, Remmers E, Shah N, McCoy JP, Tsang JS (2016) Systematic analysis of cell-to-cell expression variation of T lymphocytes in a human cohort identifies aging and genetic associations. Immunity 45(5):1162–1175
15. Finak G, Langweiler M, Jaimes M et al (2016) Standardizing flow cytometry immunophenotyping analysis from the human immunophenotyping consortium. Sci Rep 6:20686
16. Saeys Y, Gassen SV, Lambrecht BN (2016) Computational flow cytometry: helping to make sense of high-dimensional immunology data. Nat Rev Immunol 16(7):449–462
17. Aghaeepour N, Finak G; FlowCAP Consortium et al (2013) Critical assessment of automated flow cytometry data analysis techniques. Nat Methods 10(3):228–238
18. Hu Z, Jujjavarapu C, Hughey JJ et al (2018) MetaCyto: a tool for automated meta-analysis of mass and flow cytometry data. Cell Rep 24(5):1377–1388
19. Reisman BJ, Barone SM, Bachmann BO, Irish JM (2021) DebarcodeR increases fluorescent cell barcoding capacity and accuracy. Cytometry A 99(9):946–953
20. Perfetto SP, Chattopadhyay PK, Lamoreaux L, Nguyen R, Ambrozak D, Koup RA, Roederer M (2010) Amine-reactive dyes for dead cell discrimination in fixed samples. Curr Protoc Cytom Chapter 9:Unit 9.34
21. Perfetto SP, Chattopadhyay PK, Lamoreaux L, Nguyen R, Ambrozak D, Koup RA, Roederer M (2006) Amine reactive dyes: an effective tool to discriminate live and dead cells in polychromatic flow cytometry. J Immunol Methods 313(1–2):199–208
22. Johnson S, Nguyen V, Coder D (2013) Assessment of cell viability. Curr Protoc Cytom Chapter 9: Unit9.2
23. Lekishvili T, Campbell JJ (2018) Rapid comparative immunophenotyping of human mesenchymal stromal cells by a modified fluorescent cell barcoding flow cytometric assay. Cytometry A 93(9):905–915
24. Krutzik PO, Clutter MR, Trejo A, Nolan GP (2011) Fluorescent cell barcoding for multiplex flow cytometry. Curr Protoc Cytom Chapter 6:Unit 6.31

# Chapter 8

# Average Rheological Quantities of Cells in Monolayers

**Santanu Kumar Basu, Haider Dakhil, and Andreas Wierschem**

## Abstract

The method of cell monolayer rheology enables quantifying average rheological properties of cell in a single experimental run of few millions cells together in a single layer. Here we describe step-by-step procedure as to how to employ a modified commercial rotational rheometer to run rheological measurement and detect average viscoelastic properties of cells while maintaining the necessary precision level at the same time.

**Key words** Cell rheology, Linear viscoelasticity, Narrow-gap rheometry, Cell monolayer

## 1 Introduction

Rheology is the study of deformation and flow of material under stress. It deals with the measurement of viscoelasticity of solid, fluid, and viscosity of fluids [1, 2].

Rheological properties of cells are important for physiological functions such as cell migration, division, contraction, adhesion, or differentiation.

Various techniques are available for single cell studies such as atomic force microscopy, micropipette aspiration, microplate rheometer, magnetic bead cytometry, and particle tracking microrheology [3, 4]. Cells show stiffness variation by an order of magnitude based on stages of cell cycle, shape, structure, and protein expression [5, 6].

Therefore, to find out the impact of drugs on the viscoelastic properties of cells, it is important to determine the average viscoelastic properties of cells, which can be achieved by studying the average viscoelastic properties of a large number of cells as a function of the cell coverage [8], seeded in a monolayer between rheometer plates in the parallel plate configuration mode [7].

With the help of a commercial rheometer, rheology of living animal cells is a difficult task, as systematic error in the gap width ranges from 25 to 75 μm [9–11], whereas the average height of

Oliver Friedrich and Daniel F. Gilbert (eds.), *Cell Viability Assays: Methods and Protocols*, Methods in Molecular Biology, vol. 2644, https://doi.org/10.1007/978-1-0716-3052-5_8,

animal cells ranges below 25 μm. To improve the gap width precision, we have modified commercial rotational rheometers significantly reducing the variation to less than ±1 μm [8, 12, 13].

## 2 Materials

We provide the protocol for adherent cells, for example, HeLa or MCF7 cells that can be adhered on to the glass surface with extracellular matrix proteins like fibronectin.

1. Culture medium for MCF7 cells: RPMI 1640 with sodium pyruvate, 10% fetal bovine serum (FBS), 100× MEM nonessential amino acids, 10 μg/mL human insulin. Sterile filtered through 0.2 μm filter (*see* **Note 1**).
2. Culture medium for HeLa cells: RPMI 1640 without glutamine, 10% FBS, 4 mM L-Glutamine, 1% Penicillin-Streptomycin. Sterile filtered through 0.2 μm filter (*see* **Note 1**).
3. PBS−−: Dulbecco's Phosphate Buffered Saline without $MgCl_2$ and $CaCl_2$.
4. PBS++: Dulbecco's Phosphate Buffered Saline with $MgCl_2$ and $CaCl_2$.
5. 30 μg/mL Fibronectin solution: 1.7 μL Fibronectin added to 55 μL PBS++.
6. Trypsin-EDTA solution.
7. Hemocytometer.
8. 5 mM Calcein AM stock solution: 10 μL DMSO added to 50 μg of Calcein AM.
9. 30 μM Propidium iodide (PI) solution: 20 μL of 1 mg/mL PI solution added to 980 μL of culture medium with cells.
10. T 25 tissue culture flask.
11. Digital b/w camera equipped with a filter block for fluorescence imaging and with a 5× objective lens. The filter block contains light-emitting diode (LED) 490 nm, excitation filter [Center Wavelength (CWL) = 460 nm and Full Width at Half Maximum (FWHM) = 60 nm], emission filter (CWL 578 nm, FWHM 105 nm), and FITC Dichroic Filter (Reflection Band = 470–490 nm, Transmission Band = 508–675 nm) are incorporated in the lens tube for fluorescence imaging with Calcein AM and Propidium Iodide (*see* **Note 2**).
12. Confocal interferometric sensor: Working distance: 20 mm, measuring range: 90 μm.
13. Rotational rheometer, e.g., MCR 501 from Anton Paar.

14. Piezoelectric actuators.
15. Camera.
16. Two glass plates one with a diameter of 75 mm and evenness of at least $\lambda/4$ and the other glass plate with a diameter of 50 mm and evenness of $\lambda/4$, where $\lambda$ is the testing wavelength (633 nm).
17. Lens cleaning tissue.
18. Acetone (100%).
19. Sodium dodecyl sulfate (SDS).
20. Ethanol (100%).

## 3 Methods

All procedures shall be carried out at room temperature unless otherwise specified (*see* **Note 3**).

### 3.1 Cell Preparation

1. Pre-warm the Trypsin-EDTA solution, PBS−− and growth medium to 37°C.
2. Remove the culture medium from the culture flask (*see* **Note 4**).
3. Wash cells with 1 mL of PBS−− and discard the PBS−−.
4. Add 0.5 mL/10 cm$^2$ culture flask area of pre-warmed Trypsin-EDTA to the cells.
5. Incubate for 2–3 min at room temperature to let the cells detach from its inner surface.
6. When at least 90% of the cells are detached and floating, add 3 mL of fresh culture medium with FBS to the culture flask (*see* **Note 5**).
7. Transfer the cell suspension into a 15 mL conical tube and centrifuge for 5 min at 180 × g to separate the cells from the suspension.
8. Discard the supernatant and add 1 mL of culture medium with the cells and mix thoroughly.
9. Count the cells with a hemocytometer under the microscope to quantify the amount of cells needed in an individual rheological experiment.

### 3.2 Cell Staining

1. Remove the cell culture medium from the flask and wash the cells once with PBS−− to remove any remaining media.
2. Add 1 μL of Calcein AM stock solution per 1 mL of fresh medium. This yields a concentration of 5 μM (*see* **Note 6**).

3. Incubate the cells in the T 25 flask with 1 mL of the fresh medium with Calcein AM in the incubator at a temperature of 37 °C for about 1 h to let the agent pass through the cell membrane.
4. After 1 h of incubation, cells are stained with Calcein AM. Follow the procedure of Subheading 3.1 for trypsinization, centrifugation and mixing with medium.
5. Store the stained cells in suspension in a centrifuge tube inside the incubator at 37 °C. These cells are used to seed on the plates of rheometer.
6. To identify dead cells, add 20 μL of 1 mg $mL^{-1}$ Propidium Iodide solution to 980 μL of cell suspension. In presence of Propidium Iodide, dead cells will start emitting fluorescence signal immediately.

### 3.3 Cell Monolayer Rheometry with a Modified Narrow-Gap Rheometer

1. Replace the lower metal plate of the rheometer with a glass plate having a diameter of 75 mm and evenness of at least λ/4 and glue the rotating measurement head (diameter with 25 mm) with a glass plate having a diameter of 50 mm and evenness of $\lambda/4$, where $\lambda$ is the wavelength for testing the evenness (633 nm) *see* Fig. 1 [9–11] (*see* **Note 7**).

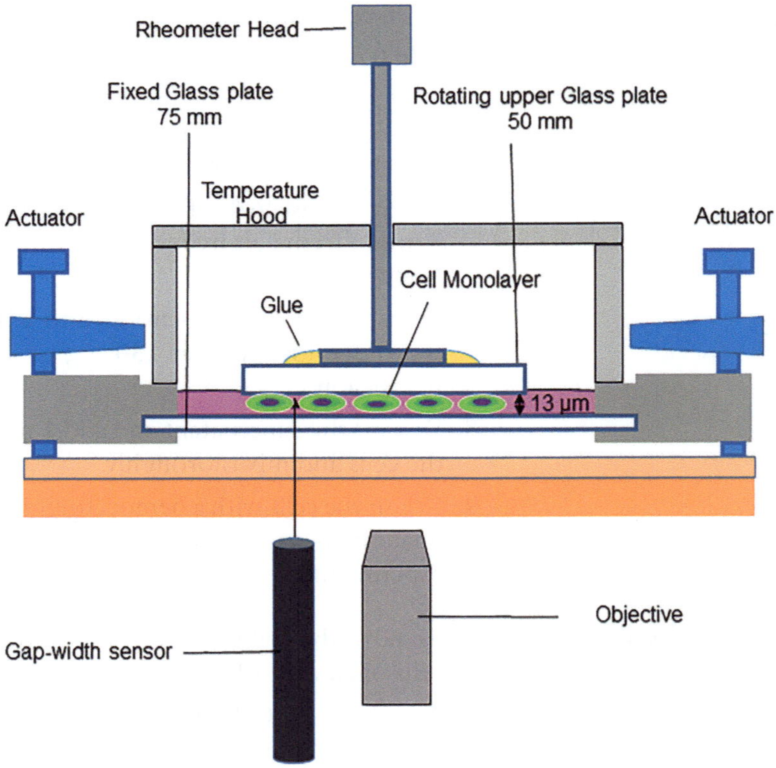


**Fig. 1** Schematic representation of the narrow-gap rheometer setup. A confocal interferometric sensor detects the gap width between the upper and lower plates; the cells are viewed with an objective and camera

2. The stationary lower glass plate stationed perpendicular to the rotation axis onto a tripod, which is fixed on the rheometer table. Adjust the tripod moving up and down by three actuators and a confocal interferometric sensor, which altogether are instrumental in making the plates parallel.

### 3.4 Alignment of Upper and Lower Plates in Narrow-Gap Rheometer

1. Clean both the upper and lower glass plates carefully with cleaning reagents e.g., ethanol (absolute alcohol), 99% acetone and 2% SDS and dry it with normal tissue paper initially and lens cleaning tissue paper finally. Place the upper glass plate atop the lower one and observe from top to count the number of optical fringes, which will indicate whether the plates are really clean or not. Clean plates will generate 10 optical fringes or less.
2. Once both plates are cleaned, check the intensity of signal of the interferometric sensor keeping the upper plate in contact with lower one. Intensity has to be 50–70%.
3. Glue the upper plate with the measurement head of the rheometer using a dissolvable adhesive like UV curing glue [12].
4. Connect the glass plate and measurement head with the rheometer. Do the zero gap. Lift the upper plate by about 20 μm.
5. With the help of the confocal interferometric sensor, measure the gap width between the plates near the edge of the smaller plate at three different positions, forming a triangle while the upper plate will keep rotating at a constant speed of 0.04 $s^{-1}$ to measure the variation in gap width in time.
6. Since the upper plate is inclined with the rotating axis of rheometer, the gap width varies periodically during rotation. Since the lower plate is not perfectly perpendicular to the rheometer axis, the mean value of the gap width is different at each position of the sensor. *See* Fig. 2a.
7. Adjust the lower plate with three actuators that is shown in Fig. 1 to minimize deviations in the mean gap width. *See* Fig. 2b.
8. Stop the rotation of the upper plate with measurement head. Disconnect it from the rheometer. Dislodge the upper plate from the measurement head by leaving it in acetone at least 5–10 min to dissolve and soften the adhesive and subsequent exposure to hot air for a short period.
9. Place the upper plate on the leveled lower plate, and glue it to the measurement head.
10. Check parallelism again with the sensor to achieve undulations at each location with an amplitude of about 1 μm, *see* Fig. 2c.
11. Press zero-gap position with the help of Rheometer software to bring down the upper glass plate in contact with the lower one.

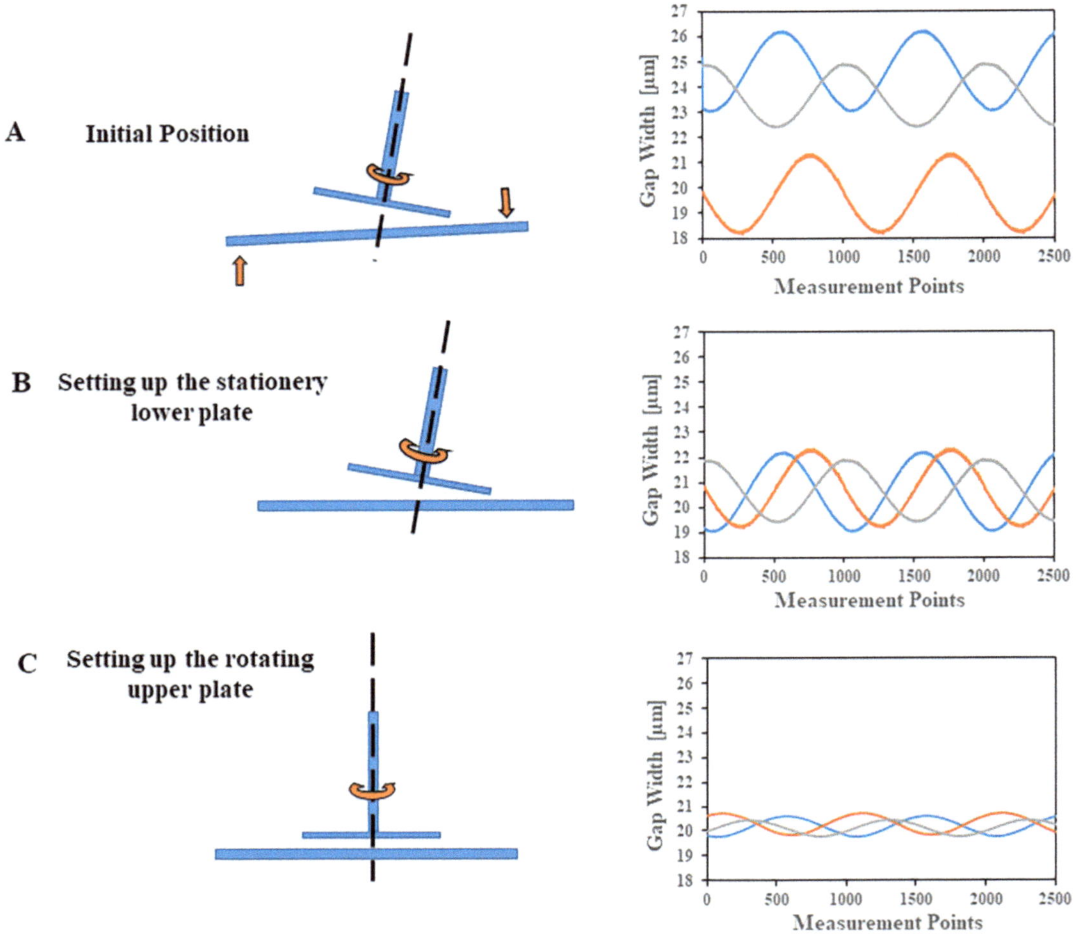


**Fig. 2** Setting up the rheometer disks. The upper disk rotates slowly, while the gap width is detected with a sensor at three different locations: before adjustment (**a**), after adjusting the lower plate (**b**), and after final adjustment of the upper disk (**c**)

12. Bring down the upper plate to the zero-gap prior to each experiment. Find out the gap width with the sensor as a three-point average, and then, adjust the lower plate to make sure the rheometer gap distance is identical with sensor reading. This significantly reduces the error in zeroing.

If the upper plate is already properly aligned, one may start directly with **step 10** of alignment of the narrow-gap rheometer.

### *3.5 Coating the Rheometer Plates*

1. In order to obtain a coating of adhesive protein on both the upper and lower glass plates of the rheometer, add fibronectin solution in the middle of the lower plate, and bring down the upper plate to a gap width of 20 µm so that fibronectin solution may stick to both the plates together at the same time.

2. Leave fibronectin solution between the glass plates forl h at a gap width of 20 μm.
3. After 60 min of coating, lift the upper plate carefully, and rinse the surface of both the plates three times with PBS++.

### 3.6 Cell Seeding on to the Lower Glass Plate of the Rheometer

1. Before the plates dry out of PBS++, seed the cell suspension on to the center of the lower glass plate.
2. Immediately after that, bring the upper plate down with high speed till the upper plate touches the upper layer of medium with cell suspension. Reset the normal force at this point. As soon as, upper plate touches the medium with cells, minimize the lowering speed of the upper plate to minimum (1 μm/s).
3. Compress the cells in between the upper and lower plate at a gap width that is close to the average height of the nucleus of the cells without any damage, and at the same time, observe the normal force range, which should be 2 to 3 N.
4. Leave the cells in between plates for 1 h to 1 h and a half so that cells will attach and adhere both the upper and lower plates.
5. After about 90 min of adhesion, raise the upper plate slowly to the gap width, at which the normal force would be close to 0 N and where oscillatory shear can be applied to cells.

### 3.7 Cell Coverage Detection

1. Capture 50 fluorescence images close to the edge of the upper plate with the camera attached with the fluorescence microscope (*see* **Note 8**).
2. Determine the relative area of gap between the plates, covered by the cells with MATLAB code (*see* **Note 9**).

### 3.8 Measurement of Viscoelastic Properties of Cells

From the oscillatory shear measurement, viscoelastic properties of the cell monolayer is well-understood. The upper glass plate rotates to create oscillatory movement around the axis while the cells are attached in between the upper and the lower glass plates of the rheometer.

In amplitude sweep, the deflection of the upper glass plate is increased step-wise, while keeping the frequency constant. On the contrary, in case of frequency sweep, frequency keeps changing, while amplitude remains constant.

From the amplitude sweeps, the linear viscoelastic region is determined. This indicates the range of lowest strain values, at which the storage ($G'$) and loss moduli ($G''$) remain constant.

The storage modulus $G'$ refers to the elastic response or energy stored in elastic structure of the sample, and loss modulus $G''$ refers the viscous part or the amount of energy dissipated in the sample. In Fig. 3, the linear viscoelastic range of the cell monolayer is below a strain amplitude of 1% (*see* **Note 10**).

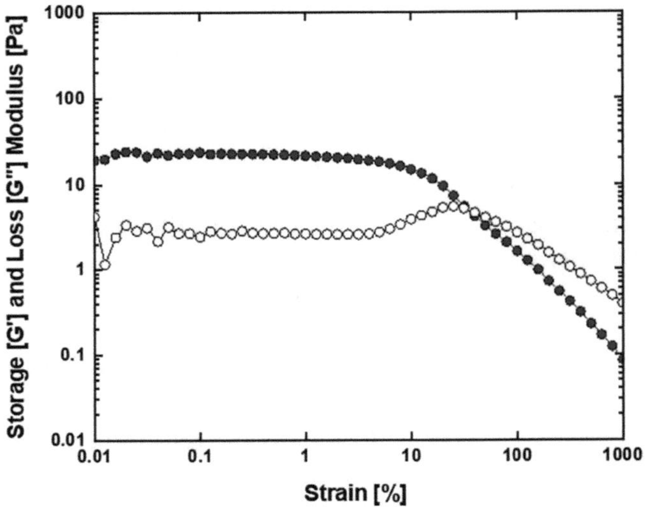


**Fig. 3** Amplitude sweep of Hela cells. Closed and open symbols indicate the storage and the loss modulus, respectively

1. Amplitude sweep: Carry out the amplitude sweep in the controlled deformation mode at strain amplitudes between 0.01% and a strain amplitude where storage and loss moduli start deviating from plateau values (typically between 1% and 10%). Fix the frequency (typically 1 Hz).
2. Frequency sweep: Carry out the frequency sweep in a controlled deformation mode at frequencies that typically range between 1 and 30 Hz. Fix the strain amplitude at a value well within the linear viscoelastic range, determined with the amplitude sweeps (a typical value is 0.1%).

Figure 3 shows the storage and the loss modulus of the cell monolayer, $G'$ and $G''$, respectively. Average cell moduli $G'_{\mathrm{Cell}}$ and $G''_{\mathrm{Cell}}$ can be determined from the measured moduli and the mean cell coverage close to gap rim, $C$:

$$G'_{\mathrm{Cell}} = \frac{G' - (1 - C)G'_{\mathrm{Medium}}}{C} \tag{1}$$

Figure 4 shows the average storage modulus of different types of cells in the linear viscoelastic regime. Standard deviation is about 20%.

## 4 Notes

1. Lid and neck of medium bottles must be kept clean and dry, especially after warming-up in the water bath, to avoid contamination.

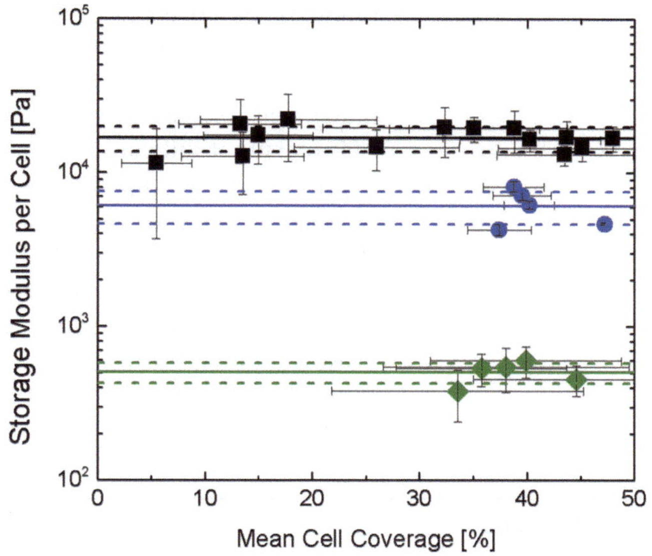


**Fig. 4** Storage modulus per cell of mouse fibroblasts (black squares), human fibroblasts (blue circles), and HeLa cells (green diamonds). Solid lines show the mean modulus; thin dashed lines indicate the range within standard deviation between the measurements [8] (License no. 3898181445518, Springer, Rheologica Acta)

2. Camera magnification: Compromise between imaging large areas to obtain good statistics for the cell coverage and sufficient resolution for precisely determining the cell size.
3. Before using the safety cabinet, be sure it is clean and sterile.
4. While preparing the cells for experiments, wear powder-free Nitrile gloves to avoid any infections.
5. FBS neutralizes the function of trypsin.
6. Live cell staining with fluorescent agents like Calcein AM is carried out to optically detect cells covering the area in the gap between the rheometer plates with a home built fluorescence microscope. Cells are stained in dark.
7. Cell monolayer rheometry, which requires adhesion of a large number of living cells in a monolayer between two parallel glass plates, allows studying the average rheological properties of cells. To obtain reproducible results, this method demands the gap-width variation between the two plates to be a fraction of the average cell height, which is not possible using a commercial rheometer that has the gap-width variation of about ±25–70 μm. The commercial one had been modified to be narrow-gap rheometer having a good precision.
8. Camera magnification: Compromise between imaging large areas to obtain good statistics for the cell coverage and sufficient resolution for precisely determining the cell size.

9. Optimal cell coverage in the monolayer is between 30% and 50%. At lower concentrations, the rheological signal may be rather weak and the scatter in local cell coverage rather large. At higher cell concentrations, cell agglomeration becomes significant making it difficult to obtain a monolayer.

10. To study cells at minimum pre-stress, the final gap width may be set to the average height of the cells, or the normal force of the rheometer may be zeroed with the measurement head initially touching the upper surface of the cell culture medium. When bringing down the measurement head, compressed cells yield a positive normal force. Minimum average pre-stress is obtained with the cells at a gap width where the normal force is zero again.

## Acknowledgments

HD gratefully acknowledges financial support by Deutscher Akademischer Austauschdienst (DAAD), Germany, and the Ministry of Higher Education & Scientific Research, Iraq.

### References

1. Chhabra RP, Richardson JF (2008) Non-Newtonian flow and applied rheology, 2nd edn. Butterworth-Heinemann, Oxford, pp 1–55
2. Steffe JF (1996) Rheological methods in food process engineering. Freeman Press, East Lansing
3. Verdier C, Etienne J, Duperray A et al (2009) Review: rheological properties of biological materials. Comptes Rendus Physique 10:790–811
4. Kollmannsberger P, Fabry B (2011) Linear and nonlinear rheology of living cells. Annu Rev Mater Res 41:75–97
5. Weissman-Shomer P, Fry M (1975) Chick embryo fibroblasts senescence in vitro: pattern of cell division and life span as a function of cell density. Mech Ageing Dev 4:159–166
6. Cai P, Mizutani Y, Tsuchiya M et al (2013) Quantifying cell-to-cell variation in power-law rheology. Biophys J 105:1093–1102
7. Fernández P, Heymann L, Ott A et al (2007) Shear rheology of a cell monolayer. New J Phys 9:419–447
8. Dakhil H, Gilbert DF, Malhotra D et al (2016) Measuring average rheological quantities of cell monolayers in the linear viscoelastic regime. Rheol Acta 55:527–536
9. Davies GA, Stokes JR (2005) On the gap error in parallel plate rheometry that arises from the presence of air when zeroing the gap. J Rheol 49:919–922
10. Davies GA, Stokes JR (2008) Thin film and high shear rheology of multiphase complex fluids. J Non-Newtonian Fluid Mech 148:73–87
11. Pipe CJ, Majmudar TS, Mckinley GH (2008) High shear rate viscometry. Rheol Acta 47:621–642
12. Dakhil H, Wierschem A (2014) Measuring low viscosities and high shear rates with a rotational rheometer in a thin-gap parallel-disk configuration. Appl Rheol 24:63795–63801
13. Dakhil H, Basu SK, Steiner S, Gerlach Y, Soller A, Pan S, Germann N, Leidenberger M, Kappes B, Wierschem A (2021) Buffered λ-DNA solutions at high shear rates. J Rheol 65:159–169

# Chapter 9

# Assaying Spontaneous Network Activity and Cellular Viability Using Multi-Well Microelectrode Arrays

**Seline S. Choo, Jackson Y. Keever, Jasmine Brown, Jenna D. Strickland, and Timothy J. Shafer**

## Abstract

Microelectrode array (MEA) technology is a neurophysiological method that allows for the measurement of spontaneous or evoked neural activity to determine chemical effects thereon. Following assessment of compound effects on multiple endpoints that evaluate network function, a cell viability endpoint in the same well is determined using a multiplexed approach. Recently, it has become possible to measure electrical impedance of cells attached to the electrodes, where greater impedance indicates greater number of cells attached. This would allow rapid and repeated assessments of cell health as the neural network develops in longer exposure assays without impacting cell health. Typically, the lactate dehydrogenase (LDH) assay for cytotoxity and CellTiter-Blue® (CTB) assay for cell viability are only performed at the end of the chemical exposure period because these assays involve lysing of the cells. Procedures describing the multiplexed methods in acute and network formation screening are included in this chapter.

**Key words** Microelectrode array (MEA), Neural network formation, Electrical impedance, Cytotoxicity, Multiplexed assays, Viability, Alamar blue, Lactate dehydrogenase

## 1 Introduction

Multi-well microelectrode array (mwMEA) systems have increased throughput of traditional MEAs, making them an effective in vitro screening tool to prioritize large sets of compounds at multiple concentrations. This phenotypic approach can be applied to drug discovery and safety screening [1–4], acute [5–9], chronic, and developmental neurotoxicity [10–12] and cardiac screening [13]. Currently, thousands of compounds lack sufficient toxicity data, especially as it relates to the nervous system and its

JPB was supported by Student Services Contracts EP-14-D-000101.

Oliver Friedrich and Daniel F. Gilbert (eds.), *Cell Viability Assays: Methods and Protocols*, Methods in Molecular Biology, vol. 2644, https://doi.org/10.1007/978-1-0716-3052-5_9,

development. Further, neurotoxicity, including seizure liability, is a common reason for the failure of drugs at the clinical trial stage of drug development or addition of "black box" FDA warnings for approved drugs. MEA high throughput chemical screening has been proposed as one approach to address these issues [14, 15]. The advancement of a multiplexed approach allowing for simple and rapid characterization of compound effects on both neurophysiological and cellular viability endpoints within the same network provides a method to differentiate between compound-induced changes in neural activity and overall reductions in cell health [16]. The ability to determine if specific effects of a compound on the neural activity (e.g., changes in network firing rates) and changes in cell viability occur simultaneously remains an important aspect of in vitro screening for neurotoxicity.

In this chapter, several different multiplexed methods are outlined for evaluating cell viability in conjunction with MEA recordings following either acute or prolonged (e.g., during network formation) exposure to compounds. In the examples in this chapter, a mixed primary rat cortical culture composed of both inhibitory and excitatory neurons and glial cells is utilized and optimized using Axion Biosystems 48-well classic or CytoView MEA plates. However, these methods may be adapted for other MEA platforms (e.g., 6-, 24-, or 96-well platforms) or cell types (e.g., cardiac cells; IPSC-derived rodent, or human neurons). Using a multiplexed screening approach, we demonstrate straightforward methods for simultaneous determination of compound effects on both neural network function and cell viability.

Comprehensive procedures are described for assessment of compound neurotoxic effects on neural network function following acute exposure or exposure throughout network formation. These procedures include culturing, recording, and viability assessment of cells on MEA plates. To assess changes in neural activity, primary rat cortical cultures were grown on Classic or CytoView MEA plates from Axion Biosystems. Changes in spontaneous network activity in response to treatment were monitored for both acute (~40 min on day in vitro (DIV) 13) and network formation exposures (~15 mins on DIV 5, 7, 9, and 12). Cell viability can be assessed from the same well by measuring electrical impedance (if the CytoView plate is used), lactate dehydrogenase (LDH) release, and metabolic activity (CellTiter-Blue®; CTB). Modifications to each of the viability assays were needed for both acute and network formation assays. While facilitating rapid determination of both changes in neural network function and viability, these methods also serve to reduce cost, time, and animal use.

## 2 Materials

The materials listed below are referenced in multiple procedures. Those specific to a certain procedure are listed in their respective materials section. All procedures in this chapter use sterile technique and are done in a biological safety cabinet with the exception of the CTB and LDH assays described later in this chapter.

1. Biological Safety Cabinet (BSC).
2. 37 °C water bath.
3. Cell Culture Incubator (37 °C +/−1 °C, 5% $CO_2$ +/−1%, 95% humidity).
4. Vacuum flasks/trap.
5. Vortex-style mixer.
6. Classic MEA plates (Axion M768-KAP-48) (*see* **Note 1**) or CytoView MEA plates (Axion M768-tMEA-48W) with viable neural cultures.
7. Pipet Aid and disposable pipettes.
8. Single channel pipette.
9. Multi-channel pipette.
10. Reagent reservoir.
11. Sterile pipette tips sized to fit pipettes.
12. 70% ethanol.
13. pH meter.
14. Stir plate and stir bar.
15. Balance.
16. Beakers (500 mL, 1 L, or other appropriate size).
17. Graduated cylinder.
18. Bottle top filter (Cellulose acetate) (0.22 μm).
19. NB/B27: 500 mL of 1× Neurobasal-A Medium, 10 mL of 50× B-27 Supplement, 5 mL of 100× GlutaMax, 5 mL of 100× Pen-Strep, pH adjusted to 7.4, sterile filtered.
20. Cortical Medium: 500 mL of 1× DMEM with GlutaMax, 55 mL of heat inactivated Horse Serum, 5 mL of 1 M HEPES, 5 mL of 100× Pen-Strep, pH adjusted to 7.4, sterile filtered.
21. Laminin: 20 μg added to each mL of seeding stock.
22. 0.05% Polyethyleneimine (PEI) solution: 100 mL of deionized water, 1.19 g of HEPES, 0.1 g of 50% (w/v) in $H_2O$ (*see* **Note 2**), pH adjusted to 8.0, sterile filtered.
23. Long Evans rat pups less than 24 h old.

### 2.1 MEA Recording

1. Axion Integrated Studio (AxIS) software v3.4.15 or later on a computer workstation.
2. Axion Biosystems MaestroPro 768-channel amplifier with $CO_2$ setup.

### 2.2 Network Formation Assay and Acute Screening Assay

1. Sterile polypropylene 96-well dosing plate.
2. Stock compounds.
3. Axion Integrated Studio (AxIS) software v3.4.15 or later.
4. Axion Biosystems MaestroPro 768-channel amplifier with $CO_2$ setup.

### 2.3 Viability: CellTiter-Blue®

1. CellTiter-Blue® Cell Viability Assay kit.
2. BMG FLUOstar OPTIMA Fluorescence/Luminescence Microplate Reader and Optima v2.20 R11 software or equivalent (excitation and emission wavelengths of 544 and 590 nm, respectively).
3. 96-well white opaque plate.
4. Test material: exposed cultures in MEA plates.
5. Plate map.

### 2.4 Viability: LDH Release (Acute)

1. CytoTox 96® Non-Radioactive Cytotoxicity Assay kit.
2. Molecular Devices VERSAMax plate reader and SoftMax Pro v5.3 software or equivalent (Absorbance 490 nm).
3. Clear 96-well plates with flat bottom.
4. Test material: exposed cultures in MEA plates.
5. Plate map.

### 2.5 Viability: Total Cellular LDH (Network Formation)

1. CytoTox 96® Non-Radioactive Cytotoxicity Assay kit.
2. Molecular Devices VERSAMax plate reader and SoftMax Pro v5.3 software or equivalent (Absorbance 490 nm).
3. Phosphate Buffered Saline, pH 7.4.
4. Clear 96-well plates with flat bottom.
5. Test Material: exposed cultures in MEA plates.
6. Plate map.

### 2.6 Viability: Impedance Measurement (See Note 3)

1. Axion Integrated Studio (AxIS) software v3.4.15 with viability module.
2. Axion Biosystems MaestroPro 768-channel amplifier with $CO_2$ setup.
3. Exposed culture in CytoView MEA plates.

## 3 Methods

### 3.1 Coating MEA Plates with Poly-Ethyleneimine (PEI)

Prior to culturing cells on MEAs, the surface of the wells must be coated with a substrate that facilitates attachment of the cells. Our protocols use PEI, but other substrates may be more appropriate depending on the cell type.

1. Add 150 μL of PEI Solution to each well of a 48-well MEA plate, ensuring that the bottom of the well is completely covered (if needed, gently tap the perimeter of the plate for full coverage).
2. Incubate the plate in a humidified cell culture incubator at 37 °C for 1 h.
3. Remove PEI from wells by aspiration, and rinse thoroughly with 500 μL sterile water for three times (*see* **Note 4**).
4. Allow plates to air dry in the BSC with the lid off. Place the lid in the BSC with the interior side facing up to ensure sterility.
5. Once dry, plates can immediately be used for cell culture or taped closed and refrigerated until they are ready for use (can be refrigerated up to 7 days after coating).

### 3.2 Culturing on MEA Plates

Primary cortical neural cultures are prepared from the neocortices of LE rat pups and is detailed in [5] with modifications. All procedures involving animals were approved by the US EPA-RTP Institutional Animal Care and Use Committee. Both classic and CytoView MEA plates are pre-coated with 0.05% polyethyleneimine (PEI) for cell attachment (refer to Subheading 3.1). For classic 48-well MEA plates, cortical cells are plated at a seeding density of $1.5 \times 10^5$ cells/well, whereas the seeding density is $1.2 \times 10^5$ cells/well for CytoView 48-well MEA plates. Cells are administered via a 25 μL (classic MEA) or 10 μL (CytoView MEA) cortical medium/laminin drop directly onto the microelectrode array (*see* **Notes 5** and **6**). The plates are transferred to a 37 °C incubator for 2 h, and then the volume of each well is adjusted to 500 μL with NB/B27 media. Viable cultures form complex neural networks over time and can remain stable for several weeks in culture. Neural firing increases in synchrony and bursting activity as the culture ages, and this activity can be quantified [17]. Plates can either be immediately dosed following seeding and monitored over several days (refer to Subheading 3.4) or allowed to form neural networks first and then acutely exposed to compound for a shorter period of time (refer to Subheading 3.5).

### 3.3 MEA Recording

1. Turn on the MaestroPro and open AxIS Navigator.
2. Once the program has connected to the MaestroPro and initialized, set the heater to 35 °C (*see* **Note 7**).

3. MaestroPro configuration should be set as follows: Neural Spikes, 1000× gain, and median referencing is selected.
4. Add the Digital Filter to the MaestroPro and use the default settings: High Pass Filter = Butterworth, High Pass Cutoff Frequency = 200 Hz for acute assay, and 0.1 Hz for network formation assay; Low Pass = Butterworth, Low Pass Cutoff Frequency = 3000 Hz for acute assay, and 5000 Hz for network formation assay.
5. Add Spike Detector to Digital Filter: Set the standard deviation to 6× and select Adaptive Threshold Crossing method.
6. Add Burst Detection: Use the Interspike Interval Threshold method of detection and make sure that the settings are as follows: Max ISI = 100 ms, Minimum number of spikes = 5, Minimum participating electrodes = 25%, Mean Firing Rate Estimation = 10 s and Synchrony = 20 ms. Check the box "Enable Network Burst Detection."
7. Add Statistics Compiler (*see* **Note 8**).
8. For experimental recordings, "AxIS Raw" files are collected (*see* **Note 9**). Additionally, you may also wish to record other files while collecting the .raw file (*see* **Note 10**).
9. Once the heater has reached 35 °C, place the MEA plate to be recorded into the MaestroPro. Ensure that the plate is in the correct position, with the notched corner on the upper left side.
10. Click play and ground any electrodes per well with noise over 4 μV (*see* **Note 11**).
11. Record for the desired length of time. Length of recording time is determined by the investigator. In AxIS Navigator, recordings can be scheduled for the future to allow for multiple delayed timed recordings. After the recording is completed, click the stop icon, and return the MEA plate to the incubator.

### 3.4 Network Formation Assay

This procedure begins with the dosing of cortical cultures in triplicate MEA plates 2 h after they are seeded (refer to Subheading 3.2). Test compounds are prepared at 1000× the final desired testing concentration in the appropriate solvent. In other words, multiple 1000× concentrations are prepared which typically range from 30 to 30,000 μM. This results in a final vehicle concentration of 0.1% by volume. A dosing plate (see Fig. 1) is prepared at 50× the final concentration (a 1:20 dilution of the 1000× stock in NB/B27) according to the plate map (*see* Fig. 2). Compounds are tested at a range of concentrations (typically 0.03–30 μM). The 50× stock is used to dose the 48-well MEA plate. Addition of the 50× stock into the MEA plate is a 1:50 dilution resulting in the desired final concentration (*see* **Note 12**).

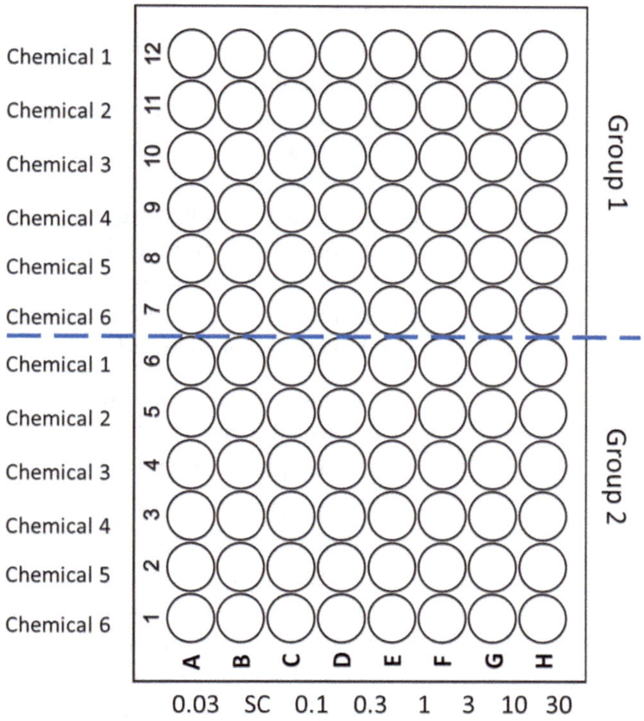


**Fig. 1** Example of dosing plate for network formation assay. One dosing plate can be used for two chemical sets. Each chemical set contains six compounds of seven concentrations half log apart and a solvent control (e.g., DMSO, $H_2O$, EtOH, DMSO:EtOH)

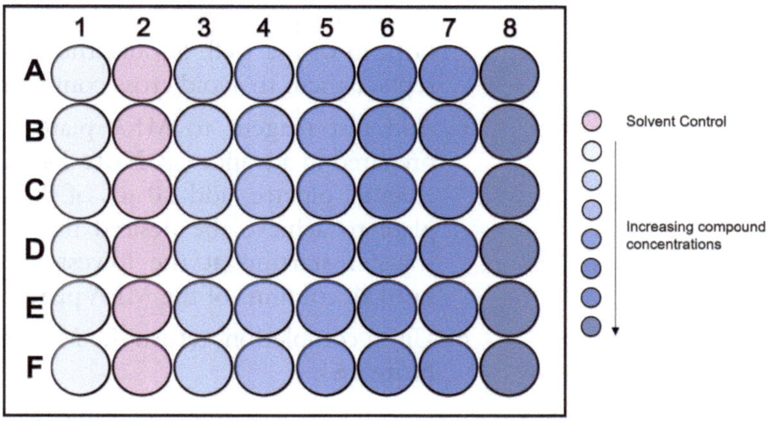


**Fig. 2** Example of 48-well MEA plate map for network formation assay. Each row contains one test compound with increasing concentration from left to right and solvent controls on the second column

A 15-min neural activity recording are taken with electrical impedance measurement on DIV 5, 7, 9, and 12 to track the culture development. Increasing spiking and bursting should be observed between DIV 5 and 12 as the neural networks develop. The average number of active electrodes should be 12 and 16, and the network should be bursting synchronously by DIV 12 in the solvent control wells (e.g., dI $H_2O$, EtOH, DMSO, etc.). Cells experience a full media change and subsequent dosing on DIV 5 and 9. After neural activity recording with electrical impedance measurement on DIV 12, CellTiter-Blue® assay and Total Cellular LDH assay are performed.

1. Aliquot ~10 mL (for two sets of chemicals; each set contains six chemicals) of NB/B27 media in preparation for the 96-well dosing plate. Keep the aliquot in the 37 °C water bath until use.
2. Rotate the 96-well dosing plate so that the letters are horizontal at the top of the plate (H to A from left to right) and the numbers are vertical (1–12 from top to bottom) on the right side of the plate. The first six rows (48 wells) are used for the first set of chemicals and the lower six rows are used for the second set of chemicals.
3. Pipette 5 μL of stock compound into the appropriate well as indicated by the 96-well dosing plate map, changing the pipette tips each time.
4. Add 95 μL of warmed (37 °C) NB/B27 media to each well using a multichannel pipette to achieve a 1:20 dilution. Dispense the media down the side of the well slowly, and do not allow the pipette tips to touch the media after it is added to the bottom of the well. Should the pipette tips touch the media, replace them to avoid cross contamination (*see* **Note 13**).
5. Add test reagent to MEA plate(s) using the 50× dilution prepared in advance (*see* **Note 14**). Using an adjustable multichannel pipette, add 10 μL of the 50× stock into the MEA plate to achieve the desired final concentration. Repeat the transfer starting at the lowest concentration and continue until all columns of the MEA plate have been dosed.
6. Upon completion, return the MEA plates to the incubator (*see* **Note 15**).
7. On DIV 5, 7, 9, and 12, using the MaestroPro MEA system, allow the 48-well MEA plate to sit in the MaestroPro for 20 min prior to recording (*see* **Note 16**).
8. Record 15 min of neural activity along with electrical impedance measurement (refer to Subheading 3.9). The following file types are collected for this assay: AxIS Raw (.raw), Advance Metrics (.csv), and Spike List (.csv). Return plates to incubator after completion.

9. After recording on DIV 5 and 9, aliquot enough NB/B27 media for the media change (25 mL of NB/B27 media is needed for one 48-well MEA plate) and place the aliquot in the water bath until it has warmed to 37 °C (approximately 30 min for 75 mL). Add in a few extra mL to account for transfer loss.
10. After media has reached 37 °C, remove it from the water bath. Pipette 25 mL of media into a sterile reservoir inside the Biological Safety Cabinet.
11. Using the vacuum trap, gently aspirate most of the media from two columns of wells at a time, ensuring that a thin film is left covering the cells.
12. Gently add 490 μL of fresh media back into the wells using an adjustable multi-channel pipette (*see* **Note 17**).
13. Repeat these **steps 10–12** for any subsequent plates. Upon completion, return the MEA plates to the incubator.
14. After media change, repeat **steps 1–6** to dose the MEA plates.
15. On DIV 12, perform CellTiter-Blue® assay and Total Cellular LDH assay (refer to Subheadings 3.6, 3.7, and 3.8).

### 3.5 Acute Assay

This procedure begins with primary cortical neurons cultured on 48-well MEA plates (refer to Subheading 3.2). The cultures experience a full media change on DIV 5 and 9 without dosing of test chemicals. Acute MEA plates should be checked on DIV 9 or 12 before feeding to determine the number of active electrodes per each well. The average number of active electrodes per well should be between 12 and 16 (each well contains 16 electrodes). Wells that do not have at least ten active electrodes should not be used for acute experiments. Plates with 50% or more of the wells with less than 10 active electrodes should not be used at all. On DIV 12, a half media change is performed on plates that will be exposed to chemicals on DIV 13, while a full media change is performed on plates that will be exposed on DIV 15. On DIV 13 or DIV 15, depending on the experimental design, a recording of baseline activity is made for 40 min. The cultures are then acutely exposed to chemicals, and a second 40-min assessment of activity is recorded. After the second recording, three cytotoxicity assays are performed. More than 90% viability should be observed in control wells. Electrical impedance can be measured using the MaestroPro system during baseline activity and dosed activity recording. After the dosed activity recording, LDH Release assay and CellTiter-Blue® assay are performed.

1. On DIV 5 and 9, aliquot NB/B27 media for the media change and place the aliquot in the water bath until it has warmed to 37 °C.

2. After media has reached 37 °C, remove it from the water bath. Pipette 25 mL of media into a sterile reservoir inside the Biological Safety Cabinet.
3. Using the vacuum trap, gently aspirate most of the media from two columns of wells at a time, ensuring that a thin film is left covering the cells.
4. Gently add 500 μL of fresh media back into the wells using an adjustable multichannel pipette.
5. Repeat these **steps 1–4** for any subsequent plates. Upon completion, return the MEA plates to the incubator.
6. On DIV 12, Repeat **steps 1–5** for plates that will be exposed to chemicals on DIV 15. For plates that will be dosed on DIV 13, remove 250 μL of spent media from each well using multichannel pipette, and add 250 μL of fresh media back into wells.
7. On the day of the chemical exposure, using the MaestroPro system, allow the 48-well MEA plate to equilibrate in the chamber for 20 min prior to recording.
8. After 20 min, record a 40-min neural activity baseline with electrical impedance measurement. The following file types are collected for this assay: AxIS Raw (.raw) and Advance Metrics (.csv). MEA recordings are performed on DIV 13 or DIV 15, depending on experimental design (refer to Subheading 3.3).
9. Aliquot ~25 mL of NB/B27 media in preparation for the 96-well deep-well dosing plate. Keep the aliquot in the 37 °C water bath until use.
10. During the baseline recording, prepare the 10× dosing plate according to the plate map.
11. Dosing solutions are prepared by making a 1:100 dilution of stock compound in NB/B27 (5 μL stock compound and 495 μL NB/B27).
12. Prepare the solvent control and positive controls (e.g., 25 μM Tetrodotoxin and 1 μM Picrotoxin) in the first column. See Fig. 3 for an example of a typical plate map for an acute experiment.
13. After recording the 40-min baseline, slide open the door of MaestroPro by pushing it back. This will open the MaestroPro lid without the machine lifting the MEA plate up. Next, remove the cover of the MEA plate. Gently pipette 55 μL of dosing solution into the corresponding wells on the MEA plate to achieve a 1:10 dilution.
14. Replace the cover and slide the lid of MaestroPro closed. Record the dosed activity for 40 min after a 20-min equilibration. Take impedance measurement after that.

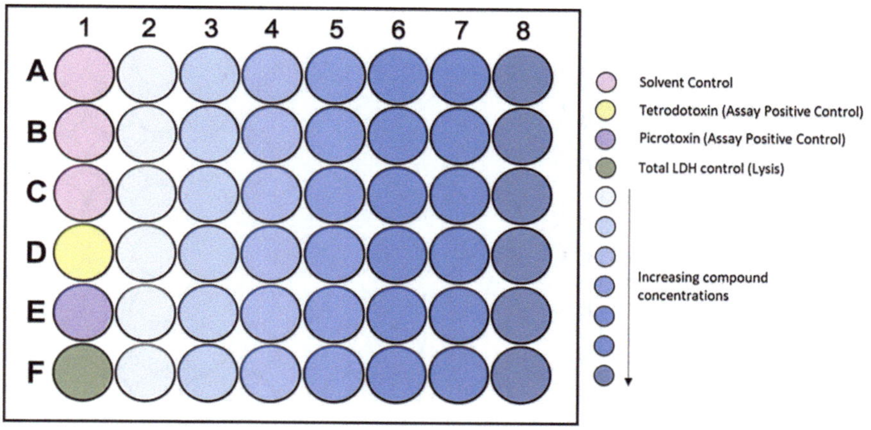


**Fig. 3** Example of 48-well MEA plate map for acute assay. Each row contains one test compound with increasing concentration from left to right. Solvent controls and assay positive controls are in the first column. Well F1 will be added with lysis solution in the LDH assay to serve as total LDH control

15. Remove MEA plate from MaestroPro, and perform LDH release assay and then CellTiter-Blue® assay (refer to Subheadings 3.6 and 3.7).

### 3.6 Viability: CellTiter-Blue® Assay

The CellTiter-Blue® Cell Viability Assay measures the fluorescence of resorufin converted from resazurin after it has undergone reduction reaction by metabolically active cells. This assay is used to determine the viability of cells following both acute and network formation exposure to chemicals. As written, this procedure has only been optimized for primary cortical cultures seeded in MEA plates.

1. Prepare a 1:6 dilution of the CellTiter-Blue® reagent by adding warm (37 °C) NB/B27 media (*see* **Note 18**). For example, add 5.5 mL of CellTiter-Blue® to 27.5 mL of media for three MEA plates.
2. Using the vacuum trap, gently aspirate most of the treatment media from two columns of wells at a time, ensuring that a thin film is left covering the cells.
3. Slowly add 200 μL fresh CellTiter-Blue® solution to each well.
4. Add 150 μL of CellTiter-Blue® solution to three wells on a 96-well opaque plate (CellTiter-Blue® transfer plate) for blanks.
5. Place both the MEA plate and the transfer plate in the incubator for 60 min, ensuring to gently shake the MEA plate every 15 min (*see* **Notes 19** and **20**).
6. At the end of the incubation period, transfer 150 μL from each well of the MEA plate to the corresponding wells of the transfer plate. See Fig. 4 for a typical CellTiter-Blue® transfer plate map.

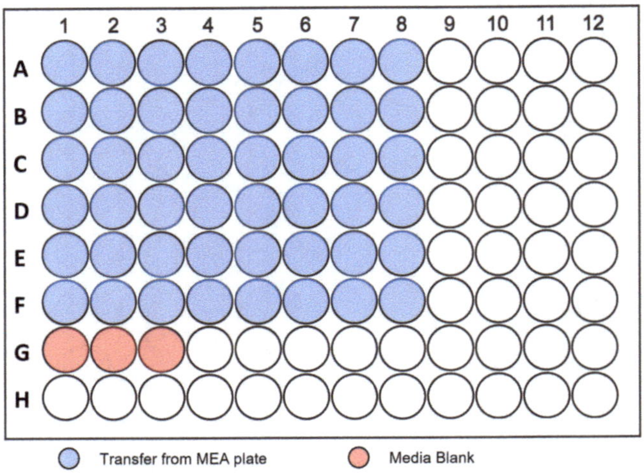


**Fig. 4.** A 96-well CellTiter-Blue® transfer plate map. After the 1-hour incubation period, 150 μL of spent CellTiter-Blue® is transferred from each well of the 48-well MEA plate to corresponding wells on a 96-well opaque plate (transfer plate). G1–G3 serves as media blank

7. Allow the transfer plate to come to room temperature, and lance any large bubbles using a syringe needle (*see* **Note 21**).
8. Measure the fluorescence using the BMG FLUOstar OPTIMA Fluorescence/Luminescence Microplate Reader (*see* **Note 22**).
9. Calculate the blank-corrected fluorescence by deducting mean of three media blank fluorescence values from each transfer plate from the raw fluorescence value.
10. Calculate the percent of control using the following formula:

$$\textbf{Percent of Control} = 100 \times \frac{\text{Blank corrected fluorescence value}}{\text{Mean of solvent controls blank corrected fluorescence value}}$$

### 3.7 Viability: LDH Release Assay

This procedure has been employed to measure LDH Release Assay from cortical cells around DIV 13 to 15 exposed acutely to a variety of compounds. Specifically, this assay is an absorbance method to measure the amount of LDH release by the cells into surrounding media after acute exposure to chemicals and is performed after the 40-min treated neural activity recording for the acute assay.

1. Thaw assay buffer and substrate mix at room temperature, and avoid excessive exposure to light. Prepare the reconstituted substrate buffer by adding 12 mL thawed assay buffer at room temperature to a vial of substrate mix. Invert mixture a few times to ensure the substrate mix has dissolved completely.
2. Prepare the positive control by making a 1:250 and 1:2500 dilution of purified LDH (provided by Promega's CytoTox 96® Non-Radioactive Cytotoxicity Assay kit) in NB/B27 on a LDH transfer plate (clear 96-well plate).

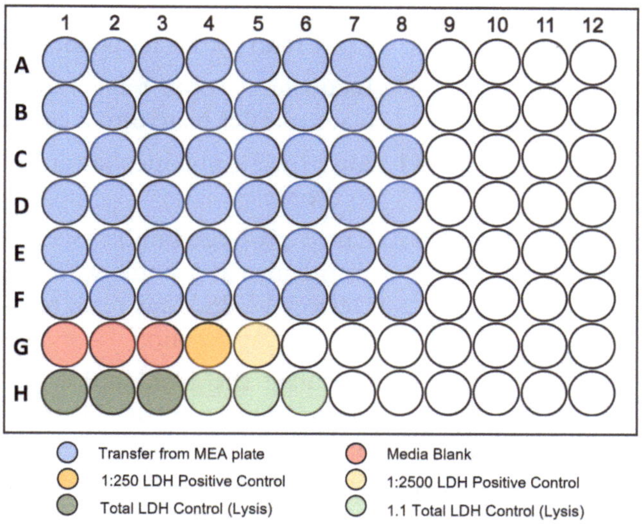


**Fig. 5** A 96-well LDH release/Total Cellular LDH transfer plate map. Cell lysate from network formation assay or NB/B27 media from acute assay are transferred from each well of the 48-well MEA to the corresponding wells on a 96-well transparent plate (transfer plate). G1–G3 serves as dilute lysis solution blank or media blank. G4 and G5 are LDH positive control prepared from the kit. Total LDH control and 1:1 total LDH control (Row H) are only used in the LDH Release Assay for acute assay

3. To prepare the 1:250 dilution, add 1 μL of the purified LDH into 249 μL of NB/N27 media and mix thoroughly. To prepare the 1:2500 dilution, transfer 20 μL of 1:250 dilution to 180 μL of NB/B27 media and mix thoroughly. Then transfer 50 μL to the appropriate well of the LDH transfer plates (*see* **Note 23**). See Fig. 5 for a typical LDH release/Total Cellular LDH transfer plate map.
4. Prepare media blanks by adding 50 μL of NB/B27 media into three wells of the LDH release transfer plate (*see* **Note 24**).
5. After the 40-min treatment recording, remove 50 μL of media from each well of the treated MEA plate and transfer to the corresponding wells on the LDH transfer plate.
6. Add 20 μL of lysis solution provided by Promega's CytoTox 96® Non-Radioactive Cytotoxicity Assay kit to Well F1 on the MEA plate; mix well and transfer well content to a microcentrifuge tube (this will be used as total cellular LDH control in analysis) (*see* **Note 25**).
7. After spinning the microcentrifuge tube briefly, transfer 50 μL of the media from the tube to three wells on LDH plate.
8. Transfer 25 μL from the tube to another three wells, and dilute 1:1 with 25 μL of media.

9. Add 50 μL of reconstituted substrate mix to each well of the LDH release transfer plate.
10. Incubate at room temperature for 30 min in the dark (e.g., place transfer plate into a closed drawer).
11. After the 30-min incubation period, add 50 μL of stop solution to the wells.
12. Lance any large bubbles with a syringe needle, and record the absorbance at a single wavelength of 490 nm with Molecular Devices VERSAMax plate reader within 1 hour after the addition of stop solution.
13. Calculate the blank-corrected LDH OD by deducting the mean of three media blank OD of each transfer plate from the raw LDH OD.
14. Calculate total cellular LDH control (refer to **step 6**) by taking the mean of blank-corrected 1:1 total LDH control dilution OD (*see* **Note 26**).
15. Calculate the percent LDH release in each well using the following formula:

$$\%\ \text{Release} = 100 \times \frac{(\text{Blank corrected released LDH OD})}{(\text{Blank corrected total LDH control OD})}$$

### 3.8 Total Cellular LDH Assay

This procedure has been employed to measure Total Cellular LDH from cortical cells exposed to a variety of compounds over a period of 12 days in vitro. Specifically, this assay is an absorbance method to determine the amount of LDH released from lysed cells remained after long-term exposure to chemicals in a culture and is performed after the DIV 12 recording for the network formation assay. Cells must be lysed to measure total cellular LDH. Hence, this assay is performed following CellTiter-Blue® Assay detailed in Subheading 3.6.

1. During the 15-min MEA recording on DIV 12, prep the LDH transfer plates (clear 96-well plate) by adding 25 μL of PBS to 48 wells.
2. Prepare dilute lysis solution by adding 1.5 mL of lysis solution to 75 mL PBS. This is enough for three 48-well MEA plates.
3. Prepare media blanks by adding 25 μL of dilute lysis solution and 25 μL of PBS into three wells (G1-G3) (*see* **Note 27**).
4. Prepare the LDH positive control by making 1:250 and 1:2500 dilution of purified LDH (provided by Promega's CytoTox 96® Non-Radioactive Cytotoxicity Assay kit).
5. To prepare the 1:250 dilution, add 1 μL of purified LDH to 249 μL of dilute lysis solution and mix thoroughly. To prepare the 1:2500 dilution, transfer 20 μL of 1:250 dilution into

180 μL of dilute lysis solution and mix thoroughly. Then, transfer 50 μL of the positive controls to the appropriate wells (refer to Fig. 5 in Subheading 3.7).

6. Aspirate the remaining CellTiter-Blue® from the 48-well MEA plate and add 500 μL of dilute lysis solution from step 2 to each well.
7. Incubate the plate at 37 °C for 30 min.
8. After the 30-min incubation period, thoroughly mix each column by pipetting up and down. Remove 25 μL from each well of the MEA plate and transfer to the corresponding well on the total cellular LDH transfer plate.
9. Add 50 μL of reconstituted substrate mix to each well of the LDH plate.
10. Incubate at room temperature for 30 min in the dark (closed drawer).
11. After the 30-min incubation period, add 50 μL of stop solution to each well (*see* **Note 28**).
12. Pop any large bubbles with a syringe needle and record the absorbance at a single wavelength of 490 nm using Molecular Devices VERSAMax plate reader within 1 hour after the addition of stop solution (*see* **Note 29**).
13. Calculate the blank-corrected LDH OD by deducting the mean of three media blank OD of each transfer plate from the raw LDH OD.
14. Calculate the percent of control in each well using the following formula:

$$\textbf{Percent of Control} = 100 \times \frac{\text{Blank corrected total LDH OD}}{\text{Mean of solvent controls blank corrected total LDH OD}}$$

### 3.9 Viability: Electrical Impedance

Impedance measurement provides a nondestructive, label-free, and real-time assessment of electrical resistance (*see* **Notes 30** and **31**). It is enabled by using the viability module within the MaestroPro MEA system from Axion Biosystems. This procedure is quick and simple to conduct and can be performed at any timepoint of the assay, as needed.

1. Turn on the MaestroPro, open AxIS Navigator software, and set the heater to 35 °C once the program has connected to the MaestroPro.
2. Switch the MaestroPro configuration into the viability configuration:

   "Configuration" → "Neural + Viability Real-Time" → "Spontaneous + Viability"

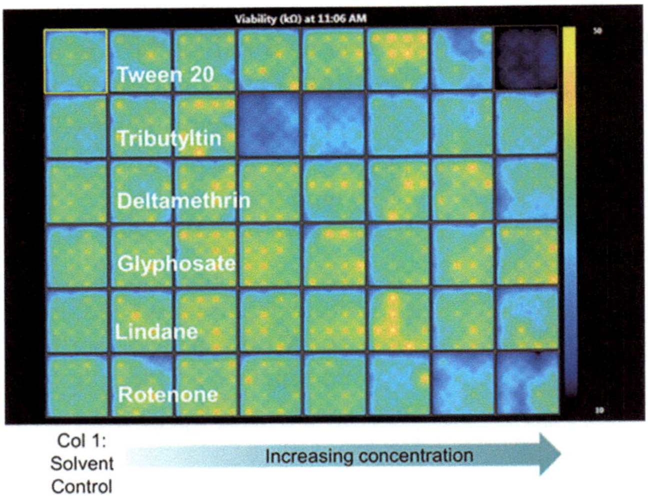


**Fig. 6** Example of impedance activity map 72 h after acute treatment as shown on AxIS Navigator. Tween 20 (Row 1) shows high cytotoxicity effect at the highest concentration and some cytotoxicity effect in the second highest concentration. Tributyltin (Row 2) and rotenone (Row 6) shows cytotoxicity effect over a few concentrations range. Deltamethrin (Row 3) shows slight cytotoxicity effect at the highest concentration. Glyphosate and lindane (Row 4, 5) shows no cytotoxicity effect

3. Add Digital Filter, Spike Detector, Burst Detector, and Statistic Compiler as well. Within the Statistic Compiler settings, check the box "Viability Data."
4. Place the CytoView MEA plate in the MaestroPro system. If you are taking only the impedance measurement, there is no need for the 20-min equilibration time. If you wish to take a neural activity recording along with the impedance measurement, then allow the plate to sit in the machine for 20 min.
5. Press play, and the machine will automatically acquire impedance measurement before moving on to neural activity recording. The process takes around a minute, and at the conclusion of the measurement, an impedance map (*see* Fig. 6) will appear.
6. Press record and record for the desired length of time.
7. Viability data can be accessed in the Statistic Compiler Advance Metrics file.
8. Calculate the blank-corrected impedance value by deducting blank impedance value (refer to Subheading 3.10) from raw impedance value of each well.
9. Calculate the percent of control in each well using the following formula:

$$\textbf{Percent of Control} = 100 \times \frac{\text{Blank corrected impedance value}}{\text{Mean of solvent controls blank corrected impedance value}}$$

### 3.10 Viability: Electrical Impedance (Establishing Media Blank Values)

Having blank wells on each CytoView MEA plate would not be cost efficient for screening purposes. Hence, to blank-correct raw impedance values, a set of blank values are collected. Considering the range and variability, the mean or median will be dedicated as the impedance blank value. The range should be 10–13 kΩ. The mean of blank raw impedance values that we obtained is 12.107 kΩ.

1. Coat three CytoView MEA plates with PEI solution as the procedure mentioned in Subheading 3.1.
2. Once the MEA plates are dry, add 10 μL of cortical medium/laminin drop onto the electrode array, and let the plates sit in the 37 °C incubator for ~2 h. Then, add 490 μL of warm NB/B27 media into each well (*see* **Note 32**).
3. Take impedance measurement using the Viability module in MaestroPro system (*see* **Note 33**).
4. Using the raw values obtained, determine the mean or median to be set as blank value.

## 4 Notes

1. The Classic MEA plate has been discontinued in favor of the CytoView MEA plate. However, these approaches are easily adapted to other plate types.
2. PEI comes as 50% (w/v) in $H_2O$. Therefore, when weighing out PEI, only half of the weight is actually PEI and so one must weigh twice as much.
3. The ability to make impedance measurements on MEA devices from other manufacturers will depend on the capabilities of their equipment and may or may not be equivalent to the assay described in this chapter.
4. PEI is toxic to cells, so PEI solution should not be allowed to dry on the plate prior to rinsing three times.
5. The drop should bead up on the surface in the center of the well and not spread.
6. The seeding protocol for CytoView MEA plates is still being optimized. As CytoView plates have lower surface tension, a smaller seeding drop is less likely to spread.
7. The optimal temperature for the MaestroPro can vary from 35 °C to 37 °C. It is recommended to run a preliminary experiment to determine the temperature where the overall cell activity in a plate is optimal.
8. Steps 3–7 describe the default settings for neural recordings. The AxIS Software will open with the most recently used settings as a default. Thus, if the previous recording was from cardiomyocytes, the settings will be different. The

experimenter should check these prior to recording to be sure they are correct. AxIS also allows saving configurations, which can be loaded upon starting an experiment. This will ensure that the correct settings are used and is recommended, but not required.

9. Always record the .raw file when conducting an experiment. This file can be replayed later to generate other file types in AxIS. The .raw files are very large and are around 1 GB per minute of recording for a 48-well plate format. Depending on the number of plates and the duration of recording, check the hard drive on the computer to make sure that sufficient storage space is available. Transfer files to an external hard drive if necessary.

10. The most used files are as follows:

    *Spike Counts*: This file lists number of spikes per second and the total number of spikes for each well and each electrode.

    *Spike List*: This file lists every spike that occurred on a plate in chronological order, noting the electrode it occurred on and the voltage amplitude.

    *Advanced Metrics*: This file type contains a variety of end-point data calculated for each well and electrode, relating to spiking, bursting, and impedance measurements (cell viability parameters).

11. Typical root mean squared (rms) noise levels should be around 2–4 μV, so the experimenter can use their best judgment on grounding electrodes with rms noise > 4 μV. Refer to AxIS Navigator User Guide Version 3.5 Section 10.2 Identifying Neural Spikes for more information.

12. The 50× dosing plate should be prepared while the cells are attaching (2 h; refer to Subheading 3.2). Time the preparation of the dosing plate so that the cells have attached at the same time or prior to the completion of the dosing plate.

13. Pipette tips only need to be changed when dosing a new plate if you are dosing from a low to high concentration.

14. The chemical treatments are rotated on different rows on the triplicates MEA plates, that is, chemical 1 on first row on plate 1, second row on plate 2, and third row on plate 3 to reduce edge effects and any other factors.

15. After dosing on DIV 0, allow MEA plates to remain in the incubator of the laboratory in which they were plated overnight before moving them to another location. Cells dosed on DIV 5 and 9 can immediately be transferred to another incubator after dosing if necessary.

16. It was previously determined [5] that there is a period of instability in mean firing rate that occurs when moving plates from the incubator into the MaestroPro.

17. Add media slowly to avoid washing the cells off the electrodes. When adding media, slowly dispense it against the side of the well for better control.
18. Adjustment of this dilution may be necessary depending upon the cell type and density used.
19. Plates are gently shaken every 15 min to ensure that all of the reagent in a well can be reduced by the cells. Since cells are seeded in the center of a single well, the reagent localized directly over the cells tends to be reduced more rapidly.
20. The rate that the CellTiter-Blue® is reduced is dependent on cell type and density.
21. Bubbles will interfere with the accuracy of the reading.
22. Acceptable values for negative control wells in this assay (using white opaque plates and the appropriate gain corresponding to this plate type) are between 26,000 and 40,000. Values outside of this range should be investigated by ensuring that the correct gain is used and checking for air bubbles. Negative control well values below 26 thousand could indicate that some cells may have been washed off during refeeding. Blank wells should be between 6 and 10 thousand but can vary depending on the combination of plate type and gain setting.
23. The kit contains purified LDH, which can be used to verify that the kit itself is working within specifications and/or develop a standard curve from which to quantify LDH. Values should range between 2.5 and 4.0 optical density (OD) for both the acute and network formation assays.
24. NB/B27 media blank absorbable values should range between 0.05 and 0.08 OD units. If the OD value is greater than 0.08, ensure there are no air bubbles.
25. For the LDH release assay, a total cellular LDH control (lysis buffer) for inducing LDH release from cells is included on every plate. Acceptable values for the positive control (lysis buffer) range from 1.9 to 3.0 OD.
26. The total cellular LDH control OD should be roughly two times the 1:1 total cellular LDH OD.
27. Dilute lysis solution media blank absorbable values should range between 0.03 and 0.06 OD units. If the OD value is greater than 0.06 OD, ensure there are no air bubbles.
28. For the total cellular LDH assay, after incubation and when stop solution is added, solvent control wells should appear dark red. Wells that have little red indicate cell death in this assay. Negative control (e.g., DMSO only) wells should range between ~0.7 and 1.5 OD.

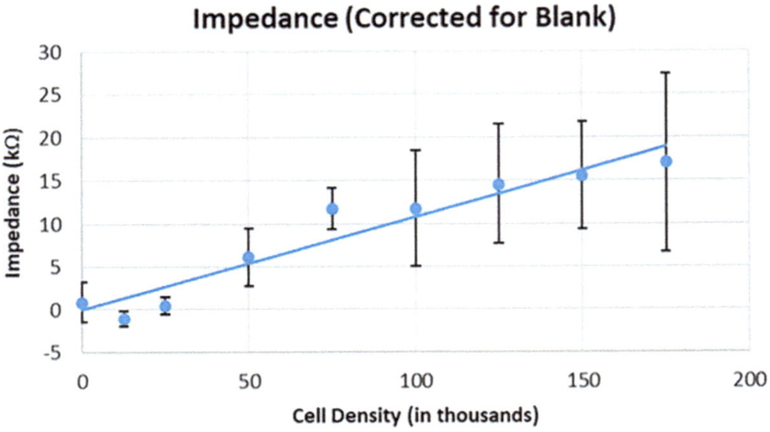


**Fig. 7** Graph plotted with cell density on the x-axis and blank-corrected impedance value on the y-axis. The graph showed increasing impedance with increasing cell density

29. Following determination of the total LDH in each well, the data can be used to (1) determine viable cells in treated wells compared to control wells or (2) perform quality checks of the data. Quality can be determined by comparing LDH values in surrounding wells on the same plate in addition to the corresponding well (same compound and concentration) in the other plates from that culture.
30. The viability module has to be installed prior to any measurement. Note that this measurement can only be taken when the cells are seeded on CytoView MEA plates.
31. Impedance is the total amount of opposition a circuit presents to electric current, and it can serve as a surrogate measure of cell viability as it reflects the number of cells attached to the electrodes. It should be recognized that multiple factors can influence impedance, including the number of cells (*see* Fig. 7).
32. It is important to fully wet the electrodes as partially wet electrodes will give inconsistent and unreasonably high values. Bubbles are likely to form during the wetting process. As long as the bubbles are not on the electrodes (some might form on the bottom edge of the well, as observed under a microscope), they should not interfere with impedance measurement. Gently tapping the plate can help dislodge bubbles, or the media can be triturated a few times using a pipette.
33. It is recommended to let the plate sit in the incubator overnight before taking impedance measurements. This will give lower and more consistent values. However, it is acceptable to take impedance values right after the 2-hour incubation.

## Acknowledgments

Preparation of this document has been funded by the US Environmental Protection Agency. This document has been subjected to review by the Center for Computational Toxicology and Exposure and approved for publication. Approval does not signify that the contents reflect the views of the Agency, nor does mention of trade names or commercial products constitute endorsement or recommendation for use.

### References

1. Bradley JA, Luithardt HH, Metea MR, Strock CJ (2018) In vitro screening for seizure liability using microelectrode Array technology. Toxicol Sci 163(1):240–253
2. Fan J, Thalody G, Kwagh J, Burnett E, Shi H, Lewen G, Chen SJ, Levesque P (2019) Assessing seizure liability using multi-electrode arrays (MEA). Toxicol In Vitro 55:93–100
3. Tukker AM, Van Kleef R, Wijnolts F, De Groot A, Westerink R (2020) Towards animal-free neurotoxicity screening: applicability of hiPSC-derived neuronal models for in vitro seizure liability assessment. ALTEX 37(1):121–135
4. Zwartsen A, Olijhoek ME, Westerink R, Hondebrink L (2020) Hazard characterization of synthetic cathinones using viability, monoamine reuptake, and neuronal activity assays. Front Neurosci 14:9
5. Valdivia P, Martin MT, Houck K, Lefew WR, Ross J, Shafer TJ (2014) Multi-well microelectrode array recordings detect neuroactivity of ToxCast compounds. Neurotoxicology 44: 204–217
6. Nicolas J, Hendriksen PJM, van Kleef RGDM, de Groot A, Bovee TFH, Rietjens IMCM, Westerink RHS (2014) Detection of marine neurotoxins in food safety testing using a multielectrode array. Mol Nutr Food Res 58:2369–2378
7. Strickland JD, Martin MT, Richard AM, Houck KA, Shafer TJ (2018) Screening the ToxCast phase II libraries for alterations in network function using cortical neurons grown on multi-well microelectrode array (mwMEA) plates. Arch Toxicol 92(1):487–500
8. Kosnik MB, Strickland JD, Marvel SW, Wallis DJ, Wallace K, Richard AM, Reif DM, Shafer TJ (2020) Concentration-response evaluation of ToxCast compounds for multivariate activity patterns of neural network function. Arch Toxicol 94(2):469–484
9. Tukker AM, Bouwman L, van Kleef R, Hendriks HS, Legler J, Westerink R (2020) Perfluorooctane sulfonate (PFOS) and perfluorooctanoate (PFOA) acutely affect human $\alpha_1\beta_2\gamma_{2L}$ $GABA_A$ receptor and spontaneous neuronal network function in vitro. Sci Rep 10(1):5311
10. Brown JP, Hall D, Frank CL, Wallace K, Mundy WR, Shafer TJ (2016) Editor's highlight: evaluation of a microelectrode array-based assay for neural network ontogeny using training set chemicals. Toxicol Sci 154(1):126–139
11. Frank CL, Brown JP, Wallace K, Mundy WR, Shafer TJ (2017) From the cover: developmental Neurotoxicants disrupt activity in cortical networks on microelectrode arrays: results of screening 86 compounds during neural network formation. Toxicol Sci 160(1):121–135
12. Shafer TJ, Brown JP, Lynch B, Davila-Montero S, Wallace K, Friedman KP (2019) Evaluation of chemical effects on network formation in cortical neurons grown on microelectrode arrays. Toxicol Sci 169(2):436–455
13. Zwartsen A, de Korte T, Nacken P, de Lange DW, Westerink R, Hondebrink L (2019) Cardiotoxicity screening of illicit drugs and new psychoactive substances (NPS) in human iPSC-derived cardiomyocytes using microelectrode array (MEA) recordings. J Mol Cell Cardiol 136:102–112
14. Johnstone AF, Gross GW, Weiss DG, Schroeder OH, Gramowski A, Shafer TJ (2010) Microelectrode arrays: a physiologically based neurotoxicity testing platform for the 21st century. Neurotox 31:331–350
15. Roberts R, Authier S, Mellon RD, Morton M, Suzuki I, Tjalkens RB, Valentin JP, Pierson JB (2021) Can we panelize seizure? Toxicol Sci 179(1):3–13. https://doi.org/10.1093/toxsci/kfaa167. PMID: 33165543

16. Wallace K, Strickland JD, Valdivia P, Mundy WR, Shafer TJ (2015) A multiplexed assay for determination of neurotoxicant effects on spontaneous network activity and viability from microelectrode arrays. Neurotoxicology 49:79–85

17. Cotterill E, Hall D, Wallace K, Mundy WR, Eglen SJ, Shafer TJ (2016) Characterization of early cortical neural network development in multiwell microelectrode array plates. J Biomol Screen 21(5):510–519

# Chapter 10

# Quantitative Live-Cell $Ca^{2+}$ Imaging During Isotropic Cell Stretch

**Ulrike Schöler, Anna-Lena Merten, Sebastian Schürmann, and Oliver Friedrich**

## Abstract

Cell viability of many cell types strongly relies on their ability to adjust to mechanical conditions and alterations. Cellular mechanisms for sensing and responding to mechanical forces and pathophysiological variations in these processes have become an emerging research field in recent years. An important signaling molecule involved in mechanotransduction as in many cellular processes is $Ca^{2+}$. New experimental methods to probe cellular $Ca^{2+}$ signaling live under conditions of mechanical stimulation facilitate new insights into previously overlooked aspects of mechanical regulation of cells.

Here, we describe a protocol for using $Ca^{2+}$ imaging in combination with a cell stretching device, the *IsoStretcher*. Cells grown on elastic membranes can be isotopically stretched in-plane, and their intracellular $Ca^{2+}$ level can be accessed online on the single cell level using fluorescent calcium indicator dyes. We show a protocol for functional screening of mechanosensitive ion channels and related drug screenings using BJ cells, a foreskin fibroblast cell line that strongly reacts to acute mechanical stimulation.

**Key words** Cell stretching, *IsoStretcher*, Calcium imaging, Mechanotransduction, Mechanosignaling, Mechanobiology, Image processing

## 1 Introduction

Mechanotransduction has become a subject of intense research over the past decades. In 2021, the Nobel Prize for Physiology or Medicine was awarded for identifying the mechanosensitive ion channels Piezo1 and Piezo2 [1]. Since the discovery of the Piezo channels, they have been found to be important mechanotransducers in a variety of cells including cardiomyocytes [2], endothelial cells [3], and fibroblasts [4].

---

**Supplementary Information** The online version contains supplementary material available at https://doi.org/10.1007/978-1-0716-3052-5_10.

Oliver Friedrich and Daniel F. Gilbert (eds.), *Cell Viability Assays: Methods and Protocols*, Methods in Molecular Biology, vol. 2644, https://doi.org/10.1007/978-1-0716-3052-5_10,

Fibroblasts are cells that are present throughout the body and play import roles in mechanosensing and mechanotransduction. Their ability to adapt to mechanical forces is important for cell viability. Fibroblasts are an important population in the cardiovascular system and are involved in cardiac remodeling [5]. Ao et al. used a microfluidic device to mechanically stimulate normal tissue-associated fibroblasts and showed that stretching triggers complex biochemical signaling events [6]. α- and β-actin redistribution and cytoskeletal remodeling was observed when connective tissue was stretched ex vivo [7].

$Ca^{2+}$ is an important cellular messenger involved in signal transduction. Mechanotransduction is the conversion of mechanical stimuli into biological signals and is, for instance, mediated via mechanosensitive ion channels, such as the calcium permeable ion channel Piezo1 [1].

Here, we describe a system for assessing mechanotransduction via a combination of a cell stretcher, the *IsoStretcher*, and an imaging system to measure intracellular $Ca^{2+}$ levels using a non-ratiometric cell permeable $Ca^{2+}$ sensitive dye, Cal-520 AM. The dye emits brighter fluorescence in the $Ca^{2+}$ bound state, and therefore, the fluorescence intensity represents the $Ca^{2+}$ level within the cells. The *IsoStretcher* is a device that applies radial stretch to cells attached to an elastic PDMS (Polydimethylsiloxane) substrate [8, 9]. The *IsoStretcher* is optimized for microscopy applications and allows for live cell imaging as it can easily be mounted onto commercial microscopes.

We show an exemplary experimental procedure to access $Ca^{2+}$ signaling in BJ cells, a fibroblast cell line established from skin taken from normal foreskin of a neonatal male human. In our experiments, the BJ fibroblasts respond to stretching the substrate with a strong increase in intracellular calcium levels. Therefore, this cell line represents a well-suited system for investigating mechanosignaling pathways and testing potential inhibitors of mechanosensitive ion channel function. In this chapter, we demonstrate our method step-by-step, using gadolinium ions ($Gd^{3+}$), which are known to inhibit mechanosensitive ion channels and thus reduce the increase in intracellular calcium levels.

Since some cells might be damaged by the mechanical overload, we provide a procedure to identify the cells with non-intact membranes at the end of each experiment: to assess the integrity of the cells' membrane, the $Ca^{2+}$ ionophore ionomycin is added to the cells after stretch. Cells with intact membranes show a strong increase in the intracellular $Ca^{2+}$ level, which is reflected by a strong increase in the fluorescence intensity of the $Ca^{2+}$ indicator.

## 2 Materials

### 2.1 Cell Culture

1. BJ cells (e.g., from American Type Culture Collection (ATCC)) or any other type of adherent cells (*see* **Note 1**).
2. BJ cell growth medium: Minimum essential medium (MEM), 10% fetal bovine serum (FBS), 1× nonessential amino acids (NEAA), 1 mM sodium pyruvate.
3. Dulbecco's phosphate-buffered saline (DPBS), no calcium, no magnesium.
4. Trypsin-EDTA solution (0.05% Trypsin).
5. T75 tissue culture flasks.
6. 15 mL centrifuge tubes.

### 2.2 Cell Stretching Device

This protocol describes our procedure for stretching adherent cells and performing live calcium imaging during stretching. For our cell stretching experiments, we use the *IsoStretcher*, a custom-built cell stretching device developed at the Institute of Medical Biotechnology of Friedrich-Alexander-University Erlangen-Nürnberg [8, 9]. The *IsoStretcher* can be used to apply radial mechanical strain onto cells attached to a PDMS chamber. All volume data and cell counts refer to the custom-built *IsoStretcher* chamber which has a growth surface of 1.86 $cm^2$ and can hold a maximum volume of 750 μL. The *IsoStretcher* comes with custom-built accessories that simplify chamber handling. All stretcher components listed here are shown in Fig. 1. Depending on the experimenter's preference, also other cell stretching systems that are suitable for live cell imaging during stretch can be employed (*see* **Note 2**).

1. Cell stretching device: *IsoStretcher* (*see* Fig. 1a).
2. PDMS chambers for the *IsoStretcher* (mixing ratio: nine-parts base elastomer, one-part curing agent, *see* **Note 3**, *see* Fig. 1b).
3. Transfer frames (*see* Fig. 1d) and corresponding transfer punch (*see* Fig. 1c): The PDMS chambers can be mounted on transfer frames before seeding the cells. Using the transfer punch, the PDMS chambers can be easily transferred onto the *IsoStretcher* directly before the experiment with minimal disturbance to the cells (*see* **Note 4**).
4. *IsoStretcher* stepper motor control unit (*see* Fig. 1e).
5. *IsoStretcher* control software (*see* Fig. 1f).

### 2.3 Buffers and Solutions

1. Sulfo-SANPAH solution: 0.5 mg/mL dissolved in sterile MilliQ Millipore water.
2. Fibronectin solution, supplied as sterile filtered solution, 1 mg/mL.

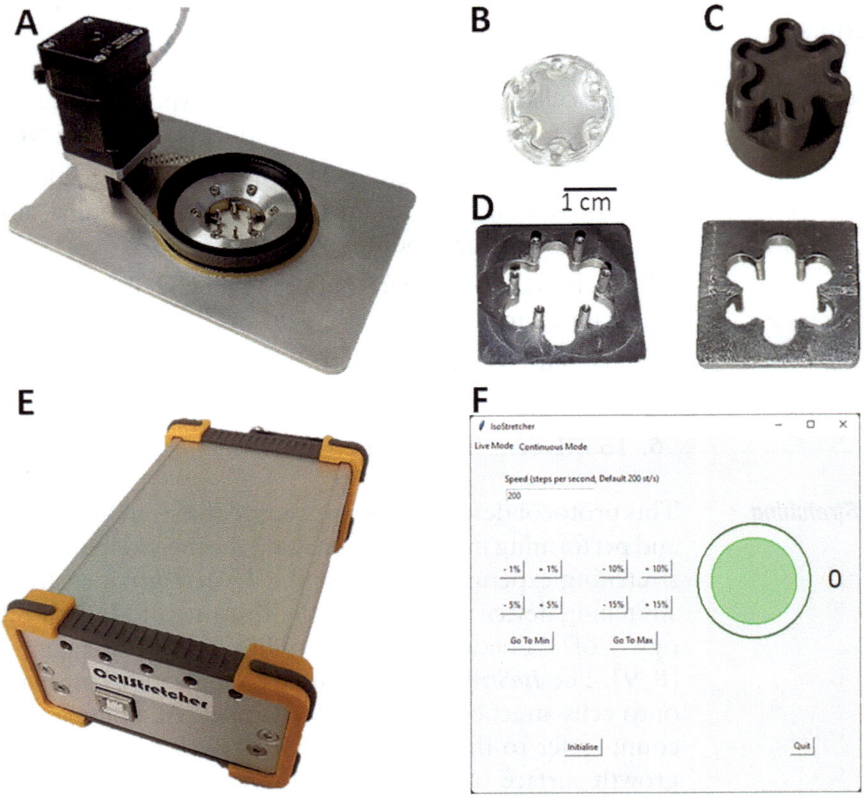


**Fig. 1** Components of the *IsoStretcher* system. (**a**) the *IsoStretcher*, (**b**) transparent PDMS chamber, (**c**) transfer punch, (**d**) transfer frame (left: view from below, right: view from above), (**e**) *IsoStretcher* control box, (**f**) *IsoStretcher* control software. (**b–d**) Scale bar = 1 cm

3. Physiological salt solution (PSS): 140 mM NaCl, 2.5 mM KCl, 1.8 mM $CaCl_2$, 1.0 mM $MgCl_2$, 20 mM HEPES, and 5.5 mM D-Glucose in water. pH adjusted to 7.4. Sterile filtered (0.2 μm filter) after dissolving.
4. Cal-520 AM stock solution (*see* **Note 5**): 1 mM Cal-520 AM in DMSO.
5. Gadolinium(III)-chloride hexahydrate stock stolution: 1 M Gadolinium(III)-chloride hexahydrate in MilliQ Millipore water.
6. Ionomycin calcium salt stock solution: 1 mM ionomycin calcium salt in DMSO.
7. Fluorescein solution: 1 M fluorescein in DMSO.

### *2.4 Laboratory Equipment*

1. Incubator (37 °C, 5% $CO_2$).
2. Water bath (37 °C).
3. Inverse phase-contrast microscope.
4. Hemocytometer.

5. Laminar flow hood.
6. UV lamp (UV lamp of the laminar flow hood is sufficient).
7. Microscope slide.
8. Cover slip.

#### 2.5 Fluorescence Microscope

1. Inverted epifluorescence microscope.
2. FITC filter set (*see* **Note 6**).
3. Long working distance objective (~20× magnification, *see* **Note 7**).
4. Camera.
5. Image acquisition software.

#### 2.6 Image Processing with Fiji

Fiji is a distribution of ImageJ and ImageJ2. The open source image processing package comes with numerous preinstalled plugins useful for the analysis of biological image data [10].

#### 2.7 Data Processing Software

Matlab, Python, or Excel (*see* **Note 8**).

## 3 Methods

All preparatory experimental steps and all cell culture work should be performed under sterile conditions in a laminar flow cabinet. The experimental steps of functional imaging can be performed under non-sterile conditions.

Refer to Fig. 2 for a visualization of the steps of the chamber preparation, of the BJ cell culturing, and of the steps of the cell seeding to the PDMS chambers.

#### 3.1 Preparation and Surface Modification of the PDMS Cell Stretching Chambers

1. Place the PDMS cell stretching chambers in a sterile tissue culture dish. Make sure that the PDMS chambers have been autoclaved and dried before starting the coating procedure.
2. Add 330 μL of Sulfo-SANPAH (0.5 mg/mL diluted in sterile MilliQ Millipore water) to each chamber and expose to UV light for 10 min. Upon exposure to UV light, the solution changes its color from orange to a dark red.
3. Remove the Sulfo-SANPAH solution and wash the inside of the chambers three times with 500 μL of DPBS.
4. Mount the PDMS chambers on the transfer frames. Make sure that the transfer frames have been sterilized before use.
5. Add 225 μL of Fibronectin solution (10 μg/mL diluted in DPBS) to each chamber and place inside the incubator (37 °C) for 3 h (*see* **Note 9**).

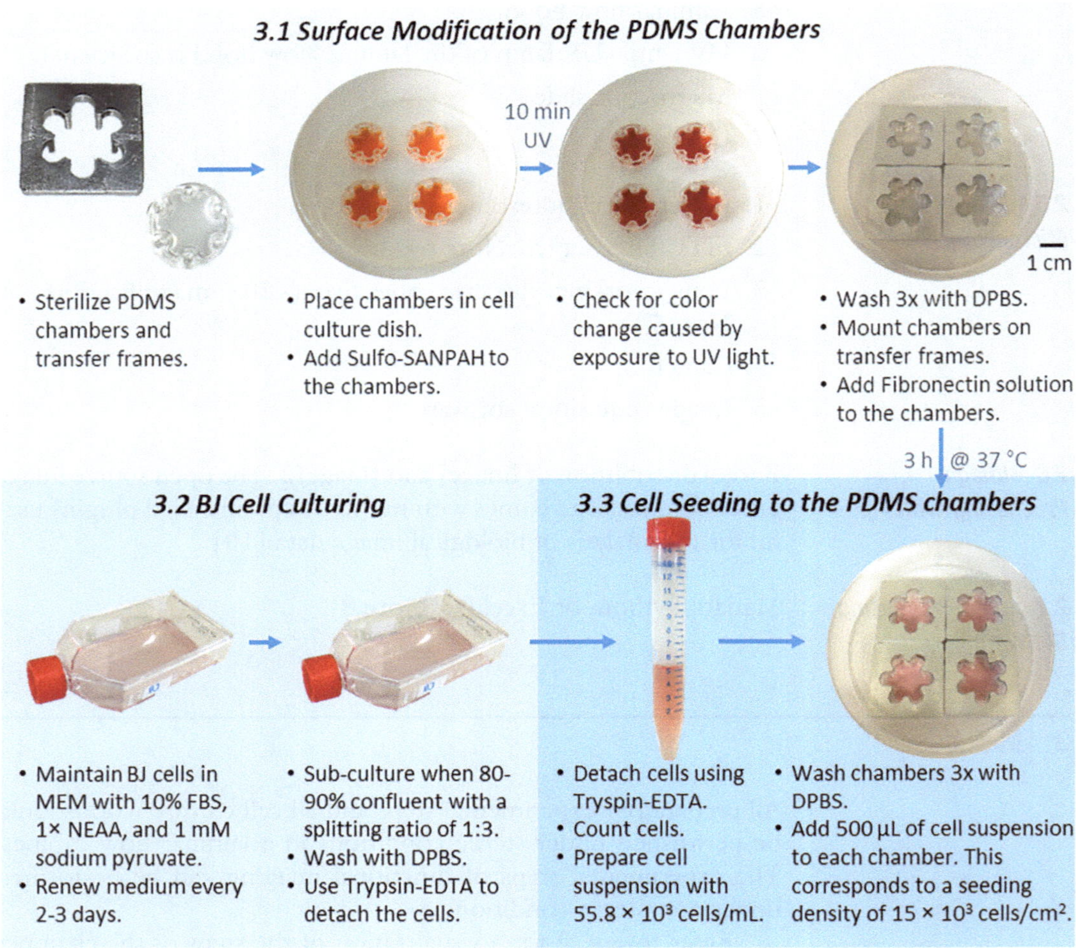


**Fig. 2** Step-by-step visualization of the procedure for PDMS chamber coating, BJ cell culture, and cell seeding. The PDMS chambers are placed inside a cell culture dish. After the Sulfo-SANPAH solution has been added to the chambers, the chambers are exposed to UV light, which causes a color change of the Sulfo-SANPAH solution and covalent binding of the Sulfo-SANPAH to the PDMS chamber. The Sulfo-SANPAH solution is aspirated, and the remaining Sulfo-SANPAH residues are removed by three washing steps with DPBS. After 3 h of incubation with the Fibronectin solution, the chambers are again washed three times with DPBS. BJ cells are routinely cultured and sub-cultivated. For cell seeding, the cells are detached from the flask, a cell suspension with defined cell concentration is prepared and added to the coated chambers

6. Remove the Fibronectin solution, and wash the inside of the chambers three times with 500 μL of DPBS.

### 3.2 BJ Cell Culturing

1. Maintain BJ cells in MEM supplemented with 10% FBS, 1× NEAA and 1 mM sodium pyruvate.
2. Renew medium every 2–3 days.
3. Sub-culturing procedure (volumes are given for a T75 flask; increase or decrease volumes for culture vessels of other sizes):

Passage the cells when approximately 80–90% confluent. Aspirate the old medium from the flask, briefly rinse the cells with 5 mL of DPBS, remove the DPBS, and add 2 mL of the Trypsin-EDTA solution to the flask. Gently rock the flask to ensure that all cells are covered by Trypsin-EDTA. Then aspirate the Trypsin-EDTA solution before the cells start to detach (*see* **Note 10**). Place the flask in the 37 °C incubator to facilitate dispersal for 5 min. Observe the cells under an inverted microscope until the cell layer is dispersed. Add 9 mL of complete growth medium to the flask. Separate the cells by carefully pipetting up and down, and transfer 3 mL of the cell suspension to a new T75 flask. This corresponds to a sub-cultivation ratio of 1:3. Add 12 mL of fresh cell growth medium to the T75 flask containing the BJ cells and mix by gentle rocking.

### 3.3 BJ Cell Seeding to the PDMS Chambers (One Day Before Cell Stretching)

1. One day before the cell stretching experiments (*see* **Note 11**), rinse the cells with 5 mL of DPBS; then, detach the BJ cells by covering the cells with 2 mL of Trypsin-EDTA which is immediately removed before the cells start to detach (*see* Subheading 3.2, **step 3** and **Note 12**).
2. Observe the cells closely with your cell culture phase contrast microscope. After detachment of the cells, add 5 mL of complete growth medium to the flask, and transfer the cells to a 15 mL centrifuge tube.
3. Count the cells using a hemocytometer.
4. In a second centrifuge tube, prepare a cell suspension with known cell concentration to be added into the PDMS chambers. Seed cells with a seeding density of $15 \times 10^3$ cells/cm$^2$ to the chambers. With a growth area of 1.86 cm$^2$ per chamber, this corresponds to $27.9 \times 10^3$ cells per chamber. To do this, adjust the cell concentration of the second cell suspension to $55.8 \times 10^3$ cells/mL. Prepare 500 μL of cell suspension for each chamber and in addition 500 μL of cell suspension as pipetting reserve.
5. Mix your cell suspension thoroughly, and add 500 μL of the cell suspension to each PDMS chamber coated as described in Subheading 3.1. Take care to use the same number of cells in each chamber, as cell behavior may change at different cell densities (*see* **Note 13**).

### 3.4 Preparation of the Cal-520 Staining Solution and of the Test Chemicals

Prepare stock solutions of your chemicals before first use as stated in Subheading 2.3. Aliquot the stock solutions in suitable quantities, and store them at 4 °C or −20 °C as specified by the manufacturer until further use. Avoid repeated freeze-thaw cycles. Prepare working solutions directly prior to use and do not store working solutions. Make sure all working solutions are brought to room temperature before use.

In the following steps, the required volume of working solution per chamber will be stated. Always allow for an appropriate pipetting reserve when preparing the working solutions.

1. Cal-520 staining solution: 5 μM Cal-520 AM (1 mM stock solution) in PSS, 300 μL per chamber.
2. Gadolinium(III)-chloride hexahydrate working solution: 400 μM Gadolinium(III)-chloride hexahydrate (1 M stock solution) in PSS, 100 μL per chamber.
3. Ionomycin calcium salt working solution: 45 μM ionomycin calcium salt (1 mM stock solution) in PSS, 50 μL per chamber.

### 3.5 Functional $Ca^{2+}$ Imaging During Global Stretch

The experimental procedure and an exemplary plot of the results are shown in Fig. 3.

1. One day after cell seeding (*see* **Note 11**), the cells should be firmly attached to the Fibronectin-coated PDMS chambers (*see* **Note 9**). Check the cells with a standard cell culture microscope. Use the chambers only when the cells have spread nicely on the surface (*see* **Note 14**).
2. Stain a maximum of three chambers at a time, as the cells are kept at room temperature after staining and may be affected by too long periods at room temperature.
3. Aspirate the growth medium from the chambers.
4. Incubate the cells in 300 μL staining solution for 30 min inside the incubator at 37 °C, followed by incubation at room temperature for 30 min. Cover the cells with a light-proof box lid to protect from light.
5. After staining, remove the staining solution from the PDMS chambers and replace with 300 μL of fresh PSS.
6. Turn on the microscope and start the microscope software. Connect the *IsoStretcher* (or the stretcher hardware of your choice) to your PC, start the control software, and initialize the device.
7. Transfer one chamber from the transfer frame to the *IsoStretcher* using the transfer punch. A visualization of this procedure is shown in Fig. 4.
8. Place the stretching device on the microscope stage. Bring the center of the chamber into the field-of-view by moving the xy-stage of your microscope (*see* **Note 15**).
9. Add another 100 μL of PSS for the *Control* experiments or 100 μL of Gadolinium(III)-chloride hexahydrate working solution (400 μM) for the *Blocking* experiments to the PDMS chamber. With a resulting volume of 400 μL (*see* **Note 16**) in the chambers, this will result in a concentration of 100 μM $Gd^{3+}$ in the chambers used for the *Blocking* experiments. Incubate for 5 min.

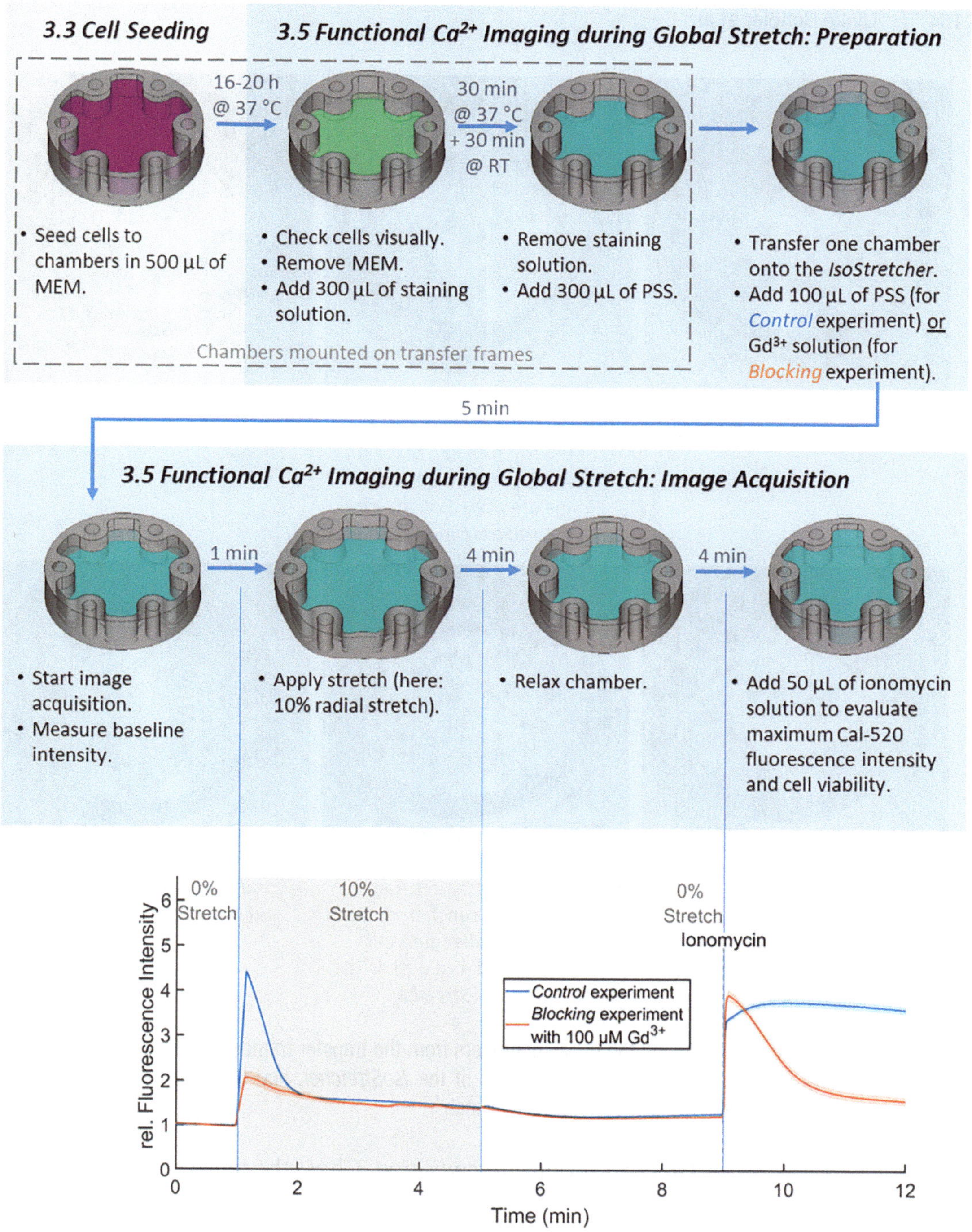


**Fig. 3** Schematic representation of the workflow of the $Ca^{2+}$ imaging experiments using the *IsoStretcher* and exemplary time curves. In the upper part of the illustration, the pipetting and stretching steps of an exemplary stretching experiment are shown. Below, the relative Cal-520 integrated density over time for two exemplary stretching experiments can be seen. The BJ cells in one chamber (orange) were exposed to a blocker of mechanosensitive ion channels (*Blocking* experiment), $Gd^{3+}$, while the second chamber serves as a control chamber (*Control* experiment, blue). While the cells in the *Control* experiment strongly react to 10% radial stretch with a strong increase in fluorescence intensity of the $Ca^{2+}$ indicator dye, only a very low increase in fluorescence intensity upon stretch was observed in the *Blocking* experiment. In both experiments, the cells strongly react to ionomycin. While the Cal-520 fluorescence intensity remains high in the *Control* cells, the fluorescence intensity quickly drops in the *Blocking* experiment after an initial rise

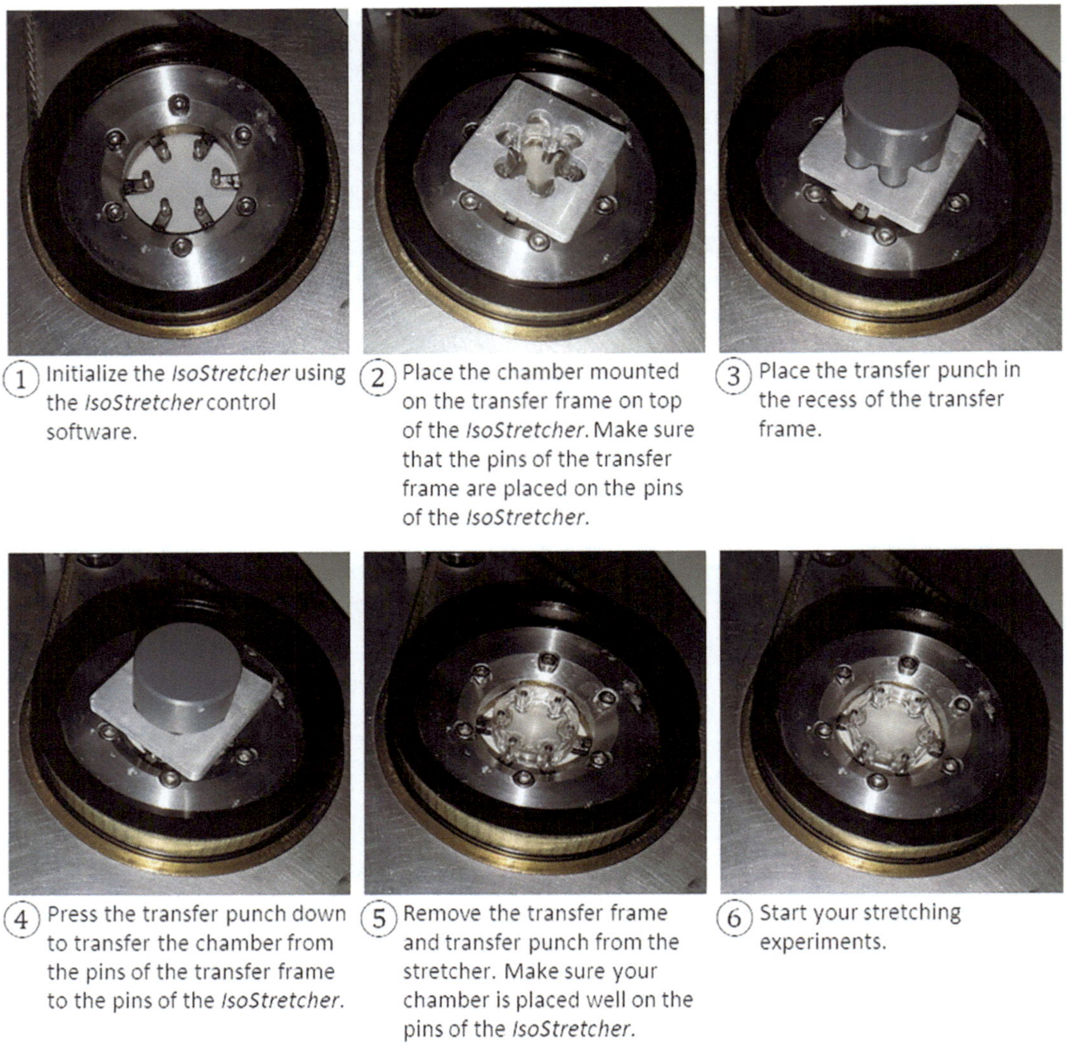


**Fig. 4** Steps involved in transferring the PDMS chambers from the transfer frames onto the *IsoStretcher*. The pins of the transfer frames are placed onto the pins of the *IsoStretcher*, and then, the PDMS chamber is transferred onto the *IsoStretcher* using the transfer punch

10. Start the image acquisition. Chose the time-lapse/video mode. Set the frame rate to one frame per second. Choose an appropriate exposure time (max. 1 s, *see* **Note 17**). Record the baseline signal of the non-stimulated cells for 1 min.
11. Stretch the cells to 10% radial stretch (*see* **Note 18**). Use your microscope software to quickly refocus. Hold the stretch for 4 min.
12. Go back to 0% stretch, refocus, and record for another 4 min.
13. Add 50 μL of the ionomycin calcium salt working solution (45 μM) to the chamber. The solution is diluted 1:9, and therefore, the resulting ionomycin concentration inside the chamber is 5 μM. Acquire the signal for another 3 min.

14. After the acquisition is finished, use tweezers to remove the chamber from the *IsoStretcher* by gently pressing against the edge of the chamber from below and pushing the chamber upwards piece by piece, changing the position of the tweezers along the edge of the chamber several times.
15. Proceed with the next chamber (*see* **Note 19**).

### 3.6 Acquisition of the Images Required for Image Correction

The image processing routine that will be described in Subheading 3.7 includes basic image correction steps. A dark image and a flat-field image are required to perform these image corrections. As long as your microscope setup is not changed, you can reuse your correction images.

1. Dark image: set up the microscope in a way that no light reaches the camera (for example set the light path on eyepiece). Choose the exposure time you will also use for the acquisition of your $Ca^{2+}$ imaging data. Acquire a time lapse of 100 images. Calculate the average of those images using Fiji (Image → Stacks → Z Project... → Average Intensity). Save this image for later use.
2. Flat-field image [11]: to correct for the uneven illumination, acquire a total of >10 images consisting of time lapses on several positions of a uniform fluorescent sample. This uniform fluorescent sample can for example be a droplet of a fluorescein solution (1 mM fluorescein in $H_2O$, dilute 1 M stock 1:1000) between a microscope slide and a coverslip. Subtract the corresponding dark image from the obtained images. Calculate the median of those images using Fiji (Image → Stacks → Z Project... → Median). Press "M" to measure the mean intensity of the resulting median image. To obtain an image with a mean intensity of 1, convert the median image to float type (Image → Type → 32-bit) and divide it by its mean intensity value (Process → Math → Divide... → Value: "Mean of the median image"). Save this image for later use.

### 3.7 Image Processing Using FIJI Software

For a quick screening of the results, it might be sufficient to evaluate the mean intensity of the whole field-of-view over time (*see* **Note 20**). For in-depth analysis, it might be necessary to analyze the data on a single-cell level. Since the field-of-view inevitably shifts slightly upon stretching/relaxation of the chamber, and the cell areas become larger/smaller, it is necessary to identify the location of the same cell in each of the different phases (before stretching/during stretching/after stretching). For this purpose, pairwise image registrations are performed, which yield transformation files describing the operations necessary to superimpose the "before stretch" image, the "during stretch" image, and the "after stretch" image. While these transformation files could then be used to transform the time lapse image data, in this protocol, the transformation files are used to transform the region of interest (ROI)

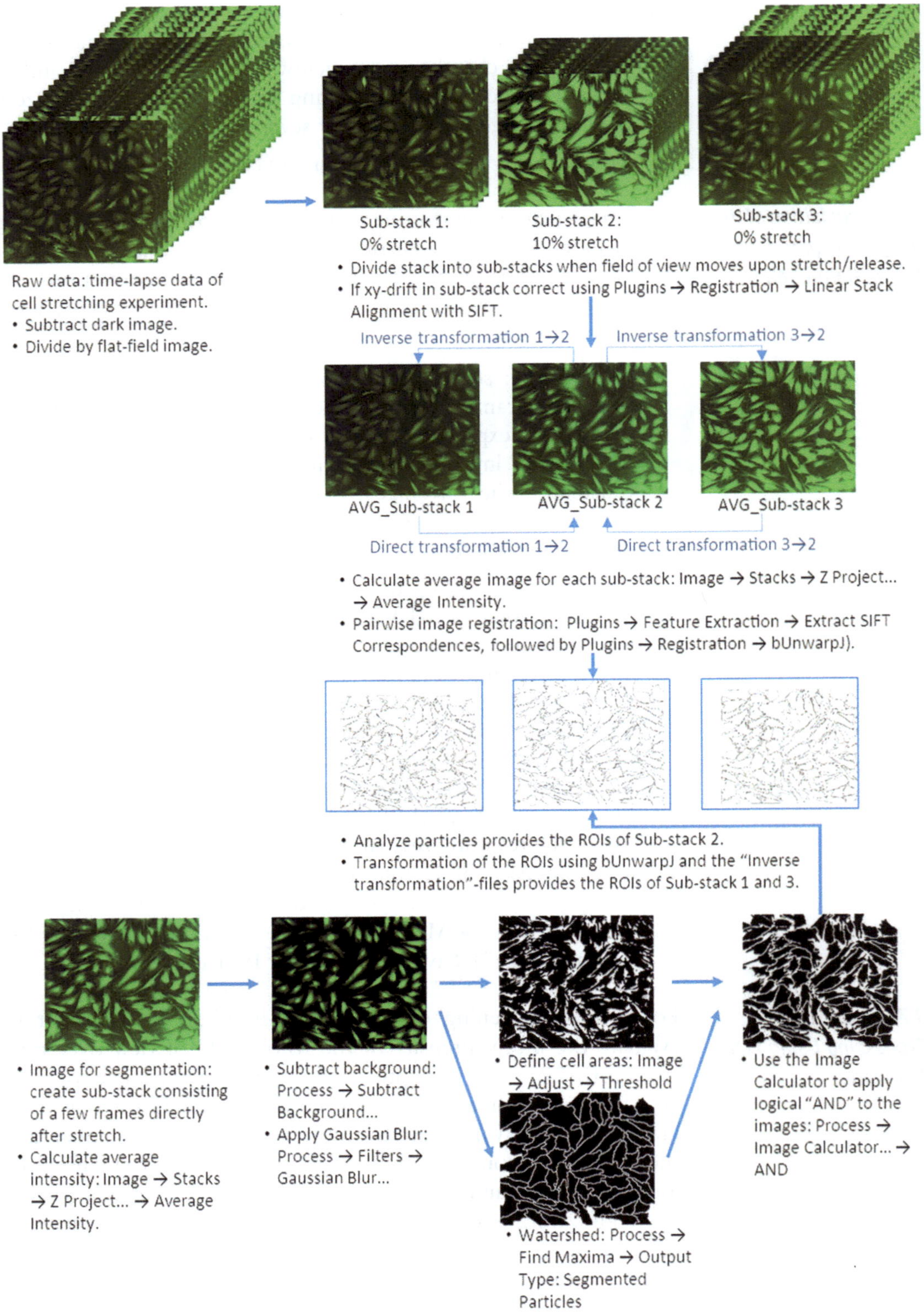


**Fig. 5** Visualization of the image processing steps involved in the ROI drawing and transformation. The time-lapse data are first corrected using the correction images. The corrected data (scale bar = 100 μm) are split into sub-stacks. The average for each sub-stack is calculated. Those average images can now be used to

files to avoid changing the image data as much as possible. The ROIs (which reflect the outlines of the cells) are determined in one of the phases and then transformed to match the outlines of the cells in each of the other phases. A visualization of the workflow for image registration, cell segmentation, and ROI transformation is shown in Fig. 5.

1. Import the time-lapse image acquired with your microscope software using the ImageJ Bio-Formats plugin.
2. Save the metadata file containing the time stamps for later use.
3. Perform basic image corrections: subtract the dark image from each image of the time lapse. Divide the resulting image stack by the flat-field image. Save the resulting image stack and use this for further analysis.
4. Split the time-lapse stack into several sub-stacks: split the recording when the field-of-view changes due to stretch or release (*see* **Notes 21** and **22**).
5. If xy-drift occurs within one sub-stack use the ImageJ plugin "Linear Stack Alignment with SIFT" (*see* **Note 23**) to align the images (Plugins → Registration → Linear Stack Alignment with SIFT).
6. For each (corrected) sub-stack calculate the z-Projection "Average Intensity" (Image → Stacks → Z Project... → Average Intensity).
7. Choose the sub-stack of images with the best signal-to-noise ratio. This does not necessarily have to be one of the sub-stacks created in **step 4**, but can, for example, only include the first ten frames of the stretch application (after refocusing). Create an average intensity projection of this sub-stack, use the "Subtract Background" operation (Process → Subtract Background...), apply Gaussian Blur (Process → Filters → Gaussian Blur...) and use the resulting image for cell segmentation. Duplicate the image. Use one duplicate to define the cell areas by setting an appropriate threshold (Image → Adjust → Threshold). To separate touching cells, use the second duplicate of the image and apply the "Find Maxima" tool which applies a Watershed algorithm to the image (Process → Find Maxima → Output Type: Segmented Particles). Apply the logical "AND" to the thus

**Fig. 5** (continued) apply the ImageJ plugin "bUnwarpJ" to perform image registration and save the transformation files. For image segmentation, another average image is created: choose the images where the cells are strongly fluorescent directly after stretch, and calculate the average image of this sub-stack. Define the cell areas by setting a threshold and separate touching cells by applying a Watershed algorithm using the "Find Maxima" tool. The transformation files created with "bUnwarpJ" can then be used to define the ROIs of the other sub-stacks

obtained "Threshold" image and the "Segmented Particles" Image (Process → Image Calculator... → "AND") (*see* **Note 24**). Save this image.

8. Perform a pairwise feature extraction on the "Average Intensity" images created in **step 6** via the ImageJ plugin "Extract SIFT Correspondences" (Plugins → Feature Extraction → Extract SIFT Correspondences), followed by an image registration using the ImageJ plugin "bUnwarpJ" (Plugins → Registration → bUnwarpJ) [12]. Set the Landmark weight of the "bUnwarpJ" plugin to 1, the Target Image is always the sub-stack you used for cell segmentation, the Source Image is one of the other sub-stacks, respectively. Enable "Save Transformations". Save the files obtained from the plugin. In the further course of this protocol, you will need the "inverse transformation" files and the third slice ("Warped Source Mask") of the "Registered Source Image" files.
9. Exclude the parts of the image that are not visible in all sub-stacks. To do so, apply logical "AND" to all "Warped Source Mask" images. This is your "Restriction Image." Apply logical "AND" to your "Restriction Image" and the image obtained from **step 7**.
10. Use the "Analyze Particles" tool; add the ROIs to the ROI manager (Analyze → Analyze Particles... → Add to Manager).
11. The ROIs obtained from the cell segmentation can now be transformed using the "inverse transformation" files obtained from the bUnwarpJ plugin. For this step, the ImageJ-macro file "Transformation_of_ROIs.ijm" provided as Electronic Supplementary Material can be used (Plugins → Macros → Run...). The macro requires as input the average image of the sub-stack for which the ROIs should be transformed, the corresponding "inverse transformation" file, and the ROI set obtained from **step 10**. The output of the macro is the transformed ROI set and a log file containing the input and output details.
12. Specify which measurements will be recorded. Enable "Integrated Density" (Analyze → Set Measurements... → Integrated density).
13. For each sub-stack, use the corresponding set of ROIs and use "Multi Measure" to measure the integrated density of the fluorescence signal per ROI (ROI Manager → More >> → "Multimeasure").

### 3.8 Data Analysis

1. Use the data processing software of your choice to import the results file obtained from Fiji.
2. For each ROI, calculate the normalized $Ca^{2+}$-dependent fluorescence intensity by dividing each Integrated Density value of

this ROI by the mean Integrated Density value during the first minute (the baseline signal).

3. Extract the time stamps from the metadata.
4. Plot the normalized Integrated Density over time.

### 3.9 Accessing Cell Viability Based on the Response to Ionomycin

We observed that during our experiments isolated cells may die off due to mechanical overload. In order to identify these in an automated way, one can take advantage of the response of the cells upon ionomycin addition. In cells with an intact membrane, the calcium ionophore leads to an increase in the intracellular $Ca^{2+}$ level and thus to an increase in the Cal-520 fluorescence intensity. In contrast, no increase in Cal-520 fluorescence intensity is observed in damaged cells, most likely because the dye molecules diffuse out of the cells through the damaged membrane. An exemplification is shown in Fig. 6.

1. Open the file containing the single cell analysis data in the data processing program of your choice.
2. For each cell, extract the maximum value of the relative integrated density after addition of the ionomycin calcium salt working solution.
3. Choose an appropriate threshold to distinguish between cells reacting with an increase in relative integrated density after addition of ionomycin and cells not reacting (*see* **Note 25**).
4. For each cell, check if the maximum value of the relative integrated density is larger than the threshold value.
5. If you wish, you can now exclude the cells that are potentially no longer viable due to mechanical overload from further analysis or separate the cells into two groups for further analysis.

## 4 Notes

1. Depending on your research question, other types of adherent cells may be used. Adjust the culture and handling solutions according to your standard protocol for culturing the respective cells.
2. For this protocol, it is crucial that the cell stretcher you wish to use is suitable for live cell imaging. The *IsoStretcher* fits on most commercially available inverted epifluorescence microscopes. The PDMS chambers used with the *IsoStretcher* have been optimized for microscopy: they have a thin, smooth, and transparent chamber bottom and a low z-displacement during stretching [8, 9]. If you decide to use a cell stretching device other than the *IsoStretcher*, be sure to match the volume

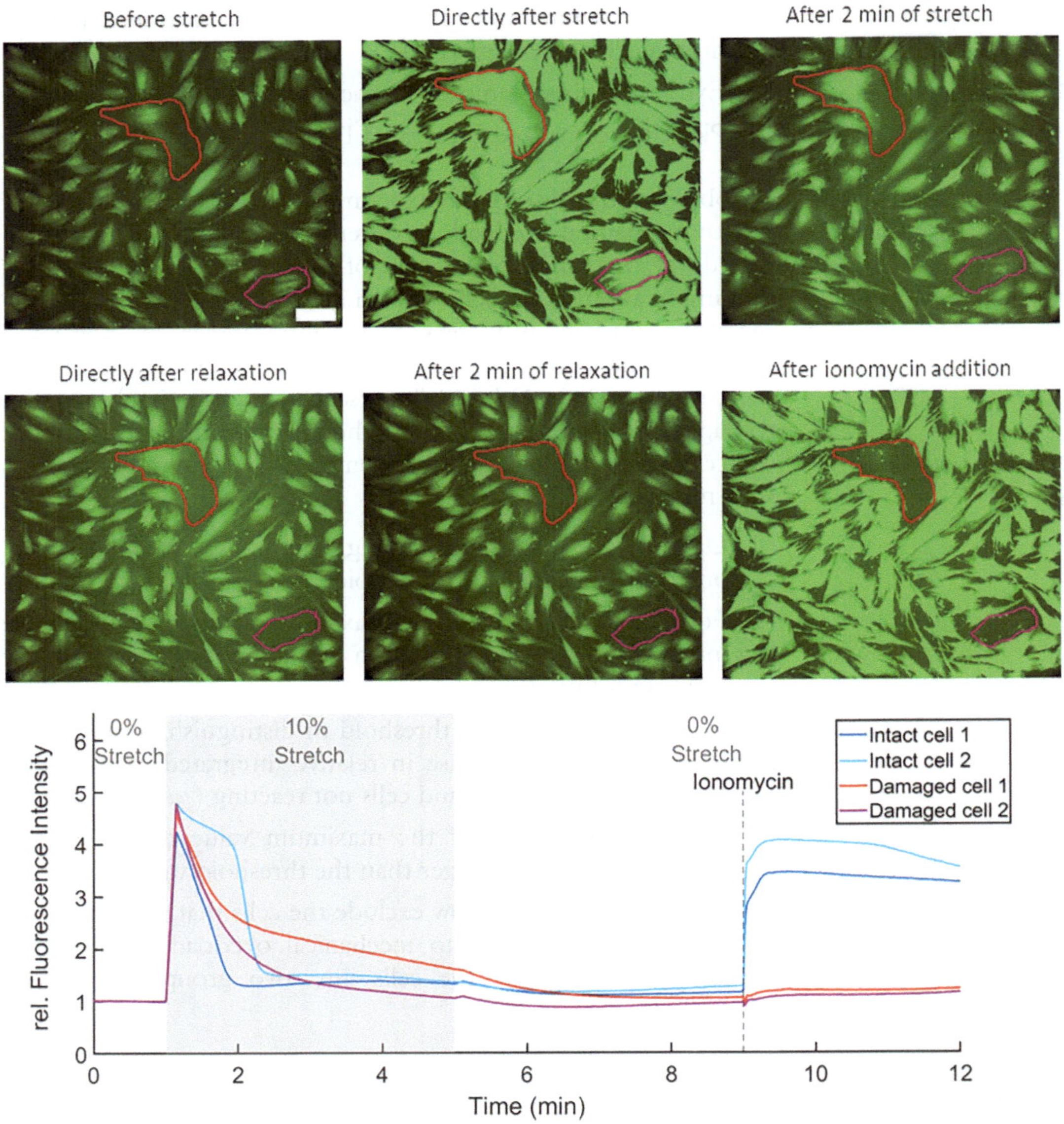


**Fig. 6** Accessing cell viability based on the response to ionomycin. Upon stretch, some single cells die off due to mechanical overload. If those cells should be excluded from the analysis or analyzed separately, the (lack of) increase in fluorescence intensity upon addition of ionomycin can be used to identify cells that have died off during the course of the experiment as they were already $Ca^{2+}$ overloaded due to compromised cell membrane integrity. In the upper part of the figure, two BJ cells are highlighted that lost membrane integrity during the experiment. In the lower part of the figure, the exemplary time course of four single cells is shown. The blue curves show exemplary time curves of intact cells, while the red and pink curve show the integrated density over time of the cells highlighted in the upper part of the figure that lost their membrane integrity. Scale bar = 100 μm

of substances and the number of cells to the volume and area of the substrate of your respective cell stretching system.

3. By varying the ratio of PDMS base to curing agent, the stiffness of the PDMS chambers can be tailored. If the amount of curing agent used is decreased, the stiffness of the PDMS chambers decreases [13].
4. The transfer frame serves as a holder to prevent stimulation/ movement of the cells during chamber handling before starting the experiments. The transfer punch is a helpful tool to transfer the PDMS chamber from the transfer frame to the *IsoStretcher*. On other cell stretching systems, you might find other solutions to avoid accidental mechanical stimulation of the cells during experiment preparation.
5. We use Cal-520 AM which is a fluorogenic calcium-sensitive dye with an improved signal-to-noise ratio and intracellular retention compared to Fluo-3 AM and Fluo-4 AM. However, if you prefer to use another calcium-sensitive dye, such as Fluo-3 AM or Fluo-4 AM, this is possible after optimization of the staining protocol. Usage of probenecid, an organic anion transporter inhibitor, might also be beneficial.
6. The FITC filter set is appropriate for imaging cells stained with Cal-520 AM, Fluo-3 AM, or Fluo-4 AM. If you wish to use a dye other than those listed, check which filter set is required.
7. We use a Nikon CFI S Plan Fluor ELWD 20XC. This objective has a long working distance and a correction ring for cover slip correction that allows adjustment to the thickness of the PDMS substrate.
8. For further processing and visualization of the data received from Fiji, programs such as Matlab or Python are recommended. Depending on the application, Excel may also be sufficient, but difficulties may arise due to the size of the data sets.
9. In our hands, our coating protocol with Sulfo-SANPAH and fibronectin worked for a wide range of cells. However, for other cell types, it might be beneficial to test other extracellular matrix (ECM) proteins.
10. According to the ATCC sub-cultivation protocol, the BJ cells are kept in the Trypsin-EDTA solution until the cell layer is dispersed. Then, the Trypsin-EDTA solution is neutralized by serum-containing medium, and the obtained cell suspension is transferred to the new flask. However, we find it useful to remove most of the Trypsin-EDTA solution immediately after the cells have been covered with it and have not yet detached, since the remaining Trypsin is sufficient to detach

the cells and is thus almost completely removed from the medium without the need to centrifuge the cell suspension.

11. While we allowed the BJ cells to grow and attach to the chambers overnight, other attachment times may be appropriate for other cell types and experimental designs. These attachment times can vary from several hours to several days.
12. It is especially important to avoid over-trypsinization at this point, as this might damage the cells and hinder proper adhesion to the PDMS chamber coated with Fibronectin.
13. We observed in our experiments that BJ cells are less responsive to stretch at very high cell densities.
14. Depending on the cell handling, the cells may have problems spreading on the surface of the PDMS chambers even if they proliferate well in your standard cell culture flasks. We have observed that medium that is too old can be a problem for the cells when seeded into the PDMS chambers, although they can still handle it when seeded into cell culture flasks.
15. Since the *IsoStretcher* stretches the PDMS chambers radially, it is important to bring the center of the chamber into the field-of-view to see the same cells before and during stretching. The further away from the center one chooses the field-of-view, the stronger the movement of the cells in the xy-plane will be upon stretch, and the more cells will move out of the field-of-view upon stretch. Depending on the cell stretching system (e.g., when using a linear stretch), you will need to optimize, which initial field-of-view is most appropriate to observe as many cells as possible throughout the experiment.
16. It might be tempting to reduce the volume of physiological salt solution inside the PDMS chamber to reduce the volume of expensive test drugs needed to reach a certain concentration. However, we do not recommend to have a volume smaller than 400 μL inside the chambers at the time of stretch as the liquid level changes during the stretching process. In the worst case, the cells could, therefore, end up being exposed to air during stretch.
17. Before performing first stretch experiments, optimize the exposure time needed with your calcium indicator dye. Find a trade-off between a sufficient signal-to-noise ratio and minimal bleaching. As the mean intensity/integrated density of whole cells will be evaluated, a low spatial resolution is sufficient. Therefore, it can also be beneficial to use binning. For example, on our setup we use an exposure time of 500 μs and $2 \times 2$ binning. Make sure that your image software switches off the fluorescence lamp while the camera is not recording.

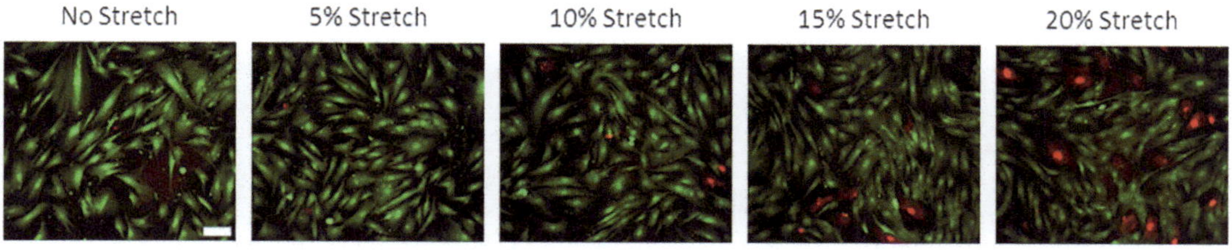


**Fig. 7** Live/Dead-Imaging of BJ cells subjected to different extents of stretch. Calcein-AM (green) stains live cells and EthD-III (red) stains dead cells. With increasing extent of stretch applied to the cells, more dead cells are seen in the chambers. Scale bar = 100 μm

18. Depending on the type of cells, experimental procedure, and scope, different levels of stretch might be appropriate. Note that too high strain load and assumed subsequent membrane stress can cause cell death. Figure 7 shows BJ cells that were seeded to PDMS chambers, incubated overnight, placed in PSS, and subjected to different extents of radial stretch for 4 min. After stretch, the cells were stained using PromoKine's Live/Dead Staining Kit II. The kit consists of Calcein-AM that stains live cells green and EthD-III that stains dead cells red. After 30 min of incubation with the dyes at room temperature, the cells were imaged using a fluorescence microscope. It can be seen that with increased stretch extent applied, more dead cells can be observed inside the chambers. Therefore, we chose 10% radial stretch for the experiments with BJ cells. However, for other cell types that are less susceptible to mechanical overstimulation, it may be useful to apply higher stretch. So far, we have not observed cell death due to stretch to this extent in other cell types.
19. Always perform corresponding substance testing and control experiments on the same day, as there might be variability in cell response from day to day. If you stain several chambers at the same time, make sure that you switch between conditions randomly and do not always image one condition first and then the other condition, as the time between staining and performing the experiment could also have an influence.
20. Upon stretch, there are two superimposing effects occurring when evaluating the mean intensity. The first effect is that the mechanical stimulation causes an activation of mechanosensitive ion channels resulting in an elevation of the calcium level in the cytoplasm resulting in brighter fluorescence of the calcium indicator dye. At the same time, the area of the chamber becomes larger while the amount of dye in the chamber stays the same. This means that the fluorescence intensity per area declines. To visualize this effect, we stained BJ cells with CellTracker Green CMFDA, a fluorescent dye whose fluorescence intensity should not be affected by changes in substance

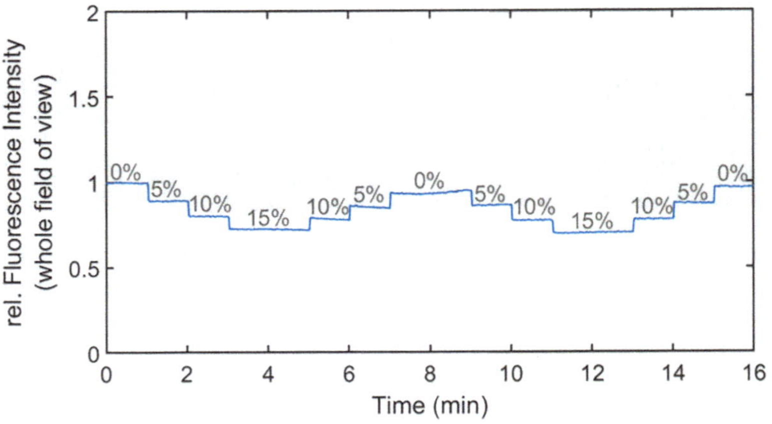


**Fig. 8** Visualization of the problem of evaluating the mean fluorescence intensity within the whole field-of-view when stretch is applied. BJ cells were stained with CellTracker Green CMFDA, a fluorescent dye. The extent of stretch was increased/reduced stepwise. With increasing chamber area, but constant dye particle concentration, the number of dye particles per area decreases upon stretch, and therefore, so does the mean intensity

concentrations such as the intracellular $Ca^{2+}$ level. As shown in Fig. 8, a stepwise decrease in fluorescence intensity can be observed when stretch is applied and a stepwise increase in fluorescence intensity when the chamber is relaxed again. For cells showing a very low response to stretch, the decrease in fluorescence intensity caused by the second effect might even be larger than the increase caused by the change in the intracellular calcium level. To overcome this issue, we usually analyze the integrated density which integrates all intensity values of one ROI instead of averaging them.

21. The data shown in Fig. 3 were split into three sub-stacks: before stretch (0–1 min), 10% stretch (1–5 min), after stretch (5–12 min).

22. When the chamber is stretched/relaxed, there is usually a z-shift. You may need some frames to refocus. We usually exclude these frames from the analysis. However, it is important to note that the strongest Cal-520 fluorescence signal (reflecting the highest intracellular $Ca^{2+}$ level) is measured within a few milliseconds to seconds after the chamber is stretched, followed by a decline that can last up to a few minutes. Thus, if you refocus too slowly, you may miss the maximum signal.

23. Note that the scale gets lost when you use the ImageJ plugin "Linear Stack Alignment with SIFT". Therefore, to avoid later confusion, it is best if you redefine the scale for your corrected image right after you applied the plugin.

24. The procedure is also explained in a tutorial available at the Melbourne Advanced Microscopy Facility and is clearly explained there [14, 15].
25. For the data shown in Fig. 5, a threshold value of 2 would be appropriate.

## Acknowledgments

This work was supported by the German Research Foundation (DFG) projects 383071714 (FR 2993/23-1), the Bavarian Equal Opportunities Sponsorship—Realisierung von Chancengleichheit von Frauen in Forschung und Lehre (FFL)—Realization Equal Opportunities for Women in Research and Teaching, and the Erlangen Graduate School in Advanced Optical Technologies (SAOT) by the German Research Foundation (DFG) in the framework of the German excellence initiative. This work is part of a PhD thesis (US and ALM).

### References

1. Coste B, Mathur J, Schmidt M et al (2010) Piezo1 and Piezo2 are essential components of distinct mechanically activated cation channels. Science 330(6000):55–60. https://doi.org/10.1126/science.1193270
2. Jiang F, Yin K, Wu K et al (2021) The mechanosensitive Piezo1 channel mediates heart mechano-chemo transduction. Nat Commun 12(1):869. https://doi.org/10.1038/s41467-021-21178-4
3. Wang S, Chennupati R, Kaur H et al (2016) Endothelial cation channel PIEZO1 controls blood pressure by mediating flow-induced ATP release. J Clin Invest 126(12):4527–4536. https://doi.org/10.1172/JCI87343
4. Emig R, Knodt W, Krussig MJ et al (2021) Piezo1 channels contribute to the regulation of human atrial fibroblast mechanical properties and matrix stiffness sensing. Cell 10(3):663
5. Fujiu K, Nagai R (2014) Fibroblast-mediated pathways in cardiac hypertrophy. J Mol Cell Cardiol 70:64–73. https://doi.org/10.1016/j.yjmcc.2014.01.013
6. Ao M, Brewer BM, Yang L et al (2015) Stretching fibroblasts remodels fibronectin and alters cancer cell migration. Sci Rep 5(1):8334. https://doi.org/10.1038/srep08334
7. Langevin HM, Storch KN, Cipolla MJ et al (2006) Fibroblast spreading induced by connective tissue stretch involves intracellular redistribution of alpha- and beta-actin. Histochem Cell Biol 125(5):487–495. https://doi.org/10.1007/s00418-005-0138-1
8. Schürmann S, Wagner S, Herlitze S et al (2016) The *IsoStretcher*: an isotropic cell stretch device to study mechanical biosensor pathways in living cells. Biosens Bioelectron 81:363–372. https://doi.org/10.1016/j.bios.2016.03.015
9. Guo Y, Merten AL, Schöler U et al (2021) In vitro cell stretching technology (*IsoStretcher*) as an approach to unravel Piezo1-mediated cardiac mechanotransduction. Prog Biophys Mol Biol 159:22–33. https://doi.org/10.1016/j.pbiomolbio.2020.07.003
10. Schindelin J, Arganda-Carreras I, Frise E et al (2012) Fiji: an open-source platform for biological-image analysis. Nat Methods 9(7):676–682. https://doi.org/10.1038/nmeth.2019
11. Model MA, Burkhardt JK (2001) A standard for calibration and shading correction of a fluorescence microscope. Cytometry 44(4):309–316. https://doi.org/10.1002/1097-0320(20010801)44:4<309::AID-CYTO1122>3.0.CO;2-3
12. Arganda-Carreras I, Sorzano COS, Marabini R et al (2006) Consistent and elastic registration of histological sections using vector-spline regularization. In: Computer vision approaches to medical image analysis. Springer, Berlin, Heidelberg, pp 85–95

13. Kah D, Winterl A, Přechová M et al (2021) A low-cost uniaxial cell stretcher for six parallel wells. HardwareX 9:e00162. https://doi.org/10.1016/j.ohx.2020.e00162
14. Cheung S, Shobhana K (2022) FIJI for Quantification Workshop: Fundamentals of Image Quantification. https://microscopy.unimelb.edu.au/optical-microscopy/capabilities/fiji-workshop-resources. Accessed 20 Jul 2022
15. McMillan P (2020) FIJI for Quantification Workshop: Fundamentals of Image Quantification. https://microscopy.unimelb.edu.au/optical-microscopy/capabilities/fiji-workshop-resources. Accessed 20 Jul 2022

# Chapter 11

# Assessment of Cell Viability in Electrically Excitable Muscle Cells Through Intact Twitch Stimulation

**Julian Bauer, Stewart I. Head, and Oliver Friedrich**

## Abstract

Muscle cells (i.e. skeletal muscle fibers) are fully viable and functional when their excitation-contraction (EC) coupling machinery is intact. This involves intact membrane integrity with polarized membrane, functional ion channels for action potential generation and conduction, an intact electro-chemical interface at the level of the fiber's triad, followed by sarcoplasmic reticulum $Ca^{2+}$ release, and subsequent activation of the chemico-mechanical interface at the level of the contractile apparatus. The ultimate end result is then a visible twitch contraction upon a brief electrical pulse stimulation. For many biomedical studies involving single muscle cells, intact and viable myofibers are of utmost importance. Thus, a simple global screening method that involves a brief electrical stimulus applied to single muscle fibers and assessment of visible contraction would be of high value. In this chapter, we describe step-by-step protocols to (i) obtain intact single muscle fibers from freshly dissected muscle tissue using an enzymatic digestion procedure and (ii) provide a workflow for the assessment of twitch response of single fibers that can be ultimately classified as viable. For this, we have prepared a unique stimulation pen for which we provide the fabrication guide for do-it-yourself rapid prototyping to eliminate the need for expensive specialized commercial equipment.

**Key words** Skeletal muscle, Single fiber, Twitch stimulation, Electrical stimulator, Enzymatic dissociation, Contraction

## 1 Introduction

For many research activities in basic research, clinical, and pharmacology applications, the use of single intact and viable cells is of utmost importance. For instance, for high-throughput/high-content drug screening or in oncology research, one of the starting points aims to generate a large batch of viable single cells that are being monitored over time concerning cell viability, for example, in response to various interventions. The assessment of 'cell viability' in most of those settings, however, not only must be recognized as

**Supplementary Information** The online version contains supplementary material available at https://doi.org/10.1007/978-1-0716-3052-5_11.

Oliver Friedrich and Daniel F. Gilbert (eds.), *Cell Viability Assays: Methods and Protocols*, Methods in Molecular Biology, vol. 2644, https://doi.org/10.1007/978-1-0716-3052-5_11,

the desirable readout but also as an undesirable intervention in itself, as most viability assays, for example, involving fluorescent probes and markers, will ultimately interfere with or even destroy the investigated cells for the sake of a single time point viability measurement [1, 2]. Thus and moreover, this also prevents the precise assessment of viable cells to start with. Consequently, in most studies, the starting cell population (e.g. obtained from cell culture or extraction of primary cells from tissues) is simply assumed to consist of a high percentage of viable input cells with the study design relying on sufficient statistical rigor to eliminate dead cell subpopulations from the analysis. In that regard, it would be desirable to actually ascertain the viability and functionality of every single cell prior to downstream experiments, using simple and non-invasive methodologies. Those would then also allow to selectively pick viable cells for subsequent processing and experimentation and sort out non-viable or defective cells more easily. Currently, the viability and functionality of each cell in a starting cell population is mostly assumed rather than ascertained. In best-case scenarios, the mere morphology of cells in a light microscopic assessment is used to qualitatively extrapolate the overall viability of cells within a batch from a given field of view [3]. Electrically excitable cells, like muscle and neuronal tissue, share the unique feature that their viability and functionality is linked to an intact cell membrane that contains delicate sets of ion channels required to convey electrical signals into subsequent information cascades. In particular, in the case of muscle, this signaling cascade consists of a sequential electro-chemical-mechanical coupling where sufficiently large membrane depolarization through an action potential will be sensed by voltage-gated ion channel sensors (the so-called Dihydropyridine receptors, DHPR) in the transverse tubular system. These sensors directly open a $Ca^{2+}$ release channel on the membrane of the internal sarcoplasmic reticulum (a $Ca^{2+}$ ion storage organelle) to result in a transient increase in myoplasmic $Ca^{2+}$ levels [4–6].

This will then activate the contractile apparatus, that is, the acto-myosin filaments then start to slide against each other [7] (*see* Fig. 1). Hence, the end result will be a visibly strong contraction and relaxation, also known as a twitch, in case of a short-pulsed (~ms or less) voltage stimulus [8, 9]. In this regard, it does not matter whether the electrical stimulus is an intrinsic one conveyed by a nerve or an external one induced by external field stimulation, the DHPR receptor will simply detect membrane potential changes [10, 11]. Of note, a 'viable' muscle cell (or muscle fiber) will have to present intact signal conversion at each of those steps, and disturbances of those may ultimately result in contractile failure and weakness as an integral part of muscle disease, or simply indicating a damaged muscle fiber. Such damage, for example, can be the result of trauma during single fiber isolation and dissection

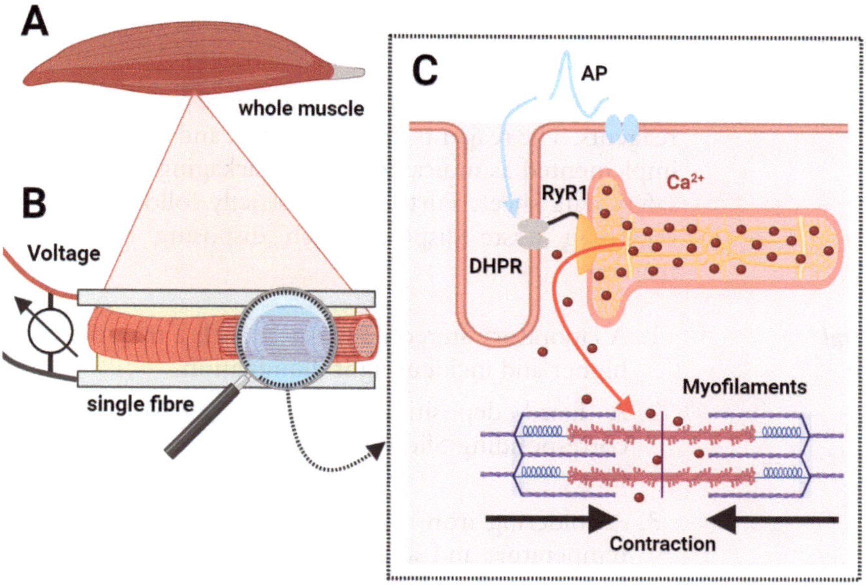


**Fig. 1** Muscular contraction through extracellular electrical stimulation. Skeletal muscle (**a**) exhibits a highly conserved modular hierarchical structure, where single muscle fibers are the smallest functional cellular unit (**b**). Physiologically, muscle contraction is initiated by an electrochemical nervous impulse which leads to membrane depolarization and, subsequently, to the induction of action potentials. This electrical impulse can be mimicked by extracellular electrodes which generate an electric field around the muscle cell (**b**). Eventually, the membrane depolarization, whether artificially (through external field stimulation) or intrinsically (through nerve input) induced, will be sensed by the voltage-sensitive DHPR receptors which (if the stimulus is sufficiently large) trigger a signaling cascade that results in a transient increase in myoplasmic $Ca^{2+}$ concentration. The elevated calcium concentration will ultimately enable the cross-bridge-cycling of the myofilaments and, therefore, lead to muscle contraction by activating the contractile apparatus (**c**). (The figure was created with BioRender.com)

from the whole organ, and it is, therefore, important to ascertain 'viability' through this relatively easily applicable methodology. However, neither the single fiber isolation nor the electrical testing are trivial and are usually confined to a few 'expert labs' worldwide [9, 12]. For the electrical stimulation, also often expensive and bulky equipment is advertised, which could be much cheaper and more versatile provided using the DOY (do-it-yourself) approach from the *MakerBot* and rapid prototyping community [13]. To broaden the applicability of single fibre viability assessment to a wider community, in this chapter, we present a step-by-step guide to (i) explain the muscle dissection and enzymatic single fiber isolation procedure for short muscle fibers of murine *interossei* toe muscles. This methodology aims to (i) yield large batches (thousands) of viable single fibers from defined input muscle tissue and, (ii) we present a fabrication guide alongside STL files to make-yourself an ergonomic, handheld electrical stimulation pen to quickly assess fiber contractility upon brief field stimulation under a preparation microscope. Finally, the workflow to assess fiber viability will be presented by examples.

## 2 Materials

All solutions are prepared using ultrapure water and analytical grade reagents. The reagents must be stored, and safety measures must be implemented as indicated on the packaging or the corresponding safety data sheet. Furthermore, strictly follow federal regulations regarding waste disposal when disposing of excess or waste material.

### 2.1 General Equipment

1. A laboratory stereo microscope with a magnification of 8× or higher and incident light illumination.
2. A fused deposition modeling (FDM) 3D printer with corresponding Slicer-Software to create 3D models from '.stl' files.
3. A soldering iron, in the ideal case with adjustable soldering temperature and suited for the use with leaded tin solder.
4. A 'helping hands' holding device to facilitate the soldering process of the circuit board inside of the stimulation pen.
5. A metal saw or a *Dremel driver* with a cutting disc to manufacture the electrode mounting tubes (*see* Subheading 3.2).
6. One sheet of sandpaper with a grain size of at least 500 grit.
7. A hole puncher or a pair of punch pliers for holes with a diameter of 5–6 mm.
8. Two finish taps without lead taper, one with thread diameter M4 and the other one with M3.
9. A vacuum desiccator with a corresponding vacuum pump.
10. A thermocycler with a heating block suited for Eppendorf tubes with a capacity of 2 mL.
11. Adjustable mechanical air displacement pipettes with displacement volume ranges of 100–1000 μL and 20–200 μL.

### 2.2 Reagents and Solutions

1. Ringer's solution: 145 mM NaCl, 5 mM KCl, 2.5 mM $CaCl_2$, 1 mM $MgCl_2$, 10 mM HEPES, and 10 mM glucose. Adjust pH to 7.4 using KOH. The solution can be prepared in large quantities and can be stored at −20 °C until usage.
2. Polydimethylsiloxane (PDMS) for PDMS-coated Petri dishes: A two-component PDMS crafting kit, consisting of a pre-polymer base component and a cross-linking curing agent (e.g. *SYLGARD* 184 Silicone Elastomer Kit).
3. Collagenase solution: Ringer's solution (*see* **item 1** and **Note 1**), 1.07 mg/mL collagenase Type 1 CLS 1.

### 2.3 Muscle Cell Stimulation Pen (ClickTwitch)

The mechanical and electronic components in this materials list are off-the-shelf parts and should be readily available at any hardware or electronics retailer.

1. 75 to 100 g of polylactide (PLA) 3D-printing filament.
2. A 9 V block battery to serve as a voltage source and a corresponding 9 V battery clip.
3. A unipolar miniature short-stroke push button (e.g. type B3F-1050) with a suitable cap. The cap height can be chosen according to personal preference.
4. Four sinter glass diodes of the type BYV 27.
5. Two 10 nF ceramic capacitors.
6. One 0.1 μF electrolyte capacitor.
7. One 25 kΩ miniature potentiometer to allow for pulse duration adjustment.
8. Two 100 kΩ resistors.
9. One 270 Ω resistor.
10. One Ø 3 mm low-current LED (e.g. type TLLG4400) in any available colour.
11. One Metal Oxide Semiconductor Field-Effect Transistor (e.g. type IRLIZ44N).
12. An inverting Schmitt trigger with six individual inverters (type 40106N).
13. Heat-shrink tubing in different diameters and lengths.
14. Two syringe cannulas of the type G21 and a length of 50 mm.
15. Approximately 21–25 mm of platinum wire which will serve as material for the excitation electrodes.
16. Roughly 1 m of stranded wire with a surface area of 1.5 $mm^2$.
17. One hole grid board with a hole distance of 2.54 mm.
18. Two disc magnets with a diameter of 4–5 mm and a height of approximately 1 mm.
19. Six countersunk head screws à M3 × 12 mm and, in case the stimulation pen is intended to be used with the wristband attached to it, two additional countersunk head screws à M4 × 10 mm are needed.
20. A velcro cable binder, preferably with a length of at least 25 cm and a width of at least 25 mm. This klettbinder will solely function as a wristband; thus, it is possible to substitute it with a wristband of any personal preference.
21. Duct tape to temporarily fix electronic components into place during assembly.

22. A pair of needle-nose pliers to facilitate the assembly of the device.
23. A small amount of instant glue.

### 2.4 Muscle Dissection

1. A laboratory mouse or preferably a mouse hind limb obtained through tissue sharing (*see* Subheading 3.3).
2. A pair of microsurgery Vannas Spring Scissors (e.g. eye surgery scissors), in the ideal case with a 3- to 4-mm cutting-edge.
3. Two pairs of type 55 Dumont micro-dissecting forceps.
4. One pair of BeeBee Bone Scissors, ideally with a cutting-edge of approximately 15 mm.
5. A selection of Minutien pins with an ideal length of approximately 1 cm.

### 2.5 Enzymatic Single Muscle Cell Bioseparation

1. PDMS-coated Petri dishes (*see* Subheading 3.1), in which the single muscle cell bioseparation can be performed.
2. Disposable plastic transfer pipette of which the tip is cut off and fire-polished at the enlarged tip opening to prevent fiber damage by shearing them against sharp edges during trituration (*see* **Note 2**).
3. An Eppendorf tube with a capacity of 2 mL.

### 2.6 Cell Viability Assessment Through Electrical Stimulation

1. An experimental chamber (e.g. a one-well chamber slide), of which the bottom ideally can be coated with a thin layer (0.1–0.5 mm) of PDMS to avoid cell attachment (*see* **Note 3** and Subheading 3.1).
2. A muscle stimulator with adjustable electrode distance which is capable to create electric field strengths of 45 to 90 V/cm at electrode distances of approximately 1 to 2 mm and pulse durations of ~2 ms (*see* **Note 4**).

## 3 Methods

### 3.1 PDMS Coating of Petri Dishes

The prefabrication of coated Petri dishes is a preliminary step to the muscle dissection and the enzymatical muscle single fiber dissociation. Therefore, it should ideally be performed at least 2 days prior to the muscle experiments.

1. Calculate the needed volume or weight of PDMS according to the amount of desired coated Petri dishes, their diameter, and the PDMS bottom layer height of 3–4 mm.
2. Take the two-component PDMS crafting kit, and mix the base component and the cross-linker in a ratio of 10:1.

3. Once the components are homogenously mixed, pour the liquid PDMS into the Petri dishes until a layer of approximately 3–4 mm thickness covers the bottom of the dish.
4. Put a lid on the dishes to avoid entrapment of particles or contamination, and cure the layer for 48 h at room temperature (*see* **Note 5**).

### 3.2 Muscle Cell Stimulation Pen (ClickTwitch) (Optional Step If No Commercial Pulse Stimulator Is to Be Used and DOY-Approach Is Pursued)

The assembly of the *ClickTwitch* muscle cell stimulation pen requires some basic understanding of 3D printing, circuit-board soldering, and manual manufacturing. In case you encounter any difficulties, consult your local electronics or mechanical manufacturing workshop. Nevertheless, the device is designed to be assembled and operated by anyone, and the specific skill set to do so can be easily acquired along the way of manufacturing it. A detailed set of explosion drawings of the stimulator is provided in the electronic supplementary material, to even further facilitate the manufacturing and assembly process.

1. Load the '.stl'-files into the Slicer-Software to convert the 3D models into G-Code and virtually arrange the models on the printing platform of the 3D printer.
2. Print the 3D models, based on the G-Code from the Slicing-Software, using the 3D printer and the PLA filament.
3. Use the thread taps to recut the threads of all components to ensure easy turning of the mounting screws during assembly (*see* the supplementary assembly manual to identify the threaded holes).
4. Solder the circuit board according to the wiring diagram given in step 4 of the supplementary assembly manual (*see* **Note 6**).
5. Cut off the tip and the plastic fitting of the syringe cannulas, grind or cut them to a length of roughly 15 mm, and remove the burr as well as possible with a piece of fine sandpaper. These tubes will become electrode mountings.
6. Skin the ends of the electrode connection cables to a length of about 5 mm and insert the endings into the syringe cannula tubes. Subsequently, use the needle-nose pliers to create a crimp connection between the cables and the tubes (*see* steps 8–10 in the supplementary assembly manual).
7. Place the circuit board into the circuit housing and bolt on the circuit board cover. While doing so, ensure that the cable of the battery clip is guided through the gap in the sidewall of the battery cavity, and the pen connection cable runs through its desired opening in the circuit board cover (*see* step 7 in the supplementary assembly manual).
8. Place the electrode mounting tubes into their corresponding grooves in the lower part of the pen, and fix them into place

with a drop of instant glue (*see* steps 8–10 in the supplementary assembly manual).

9. Insert the push button into its desired slot in the lower part of the pen, and guide all internal cables through the cable outlet at the end of the pen (*see* steps 8–10 in the supplementary assembly manual).
10. Place the upper part of the pen on the lower half, and complete the pen assembly by bolting it onto the lower part with two countersunk head screws (*see* steps 8–10 in the supplementary assembly manual).
11. Punch two holes into the Velcro cable binder, using the hole puncher or the pair of punch pliers (*see* **Note 7**).
12. Attach the cable binder to the bottom of the circuit housing with the two M4 screws (*see* **step 12** in the supplementary assembly manual).
13. Connect the 9 V block battery to the battery clip, place it inside the battery cavity, and close the battery cover by screwing it onto the lower part of the housing with a screw (*see* step 13 in the supplementary assembly manual).
14. Cut the platinum wire to a length of your liking, bend it according to its desired application, and insert it into the electrode mounting tubes (*see* step 14 in the supplementary assembly manual and **Note 8**).
15. Insert the magnets into their desired holes in the pen-holder brackets, and fix them into place with a drop of superglue (*see* step 15 in the supplementary assembly manual).
16. The *ClickTwitch* muscle stimulation pen is now ready for operation (*see* step 16 in the supplementary assembly manual).

### 3.3 Muscle Dissection

Animal husbandry and the euthanasia of laboratory animals must be conducted in compliance with the respective local Animal Welfare laws. Following sacrifice of the animal, the hind limbs can be cut off and stored in Ringer's solution at 4 °C to preserve tissue vitality for several hours. The muscle of choice for this experimental protocol are the murine *musculi interossei plantares* (MIP) which consist of four independent muscles located plantar of the metatarsal bones.

1. Severe the hind paw from the cut-off murine hind limb right above the ankle joint using the bone cutters (*see* Fig. 2a).
2. Place the paw inside of a PDMS-coated petri dish with its plantar surface facing upwards.
3. Use the pair of tweezers and the Minutien pins to mechanically spread the paw digits and pin the paw down onto the well bottom (*see* **Note 9** and Fig. 2b).

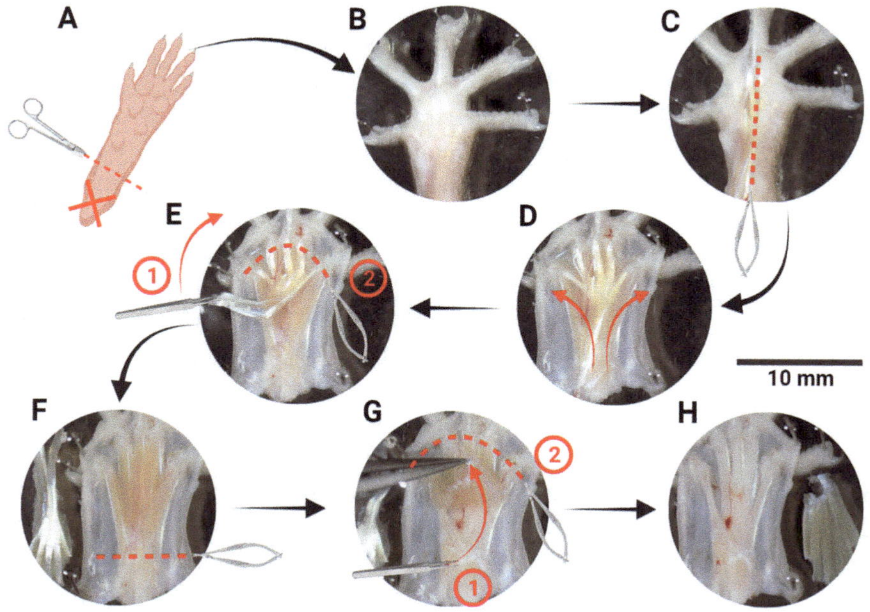


**Fig. 2** A detailed description of how to extract the *musculi interossei plantares* from a murine hind paw. After the paw is cut distally to the ankle joint and separated from the hind limb (**a**), it must be fixated onto the well bottom with its digits fanned out and its palmar surface facing up (**b**). Subsequently, the skin is opened up by a medial cut in distal direction (**c**). Once the underlying muscular plateau (FDB on top of plantar aponeurosis on top of *interossei* muscles) is exposed, the skin is spread open laterally and is fixated abaxially onto the dish bottom to facilitate access (**d**). The *flexor digitorum brevis* (FDB) including the plantar aponeurosis can now be removed by gently lifting up and pulling both in distal direction, carefully severing the underlying connective tissue and removing it by cutting the distal insertions (**e**). The *musculi interossei plantares* are now exposed (**f**) and can be removed similarly to the previous removal of upper tissue, by cutting the proximal insertions, gently pulling toward the distal end and cutting the distal muscle insertions (**g**). Due to the tight attachment of the muscles to the underlying bones, while lifting up the proximal muscles' portion, under a flat horizontal angle, the muscles need to be 'scraped off' the bone by undermining and cutting movements with the precision scissors toward the distal end. Once freed and cut off, the muscles are now ready for the enzymatic single fiber isolation (**h**). (The figure was created with BioRender.com)

4. Cut the plantar skin medially towards the distal end of the paw with a pair of micro-surgery scissors to expose the subjacent muscular layers (*see* Fig. 2c).
5. Spread the skin apart by gently pulling it sideward, severing the underlying connective tissue with the scissors as you go, and pin the skin down abaxially with Minutien pins (*see* **Note 10** and Fig. 2d).
6. The uppermost tissue layer under the skin is now the *M. flexor digitorum brevis* (FDB). Gently lift up the proximal end of the FDB tendon, including the underlying plantar aponeurosis, and pull them towards the distal end of the paw (toward the digits), while continuously cleaving the subjacent connective tissue and eventually severe the terminal tendon insertions.

This step will remove the FDB tendon including the overlying muscular plateau, thus the MIP are now visible (*see* **Note 11** and Fig. 2).

7. Severe the proximal muscle insertions with a transversal cut close to the metacarpal bones and gently pull the muscles towards the distal end of the paw while the underlying connective tissue needs to be continuously cleaved. Do this as done before during removal of the flexor tendon, this time however, with a more scraping-like cutting to free the muscle tissue from the metatarsal bones (*see* **Note 12** and Fig. 2g).
8. Severe the distal muscle insertions which are located at the metacarpophalangeal joints and remove the *musculi interossei plantares* from the paw (*see* Fig. 2h).

### 3.4 Enzymatic Single Muscle Fiber Dissociation

1. Transfer the freshly dissected *interossei* muscles into a 2 mL Eppendorf tube containing 1 mL of collagenase solution (*see* Fig. 3a).
2. Place the Eppendorf tube in a thermocycler and incubate it. In our hands, 45 min at 37 °C and 500 rpm mixing speed (*see* **Note 13** and Fig. 3b) were found to yield optimal numbers of viable fibers.
3. Gently transfer the muscles into a PDMS-coated Petri dish containing Ringer's solution using a 1000 μL air displacement pipette (*see* **Note 14** and Fig. 3c, d).
4. Use the cut-off transfer pipette (*see* **Note 2**) to carefully triturate the muscle tissue until the single muscle fibers readily detach. This procedure must be executed until a sufficient number of cells are isolated (*see* **Note 15** and Fig. 3e, f).
5. The cell suspension can now be stored at 4 °C until further usage (*see* Fig. 3g).

### 3.5 Cell Viability Assessment Through Electrical Stimulation

1. Pipette Ringer's solution into the experimental chamber to have enough fluid level for cells to sink down to the bottom.
2. Transfer approximately 30–50 μL of cell suspension into the chamber, and make sure that a sufficient number of cells was being transferred by visually examining the cells under the microscope with 6× to 8× magnification (*see* **Note 16**). By diluting the cell suspension in the fresh Ringer's volume, fiber density becomes low enough for single fibers to be approached by the relatively large stimulation electrodes.
3. Place the stimulation electrodes in the chamber, and ensure that the electrode surface is completely submerged.
4. Position the pair of electrodes around a muscle fiber of interest under visual control through the stereomicroscope, and align the axis of the parallel electrodes with the fiber axis (*see* **Note**

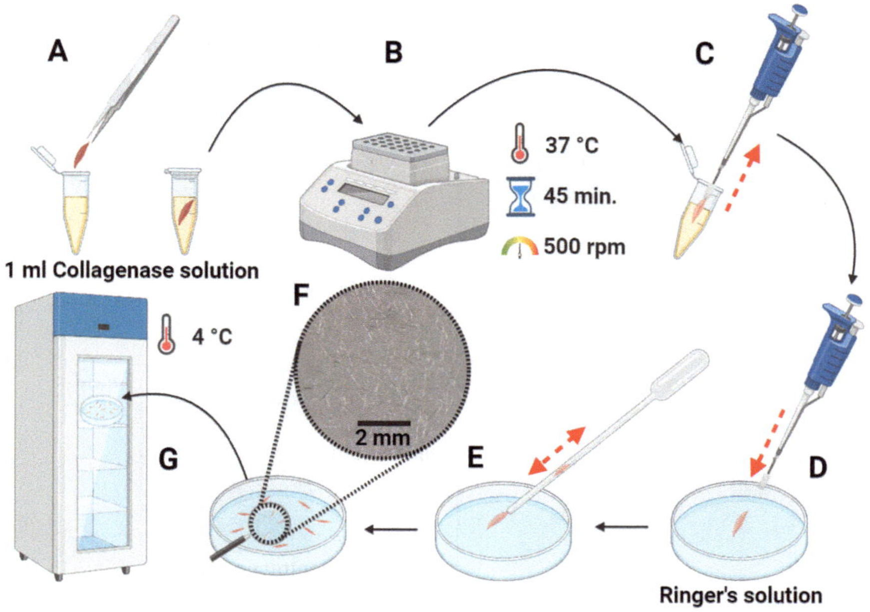


**Fig. 3** From vital muscle tissue to viable single muscle fibers through enzymatic digestion and mechanical isolation (trituration). Initially, the muscle tissue is transferred into a 2 mL Eppendorf tube containing 1 mL of collagenase solution (**a**) and is afterwards incubated for 45 min at 37 °C and 500 rpm in a thermocycler (**b**). After the incubation period, the muscle tissue is transferred via slight suction into a Petri dish filled with Ringer's solution, using a 1000 μL air displacement pipette (**c**, **d**). Next, the muscle tissue is being triturated with a transfer pipette until the single fibers become detached from the muscle through the fluid shear forces (**e**, **f**). The cell suspension can now be stored at 4 °C until further use (**g**). (The figure was created with BioRender.com)

17). The design of the stimulation pen being held between the thumb and the index fingers should allow very fine movements of the electrode tips under visual control.

5. Slowly reduce the electrode distance as much as possible (approximately 1–2 mm), but assure that there is no physical contact between the electrodes and the muscle cell.
6. Apply a square unipolar rectangular pulse with an amplitude of 9 V and a pulse duration of 2 ms (*see* **Note 18**). An electrode distance between 1 and 2 mm at a pulse amplitude of 9 V accounts for an electrical field strength between 90 and 45 V/cm.
7. If the applied electrical pulse results in a twitching movement of the fiber (*see* Fig. 4), it is obvious that the electrical excitability, and thus, the cell membrane integrity was not damaged during the enzymatic isolation and, therefore, the cell is viable and can be used for further downstream experiments. If no twitch is observed, dispose of the cell and repeat the test with a different muscle fiber (*see* **Note 19**).

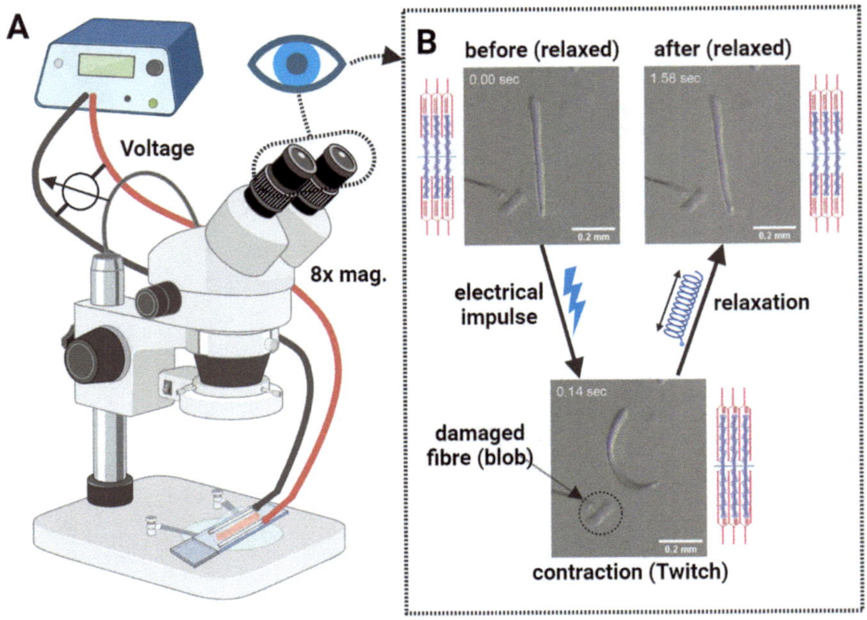


**Fig. 4** Cell viability assessment of a single muscle fiber through electric field stimulation. The viability assessment of single muscle fibers through electric field stimulation should be performed under a laboratory stereo microscope with sufficient magnification (6× or higher) and incident light illumination. Ringer's solution and a batch of cell suspension are transferred into a volumetric experimental chamber, placed under the stereomicroscope, and the electrodes are positioned around a fiber of interest under visual inspection through the stereomicroscope (**a**). Once the electrodes and the sample are adjusted and positioned, a unipolar rectangular direct current pulse (e.g., 9 V amplitude and 2 ms duration) is applied, which will result in a twitch, in case the muscle fiber was viable (**b**). (The figure was created with BioRender.com)

## 4 Notes

1. The collagenase solution is used for the enzymatic separation of single muscle fibers out of a vital murine skeletal muscle. The basis of this solution is Ringer's solution (*see* Subheading 2.2) to which then 1.07 mg/mL Collagenase Type 1 CLS 1 is added for collagen digestion. The concentration of the collagenase is based on an enzyme with an average enzymatic activity of 200 CDU/mL. In contrast to the standardized Ringer's solution, this solution should not be prepared in large quantities, as storage periods longer than 1 month at −20 °C could lead to declining enzymatic activity. This also applies to repeated freeze-thaw cycles, which should be avoided.
2. Cut off the tip of a transfer pipette right at the enlarged tip opening. The inlet of the pipette should now have an inner diameter of about 3.5 to 4 mm. To eliminate the risk of sharp-cutting edges and accompanying tissue or muscle fiber damage, the inlet must be fire-polished using a lighter. To do so, swipe

the tip of the pipette through the flame repeatedly until a visible slight constriction is building up at the inlet, and the edges are noticeably rounded off.

3. If PDMS coating is not possible, we found it to be useful to reuse experimental chambers several times over a series of experiments and to not clean them with anything else but ultrapure water. The slight residual deposits which would build up over time, tend to prevent fiber attachment to the coverslip. Therefore, the agitation and handling of the fibers were noticeably facilitated, and, in particular, fibers not sticking to the substrate will experience less trauma during the vigorous contraction upon electrical stimulation.
4. A muscle stimulation device exhibiting the required electrical parameters is provided in the electronic supplementary material, as a do-it-yourself production guide. This device is a self-developed handheld stimulation pen that allows for the evaluation of electrical excitability of muscle cells while simply incorporating off-the-shelf electronic parts and a housing that can be manufactured through FDM (Fused Deposition Modeling) 3D printing.
5. To achieve optimal results without entrapment of air bubbles inside the PDMS layer and in case the curation process of the chosen PDMS kit can be performed at room temperature, the curing procedure should be executed inside a desiccator under vacuum conditions. Additionally, the dishes should be covered with a lid during the entire curing procedure.
6. Keep in mind that the wire connection between the electrode and the push button requires some slack (approximately 20 cm), as these parts will be positioned inside the pen. Furthermore, the connection cables between the circuit housing and the pen can be covered with heat shrink tubing to give them the appearance of a single cable. To do so, make sure that the tubing is being applied before the push button is soldered onto the cables. The integrated LED serves as a functionality indicator of the circuit.
7. The positions of the punched holes along the length of the strap are variable, but they must be centered on the longitudinal axis and the hole spacing should not be less than 39 cm.
8. The shaft of the electrode wires can be bent to a slight curvature to achieve a self-locking effect inside of the electrode mounting tubes.
9. To pin the paw down securely or for proper fixation throughout the preparation procedure, it is recommended to pin down every digit individually and to apply light strain to the skin by stretching and positioning the digits in a star-shaped manner. In our hands, needle-pinning the first, third, and fifth digit is sufficient for secure fixation.

The Minutien pins can be pierced through the distal pad of the digits while pulling the digit toward the distal direction and pinning it down by inserting the pin into the PDMS layer of the Petri dish (*see* Fig. 2a). It is advantageous to punch the pin coming from the far end of the digit and penetrate the tissue under a flat angle towards the paw. That way, all needles will point away from the paw to maximize the preparation area and to avoid collision with the pins during ongoing dissection. Punching the pins vertically through the tissue is disadvantageous as it will lead to three to five needle pillars sticking out of the dish that interfere with the ongoing dissection. Flat and pointing away tissue punch is, therefore, the best way to maintain a widely accessible dissection area for scissors and tweezer handling.

10. To increase access to the subjacent muscular plateau, the skin can be removed completely at will.
11. The paw should be cut off from the hind limb distally to the ankle joint, as this will already sever the flexor tendon close to its proximal insertion and, therefore, facilitate manual access to its loose end.
12. Only grab the muscle at its tendons, to avoid muscular damage. To efficiently remove the underlying connective tissue during muscle extraction, perform a swiping motion with the scissors on top of the metatarsal bones to scrape off the muscle tissue with undermining sweep-and-cut movements.
13. The enzymatic activity of the collagenase solution and the muscles' content of connective tissue might slightly differ throughout several experiments, because of fabrication tolerances and phenotypical differences between laboratory animals. Therefore, an optimization between incubation duration and incubation temperature has to be found in a range of 30–50 min and 30 to 37 °C.
14. For the transfer, the muscle should not be sucked into the pipette tip completely but attached to it through slight suction. This step avoids unnecessary shear stress to the tissue and limits the contamination of the Ringer's solution with collagenase to a minimum, as the fluid transfer is limited to the excess volume on the muscle surface.
15. Avoid unnecessary shear stress by shearing the muscle between the PDMS coating of the petri dish and the opening of the pipette. Keep the pipette positioned at a flat angle to the dish bottom and triturate slowly and steadily for several minutes. As the Ringer's solution contains $Ca^{2+}$-ions, fibers whose sarcolemma was damaged during trituration would instantly exhibit irreversible muscle contracture, which is an advantage as those show up as debris. Only nicely tube- or banana-shaped single fibers will be likely to be viable.

16. Before transferring a batch of cell suspension into the test chamber, try to visually distinguish cell accumulations that seem vital and relaxed using the microscope with high magnification. Two visual indicators for cell viability are the shimmering gloss and transparency of the single fibers and a fiber length of approximately 0.5–1 mm. Fibers that are already contracted appear as rather spherical, and opaque blobs or milky rods and can be ignored (*see* Fig. 4).
17. Agitation of isolated single fibers should not be performed using metal tweezers, as this could lead to fiber contraction (mechanical stimulus leading to depolarization or membrane damage with $Ca^{2+}$ influx). A 20 μL pipette tip whose tip is cut off diagonally can be used instead. Another possibility is to coat the tips of a pair of tweezers with PDMS.
18. The given electrical parameters can be seen as suggested values which originate from the electrical capabilities of the provided stimulation pen. Those create electric field strengths of 45–90 V/cm at electrode distances of approximately one to 2 mm. In contrast to the non-adjustable voltage source of the stimulation pen, commercially available systems mostly allow for voltage and pulse duration configuration. Therefore, other combinations with regards to electrode distance, voltage, and pulse durations are possible. For example, in single 'mechanically skinned' fibers, electric field strength thresholds for the contractile response have been reported from ~20 V/cm, and robust contractions were elicited from 50 V/cm [10].
19. As shown in Fig. 4, muscle fibers do not always contract linearly along the fiber axis during electric field stimulation, they may rather exhibit a significant curvature during the vigorous, unloaded contraction (*see* [8]). If the cells are supposed to be used for quantitative assessment of fiber-shortening parameters, this curvature must be considered, according to the instructions given by Friedrich et al. [8]. We found out that the number of bending fibers was less when using PDMS-coated or reused experimental chambers (*see* Subheading 3.1 and **Note 3**).

## References


1. Jensen EC (2012) Use of fluorescent probes: their effect on cell biology and limitations. Anat Rec 295:2031–2036
2. Ramirez CN, Antczak C, Djaballah H (2010) Cell viability assessment: toward content-rich platforms. Expert Opin Drug Discov 5(3): 223–233
3. Chan LLY, Rice WL, Qiu J (2020) Observation and quantification of the morphological effect of trypan blue rupturing dead or dying cells. PLoS One 15(1):e0227950
4. Berchtold MW, Brinkmeier H, Müntener M (2000) Calcium ion in skeletal muscle: its crucial role for muscle function, plasticity, and disease. Physiol Rev 80(3):1215–1265
5. Allard B (2018) From excitation to intracellular Ca 2+ movements in skeletal muscle: basic

aspects and related clinical disorders. Neuromuscul Disord 28(5):394–401

6. Friedrich O, Reid MB, Van den Berghe G, Vanhorebeek I, Hermanns G, Rich MM, Larsson L (2015) The sick and the weak: neuropathies/myopathies in the critically ill. Physiol Rev 95:1025–1119

7. Lamb GD, Stephenson DG (2018) Measurement of force and calcium release using mechanically skinned fibres from mammalian skeletal muscle. J Appl Physiol 125:1105–1127

8. Friedrich O, Weber C, Wegner F, Chamberlain JS, Fink RHA (2008) Unloaded speed of shortening in voltage-clamped intact skeletal muscle fibres from wt, *mdx* and transgenic minidystrophin mice using a novel high-speed acquisition system. Biophys J 94:4751–4765

9. Head SI, Chan S, Houweling PJ, Quinlan KGR, Murphy R, Wagner S, Friedrich O, North KN (2015) Altered $Ca^{2+}$ kinetics associated with α-actinin-3 deficiency may explain positive selection for ACTN3 null allele human evolution. PLoS Genet 11(2):e1004862

10. Posterino GS, Lamb GD, Stephenson DG (2000) Twitch and tetanic force responses and longitudinal propagation of action potentials in skinned skeletal muscle fibres of the rat. J Physiol 527(1):131–137

11. Marš T, Miš K, Meznarič M, Prpar Mihevc S, Jan V, Haugen F, Rogelj B, Rustan AC, Thoresen GH, Pirkmajer S, Nikolić N (2021) Innervation and electrical pulse stimulation - in vitro effects on human skeletal muscle cells. Appl Physiol Nutr Metab 46(4):299–308

12. Goodall MH, Ward CW, Pratt SJP, Bloch RJ, Lovering RM (2012) Structural and functional evaluation of branched myofibers lacking intermediate filaments. Am J Physiol Cell Physiol 303:C224–C232

13. Schmitt MA, Friedrich O, Gilbert DF (2019) Portoporator©: a portable low-cost electroporation device for gene transfer to cultured cells in biotechnology, biomedical research and education. Biosens Bioelectron 131:95–103

# Chapter 12

# Evaluating Cellular Viability by iTRAQ Proteomic Profiling

Anne Poljak, Mark Raftery, and Patsie Polly

## Abstract

Cellular health, functionality, response to environment, and other variables affecting cell, tissue, or organ viability are reflected in the cellular proteomes and metabolomes. These "omic" profiles are in constant flux even during normal cellular functioning, to maintain cellular homeostasis, in response to small environmental changes and maintenance of optimal cell viability. However proteomic "fingerprints" can also provide insight into cellular ageing, response to disease, adjustment to environmental changes, and other variables that impact cellular viability. A variety of proteomic methods can be used to determine qualitative and quantitative proteomic change. In this chapter, we will focus on a labeling method called isobaric tags for relative and absolute quantification (iTRAQ), which is frequently used to identify and quantify proteomic expression changes in cells and tissues.

**Key words** Isobaric tags for relative and absolute quantification (iTRAQ), Proteomics, Mass spectrometry, Cell viability

## 1 Introduction

Cell viability is a measure of overall cellular or tissue health and is frequently examined in cell culture assays where the effects on cells of specific treatments are to be assessed. Cell viability may improve or deteriorate. A variety of assays can provide a measure of cell viability, including (1) permeability of the cell membrane assessed by the trypan blue dye uptake assay [1], (2) release of intracellular protein, such as lactate dehydrogenase (LDH), into the extracellular media [2–4], which can be taken as an indication of compromised cell viability or cell cytotoxicity, (3) alteration of cellular metabolism can be assessed by MTT (3-(4,5-Dimethylthiazol-2-yl)-2,5-diphenyltetrazolium bromide, a tetrazole) assay [5, 6], which may be an indicator of adversely affected cellular health, as a variety of disease states can impact cellular metabolic state, and similarly (4) cellular respiration can be assessed by Seahorse assay [7, 8]. Most of these assays give an idea of severe or late-stage effects on cells and/or tissues. Earlier and more subtle effects on

Oliver Friedrich and Daniel F. Gilbert (eds.), *Cell Viability Assays: Methods and Protocols*, Methods in Molecular Biology, vol. 2644, https://doi.org/10.1007/978-1-0716-3052-5_12,

cell viability can be assessed using "omic" techniques to assay cellular metabolites and/or biopolymers, as with proteomic profiling [9, 10]. Changes to proteomic expression profiles are highly dynamic, detectable even with relatively mild effects on cells, and can be a sensitive indicator of cellular impact, well before gross effects on cell integrity, such as compromised cell membranes, are observed.

Cell viability can be affected by diverse endogenous and exogenous stimuli, including disease, changed physiological state (e.g., diet, exercise, stress, etc.), exposure to toxins, or other stressors. Proteomics approaches can be used qualitatively, to determine which proteins have altered expression in response to specific stimuli, and also quantitatively to determine extent of expression change, expressed either in absolute terms (e.g., μg/mL) or in relative terms, such as fold-change relative to controls. Diverse sample and tissue types are amenable to proteomic analysis, including cultured cells, organelles or organoids [11–13], tissues [14], or body fluids such as serum, plasma, and CSF [9, 15].

Many different proteomic workflows exist. The classical 2D SDS-PAGE gel separation workflow is based on protein quantitative differences that are determined off-line based on 2D SDS-PAGE spot quantification, which are followed by mass spectrometric identification of the protein(s) within the spot [11, 16, 17]. More commonly, mass spectrometry is used in a MudPit approach, for both quantification and protein identification. In this approach, the entire cell or tissue extracted proteome is enzymatically digested (generally with trypsin), processed, and analyzed. Many different sample preparation approaches are available for MudPit workflows, including label-free approaches [10], labeling samples with tags such as Tandem Mass Tag (TMT) or isobaric Tag for Relative and Absolute Quantitation iTRAQ [9]. In addition, there are multiple approaches to characterize post-translational modifications, such as phosphopeptide enrichment [18–20] or to track in vivo protein–protein interactions using approaches such as BioID tagging [21]. Each of these techniques has specific advantages and limitations, and the experimental design and questions to be addressed will guide the choice of the sample preparation technique. For example, (1) the TMT approach utilizes up to 18 tags along with the need for ultra-high resolution Orbitrap instruments, which can be important in experimental designs with a higher number of sample categories, since the maximum number of tags in iTRAQ is eight. TMT kits, including the 18-plex TMT-Pro kit, are available, for example, from Thermo Fisher Scientific, (2) the number of tags available can be a limitation for all labeling approaches, when sample numbers are large (i.e., >18 samples), and in this case a label-free approach may be needed. Tribred Orbitrap instruments, such as the Lumos and Eclipse, offer the possibility to perform MS3 isolation of reporter ions allowing improved accuracy and precision for quantitative proteomics [22].

Most importantly, the analysis of the liquid chromatography tandem mass spectrometry LCMSMS data is reliant on availability of software capable of managing each specific workflow approach. A variety of public domain and commercial software is in common use. This software includes Mascot (MatrixScience), MaxQuant (freeware; [23]), Proteome Discoverer (Thermo Fisher Scientific Inc.), Scaffold (Proteome Software Inc.), Peaks (Bioinformatics Solutions Inc.), Progenesis QI (Nonlinear Dynamics, Waters Corp.), and ProteinPilot (AB Sciex LLC), to name a few. These different types of software have the capability to process mass spectrometry data to identify peptides based on LCMSMS raw data. For experiments using labels, such as iTRAQ tags, various types of software can also integrate peak areas of the labels to provide quantitative information and are often able to provide some basic statistics to compare protein expression levels across samples/biological replicates.

For untargeted data-dependent quantitative analysis of complex sample mixtures, labeling approaches may be a more robust approach for quantification. This is because multiple peptides from each protein component are independently labeled with the tags, which are all independently quantified to determine the average expression change in control versus test samples. Since multiple peptides from the same protein are quantified, there are multiple measures of expression change for each protein. The average for each set of peptides provides a confident indicator of degree of fold-change relative to comparison samples (such as controls), similarity of fold-change across technical and biological replicates, and differences between "treatment" conditions to name a few parameters.

In this chapter, we describe a commonly used workflow for preparing, processing, and analyzing iTRAQ-labeled samples (For schematic of complete workflow, see Fig. 1).

## 2 Materials

### 2.1 Solutions and Reagents

1. Dissolving buffer: 300 μL MilliQ, 30 μL 500 mM sodium bicarbonate (2.1 g $NaHCO_3$/50 mL = 500 mM), 15 μL 2% SDS (available in the iTRAQ 8-plex kit, or else can be prepared separately).
2. Reducing agent TCEP (Tris-(2-carboxyethyl) phosphine, e.g., in the iTRAQ test kit, see Subheading 2.2), 50 mM. The kit only has a single vial, so if it runs out, it can also be prepared separately (*see* **Note 1**).
3. Alkylating reagent iodoacetamide: up to 200 mM (e.g., 37 mg IAA/1 mL 50 mM ammonium bicarbonate).

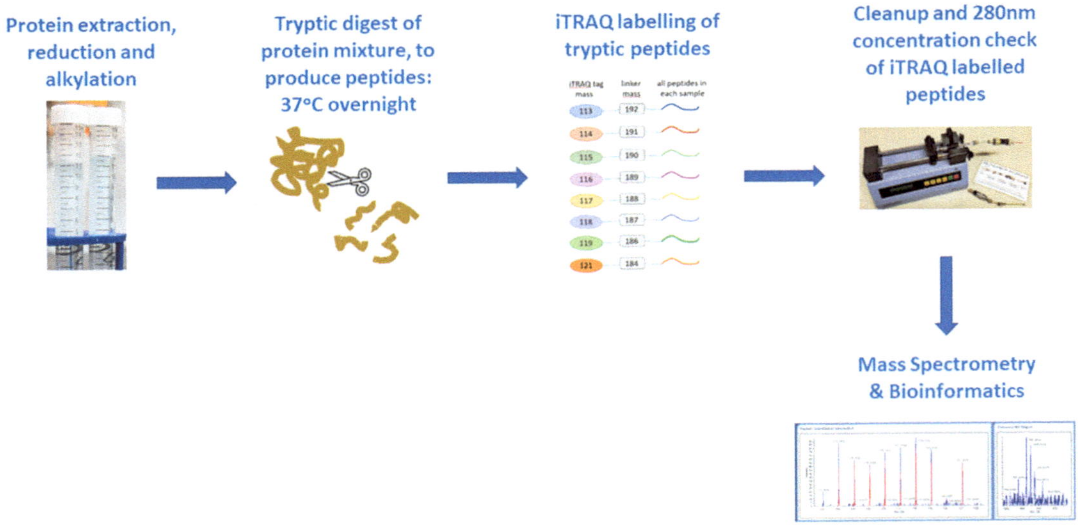


**Fig. 1** Schematic of iTRAQ workflow, showing steps from sample workup through to bioinformatics analysis

4. iTRAQ sample cation exchange load buffer: 10 mM potassium phosphate $KH_2PO_4$ in 25% acetonitrile, pH 3.0. For 100 mL volume: 0.136 g $KH_2PO_4$, 75 mL MilliQ, 25 mL $CH_3CN$, pH 3 with ~150 μL neat HCL. Store at 4 °C until use.
5. iTRAQ sample cation exchange elute buffer: 10 mM potassium phosphate $KH_2PO_4$ in 25% acetonitrile and 350 mM KCl, pH 3.0. For 100 mL volume: 0.136 g $KH_2PO_4$, 2.33 g/ 100 mL KCl, 75 mL MilliQ, 25 mL $CH_3CN$, pH adjusted to 3 with ~150 μL neat HCl. Store at 4 °C until use.
6. iTRAQ sample cation exchange clean buffer: 10 mM $KH_2PO_4$ in 25% acetonitrile and 1 M KCl, at pH 3.0. For 100 mL: 0.136 g $KH_2PO_4$, 6.66 g/100 mL KCl, 75 mL MilliQ, 25 mL $CH_3CN$, pH adjusted to 3 with ~150 μL neat HCl. Store at 4 °C until use.
7. iTRAQ sample cation exchange storage buffer: 10 mM $KH_2PO_4$ in 25% acetonitrile, pH 3.0 + 0.1% sodium azide ($NaN_3$). For 100 mL volume: 0.136 g $KH_2PO_4$, 0.1 g $NaN_3$, 75 mL MilliQ purified water, 25 mL $CH_3CN$. pH ~2–3 by adding ~150 μL neat HCl, check using either a pH meter or 0–14 pH strips. Keep adjusting buffer until it is at low pH of ~2–3. Store at 4 °C until use.
8. Sequencing grade trypsin (Promega Corp., Madison, WI) (*see* **Note 2**).

### 2.2 Test Kits

Test kits for performing iTRAQ labeling can be purchased from AB Sciex LLC (Framingham, MA), as either 4-plex or 8-plex kits.

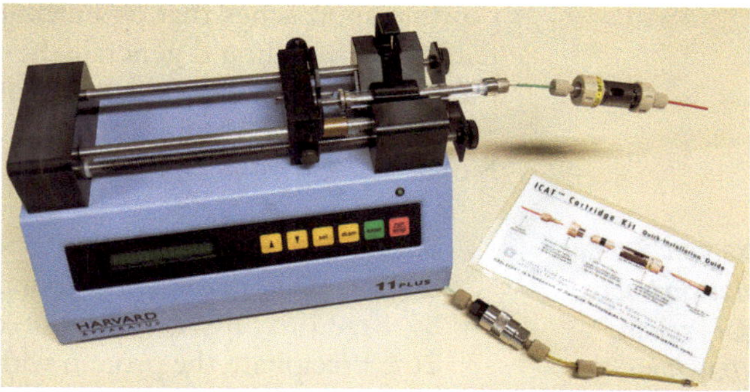

**Fig. 2** Syringe pump, with the iTRAQ macrotrap and housing for sample clean-up installed

*2.3 General Equipment and Software*

1. ICAT accessories kit: this can be kept in a 4 °C storage location, such as a fridge or walk in cold room, and consists of ICAT cartridge – cation exchange, containing 200 μL POROS 50 HS, 50 μm particle size (4.0 mm × 15 mm; Applied Biosystems/Thermo Fisher Scientific Inc., Waltham, MA), cation exchange buffers (as detailed in Subheading 2.1 above), peptide cleanup cartridge (Macrotrap, peptide, Bruker Corp., Billerica, MA).
2. Syringe pump: this is used for sample cleanup, to keep flow rate slow and consistent, and is used at a flow rate of 9.5 mL/h for the steps described in this method (see Fig. 2).
3. Centrifugal vacuum dryer: such as a SpeedVac (Thermo Fisher Scientific Inc., Waltham, MA) for sample drying and concentration.
4. Nano-spectrophotometer with the capability of taking readings at 280 nm (such as a NanoDrop or DeNovix).
5. Mass Spectrometer: such as a QStar ELITE quad TOF hybrid mass spectrometer (AB Sciex LLC, Framingham, MA) (*see* also **Note 3**).
6. Software with the capability of integrating and comparing iTRAQ data is needed, such as the Protein Pilot Software from AB Sciex LLC (Framingham, MA). Other software with the capability of processing iTRAQ tag data include Scaffold (Proteome Software Inc.), Mascot (MatrixScience), and Proteome Discoverer (Thermo Fisher Scientific Inc.) (*see* **Note 4**).

## 3 Methods

General introductory notes relating to the iTRAQ procedure: A wide range of sample extraction and total protein assay protocols exist, and the protocol provided below is just one example of many possible approaches that could be utilized. However, amine-containing buffers should generally be avoided, as they can react with the iTRAQ tags, which label protein amines, so reaction with amine-containing buffer components should be avoided (*see* **Notes 5** and **6**).

### 3.1 Example Protocol for Sample Extraction and Total Protein Assay (Day One)

1. Start with 100 μg/100 μL total protein (e.g., whole cell extract, CSF, plasma low abundance proteins, etc.) as assayed by Bradford, BCA, Lowry or other total protein assay [24–26]. Precipitate the protein with nine volumes of cold acetone (leave in −20 °C room for 1–2 h). Also refer to **Notes 7–9**.
2. Spin down the protein pellet (5 min, 10,410 g, ambient temp) and decant and discard the acetone. Allow the pellets to air dry for ~5 min.
3. Make up a solution of dissolving buffer (*see* **Note 10**). Add 20 μL of the dissolving buffer to each sample. Vortex each sample continuously for ~5 min, then sonicate each for ~2–3 min.
4. Add 2 μL of reducing agent (TCEP = Tris-(2-carboxyethyl) phosphine), which is located in the iTRAQ kit and has a concentration of 50 mM) to each sample and incubate at 60 °C for 60 min (*see* **Note 1**).
5. Add 1 μL of iodoacetamide to each sample, vortex, spin briefly in a microfuge, and incubate at ambient temperature for 10 min (*see* **Note 11**).
6. Digest with trypsin as follows: reconstitute one vial of 20 μg sequencing grade trypsin (e.g., Promega Corp., Madison, WI) in 45 μL MilliQ, and allow to sit for 2–3 min at ambient temperature to dissolve. Add 10 μL to each sample (may need more vials, if have more than four samples), and incubate overnight at 37 °C (12–16 h).
7. There should now be 100 μg total protein/103 μL (or 100 μg total protein/33 μL in the case of acetone precipitation) for each sample (i.e., 0.97–3.03 μg/μL).

### 3.2 iTRAQ 8-Plex Labeling Followed by Cation Exchange Cleanup of Sample (Day Two)

1. Spin the digested sample down to bottom of tube (brief spin in microfuge, just to collect all droplets at the bottom of the tube). Check pH with a pH strip, testing (take out just the very minimum to check pH on the strip <0.5 μL). The pH needs to have a basic pH for the labeling to work. If it is not basic, then adjust the pH by adding 2.5–4 μL of freshly prepared sodium carbonate (500 mM $Na_2CO_3$, 0.053 g/mL) until the pH is basic (*see* **Note 12**).
2. Remove the required number of iTRAQ reagents (usually stored at −20 °C), and allow them to reach ambient temperature. Spin briefly to bring solution droplets to the bottom of

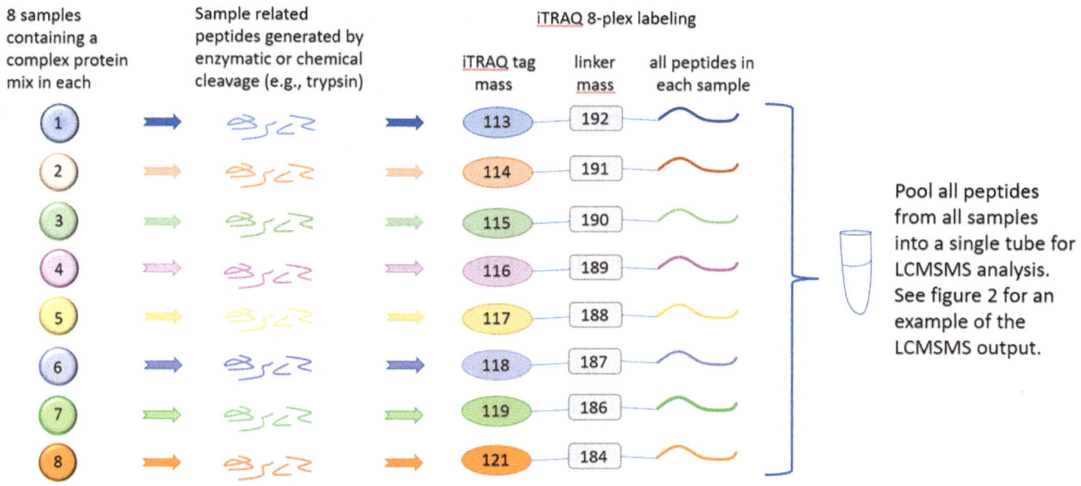


**Fig. 3** Schematic of the iTRAQ-labeling procedure. This schematic is an author created drawing, based on ABSciex technical literature and procedural notes (https:/sciex.com/products/consumables/itraq-reagent and https:/sciex.com/content/dam/SCIEX/pdf/brochures/mass-spectrometry-4375249C.pdf)

the tube. To each iTRAQ reagent, add 50 μL of isopropanol (also called; isopropyl alcohol, 2-propanol, secondary propyl alcohol, dimethyl carbinol or petrohol). Vortex and spin briefly, then transfer the whole contents of one iTRAQ reagent vial to one sample tube. Vortex, spin briefly, and incubate at ambient temperature for 1 h (see Fig. 3).

3. Combine the contents of each iTRAQ reagent-labeled sample tube into one tube, then vortex and spin briefly. Collect the ICAT accessories kit from the 4 °C storage location (i.e., cation exchange cartridge and cartridge holder, cation exchange buffers).
4. Clean-up the sample using a cation exchange cartridge as follows: set up (condition) the syringe pump (9.5 mL/h) by passing 1 mL of cation exchange "clean" buffer through it (to waste), followed by 2 mL of cation exchange "load" buffer (to waste).
5. Reduce the concentration of buffer salts and organics by diluting the sample mixture ten-fold with cation exchange "load" buffer (i.e., needs to be <10 mM salt conc). Vortex and check pH. The pH should be between 2.5 and 3.3 pH, and if not, then add some neat acetic acid until it reaches the correct pH (e.g., 400 μL sample + 4 mL cation exchange load buffer + 25 μL neat acetic acid).
6. Slowly load the diluted sample onto the cation exchange cartridge (9.5 mL/h using the syringe pump) and collect the eluate in a sample tube. For a sample volume of several mL, this step is repeated till all of the sample volume has been passed through the cartridge.

7. Inject 1 mL of cation exchange "load" buffer to wash the TCEP, SDS, calcium chloride, and excess iTRAQ reagents from the cartridge. Collect the eluate in the same sample tube as in **step 6**. Keep this sample until you verify by LCMSMS that the labeling and clean-up were successful.
8. To elute the peptides from the cation exchange cartridge, slowly inject 500 μL of cation exchange "elute" buffer, and capture the eluent into a clean 1.5 mL polypropylene tube as a single fraction.
9. Wash the undigested proteins such as trypsin from the cation-exchange cartridge by injecting 1 mL of the cation exchange "clean" buffer (divert to waste).
10. After cleaning the cartridge as in **step 9**, inject 2 mL of the cation exchange "storage" buffer. Remove the cartridge, seal the ends with the two end caps, and store at 2–8 °C (*see* **Note 13**). Clean the needle port adapter, outlet connector, and syringe with MilliQ water.

### 3.3 C18 Cleanup and Checking by MSMS to Ensure that the Labeling Has Worked

1. Dry the 500 μL of cation exchange eluant in the SpeedVac (may not be completely dry, but try to have <50 μL volume remaining), then resuspend in 500 μL of 0.2% aqueous heptafluorobutyric acid (HFBA). There is a considerable amount of salt in this sample, deriving from the elute buffer (see Subheading 3.3, **step 8**). The current C18 clean-up is performed to remove these salts, and prepare a clean peptide mixture.
2. Using the syringe pump (9.5 mL/h), prime the C18 cartridge with 800 μL–1 mL neat acetonitrile ($CH_3CN$). Use the green macrotrap from Microm.
3. Wash with 800 μL–1 mL, 50% $CH_3CN$/0.1% formic acid in MilliQ water.
4. Wash with 800 μL–1 mL, 0.2% HFBA in MilliQ water.
5. Load sample onto C18 cartridge, 9.5 mL/h (collect the load eluant—just in case of any sample losses, or if Oasis cartridge is to be used—until after MSMS check) (*see* **Note 14**).
6. Wash with 1 mL of 0.2% HFBA/MilliQ (collect the eluant—just in case of any sample losses, or if the Oasis cartridge is to be used—until after MSMS check) (*see* **Note 14**).
7. Elute with 500 μL 0.1% formic acid/50% $CH_3CN$/MilliQ. Elute with additional 200 μL neat $CH_3CN$, pooling the two eluants as a 700 μL total volume in a clean 1.5 mL polypropylene tube.

8. Dry 75 μL of the total 700 μL eluant in the SpeedVac (store the remaining volume at -80 °C either wet or dry). Once completely dry, add 25 μL of 0.05% HFBA/1.0% formic acid to dissolve. Make a 1:40 dilution by adding 1 μL sample to 39 μL 0.05% HFBA/1.0% formic and load 1 μL into the mass spectrometer (e.g., QStar Elite, 5600 hybrid Qtof system, etc.) as a test run to check on peptide abundance and iTRAQ labeling (this should be no more than 0.3–0.8 μg/1 μL on-column).
9. Mass Spectrometry of iTRAQ-labeled sample (*see* **Note 15**).

### 3.4 Tandem Mass Spectrometry of iTRAQ-Labeled Peptides Using 2D LCMSMS

1. Peptides are eluted from the cation exchange capture cartridge in a series of elution steps of progressively increasing salt concentration (e.g., multiple steps of increasing ammonium acetate concentration; 0 mM = initial sample load step, 5 mM, 10 mM, 15 mM, 20 mM, 25 mM, 30 mM, 40 mM, 50 mM, 100 mM, 250 mM, 500 mM, 1000 mM).
2. Eluted peptides are then captured onto a C18 capture cartridge, where the salt solution is washed off, and then peptides fractionated on a gradient of organic solvent run on the C18 capillary column (*see* **Note 16**).
3. If a higher duty cycle mass spectrometer is available, 2D LCMSMS can be replaced with the shorter procedure of 1D LCMSMS using a 4 h gradient (*see* **Note 17**).

### 3.5 Data Analysis: ProteinPilot

A variety of commercial and public domain software options are able to process iTRAQ data (*see* **Note 4** and Figs. 4 and 5).

## 4 Notes

1. The reducing agent TCEP (Tris-(2-carboxyethyl) phosphine, 50 mM) is provided in the iTRAQ kit; however, if it runs out before all iTRAQ reagents are consumed, it can also be prepared separately from a commercial source such as Fluka, Sigma, and so forth. In this case, it should also be made up to a concentration of 50 mM either in MilliQ or a weak buffer such as 50 mM ammonium bicarbonate. The advantage of TCEP as a reducing agent is that it has a wider working pH range than sulfhydryl containing reducing agents such as DDT or β-mercaptoethanol, which are effective at neutral to slightly basic pH. TCEP is effective as a reducing agent between pH 1.5 and 9, but can be acidic, especially when made up with water and pH should, nevertheless, be checked since subsequent steps such as IAA and trypsin work better at neutral to slightly alkaline pH.

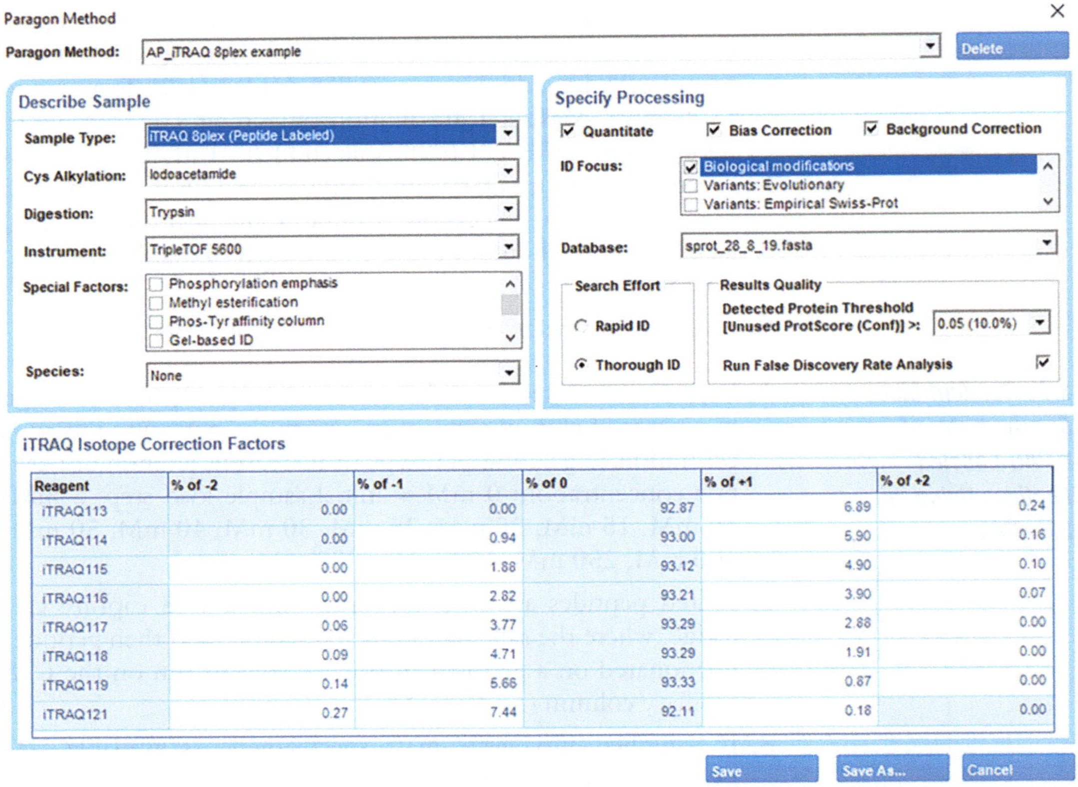


**Fig. 4** ProteinPilot v4 identification and quantification method setup options. The panels in this figure are taken from a copy of ProteinPilot v4 software licensed to the UNSW Bioanalytical Mass Spectrometry Facility

2. Sequencing grade trypsin is available from a number of commercial sources. A commonly used trypsin is available from Promega Corp. (Madison, WI) and is a modified trypsin sequence with the advantage of being minimally affected by autolysis. In this case, relatively few trypsin autolysis peaks are identified following LCMSMS, hence, minimizing detection of non-sample-related peptides.

3. Mass Spectrometer: Many different mass spectrometers with LCMSMS capabilities can be used. The procedures described here utilize a QStar ELITE quad TOF hybrid mass spectrometer (AB Sciex LLC, Framingham, MA). Consequently, to improve detection of higher protein/peptide numbers, a 2D LCMSMS approach can be used on such a model. Newer models of hybrid quad TOF mass spectrometer are now available (including 5600, 6600 TripleTOF, ZenoTOF 7600) and Orbitraps (ThermoFisher Scientific), and on such higher scanning/sensitivity/duty cycle equipment, a 1D UPLC-MSMS approach can be used, with longer columns and gradient, to improve peptide LC separation (e.g., 1.9 micron particles, 50 cm column, and 4 h gradient). Whether a newer or older

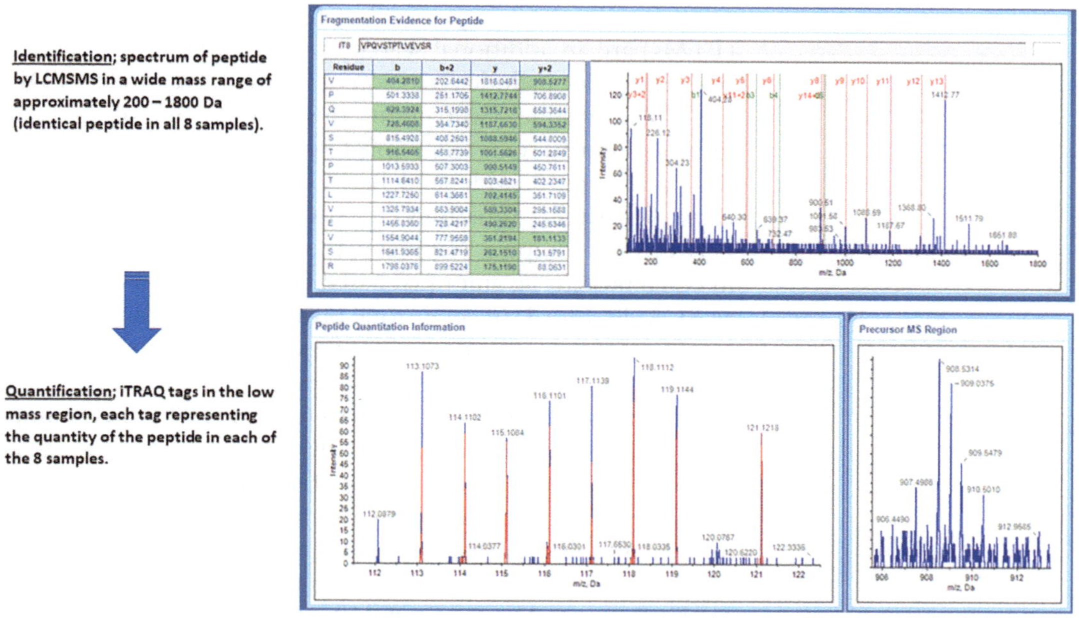


**Fig. 5** Spectrum of an iTRAQ-labeled peptide, showing MSMS peptide sequence and the spectrum of the iTRAQ 8-plex tags in the low mass range. The panels in this figure are representative data on a peptide from an 8-plex iTRAQ experiment carried out in the UNSW Bioanalytical Mass Spectrometry Facility, and processed using ProteinPilot v4

model of mass spectrometer is available, an important point is that the instrument used should have good sensitivity for the iTRAQ tags in the low mass range (*m/z* 100–150) and the higher mass range, which contains peptide sequencing ions (*m/z* 350–2000).

4. The ProteinPilot software has evolved over the last two decades with multiple versions and subversions issued from V1 to V5.0.2 (released in 2018), with new features and options available with each new version. For example, versions prior to V3 did not include background and bias correction. The version described in this chapter is based on the options available in ProteinPilot V4 (Fig. 4).

   For WIFF file types obtained on ABSciex instrumentation, the ProteinPilot software was the original platform developed for iTRAQ sample analysis (http://sciex.com/support/software-downloads) and remains one of the most powerful, though relatively easy to use, options for iTRAQ data [27]. ProteinPilot provides detailed output on the data, with the processed results providing detailed qualitative information at the protein, peptide, and spectral levels and protein quantification based on relative comparisons of integrated peak areas of the iTRAQ tags (https://sciex.com/content/dam/SCIEX/pdf/tech-notes/all/ProteinPilot-Report-TN-RUO-MKT-02-

1778-B.pdf). See Fig. 5. Post-translational modifications (PTMs) are an additional search option. Some of the features of ProteinPilot include provision of false discovery rates (FDR) [28]. Background Correction: this feature eliminates background contributions from non-target peptides, which coelute with the target peptide, and Bias Correction: this option corrects for slight variations in total protein concentration between the samples. Normally, the samples should be mixed in equal protein μg amounts, based on results from assays such as BCA, Bradford, or other total protein assay. However, there may be slight discrepancies due to assay errors, sample mixing in unequal combination, and so forth, and this is what the "auto bias correction" attempts to redress. The basic assumption underlying the procedure is that the majority of proteins have no expression level change. If a systematic change in all proteins is detected, then this is what auto bias correct fixes mathematically. Bias correction can only be used if the dataset has a minimum of 20 proteins. Less than this is too small for a median value to be meaningful. An option for manual correction also exists, but this is probably useful only if the user is aware of a known bias (e.g., deliberately added 2x as much protein for one sample as for another). Typical method setup options for a ProteinPilot (v4 software) search are shown in Fig. 4.

5. The commercial iTRAQ kits come with detailed instructions and manual, which should also be referred to for additional details and background information. Note that the procedure described in this chapter covers additional points, apart from iTRAQ labeling, such as initial protein extraction. Furthermore, some modifications to the commercial iTRAQ protocol have been included (such as use of iodoacetamide instead of MES for alkylation of cysteines, peptides clean-up procedure following iTRAQ labeling, and tryptic peptides sample clean-up using strong cation exchange and other SPE procedures).
6. The reporter masses for iTRAQ 8-plex kit reagents are 113, 114, 115, 116, 117, 118, 119, and 121—there are eight reporter masses for the iTRAQ 8-plex kit, however, the 120 reporter is not present as the immonium ion of phenylalanine is often highly abundant at $m/z = 120$ (Figs. 3 and 5).

   The iTRAQ kits can be purchased as either a '4-plex' set (i.e., four different isobaric tags are included, so that four samples can be labeled) or an '8-plex' set (i.e., eight different isobaric tags are included, so that eight samples can be labeled). This allows for a greater range of experimental designs and numbers of technical and/or biological replicates. However, it is one of the constraints of all labeling methods, that the numbers of samples that can be analyzed are limited by the numbers of labels available. Prior to protein labeling commencement, total protein must be extracted from the cells or tissues, and the amount of total protein estimated. This is because the first part of experimental normalization is based on utilizing equal amounts of total protein for each of the samples in a given labeling experiment. Once proteins are extracted, disulfides present in the proteins are reduced and alkylated so that recovery of cysteine-containing peptides can be maximized, sequence analysis by LCMSMS facilitated, and protein sequence coverage maximized. The second part of data normalization is generally carried out during the data analysis step (e.g., bias correction within the Protein Pilot software; https://sciex.com/content/dam/SCIEX/pdf/posters/ProteinPilot-Software-Overview.pdf) [27, 28].

7. If sample amounts are limiting, so that only 1–2 μL amounts are available for use, then a reading on a nano-spectrometer (such as a NanoDrop or DeNovix) can also be taken at 280 nm using the extinction coefficient of BSA or IgG.
8. Buffers containing Tris or other amine-containing buffers should be avoided, as they may react with the iTRAQ reagents, depleting the supply for peptide labeling.
9. If the sample buffer contains any components that are likely to be problematic for LCMSMS (e.g., Tris buffer is not compatible with iTRAQ due to its amine groups), then an acetone precipitation or other sample clean-up approach might be necessary. However, it is recommended that this is not used unless there is a definite buffer problem, since each sample processing step represents a potential source of sample loss. To maximize sample recovery, keep sample processing as simple as possible, and reduce the sample processing steps to just the essential ones. This approach is likely to result in maximized sample recovery. If you need to use acetone precipitation, follow **steps 1** and **2** in Subheading 3.1, otherwise check that sample pH is within the optimum for trypsin (pH = 7–9). If pH needs to be adjusted, use a few μL of 500 mM sodium bicarbonate and go to **step (5)**. The sample clean-up procedure described here involves acetone precipitation of the total protein. Alternative protein clean-up procedures can also be considered, such as buffer exchange with 3 kDa spin filters (Amicon Ultra-4, Merck Millipore, Cork Ireland), or PD10 sephadex G25 disposable desalting columns (Cytiva Marlborough, MA).
10. If the sample is not completely soluble in just the dissolving buffer, then add SDS as outlined in the materials section above.
11. The cysteine blocking reagent in the iTRAQ kit is 2-(N-morpholino)ethanesulfonic acid (MES), but iodoacetamide (IAA) is a more widely used alkylating agent, and the mass

spectrometry data acquired will, therefore, be more readily comparable to a wider number of published studies. This is specifically relevant to cysteine containing peptides.

12. This step is particularly important for the 8-plex procedure, so be sure to take great care with adjusting the pH.
13. Keep a record as to number of usages of the cartridge.
14. At **step (6)** in Subheading 3.3 above, an Oasis HLB Plus LP extraction cartridge (Waters Corp) can also be used if greater capacity is needed, or to potentially maximize peptide recovery, for example, if larger amounts of protein are present, such as >300 μg total protein. If using the Oasis cartridge, then scale up all the volumes 3–4×. Note that it may be best to initially pass the sample through the macrotrap, and pass the eluate onto the Oasis trap (in case any peptides are still present in the eluate). Therefore, following the macrotrap peptide capture with an Oasis HLB step may be the best approach to maximize recovery. To do this, simply collect the eluate and wash from the green macrotrap and pass it through a primed Oasis HLB cartridge. Use a 1 mL plastic syringe barrel. The priming/preparation of the Oasis HLB can all be done manually (i.e., by hand – not using syringe pump), just leaving the peptide load step for the syringe pump at the 9.5 mL/h flow rate. Pool the peptide eluant with the peptides collected from the macrotrap, and dry the lot, which should facilitate maximal peptide recovery.
15. A variety of mass spectrometers and LCMSMS approaches can be used to maximize peptide detection and proteome coverage. Importantly, the mass spectrometer should have good sensitivity in the low mass range (100–150 *m/z*) in which the iTRAQ labels need to be detected and quantified and have the capability to perform tandem mass spectrometry (MSMS) for peptide sequencing. If a somewhat older or less sensitive model mass spectrometer is available (e.g., QStar Elite mass spectrometer), then good peptide detection can still be achieved by sample fractionation, either on-line or off-line. For example, 2D LCMSMS has been described [13, 29], in which peptides are initially captured onto a cation exchange column in-line and in sequence with a C18 capture cartridge, followed by a C18 nano LC column.
16. This procedure can improve proteome coverage and may be particularly useful where only a slower scanning/lower duty cycle instrument is available, but it does involve more time to complete the experiment, depending on the number of salt fractionation steps included. On instruments with a faster duty cycle (e.g., 5600 or 6600 TripleTOF, ZenoTOF 7600, etc.), the 2D chromatography approach may not be necessary and could instead be substituted with a longer gradient (e.g., 2–4 h) in a 1D UPLC chromatography approach [15].

17. On an instrument with a higher duty cycle, such as a 5600 TripleTOF mass spectrometer, the additional chromatography time using a 2D LC approach can be avoided, and 1D LC using an extended gradient of 2–4 h can offer similar or better sensitivity on complex peptide mixtures [15]. In brief, the method for a 5600 system includes the following parameters: a nano HPLC system should be used, such as a Dionex UltiMate 3000 RSLCnano pump system, valve unit (Switchos), and Famos autosampler (Thermo Scientific Dionex, Waltham, USA). The resuspended iTRAQ-labeled peptides are injected onto a C18 precolumn cartridge (Acclaim PepMap 100, 5 μm 100 Å, Thermo Scientific Dionex, Waltham, USA), which is washed (10 min) and then switched inline to a capillary column (10–25 cm) containing C18 reverse phase packing material (Reprosil-Pur, 1.9 μm, 200 Å, Dr. Maisch GmbH, Ammerbuch-Entringen, Germany). Peptides are eluted using a 240 min gradient of buffer A ($H_2O$:$CH_3CN$ of 98:2 with 0.1% formic acid) to buffer B ($H_2O$:$CH_3CN$ of 20:80 with 0.1% formic acid) at a flow rate of 200 nL/min. A high voltage (2300 V) was applied to the column tip through a low volume tee (Upchurch Scientific, Oak Harbor, USA) at the column inlet and the outlet positioned ~1 cm from the TripleTOF 5600+ hybrid tandem mass spectrometer source inlet (ABSciex, Foster City, USA). Positive ions are generated by electrospray ionization and the TripleTOF 5600+ system is operated in information-dependent acquisition (IDA) mode. A time-of-flight MS survey scan is acquired ($m/z$ 375–1600, 0.4 s) and up to ten multiple-charged ions ($m/z$ 375–1250, counts >200, charge state $\geq$2+ and $\leq$ 5+) sequentially selected by Q1 for MSMS analysis. Nitrogen is used as collision gas with an optimum collision energy automatically chosen based on charge state and mass. Tandem mass spectra are accumulated for 0.3 s. LC-MSMS should be repeated at least two to three times to acquire data on analytical replicates. See Fig. 5 for typical mass spectrometry output from an iTRAQ experiment.

## 5 Conclusion

Quantitative proteomics provides a powerful approach to study effects on cell viability, providing potential insight into the mechanisms by which cell viability may be impacted under a diversity of conditions, such as disease, treatments, environmental exposures, and so forth. The iTRAQ method described in this chapter is one such quantitative proteomics approach, which in conjunction with ProteinPilot software analysis, offers the field a wider option in analyzing data to evaluate cellular viability.

## Acknowledgments

This project was supported by a UNSW Sydney Faculty of Medicine Research Grant (P. Polly). Subsidized access to the Bioanalytical Mass Spectrometry Facility within the UNSW Mark Wainwright Analytical Center is gratefully acknowledged. We thank these organizations for their financial support.

### References

1. Strober W (2015) Trypan blue exclusion test of cell viability. Curr Protoc Immunol 111:A3.B.1–A3.B.3. https://doi.org/10.1002/0471142735.ima03bs111
2. Burd JF, Usategui-Gomez M (1973) A colorimetric assay for serum lactate dehydrogenase. Clin Chim Acta 46(3):223–227. https://doi.org/10.1016/0009-8981(73)90174-5
3. Chan FK, Moriwaki K, De Rosa MJ (2013) Detection of necrosis by release of lactate dehydrogenase activity. Methods Mol Biol 979:65–70. https://doi.org/10.1007/978-1-62703-290-2_7
4. Kumar P, Nagarajan A, Uchil PD (2018) Analysis of cell viability by the lactate dehydrogenase assay. Cold Spring Harb Protoc 2018(6):pdb.prot095497. https://doi.org/10.1101/pdb.prot095497
5. Mosmann T (1983) Rapid colorimetric assay for cellular growth and survival: application to proliferation and cytotoxicity assays. J Immunol Methods 65(1–2):55–63. https://doi.org/10.1016/0022-1759(83)90303-4
6. van de Loosdrecht AA, Beelen RH, Ossenkoppele GJ, Broekhoven MG, Langenhuijsen MM (1994) A tetrazolium-based colorimetric MTT assay to quantitate human monocyte mediated cytotoxicity against leukemic cells from cell lines and patients with acute myeloid leukemia. J Immunol Methods 174(1–2):311–320. https://doi.org/10.1016/0022-1759(94)90034-5
7. Schuh RA, Clerc P, Hwang H, Mehrabian Z, Bittman K, Chen H, Polster BM (2011) Adaptation of microplate-based respirometry for hippocampal slices and analysis of respiratory capacity. J Neurosci Res 89(12):1979–1988. https://doi.org/10.1002/jnr.22650
8. Jayasena T, Poljak A, Braidy N, Smythe G, Raftery M, Hill M, Brodaty H, Trollor J, Kochan N, Sachdev P (2015) Upregulation of glycolytic enzymes, mitochondrial dysfunction and increased cytotoxicity in glial cells treated with Alzheimer's disease plasma. PLoS One 10(3):e0116092. https://doi.org/10.1371/journal.pone.0116092
9. Poljak A, Hill M, Hall RJ, MacLullich AM, Raftery MJ, Tai J, Yan S, Caplan GA (2014) Quantitative proteomics of delirium cerebrospinal fluid. Transl Psychiatry 4:e477. https://doi.org/10.1038/tp.2014.114
10. Kaur G, Poljak A, Ali SA, Zhong L, Raftery MJ, Sachdev P (2021) Extending the depth of human plasma proteome coverage using simple fractionation techniques. J Proteome Res 20(2):1261–1279. https://doi.org/10.1021/acs.jproteome.0c00670
11. Coumans JV, Gau D, Poljak A, Wasinger V, Roy P, Moens P (2014) Green fluorescent protein expression triggers proteome changes in breast cancer cells. Exp Cell Res 320(1):33–45. https://doi.org/10.1016/j.yexcr.2013.07.019
12. Williams TJ, Lauro FM, Ertan H, Burg DW, Poljak A, Raftery MJ, Cavicchioli R (2011) Defining the response of a microorganism to temperatures that span its complete growth temperature range (−2 degrees C to 28 degrees C) using multiplex quantitative proteomics. Environ Microbiol 13(8):2186–2203. https://doi.org/10.1111/j.1462-2920.2011.02467.x
13. Lim YA, Rhein V, Baysang G, Meier F, Poljak A, Raftery MJ, Guilhaus M, Ittner LM, Eckert A, Gotz J (2010) Abeta and human amylin share a common toxicity pathway via mitochondrial dysfunction. Proteomics 10(8):1621–1633. https://doi.org/10.1002/pmic.200900651
14. Shum AMY, Poljak A, Bentley NL, Turner N, Tan TC, Polly P (2018) Proteomic profiling of skeletal and cardiac muscle in cancer cachexia: alterations in sarcomeric and mitochondrial protein expression. Oncotarget 9(31):22001–22022. https://doi.org/10.18632/oncotarget.25146
15. Muenchhoff J, Poljak A, Thalamuthu A, Gupta VB, Chatterjee P, Raftery M, Masters CL,

Morris JC, Bateman RJ, Fagan AM, Martins RN, Sachdev PS (2016) Changes in the plasma proteome at asymptomatic and symptomatic stages of autosomal dominant Alzheimer's disease. Sci Rep 6:29078. https://doi.org/10.1038/srep29078

16. Coumans JV, Gau D, Poljak A, Wasinger V, Roy P, Moens PD (2014) Profilin-1 overexpression in MDA-MB-231 breast cancer cells is associated with alterations in proteomics biomarkers of cell proliferation, survival, and motility as revealed by global proteomics analyses. OMICS 18(12):778–791. https://doi.org/10.1089/omi.2014.0075
17. Coumans JV, Harvey J, Backhouse D, Poljak A, Raftery MJ, Nehl D, Katz ME, Pereg L (2011) Proteomic assessment of host-associated microevolution in the fungus Thielaviopsis basicola. Environ Microbiol 13(3):576–588. https://doi.org/10.1111/j.1462-2920.2010.02358.x
18. Ittner A, Chua SW, Bertz J, Volkerling A, van der Hoven J, Gladbach A, Przybyla M, Bi M, van Hummel A, Stevens CH, Ippati S, Suh LS, Macmillan A, Sutherland G, Kril JJ, Silva AP, Mackay JP, Poljak A, Delerue F, Ke YD, Ittner LM (2016) Site-specific phosphorylation of tau inhibits amyloid-beta toxicity in Alzheimer's mice. Science 354(6314):904–908. https://doi.org/10.1126/science.aah6205
19. Stefanoska K, Volkerling A, Bertz J, Poljak A, Ke YD, Ittner LM, Ittner A (2018) An N-terminal motif unique to primate tau enables differential protein-protein interactions. J Biol Chem 293(10):3710–3719. https://doi.org/10.1074/jbc.RA118.001784
20. Stefanoska K, Gajwani M, Tan ARP, Ahel HI, Asih PR, Volkerling A, Poljak A, Ittner A (2022) Alzheimer's disease: ablating single master site abolishes tau hyperphosphorylation. Sci Adv 8(27):eabl8809. https://doi.org/10.1126/sciadv.abl8809
21. Prikas E, Poljak A, Ittner A (2020) Mapping p38alpha mitogen-activated protein kinase signaling by proximity-dependent labeling. Protein Sci 29(5):1196–1210. https://doi.org/10.1002/pro.3854
22. Paulo JA, O'Connell JD, Everley RA, O'Brien J, Gygi MA, Gygi SP (2016) Quantitative mass spectrometry-based multiplexing compares the abundance of 5000 S. cerevisiae proteins across 10 carbon sources. J Proteome 148:85–93. https://doi.org/10.1016/j.jprot.2016.07.005
23. Cox J, Mann M (2008) MaxQuant enables high peptide identification rates, individualized p.p.b.-range mass accuracies and proteome-wide protein quantification. Nat Biotechnol 26(12):1367–1372. https://doi.org/10.1038/nbt.1511
24. Bradford MM (1976) A rapid and sensitive method for the quantitation of microgram quantities of protein utilizing the principle of protein-dye binding. Anal Biochem 72:248–254. https://doi.org/10.1006/abio.1976.9999
25. Smith PK, Krohn RI, Hermanson GT, Mallia AK, Gartner FH, Provenzano MD, Fujimoto EK, Goeke NM, Olson BJ, Klenk DC (1985) Measurement of protein using bicinchoninic acid. Anal Biochem 150(1):76–85. https://doi.org/10.1016/0003-2697(85)90442-7
26. Lowry OH, Rosebrough NJ, Farr AL, Randall RJ (1951) Protein measurement with the Folin phenol reagent. J Biol Chem 193(1):265–275
27. Shilov IV, Seymour SL, Patel AA, Loboda A, Tang WH, Keating SP, Hunter CL, Nuwaysir LM, Schaeffer DA (2007) The Paragon Algorithm, a next generation search engine that uses sequence temperature values and feature probabilities to identify peptides from tandem mass spectra. Mol Cell Proteomics 6(9):1638–1655. https://doi.org/10.1074/mcp.T600050-MCP200
28. Tang WH, Shilov IV, Seymour SL (2008) Nonlinear fitting method for determining local false discovery rates from decoy database searches. J Proteome Res 7(9):3661–3667. https://doi.org/10.1021/pr070492f
29. Evans FF, Raftery MJ, Egan S, Kjelleberg S (2007) Profiling the secretome of the marine bacterium Pseudoalteromonas tunicata using amine-specific isobaric tagging (iTRAQ). J Proteome Res 6(3):967–975. https://doi.org/10.1021/pr060416x

# Chapter 13

# Quantifying pH in Malaria Using pHluorin and Flow Cytometry

**Jeffrey Agyapong and Petra Rohrbach**


## Abstract

Intracellular pH ($pH_i$) plays a critical role in the regulation of numerous biological functions where specific pH ranges are required for optimal operation within cells. Slight pH changes can impact the regulation of diverse molecular processes, including enzymatic activities, ion channels, and transporters, which all play a role in cell functions. Methods for quantifying $pH_i$ continue to evolve and include various optical methods using fluorescent pH indicators. Here, we provide a protocol to measure $pH_i$ in the cytosol of *Plasmodium falciparum* blood stage parasites by means of flow cytometry and using pHluorin2, a pH-sensitive fluorescent protein that has been introduced into the genome of the parasite.


**Key words** Fluorescence, pHluorin2, Green fluorescent protein, *Plasmodium*, Flow cytometry, pH quantification, Intracellular pH, Acidification, Alkalization

## 1 Introduction

Most biological processes within organisms are pH-dependent, and a small change in intracellular pH ($pH_i$) can significantly affect cell viability. $pH_i$ is regulated with the help of membrane transporters and/or chemical buffering systems, such as the $Na^+$-$H^+$ ion exchanger [1] and the bicarbonate-carbonic acid ($NaHCO_3$/$H_2CO_3$) buffer. Both are biochemical regulators that cells use to respond to pH changes. Maintaining pH within a cell or an organelle provides biomolecule functionality in an optimal environment. Many organelles, such as lysosomes or endosomal vesicles, maintain an organellar pH that differs from the cytoplasmic pH, and these variances have important functional consequences for those organelles. In the unicellular malaria parasite *Plasmodium falciparum* (*P. falciparum*), for example, various plasmepsins break down proteins found in the acidic digestive vacuole of the parasite. It has

Oliver Friedrich and Daniel F. Gilbert (eds.), *Cell Viability Assays: Methods and Protocols*, Methods in Molecular Biology, vol. 2644, https://doi.org/10.1007/978-1-0716-3052-5_13,

been shown that, if the pH of this organelle is increased from 5.4 to 5.7, the activity of the aspartic protease plasmepsin IV is significantly reduced from 100% to less than 50% [2].

Investigating cellular processes under the control of pH is challenging due to the limitations of experimental techniques, especially when quantifying pH in living cells. This is even more challenging when the size of the cell is very small and/or if pH is to be measured within an organelle. Numerous fluorescent pH indicators are available, including either dual-excitation or dual-emission probes that provide ratiometric measurements. Other indicators, such as LysoSensor probes, are ideal for pH quantification in acidic compartments, and their fluorescence emission is dependent on an acidic environment. Many fluorochromes are nonspecific and cannot be targeted or compartmentalized to a specific organelle of interest. Loading live cells with these fluorochromes is straight forward; however, prudence is advised when quantifying the data. Fluorochrome-loaded cells often show fluorescence dispersed throughout the cell, and fluorescent signals may not be easily differentiated between the various regions of the cell. Furthermore, the end concentration of loaded fluorochromes is dependent on numerous factors such as loading concentrations, loading time, loading temperature, and the number of cells in solution, and all will factor into the amount of fluorescence emitted and captured. Finally, certain fluorochromes have sub-optimal fluorescence characteristics such as low brightness, high susceptibility to photobleaching, and low quantum yields [3–6]. Due to these issues, it is often favorable to select ratiometric indicators for the quantification of emission signals.

The emergence of green fluorescent protein (GFP)-based ratiometric pH indicators has become popular for overcoming many of these obstacles. Fluorescent proteins (FPs) are stably expressed within the cell and can be targeted to specific locations by the addition of signal sequences. pHluorin [7], a modified pH-sensitive GFP, was mutated to direct pH sensitivity and displays a reversible excitation ratio between pH 7.5 and 5.5, producing a ratiometric sensor with two excitation wavelengths (a major peak at 395 nm and a minor peak at 475 nm).

pHluorin has previously been used to measure pH in intact *P. falciparum*-infected red blood cells (iRBCs) under physiological conditions via confocal microscopy [5, 8]. Quantifying the pH of the cytosol and digestive vacuole of malaria parasites (Fig. 1) was possible by generating parasites transiently expressing pHluorin. This FP provides an additional advantage of monitoring compartmentalized pH due to its moderately high fluorescence at low pH and its two inversely related fluorescent emission peaks that vary with pH changes. The ratiometric nature eliminates any changes in fluorescence due to fluorochrome bleaching, changes in focal point, and/or variations in laser intensity when imaging [9]. In

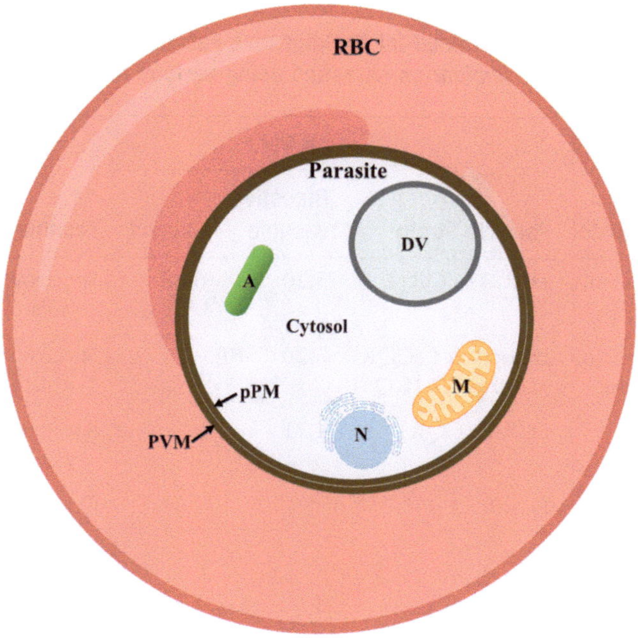


**Fig. 1** *Plasmodium falciparum* parasite within a red blood cell. The malaria parasite resides within a parsitophorous vacuolar membrane found within the RBC. Human RBCs have a concave shape and are approx. 6–8 μm in diameter and 2.5–3 μm thick. The concave shape is lost when infected with *Plasmodium*. The parasite develops within the RBC and increases in size from initially 0.5 to approx. 7 μm. Trophozoite stage parasites used in this protocol, as shown here, are approx. 3–5 μm in size. Abbreviations: A apicoplast. M mitochondrion, N nucleus, DV digestive vacuole, pPM parasite plasma membrane, PVM parasitophorous vacuolar membrane

addition, pHluorin can be specifically targeted to an organelle of interest by adding the required signal peptide, thereby overcoming the limitations found in fluorochrome loading by pure diffusion [10]. An enhanced version of pHluorin, termed pHluorin2, was developed by introducing mammalian codons and the F64L mutation that displays markedly higher fluorescence [11]. This modification allows for the detection of fluorescence even at low protein expression.

Initially, pH measurements of the malaria parasite depended on invasive techniques, where *P. falciparum*-iRBCs or isolated parasites were lysed, consequently affecting $pH_i$ homeostasis of the system [12–14]. Epifluorescence and confocal microscopy allowed pH measurements to be acquired in intact live cells (Table 1). The use of confocal microscopy to measure pH has the advantage of observing fluorescence localization within specific organelles or compartments of a cell. It also has the advantage of quantifying fluorescence within a defined region of interest (e.g., an organelle) should any additional fluorescence be located outside this region.

**Table 1**
**Steady-state pH determined for the digestive vacuole and cytosol of drug sensitive (CQS) and drug-resistant (CQR) *P. falciparum* parasites using various fluorochromes and methods**

| pH indicator | Parasite strain | pH Value: Digestive vacuole | pH Value: Cytosol | Technique | Reference |
|---|---|---|---|---|---|
| Fluorescein-dextran | FCR-3 (CQR) | 5.20 | 6.85 | Spectrofluorimetry using isolated parasites at 24 °C | [12] |
| Fluorescein-dextran | Indochina I (CQR) & FCC-2 (CQS) | 5.20–5.40 | – | Spectrofluorimetry using isolated parasites | [13] |
| Acridine orange | FCR-3 (CQR) | 4.21 | – | Spectrofluorimetry using *P. falciparum*-iRBCs | [4] |
| Fluorescein-BCECF-AM | FCR3-A2 | – | 7.34 | Confocal microscopy | [16] |
| Fluorescein-BCECF | 3D7 (CQS) | – | 7.28 | Confocal microscopy | [17] |
| Acridine orange | Dd2 (CQR)<br>HB3 (CQS) | 5.21<br>5.64 | –<br>– | Epifluorescence microscopy using *P. falciparum*-iRBCs at 37 °C | [3] |
| Fluorescein-dextran | 7G8 (CQR)<br>D10 (CQS) | 5.72<br>5.50 | –<br>– | Spectrofluorimetry using isolated parasites at 37 °C | [14] |
| Oregon green | 7G8 (CQR)<br>D10 (CQS) | 4.66<br>4.40 | –<br>– | | |
| pHluorin | Dd2 (CQR)<br>HB3 (CQS) | 5.18<br>5.17 | 7.20<br>7.03 | Confocal microscopy of intact *P. falciparum*-iRBCs at 37 °C | [8] |

Despite these advantages, the process is very time-consuming, since only one cell can be processed at a time, and the number of cells that are quantified is low (e.g., $n = 6$–9). Flow cytometry, an alternative method to detect fluorescence, has the advantage of allowing high-throughput data analysis, as thousands of cells can be quantified in a short period of time. Important here is that the emitted fluorochrome fluorescence is contained in the region of interest to be measured. When combining this technique with a targeted fluorescent protein such as pHluorin2, accurate fluorescent measurements are obtainable. In this chapter, we provide a step-by-step guide to measure pH in intracellular compartments using flow cytometry and the fluorescent protein pHluorin2 expressed in the cytosol of malaria parasites. This method can be applied to any appropriate biological system.

## 2 Materials

All solutions are prepared using Millipore-filtered water (dd$H_2O$) and stored at 4 °C if not otherwise stated. Antimalarial stock solutions should be kept at −80 °C. Solutions and consumables must be sterile. Malaria cultures should be handled under a laminar flow cabinet (Class II biological safety cabinet).

### *2.1 Cell Cultures*

1. Uninfected cells: Human $A^+$ red blood cells (RBCs).
2. Malaria parasites: stably transfected trophozoite-stage *P. falciparum*-iRBCs expressing pHluorin2 in the cytosol of the parasite (*see* **Note 1**).

### *2.2 Solutions and Reagents*

1. Complete parasite media (CPM): RPMI 1640 media, 5% Albumax I, 100 μM hypoxanthine, 20 μg/mL gentamicin.
2. Ringer's solution: 122.5 mM NaCl, 1 mM $NaH_2PO_4$, 5.4 mM KCl, 1.2 mM $CaCl_2$, 0.8 mM $MgCl_2$, 10 mM HEPES, 11 mM D-glucose. Adjust the pH of the solution to 7.4 using HCl or NaOH. Filter-sterilize using a 0.22 μm pore-size filter (*see* **Note 2**).
3. pH calibration solution: 130 mM KCl, 1.2 mM $CaCl_2$, 0.8 mM $MgCl_2$, 10 mM HEPES, 11 mM D-glucose. Filter-sterilize using a 0.22 μm pore-size filter. On the day of the experiment, adjust pH to a set of defined pH values (*see* **Note 3**).
4. Ionophore (nigericin): 10 mM nigericin in 100% DMSO (stock solution) (*see* **Note 4**).

### *2.3 General Equipment*

1. Flow Cytometer (e.g., Attune® NxT™ Acoustic Focusing Cytometer) with the following lasers and filters: lasers (violet: excitation peak at 405 nm, blue: excitation peak at 488 nm), filters (violet: 635 dichromic longpass (DLP), 415 DLP, 405/10 bandpass (BP), 555 DLP, 525/50 BP, blue: 495 DLP, 488/10 BP, 503 LP, 555 DLP, 530/30 BP).
2. Incubator with gas mixture (92% $N_2$, 5% $CO_2$, and 3% $O_2$) for malaria parasite cultures.
3. Centrifuge.
4. Water bath.
5. pH meter.
6. Petri dishes.
7. 1.5 mL Eppendorf tubes.
8. 15 mL test tubes.
9. FACS tubes.

#### 2.4 Software

1. Flow cytometry software (e.g., Attune Nxt Software).
2. Spreadsheet software, such as Numbers (Mac OS) or Excel (Microsoft).
3. Statistical software, such as Prism 9 (GraphPad software).

## 3 Methods

All steps (Fig. 2), including incubating parasites with pH calibration solutions and antimalarial drugs, are performed at 37 °C.

#### 3.1 Parasite Culturing

1. Maintain transfected parasites in 10 cm Petri dishes containing 10 mL of CPM and human $A^+$ RBCs. Continuously keep cultures at 4% hematocrit and 3–5% parasitemia (*see* **Note 1**). Incubate parasite cultures under humidified conditions at 37 °C with 92% $N_2$, 5% $CO_2$, and 3% $O_2$ [15]. Stably transfected trophozoite-stage *P. falciparum*-iRBCs expressing pHluorin2 in the cytosol of the parasite (Fig. 3a) will be used for pH quantification.
2. Subculture parasites every other day to a parasitemia of 0.5%, using pre-warmed CPM and uninfected RBCs.

#### 3.2 Parasite Preparation

An overview for the preparation of parasites is found in Fig. 2, **step 1**. More specifically:

1. From trophozoite stage *P. falciparum*-iRBC cultures, remove 10% of the culture (e.g., aliquot 500 μL from a 5 mL culture) with 5–10% parasitemia and transfer this into a 1.5 mL Eppendorf tube (*see* **Note 5**).
2. Remove culture media from *P. falciparum*-iRBC cultures by centrifuging at 600 × g for 5 min. Discard supernatant.
3. Spin-wash *P. falciparum*-iRBCs with 1 mL pre-warmed Ringer's solution at 600 × g for 2 min. Discard supernatant. Repeat.
4. Resuspend *P. falciparum*-iRBCs pellet in 1 mL of Ringer's solution, and keep in a water bath at 37 °C.

#### 3.3 pH Calibration

To calibrate pH, a summary of the steps is shown in Fig. 2, **steps 2** and **3**. The details are as follows:

1. Aliquot 10 mL of Ringer's solution into a 15 mL test tube and keep at 37 °C.
2. Aliquot 10 mL of calibration solution into eight different 15 mL tubes. Adjust pH of the solution in each tube, ranging from pH 5.0 to pH 8.5 in 0.5 pH increments (*see* **Note 3**). Add 5 μL of the ionophore nigericin stock solution to each tube

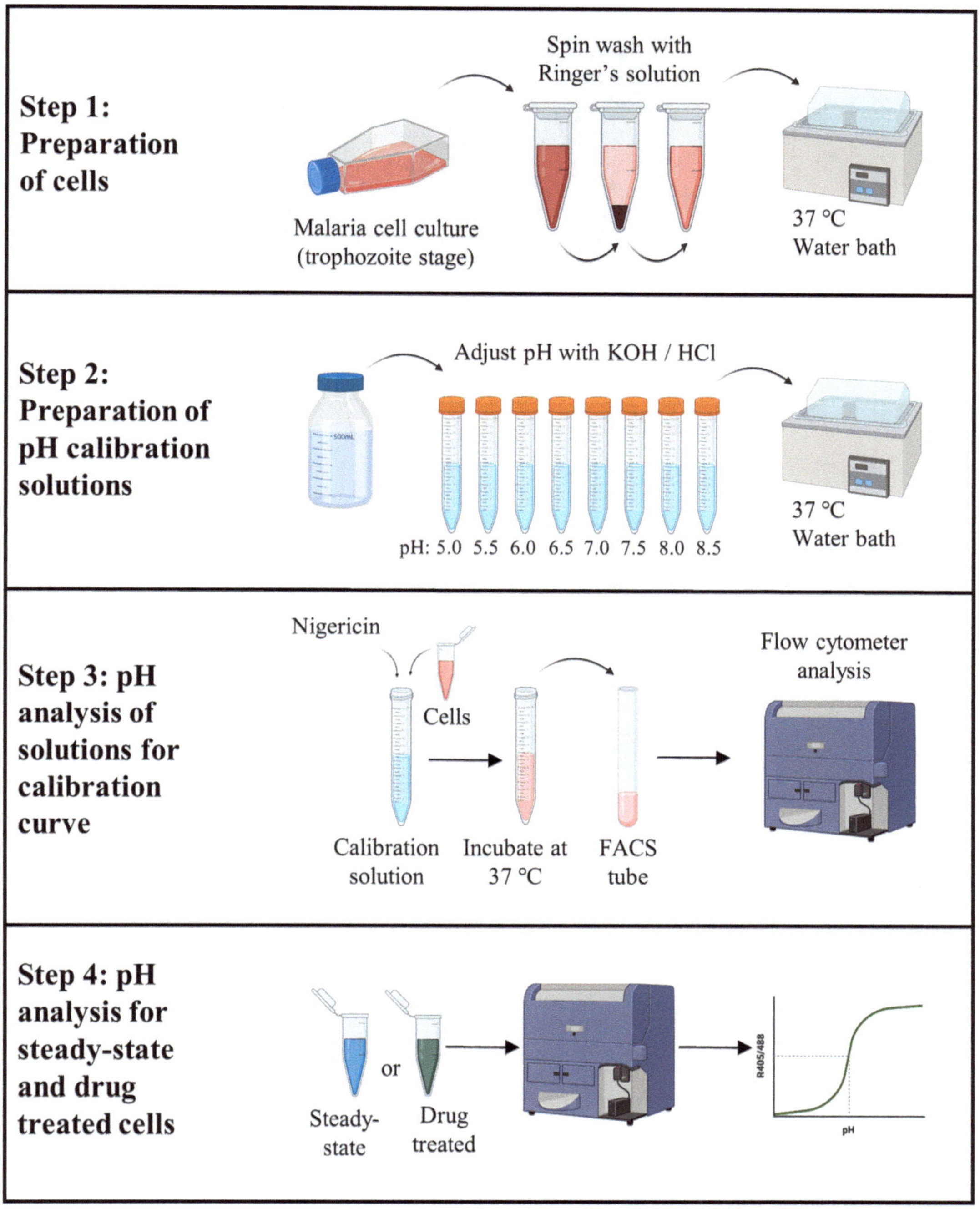


**Fig. 2** pH quantification using flow cytometry. A schematic representation of the experimental steps in chronological order. Details are found in the text

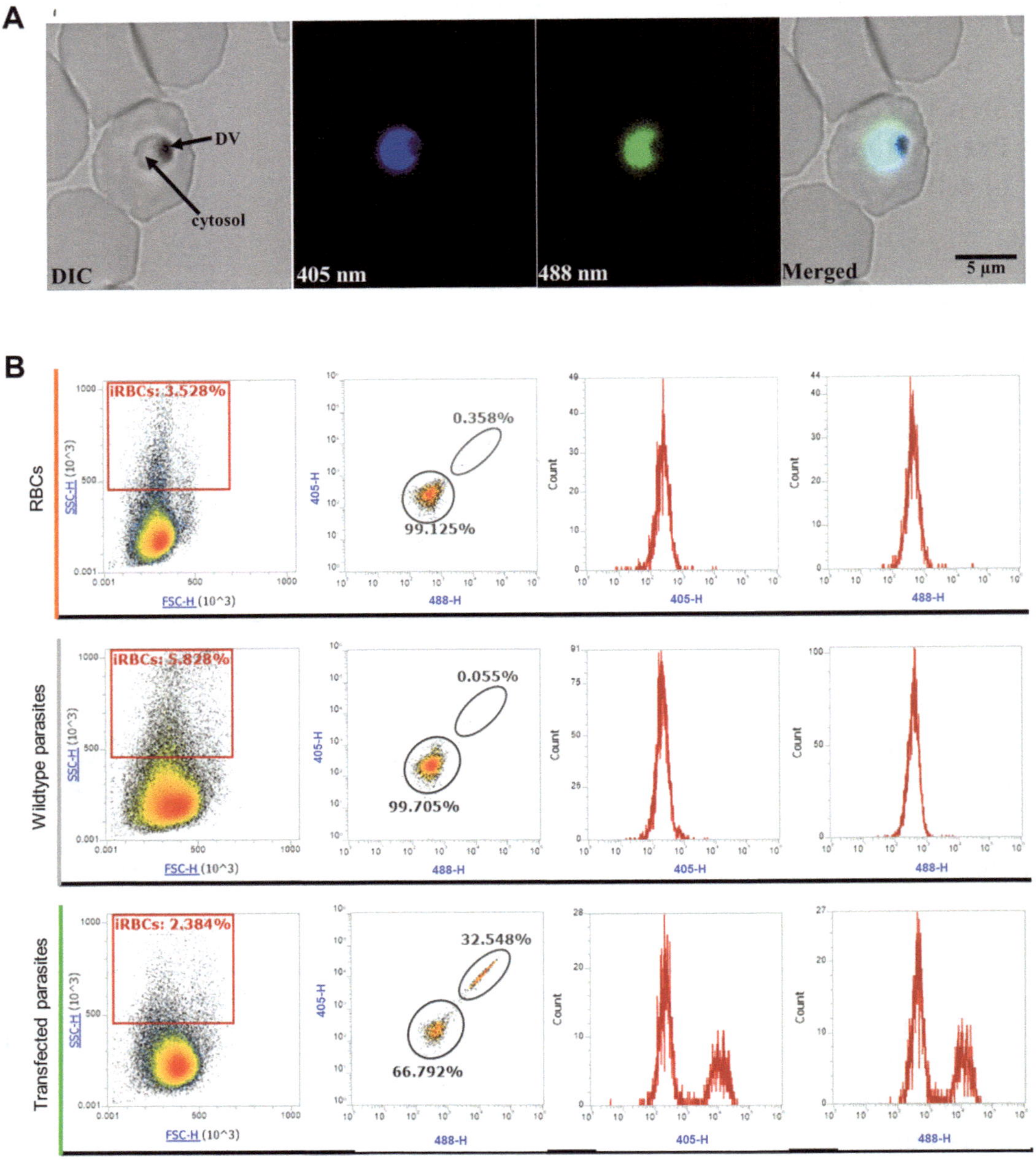


**Fig. 3** pHluorin2 emission in the cytosol of *P. falciparum*. (**a**) Live cell fluorescent images of *P. falciparum*-iRBCs transfected with pHluorin2 with signals confined to the cytosol of the parasite. pHluorin2 emission was captured using confocal microscopy after excitation with 405 nm (blue) and 488 nm (green) laser. Scale bar, 5 µm. (**b**) pHluorin fluorescence emitted from the cytosol of transfected *P. falciparum*-iRBCs using flow cytometry. Gates were set to detect emission signals from excitation with 405 nm and 488 nm. No fluorescence is detected in either channel (i.e., 405-H and 488-H) for uninfected RBCs (top row) and for wildtype parasites (middle row). Fluorescence is only detected in pHluorin-transfected parasites (bottom row) in both 405 nm (405-H) and 488 nm (488-0H) channels

(final concentration will be 5 μM). Vortex and keep tubes in water bath set at 37 °C.

3. Transfer 32.5 μL of *P. falciparum*-iRBC suspension (prepared in Subheading 3.2, **step 4**) into the 15 mL test tube containing 10 mL of either Ringer's solution or pH-set calibration solution (*see* **Notes 2** and **3**), containing 5 μM nigericin and incubate for 2 mins at 37 °C (*see* **Note 4**). This results in a cell concentration of approx. $1.3 \times 10^6$ cells/mL.
4. After incubation, mix *P. falciparum*-iRBC suspension by inverting twice and transfer 1 mL to FACS tubes to be analyzed immediately by flow cytometry.
5. Use controls described below to set up gated populations, as shown in Fig. 3b:
   - RBCs (uninfected RBCs are used to set or gate this population as blank for unstained controls): RBCs only, no parasites, that is, no *P. falciparum*-iRBCs.
   - Wild-type parasites (*P. falciparum*-iRBC cultures, that is, non-transfected parasites containing no pHluorin2 are used to set or gate iRBC population that have no pHluorin2 emitted fluorescence); pHluorin negative *P. falciparum*-iRBCs.
   - Transfected parasites (*P. falciparum*-iRBC cultures containing pHluorin transfected parasites are used to set or gate cell populations that produce a pHluorin2 emitted fluorescence): pHluorin positive *P. falciparum*-iRBCs.

### 3.4 pH Quantification

A summary to quantify the pH is shown in Fig. 2, **step 4**. In more detail:

1. Measure the emitted fluorescence from the transfected parasites at both 405 nm (labeled as 405-H) and 488 nm (labeled as 488-H) (Fig. 4). Three biological replicates are required to determine the pH value in the cytosol of the parasite.
2. To determine the steady-state pH in the cytosol of *P. falciparum*, first export the data generated from the software, labeled "statistics," to CSV files. These files can be opened using a spreadsheet software such as Numbers (Mac OS) or Excel (Microsoft).
3. Select the data under the "X mean" column (i.e., these are the mean values for the measured emitted fluorescence of a sample group) that corresponds to each of the "groups" that were prepared and measured (i.e., parasites kept at steady-state conditions and parasites at varying pH buffer conditions) (Fig. 5a) and create a new table with these values.

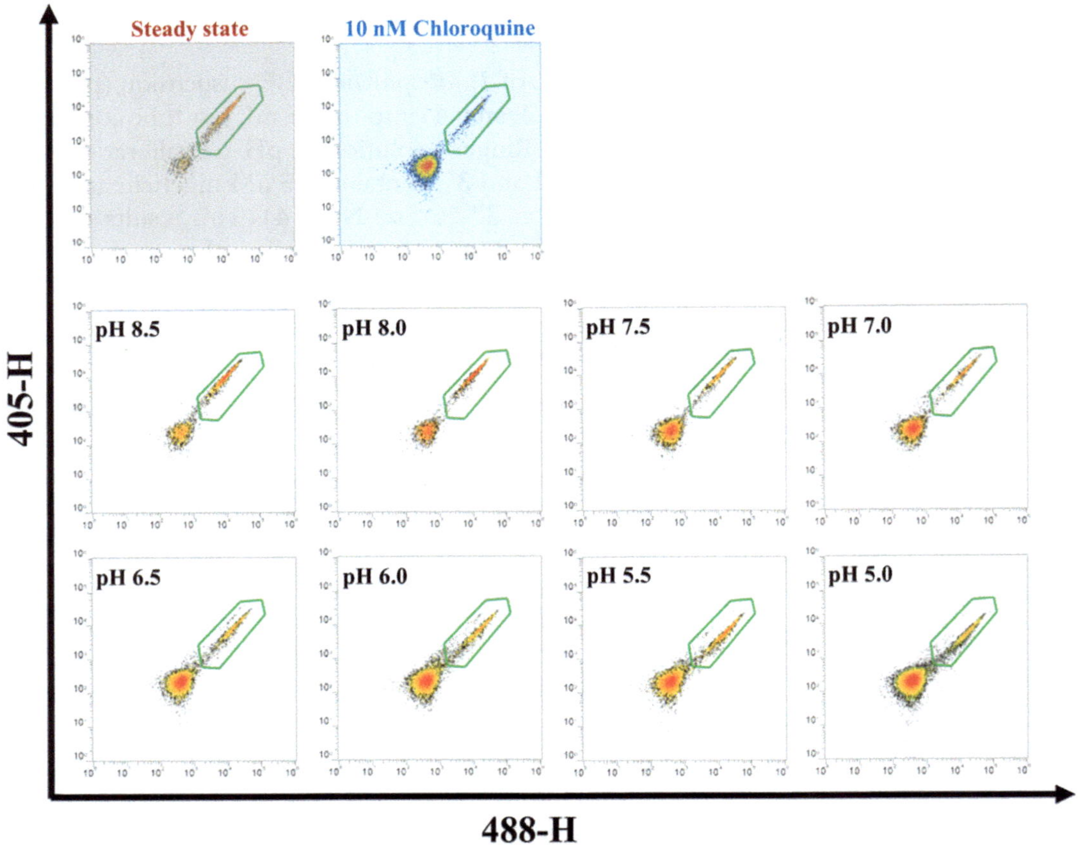


**Fig. 4** In situ calibration of pHluorin2 fluorescence to determine pH in *P. falciparum.* Fluorescence measurements of transfected parasites obtained for steady-state, drug treated (chloroquine) and the set of pH calibration solutions. Parasites were incubated in various solutions for 2 min at 37 °C. The gated cell populations, positive in both violet (405-H) and blue (488-H) channels, is seen to shift towards the bottom right of the gated quadrant as the pH of the solution changes from alkaline to acidic

4. The "group" has two "x parameters" associated with it that corresponds to the two emission wavelengths measured from exciting pHluorin with 405 nm and 488 nm: VL1-H for the measured emission from excitation with 405 nm and BL1-H for the measured emission from excitation with 488 nm (Fig. 5a). Quantify the fluorescence ratio from the emission values of the two excitation wavelengths (405 and 488 nm) by dividing the values obtained from 405 nm with 488 nm.
5. Transfer the data for the fluorescence ratio with their corresponding pH values to a statistical program such as GraphPad Prism and analyze the data by using a non-linear regression model (standard sigmoidal curve).
6. The steady state pH in the cytosol of the parasite is determined by interpolation from the ratio values or the calibration curve (Fig. 5b).

**A**

| Group | Gate | X Parameter | Count | Volume | Concentration | %Total | %Gated | X Mean | Y Mean | X Median | Y Median | X Peak | Y Peak | X SD | Y SD | X %CV | Y %CV |
|---|---|---|---|---|---|---|---|---|---|---|---|---|---|---|---|---|---|
| Homeostat (pH 7.4) | pHluorin2 | BL1-H | 485 | 50 | 9.700 | 1.322 | 1.322 | 5083 | N/A | 4532 | N/A | 3931 | N/A | 2398.53 | N/A | 47.18 | N/A |
| Homeostat (pH 7.4) | pHluorin2 | VL2-H | 485 | 50 | 9.700 | 1.322 | 1.322 | 2140 | N/A | 1894 | N/A | 2112 | N/A | 1148.85 | N/A | 53.69 | N/A |
| pH 8.5 | pHluorin2 | BL1-H | 305 | 50 | 6.100 | 0.793 | 0.793 | 5803 | N/A | 5053 | N/A | 4829 | N/A | 2984.08 | N/A | 51.42 | N/A |
| pH 8.5 | pHluorin2 | VL2-H | 305 | 50 | 6.100 | 0.793 | 0.793 | 3496 | N/A | 2808 | N/A | 2484 | N/A | 2176.80 | N/A | 62.26 | N/A |
| pH 8.0 | pHluorin2 | BL1-H | 456 | 50 | 9.120 | 1.216 | 1.216 | 6184 | N/A | 5974 | N/A | 6425 | N/A | 2745.36 | N/A | 44.39 | N/A |
| pH 8.0 | pHluorin2 | VL2-H | 456 | 50 | 9.120 | 1.216 | 1.216 | 4235 | N/A | 3558 | N/A | 4438 | N/A | 2508.08 | N/A | 59.22 | N/A |
| pH 7.5 | pHluorin2 | BL1-H | 433 | 50 | 8.660 | 1.235 | 1.235 | 5949 | N/A | 5252 | N/A | 4061 | N/A | 3160.77 | N/A | 53.13 | N/A |
| pH 7.5 | pHluorin2 | VL2-H | 433 | 50 | 8.660 | 1.235 | 1.235 | 3512 | N/A | 2779 | N/A | 1977 | N/A | 2742.36 | N/A | 78.09 | N/A |
| pH 7.0 | pHluorin2 | BL1-H | 524 | 50 | 10.480 | 1.414 | 1.414 | 6199 | N/A | 5467 | N/A | 6503 | N/A | 3495.81 | N/A | 56.40 | N/A |
| pH 7.0 | pHluorin2 | VL2-H | 524 | 50 | 10.480 | 1.414 | 1.414 | 3654 | N/A | 3285 | N/A | 2729 | N/A | 1888.39 | N/A | 51.68 | N/A |
| pH 6.5 | pHluorin2 | BL1-H | 521 | 50 | 10.420 | 1.518 | 1.518 | 6859 | N/A | 6115 | N/A | 5891 | N/A | 3669.74 | N/A | 53.51 | N/A |
| pH 6.5 | pHluorin2 | VL2-H | 521 | 50 | 10.420 | 1.518 | 1.518 | 3430 | N/A | 2805 | N/A | 2695 | N/A | 2241.69 | N/A | 65.35 | N/A |
| pH 6.0 | pHluorin2 | BL1-H | 658 | 50 | 13.160 | 2.030 | 2.030 | 5963 | N/A | 5305 | N/A | 3942 | N/A | 3591.91 | N/A | 60.24 | N/A |
| pH 6.0 | pHluorin2 | VL2-H | 658 | 50 | 13.160 | 2.030 | 2.030 | 2982 | N/A | 2553 | N/A | 1819 | N/A | 1645.57 | N/A | 55.19 | N/A |
| pH 5.5 | pHluorin2 | BL1-H | 668 | 50 | 13.360 | 2.184 | 2.184 | 6253 | N/A | 5626 | N/A | 4088 | N/A | 3252.17 | N/A | 52.01 | N/A |
| pH 5.5 | pHluorin2 | VL2-H | 668 | 50 | 13.360 | 2.184 | 2.184 | 2825 | N/A | 2473 | N/A | 1894 | N/A | 1534.63 | N/A | 54.33 | N/A |
| pH 5.0 | pHluorin2 | BL1-H | 654 | 50 | 13.080 | 1.964 | 1.964 | 6514 | N/A | 6070 | N/A | 6495 | N/A | 3656.11 | N/A | 56.13 | N/A |
| pH 5.0 | pHluorin2 | VL2-H | 654 | 50 | 13.080 | 1.964 | 1.964 | 2873 | N/A | 2499 | N/A | 1591 | N/A | 1761.60 | N/A | 61.32 | N/A |

**B**

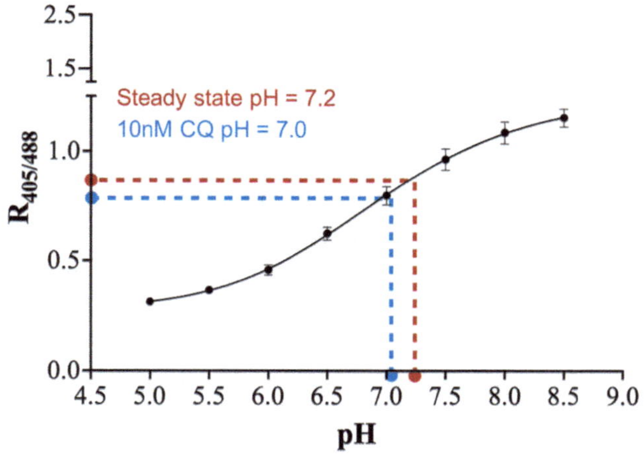


**Fig. 5** pH quantification. (**a**) Example of a spreadsheet generated by the Attune Nxt software. Data in the "X mean" column (highlighted with a red border) represent the average of the fluorescence intensities emitted by *P. falciparum*-iRBCs in a pre-determined gated population. These values are used to generate the pH calibration curve. The "group" column lists the various samples measured. The "x-parameter" lists which excitation wavelength was used, that is, BL1-H refers to 488 nm, and VL1-H refers to 405 nm. (**b**) Example of a pH calibration curve. The quantified fluorescent ratio of pHluorin transfected parasites in pH buffers ranging from 5 to 8.5 was plotted and a nonlinear regression curve was fitted. Each point represents the mean of three biological replicates with respective standard deviation. Quantified steady state pH (red dash line) of the parasite's cytosol was interpolated to be 7.2, which changes to 7.0 when chloroquine is added to the sample (blue dash line)

## 4 Notes

1. Use only synchronized trophozoite *P. falciparum*-iRBC cultures to generated data using flow cytometry. Make sure the parasitemia of the culture is approx. 2% (synchronous cultures consisting of mostly trophozoites) 48 h before the day of the experiment to obtain a parasitemia of approx. 5–10% on the day of the experiment. It is recommended to use high parasitemia for flow cytometry to obtain more accurate high-throughput data. Parasitemia [%] is determined by counting the number of *P. falciparum*-iRBCs per 1000 total cells (i.e., includes both iRBCs and uninfected RBCs) and expressed as a percentage:

   Parasitemia [%] = (number of iRBCs/1000 cells) × 100.

2. Ringer's solution contains several salts to create an isotonic solution that mimics the environment of the cell. The proportions of these salts vary from organism to organism. The Ringer's solution prepared here has been optimized for *P. falciparum*-iRBC cultures. Adjusting the pH of Ringer's solution should be done at 37 °C as temperature affects pH. After freshly preparing the Ringer's solution make sure not to dissolve the components of Ringer's solution in the total volume of $ddH_20$. For instance, if you are making 500 mL of Ringer's solution, dissolve components in 350 mL of $ddH_20$, adjust pH and then add $ddH_20$ to a total of 500 mL, put a clean stirrer into the beaker, and place it on a hot plate. Set the hot plate to 37 °C, and turn on the stirrer. It takes a while to have the temperature of the solution reach 37 °C; you may need to increase the temperature of the hot plate to speed up this process. Check the pH of the Ringer's solution while making sure that the temperature of the Ringer's solution remains constant at 37 °C. When the temperature of the solution is stable at 37 °C, adjust pH to the desired pH, in this case 7.4. Store Ringer's solution at 4 °C after the pH has been adjusted. Ringer's solution can be stored and used for a maximum of one month. It is advised to verify the pH of the solution at 37 °C on the day of use, even though it should remain constant even after storing at 4 °C.

3. For $pH_i$ calibration, a sodium-free calibration solution with varying pH is required. There is no need to initially adjust the pH of the calibration solution, as this will be done on the day the calibration solutions are used. Store the solution at 4 °C after preparation and sterilization. The pH of the aliquoted solution is adjusted on the day of experiment or not more than 24 h before the day of the experiment (it takes roughly an hour to adjust the pH for the eight aliquoted pH calibration solutions). pH adjustment should be done at 37 °C, as stated in

**Note 2**. Only use KOH (not NaOH) and HCl to adjust the pH, as sodium should be absent from the pH calibration solution.

4. Nigericin is a selective $H^+$ ionophore that exchanges $K^+$ and $H^+$ across cell membranes. It is used in live-cell imaging experiments to equilibrate intracellular pH to that of the prepared pH calibration solutions. By adding parasites to these calibration solutions, fluorescence intensity obtained from the cellular contents of the test sample can be measured and directly be linked to the set pH value of the solution. Prepare stock solutions of 10 mM nigericin in DMSO and store at −20 °C. Calibration solutions with a defined pH should contain and end concentration of 5 μM nigericin. Only add nigericin to the aliquoted calibration buffers after pH is adjusted. Limit the nigericin incubation period with the malaria parasites to a maximum of 5 min, as prolonged nigericin exposure can reduce cell viability, making it difficult to measure accurate fluorescence data in specific cellular regions. Once nigericin is added, the calibration solution can only be used on the day of the experiment. Do not store at 4 °C to use the following day.
5. Ensure that the RBCs and *P. falciparum*-iRBCs in culture are resuspended in media by gently pipetting up and down 2–3 times before transferring to an Eppendorf tube. RBCs are non-adhering cells that makes it easy to resuspend into solution after they have settled at the bottom of a Petri dish or culture flask. Cells can be incubated in Ringer's solution with or without antimalarials at 37 °C during a 4-h period. We do not recommend incubating samples beyond 4 h for flow cytometry.

## Acknowledgments

The authors gratefully acknowledge funding of the Fonds de Recherche du Quebec – Nature and Technology (FRQ-NT) and the Natural Sciences and Engineering Research Council of Canada (NSERC). The funders had no role in the study design, data collection, and analysis, decision to publish, or preparation of the manuscript.

## References

1. Putnam RW (2012) Intracellular pH regulation. Cell Physiology Book Source, Academic press; 17(4):303–321. https://doi.org/10.1016/B978-0-12-387738-3.00017-2
2. Banerjee R, Liu J, Beatty W et al (2001) Four plasmepsins are active in the plasmodium falciparum food vacuole, including a protease with an active-site histidine. PNAS 99(2):990–995. https://doi.org/10.1073/pnas.022630099
3. Dzekunov SM, Ursos LM, Roepe PD (2000) Digestive vacuolar pH of intact intraerythrocytic P. falciparum either sensitive or resistant to chloroquine. Mol Biochem Parasitol 110(1):

107–124. https://doi.org/10.1016/s0166-6851(00)00261-9

4. Ginsburg HE, Nissani E, Krugliak M (1989) Alkalinization of the food vacuole of malaria parasites by quinoline drugs and alkylamines is not correlated with their antimalarial activity. Biochem Pharmacol 38(16):2645–2654. https://doi.org/10.1016/0006-2952(89)90550-9
5. Rohrbach P (2009) Imaging ion flux and ion homeostasis in blood stage malaria parasites. Biotechnol J 4(6):812–825. https://doi.org/10.1002/biot.200900084
6. Wissing F, Sanchez CP, Rohrbach P et al (2002) Illumination of the malaria parasite plasmodium falciparum alters intracellular pH. Implications for live cell imaging. J Biol Chem 277(40):37747–37755. https://doi.org/10.1074/jbc.M204845200
7. Miesenbock G, De Angelis DA, Rothman JE (1998) Visualizing secretion and synaptic transmission with pH sensitive green fluorescent proteins. Nature 394(6689):192–195. https://doi.org/10.1038/28190
8. Kuhn Y, Rohrbach P, Lanzer M (2007) Quantitative pH measurements in plasmodium falciparum-infected erythrocytes using pHluorin. Cell Microbiol 9(4):1004–1013. https://doi.org/10.1111/j.1462-5822.2006.00847.x
9. Hess ST, Heikal AA, Webb WW (2004) Fluorescence Photoconversion kinetics in novel green fluorescent protein pH sensors (pHluorins). J Phys Chem B 108(28):10138–10148. https://doi.org/10.1021/jp0362077
10. Rohrbach P (2012) Quantitative fluorescent live cell imaging of intracellular Ca2+ and H+ ions in malaria parasites. Methods Enzymol 505:469–483. https://doi.org/10.1016/b978-0-12-388448-0.00032-2
11. Mahon MJ (2011) pHluorin2: an enhanced, ratiometric, pH-sensitive green florescent protein. Adv Biosci Biotechnol 2(3):132–137. https://doi.org/10.4236/abb.2011.23021
12. Yayon A, Cabantchik ZI, Ginsburg H (1984) Identification of the acidic compartment of plasmodium falciparum-infected human erythrocytes as the target of the antimalarial drug chloroquine. EMBO J 3(11):269–2700. https://doi.org/10.1002/j.1460-2075.1984.tb02195.x
13. Krogstad DJ, Schlesinger PH, Gluzman IY (1985) Antimalarials increase vesicle pH in plasmodium falciparum. J Cell Biol 101(6):2302–2309. https://doi.org/10.1083/jcb.101.6.2302
14. Hayward R, Saliba KJ, Kirk K (2006) The pH of the digestive vacuole of plasmodium falciparum is not associated with chloroquine resistance. J Cell Sci 119(Pt 6):1016–1025. https://doi.org/10.1242/jcs.02795
15. Trager W, Jensen JB (1976) Human malaria parasite in continuous culture. Science 193(4254):673–675. https://doi.org/10.1126/science.781840
16. Wünsch S, Sanchez CP, Gekle M et al (1998) Differential stimulation of the Na+/H+ exchanger determines chloroquine uptake in plasmodium falciparum. J Cell Biol 140(2):335–345. https://doi.org/10.1083/jcb.140.2.335
17. Wünsch S, Horrocks P, Gekle M et al (1999) Single-cell measurements of ion concentrations within the intracellular parasite. Parasitol Today 15(5):198–200. https://doi.org/10.1016/s0169-4758(99)01432-5

# Chapter 14

# Cell Membrane State, Permeability, and Elasticity Assessment for Single Cells and Cell Ensembles

**Nicolas Färber, Simon V. Neidinger, and Christoph Westerhausen**

## Abstract

The phase state and especially phase transitions of synthetic lipid membranes are known to drastically modulate mechanical membrane properties like permeability and bending modulus. Although the main transition of lipid membranes is typically detected employing differential scanning calorimetry (DSC), this technique is not suitable for many biological membranes. Moreover, often single cell data on the membrane state or order is of interest. We here first describe how to use a membrane polarity-sensitive dye, Laurdan, to optically determine the order of cell ensembles over a wide temperature range from $T = -40$ °C to +95 °C. This allows to quantify the position and width of biological membrane order–disorder transitions. Second, we show that the distribution of membrane order within a cell ensemble allows for correlation analysis of membrane order and permeability. Third, combining the technique with conventional atomic force spectroscopy allows for the quantitative correlation of an overall effective Young's modulus of living cells with the membrane order.

**Key words** Cell membrane, Lipid order, Laurdan, Fluorescence spectroscopy, Permeability, Elasticity

## 1 Introduction

From synthetic lipid membranes, it is known that the membrane state modulates nonlinearly membrane characteristics like bending energy, membrane tension, permeability, and so forth. As biological membranes are composed to a large extent of lipid molecules, the question arises whether the physical membrane properties of living cells and, in consequence, cellular functions depend as well on the lipid phase state. So far, it was shown that the cell membrane phase state measured by the determination of the lipid order influences, for example, the membrane bending rigidity, the immunological response in T cell membranes, and focal adhesion domains [1–3]. To which extent other cellular functions are regulated by the membrane phase state needs to be investigated.

In this context, we describe the technique to determine the cell membrane lipid order as measure of the phase state and provide a

Oliver Friedrich and Daniel F. Gilbert (eds.), *Cell Viability Assays: Methods and Protocols*, Methods in Molecular Biology, vol. 2644, https://doi.org/10.1007/978-1-0716-3052-5_14,

protocol to simultaneously measure the membrane properties "permeability" and "elasticity." For this, we use conventional epifluorescence microscopy/spectroscopy, the dye Laurdan for lipid order determination, and rhodamine B and Hoechst 33342 as permeability probe. However, the cargo of interest simply can be exchanged. The elastic properties of the cell membrane are measured by atomic force microscopy/spectroscopy.

Exchanging the dye Laurdan with a solvatochromic dye specific for the plasma membrane, for example, NR4A, could be an interesting modification of our protocol especially regarding the correlation of permeability and membrane order [4].

## 2 Materials

### 2.1 Solutions and Reagents

1. Laurdan solution: 2.8 mM Laurdan in dimethyl sulfoxide (DMSO). Dissolve 1 mg of Laurdan powder in 1 mL dimethyl sulfoxide (DMSO) in an amber glass vial. To speed up this procedure and to ensure full dissolution, use an ultrasonic bath at $T = 50$ °C for about 30 min. Each time after storage in the fridge, the ultrasonic treatment should be repeated to dissolve all Laurdan crystals.
2. Rhodamine B solution: 2 mM rhodamine B in phosphate-buffered saline (PBS).
3. Hoechst 33342 solution: 16 μM Hoechst 33342 in PBS buffer.
4. Trypsin-EDTA solution: 0.05%.
5. Standard PBS.

### 2.2 Cell Culture

1. Any cell line of interest.
2. Standard cell culture equipment (e.g., clean bench, incubators, sterile equipment, etc.).
3. Petri dish.
4. Coverslip.
5. Cell culture medium (DMEM with 10% FBS and 1% Penicillin-Streptomycin).
6. 4 μL Poly-L-lysine (PLL, 0.1% in water with 0.01% thiomersal) with 996 μL PBS mixture.

### 2.3 Other Equipment

1. 440 nm filter cube: 360 nm excitation, 440 nm emission.
2. 490 nm filter cube: 360 nm excitation, 490 nm emission.
3. 645 nm filter cube: 585 nm excitation, 645 nm emission.
4. 360 nm, 3 W, LED.
5. Spectrometer: minimum range 400–700 nm.

6. Cantilevers of the desired type.
7. Plasma cleaner.
8. Atomic force microscope (AFM).

## 3 Methods

### 3.1 Laurdan Staining and Incubation

1. For cell membrane staining, add the Laurdan solution to the culture medium in a volume ratio of 1:100. For example, for staining of adherent cells in a 25 $cm^2$ culture flask, add 50 μL of the staining solution to 5 mL of culture medium (for further information about influences of the staining procedure *see* **Notes 1** and **2**). To avoid cell damage by locally high DMSO concentrations, premix the 50 μL staining solution with 1 mL of the culture medium in a separate vial, and add this solution to the culture flask. Do not leave the amber staining solution vial open for too long. DMSO is hygroscopic, and therefore, it will absorb water from the air. Because of this, after long usage of the same staining solution, it will not freeze anymore in the fridge. At this point at latest, you should prepare a fresh solution.
2. After incubation for 2 h in the dark at $T = 37$ °C, remove all of the medium and rinse three times with 5 mL PBS to reduce the background signal (for further information about the staining time *see* **Note 2**). You must be very careful during this washing procedure as the cells easily detach due to the high-DMSO concentration during staining. For analysis of adherent cells, add 5 mL of PBS as imaging medium. Do not use media containing phenol red as it interferes with the Laurdan fluorescence emission signal.
3. For analysis of the cells in suspension, trypsinate with 1 mL of 0.05% trypsin-EDTA solution, and incubate for 5 min. Again, care should be taken that no phenol red or other absorbing/fluorescing agents are added before spectral analysis.

### 3.2 Permeability Assessment

Membrane permeability is assessed by the analysis of rhodamine B or Hoechst 33342 diffusion into the cytosol.

1. Add the rhodamine B solution 1:1 to the culture medium already containing Laurdan 10 min prior to the end of the Laurdan staining procedure.
2. Remove the medium containing Laurdan and rhodamine B after 10 min, and rinse the cells carefully three times with PBS as described above.

While the use of rhodamine B allows for simultaneous phase state and permeability analysis, in contrast, the uptake of Hoechst

33342 must be analyzed after phase state determination due to channel cross talk. For the permeability measurement using Hoechst 33342, add the solution 1:1 to the imaging buffer (PBS). The uptake of Hoechst 33342 can be imaged without rinsing procedure as it only fluoresces after uptake and intercalation in the DNA.

### 3.3 Spectral Analysis of Laurdan with Two Filter Cubes and Permeability Measurement

1. Mount your sample into a tempered microscope setup. *See* **Note 3** or options to control the sample temperature.
2. Record two images per region of interest using the 440 nm and the 490 nm filter cubes. For more detailed information about how Laurdan measures lipid order, *see* **Note 4**.
3. Keep the ultraviolet exposure during image acquisition as low as possible to avoid artefacts as explained in **Note 5**.
4. Record a third image for each region of interest, for example, using the 645 nm filter cube to measure membrane permeability by uptake of rhodamine B simultaneously to lipid order. Be careful about channel cross talk as discussed in **Note 6,** or measure membrane permeability and lipid order sequentially as explained in **Note** 7.

### 3.4 Spectral Analysis of Laurdan with a Spectrometer

1. Put your sample into a transparent sample container.
2. Place the sample container in a tempered sample holder with optical fiber mount.
3. Attach an optical fiber and a spectrometer.
4. Excite the sample with an ultraviolet LED and record the spectrum with a spectrometer. The difference to the data obtained using a fluorescence microscope is discussed in **Note 8**.

### 3.5 Acquisition of Elasticity Data

1. Use a new cantilever or clean a used one with a short plasma procedure followed by a buffer bath.
2. Leave the PLL-PBS mixture to act on a coverslip in a Petri dish for 1 h.
3. Rinse the incubated cover slip with PBS and cell medium.
4. Add the cell suspension and wait a few moments, for example, 5 min.
5. Carefully drain off the remaining liquid and transfer the cover slip to a new Petri dish.
6. Fill the Petri dish with cell medium or PBS, and start the AFM measurement with the desired parameters (*see* **Notes 9–14**).

### 3.6 Microscopy Data Analysis

1. Correct the images for background intensity (*see* **Note 15**).
2. Proceed with the image segmentation (*see* **Note 16**).

3. Calculate GP for each pixel (*see* **Note 17**).
4. Calculate GP for each cell (*see* **Note 18**).
5. Calculate the cell membrane permeability (*see* **Note 19**).

### 3.7 Spectrometer Data Analysis

1. Approximate the spectral data by two lognormal functions to calculate GP (*see* **Note 20**).
2. Plot GP as function of temperature (*see* **Note 21**).
3. Calculate the derivative of GP with respect to temperature to obtain thermodynamic data (*see* **Note 21**).

### 3.8 Mechanical Data Analysis

1. Subtract baseline and contact point of the AFM force curve.
2. Correct the cantilever bending.
3. Use the fit function appropriate for the tip geometry (*see* **Notes 22** and **23**).
4. Apply descriptive statistics (*see* **Note 24**).

## 4 Notes

1. Influence of environmental conditions on lipid order: The cell membrane phase state is a quantity depending on many variables such as temperature, pH, solvent content, salt concentration, lipid composition, cholesterol content, and more. Therefore, care must be taken when you prepare your own culture media to ensure that these variables do not vary significantly between the single experiments. In addition, the local environment of each cell influences the membrane phase state as well: with increasing confluence in the culture flask, the lipid order increases as well [1, 5]. In consequence, you must always ensure to work with cells of equal confluence state. Moreover, depending on the cell type, lipid order can increase or decrease with the number of the cell passages [6].Care should also be taken when comparing cells in the adherent and suspended state. During trypsination-induced cell detachment, lipid order decreases [5].
2. Influence of Laurdan incubation time and concentration: as this protocol is designed for lipid order analysis using conventional epifluorescence microscopes where the fluorescence intensity should be high to compensate for losses due to by photobleaching, the dye concentration is comparably high and, in consequence, the DMSO content as well. Both quantities again influence the lipid order: During staining, the lipid order decreases. For the abovementioned Laurdan/DMSO concentration, the staining-induced lipid order changes saturate after 2 h [5]. To avoid that small variations in the staining process

lead to apparent varations of the detected membrane-phase state, the incubation time is set to 2 h as well. If other concentrations are used, this time can be shortened or might have to be enlarged. The same holds true for other incubation temperatures: The staining-induced lipid changes might saturate later at lower temperatures.

3. Temperature control and temperature-dependent lipid order: a key factor in the context of lipid order analysis is the temperature control. On the one hand, the precise control of temperature guarantees high comparability and reproducibility among data sets. On the other hand, the analysis of the temperature dependence of the cell membrane lipid order allows for the determination of thermodynamic quantities such as cooperativity, enthalpy, and melting temperature. At the microscope, temperature can be regulated by commercially available microscope incubators, which are commonly limited to a temperature interval from room temperature to 20 K above. For wider ranges even below room temperature, custom-made setups containing heat baths and Peltier elements are necessary. Independent of the setup, an in situ temperature probe is strongly recommended, as stage incubators exhibit often long temperature equilibration times.
4. The emission spectrum of Laurdan depends on the polarity and the relaxation dynamics of its environment [7, 8]. Therefore, the emission maximum shifts from 440 nm, meaning that the lipid chain order is high and the polarity in the membrane is low due to the lack of water molecules to 490 nm. This indicates that the lipid order is low, and energy is lost due to solvent relaxation. This change can be monitored by the use of two fluorescence microscope filter cubes with the same excitation filters (360 nm) but different emission filters: 440 and 490 nm. To analyze the cell membrane lipid order, two micrographs have to be taken at the same position successively; one for each filter cube.
5. Possible artifacts: During image acquisition, photobleaching is not only an issue of intensity loss, but the ultraviolet irradiation impacts the phase state of cell membranes such as it increases lipid order. Therefore, it is necessary to record reference images if cells are imaged multiple times to distinguish between UV induced artifacts and other phase state changes. Of course, the UV effect is strongly setup dependent, but as a rough estimate, we here give exemplary empirical values for our setup. We observed at a Zeiss Axiovert 200 M epifluorescence microscope with a mercury-vapor lamp, a 10x objective and 1 s exposure time, a change of the generalized polarization (GP) as measure of lipid order of roughly $\Delta GP = 0.005$ per recorded image at the same position.

6. If another quantity, such as permeability, is analyzed by fluorescence microscopy, care must be taken to avoid channel cross talk, which could influence the lipid order measurements. In case of rhodamine B, the emission spectrum is far enough in the red compared to Laurdan so that permeability can be measured simultaneously to the lipid order analysis by the use of a third channel (645 nm filter cube). The TexasRed intensity is proportional to the uptake of rhodamine B and acts as a measure for permeability.
7. The use of Hoechst 33342 facilitates the experimental procedure as no rinsing steps are required because of the lack of background intensity. Therefore, it also allows for recording a time series of Hoechst 33342 uptake where increasing intensity again indicates a high membrane permeability. But in contrast to rhodamine B, there is spectral overlap of the Laurdan and Hoechst 33342 spectra. Therefore, the lipid order determination with the 440/490 nm filter cubes has to be done prior to the Hoechst 33342 addition. Afterwards, the fluorescence intensity of Hoechst 33342 should be measured with the 440 nm emission filter cube. By this, the permeability measurement can later be corrected for the Laurdan fluorescence intensity.
8. The use of a microscope with the mentioned filter sets for Laurdan analysis allows for spatially resolved lipid order analysis. But, as only a fraction of the spectrum is analyzed, namely the regions around 440 and 490 nm, the sensitivity to phase state changes is reduced as compared to the analysis of the whole spectrum with a spectrometer. The higher sensitivity is accompanied by the loss of spatial information. If the spectrometer is attached with an optical fiber to the microscope, the obtained spectrum is a mean value of the cell ensemble in the focal plane.
9. Atomic force microscopy: we recommend calibration before measurement on a cover glass without PLL.
10. To increase the validity of the results, it must be ensured that cells are not unintentionally measured twice and that the positioning of the cantilevers on the cell is always comparable.
11. Influence of tip geometry: depending on the desired measurement intention, the use of different cantilevers should be considered. Measuring probes with a sharp tip could damage the cell [9], but at the same time provides a high spatial resolution. Cantilevers without a tip, on the other hand, can be used very well for determining the global cell properties without putting too much strain on the cell locally.
12. Influence of repeated measurements: when recording a larger number of force curves on a particular cell, it becomes apparent

that, on the one hand, there is a significant scatter between the individual force curves, and, on the other hand, a medium-term change in the Young's modulus can be observed. Therefore, it is recommended to record a comparatively small number of force curves (approx. 20) from a single cell.

13. Influence of variance: The measurement curves vary considerably between individual cells. It is, therefore, advisable to run several parameters on one cell to allow for the analysis of relative changes of the Young's modulus.

14. Spatial measurements: while classical imaging initially only reveals information about the topography, using a measuring grid in Force Spectroscopy mode allows both forces to be measured and topography data to be collected via the set-point height. Thus, even with a conventional atomic force microscope, Young's modulus maps of cells can be created without special software. Be aware of the overall geometry.

15. Prior to the calculation of the generalized polarization (GP) (see 3.6.3) values, the 440 and 490 nm images need to be corrected for background intensity resulting from thermal detector noise, stray light, or not membrane-embedded Laurdan. If this is not done, the GP values are shifted toward zero with decreasing signal to background ratio.

16. For lipid order and permeability analysis on a single-cell level, each pixel has to be assigned to the background or a cell (see Fig. 1). Neural networks are predestined for this task, and platforms such as the python package "cellpose" are freeware, easy to use, and do not have to be trained [10].

17. After assigning each pixel to the corresponding cell, the lipid order at each pixel can be quantified using the "generalized polarization" (GP), which is defined as the difference between the 440 and 490 nm intensities normalized to their intensity-sum: $\frac{I_{440}-I_{490}}{I_{440}+I_{490}}$ [11] (see Fig. 1).

18. Cell-GP-calculation: either different regions of each single cell can be analyzed separately, or a mean GP value can be calculated as sum of all pixel GP values divided by the number of pixels. Another method for mean GP value calculation would be to calculate the mean 440/490 nm intensity of all cell pixels and use these mean values for GP calculation. The two approaches yield different results as in case of the second method; regions with lower fluorescence intensity are less weighted than regions of high-fluorescence intensity. If microscopy data are compared to spectrometer data, the second method of mean GP calculation turns out to be more comparable.

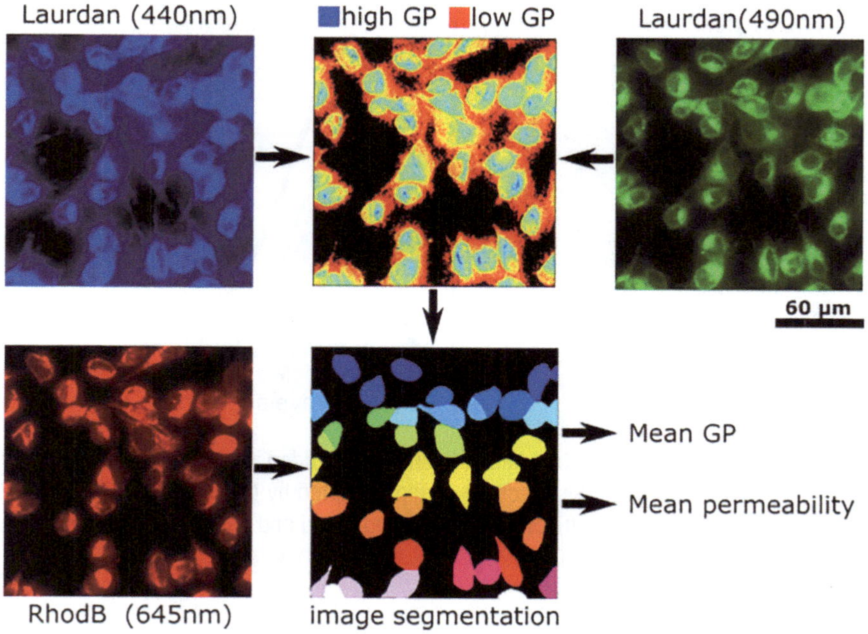


**Fig. 1** Raw and processed image data of a HeLa cell lipid order and membrane permeability assay. While the two Laurdan images captured with different filter sets allow for quantitation of lipid order measured by GP, the rhodamine B intensities provide data about the membrane permeability. Both quantities can be assigned to single cells using an image segmentation algorithm

19. In case of rhodamine B uptake measurements, the mean pixel intensity of the rhodamine B channel can be used after background correction as measure for membrane permeability (see Fig. 1). The mean intensity data of Hoechst 33342 recorded using the 490 nm filter set has to be corrected for the Laurdan intensity determined prior to the permeability assay. As a result of the microscopy image analysis, it is possible to quantify a cell membrane property, such as permeability, as function of the lipid order and test for correlation of both quantities.
20. Spectrometric data can be analyzed by the use of the two wavelength regions around 440 and 490 nm. Another more precise method that uses all of the spectral data is shown in Fig. 2: The data set is approximated by two lognormal functions centered around 440/490 nm [12]. The according fit function areas correspond to the number of ordered/disordered lipids and can be used for GP calculation.
21. Deduction of quantities of interest: as mentioned before, the temperature dependence of membrane lipid order provides insight into thermodynamic properties of the lipid system. This is illustrated in Fig. 3a: The temperature-dependent GP values of different phospholipid vesicles indicate phase transitions as a steep decrease, meaning that lipid order is strongly

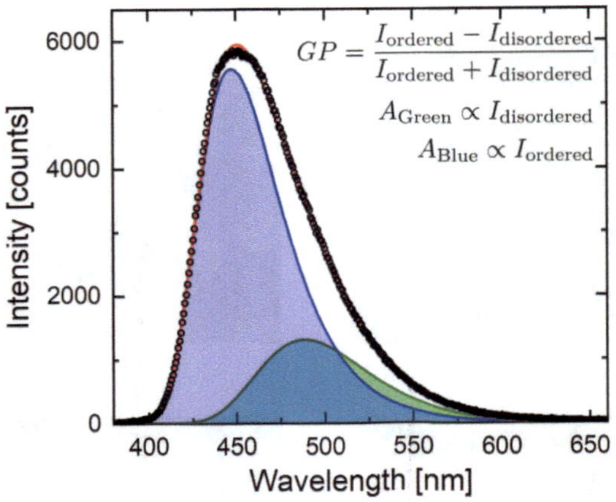


**Fig. 2** Spectral data of Laurdan embedded in the membrane of HeLa cells (black circles). The spectrum is approximated by the red line, which is the sum of two lognormal functions represented in red and blue. The blue/green areas are used for GP calculation

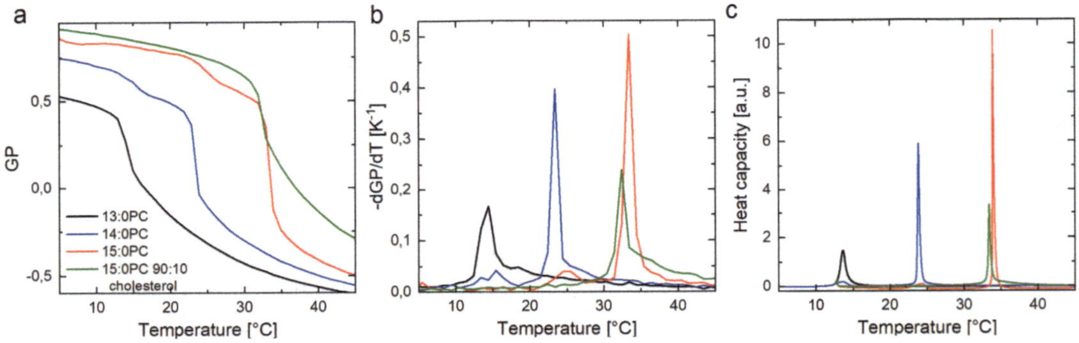


**Fig. 3** Phase state analysis of phospholipid vesicles. (**a**) GP as function of temperature of different phospholipid vesicle suspensions recorded with a spectrometer in a tempered cuvette. (**b**) Derivative of GP data in (a) with respect to temperature. The peaks indicate the phase transition of the lipid vesicles. (**c**) calorimetric data of the vesicles analyzed in (a) and (b). The peak position and the peak height of optical and calorimetric data are in good agreement

reduced at the melting temperature. The derivative of GP with respect to temperature in Fig. 3b shows a peak at the melting transition as it is known from differential calorimetry measurements. The good agreement of the optical and the calorimetric data in in Fig. 3c suggests that the Laurdan lipid order analysis can provide useful thermodynamic data such as melting temperature (peak position) and cooperativity (peak height and width) of lipid systems [5]. It is also possible to record such melting profiles in cellular membranes as shown in Fig. 4. But as the melting regime of biological membranes is markedly larger, the temperature range for GP determination has to be enlarged from −40 °C to 90 °C.

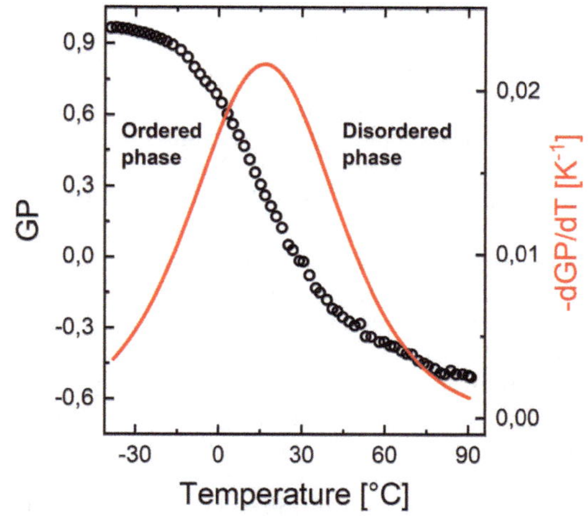


**Fig. 4** GP and the derivative of GP with respect to temperature as a function of temperature. The melting profile of HeLa cell membranes extends over a much larger temperature regime as compared to phospholipid vesicles

22. Fit limitations: please note the assumptions and limitations associated with the fit [13].
23. Analysis of viscosity: in addition to the elasticity of the cells, their viscous behavior can also be analyzed. For example, after the set-point force has been reached, a pause at constant force can be built into the procedure, and the viscous properties of the cell can be inferred from the flow behavior [14].
24. Statistics: To maximize the informative value of the data obtained, the use of appropriate statistics is recommended. For example, intra-class correlation can be used, which deals with the variance between measurements and within measurements [15].

## Acknowledgments

We would like to acknowledge financial support by "WIR Wissenstransfer Region Augsburg" for the project idea "MOlyzer—an automated membrane order analyzer." C.W. would like to acknowledge funding for the project "Physical and functional interaction mechanisms at cell membranes and vessel walls" by the University of Augsburg. N.F. would like to acknowledge financial support by the Joachim Herz foundation. S.V.N. would like to thank the Hanns-Seidel-Stiftung e.V. for its support.

## References


1. Steinkühler J, Sezgin E, Urbančič I et al (2019) Mechanical properties of plasma membrane vesicles correlate with lipid order, viscosity and cell density. Commun Biol 2:337. https://doi.org/10.1038/s42003-019-0583-3
2. Gaus K, Chklovskaia E, Fazekas De St. Groth B et al (2005) Condensation of the plasma membrane at the site of T lymphocyte activation. J Cell Biol 171:121–131. https://doi.org/10.1083/jcb.200505047
3. Gaus K, Le Lay S, Balasubramanian N et al (2006) Integrin-mediated adhesion regulates membrane order. J Cell Biol 174:725–734. https://doi.org/10.1083/jcb.200603034
4. Danylchuk DI, Moon S, Xu K et al (2019) Switchable Solvatochromic probes for live-cell super-resolution imaging of plasma membrane organization. Angew Chem Int Ed Engl 131: 15062–15066. https://doi.org/10.1002/ange.201907690
5. Färber N, Westerhausen C (1864) Broad lipid phase transitions in mammalian cell membranes measured by Laurdan fluorescence spectroscopy. Biochim Biophys Acta Biomembr 2022:183794. https://doi.org/10.1016/j.bbamem.2021.183794
6. Pokorna S, Ventura AE, Santos TCB et al (2022) Laurdan in live cell imaging: effect of acquisition settings, cell culture conditions and data analysis on generalized polarization measurements. J Photochem Photobiol B Biol 228: 112404
7. Parasassi T, Conti F, Gratton E (1986) Time-resolved fluorescence emission spectra of Laurdan in phospholipid vesicles by multifrequency phase and modulation fluorometry. Cell Mol Biol 32:103–108
8. Parasassi T, Gratton E (1992) Packing of phospholipid vesicles studied by oxygen quenching of Laurdan fluorescence. J Fluoresc 2:167–174. https://doi.org/10.1007/BF00866931
9. Chen CL, Mahalingam D, Osmulski P et al (2013) Single-cell analysis of circulating tumor cells identifies cumulative expression patterns of EMT-related genes in metastatic prostate cancer. Prostate 73:813–826. https://doi.org/10.1002/pros.22625
10. Stringer C, Wang T, Michaelos M et al (2021) Cellpose: a generalist algorithm for cellular segmentation. Nat Methods 18:100–106. https://doi.org/10.1038/s41592-020-01018-x
11. Parasassi T, De Stasio G, Ravagnan G et al (1991) Quantitation of lipid phases in phospholipid vesicles by the generalized polarization of Laurdan fluorescence. Biophys J 60: 179–189
12. Bacalum M, Zorila B, Radu M (2013) Fluorescence spectra decomposition by asymmetric functions: Laurdan spectrum revisited. Anal Biochem 440:123–129. https://doi.org/10.1016/j.ab.2013.05.031
13. JPK DP (2015) Data Processing Software Manual, Version 5.0
14. Hecht FM, Rheinlaender J, Schierbaum N et al (2015) Imaging viscoelastic properties of live cells by AFM: power-law rheology on the nanoscale. Soft Matter 11:4584–4591. https://doi.org/10.1039/c4sm02718c
15. Liljequist D, Elfving B, Roaldsen KS (2019) Intraclass correlation – a discussion and demonstration of basic features. PLoS One 14: e0219854. https://doi.org/10.1371/journal.pone.0219854

# Chapter 15

# Neutral Red Uptake Assay to Assess Cytotoxicity In Vitro

**Robim M. Rodrigues, Marth Stinckens, Gamze Ates, and Tamara Vanhaecke**

## Abstract

The neutral red uptake (NRU) assay is a cell viability assay that can be used for the assessment of compound-induced cytotoxicity. It is based on the ability of living cells to incorporate neutral red, a weak cationic dye, in lysosomes. The quantification of xenobiotic-induced cytotoxicity is expressed as a concentration-dependent reduction of the uptake of neutral red when compared to cells exposed to corresponding vehicle controls. The NRU assay is mainly used for hazard assessment in in vitro toxicology applications. Hence, this method has been incorporated in regulatory recommendations such as the OECD test guideline TG 432, in which an in vitro 3T3-NRU-phototoxicityassay is described to assess the cytotoxicity of compounds in the presence or absence of UV light.

This book chapter describes a detailed protocol to carry out the NRU assay using the human hepatoma cell line HepG2, which is frequently employed as an alternative in vitro model for human hepatocytes. As an example, the cytotoxicity of acetaminophen and acetylsalicylic acid is assessed.

**Key words** Viability assay, Neutral red uptake, HepG2, In vitro cytotoxicity

## 1 Introduction

The neutral red uptake (NRU) assay is a viability assay in which viable, undamaged cells bind the supravital dye neutral red (NR) (Fig. 1) [1]. This weak cationic eurhodine can penetrate cells by non-ionic diffusion at physiological pH. Once NR is in the cell, it accumulates in the lysosomes, where a proton gradient assures a more acidic pH allowing positive charging of NR [2]. In dead or damaged cells, NR does not accumulate in the lysosomes and can be washed away. Xenobiotics can cause alterations of the cell surface or lysosomal membrane, resulting in a decreased uptake and binding of NR. As such, the NRU assay permits the assessment of membrane permeability and lysosomal activity, making it possible to distinguish between viable, damaged, and dead cells. Cytotoxicity is expressed as a concentration-dependent reduction of the uptake of NR after exposure of the cells to the xenobiotic and

Oliver Friedrich and Daniel F. Gilbert (eds.), *Cell Viability Assays: Methods and Protocols*, Methods in Molecular Biology, vol. 2644, https://doi.org/10.1007/978-1-0716-3052-5_15,

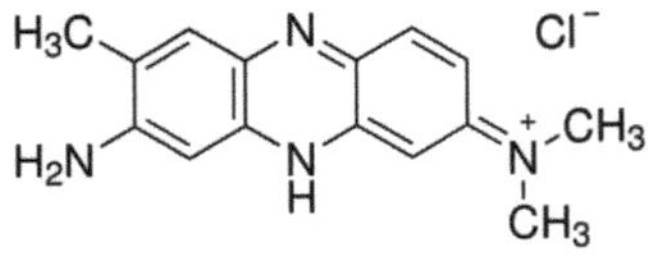

**Fig. 1** Structural formula of neutral red

comparison to cells exposed to the solvent used to dissolve the compound under investigation (i.e., the vehicle control (VC)). The NRU assay provides a sensitive, integrated signal of both cell integrity and cell growth inhibition [1].

The use of the most correct terminology when referring to "viability assays" or "cytotoxicity assays" has been questioned. Yet, both terminologies have been used simultaneously depending on the emphasis chosen in the application of the assay. Whereas cell "viability" is used when the number of living cells is quantified, "cytotoxicity" indicates rather the response of the cell as a result of a toxic insult by a xenobiotic [3]. The NRU assay quantifies a cellular function of viable cells represented by lysosomal activity and as such measures directly a viability parameter. However, when it is applied in the assessment of the toxicity of xenobiotics, it can be referred to as a cytotoxicity assay. The NRU assay has miscellaneous biological applications and is commonly used in in vitro hazard assessment of a variety of chemical substances such as pharmaceuticals and cosmetics [4, 5]. Several validation studies have been set up for the NRU assay as a test for cytotoxicity [6]. In 2000, the NRU test on Balb/c 3T3 mouse fibroblasts to assess phototoxicity was regulatory accepted in all EU member states, and in 2004, it was adopted as an official Organisation for Economic Co-operation and Development (OECD) test guideline TG 432 [7]. In 2013, the European Commission Joint Research Centre published a recommendation on the use of the 3T3-NRU assay in which it stresses the validity of the NRU in a weight-of-evidence approach to predict acute oral toxicity of chemicals as an alternative to animal testing in a regulatory setting [8]. The simplicity of the NRU assay permits automation, which improves throughput and allows fast and reliable screening of a large number of test compounds in a relatively short time span [5, 9].

For this book chapter, the NRU assay is carried out using HepG2 cells. This human hepatoma cell line was derived from a hepatocellular carcinoma of a 15-year-old Caucasian male and is widely used in hepatotoxicity studies [10]. Under adequate culture conditions, HepG2 cells display (limited) hepatocyte-like features and are therefore often utilized as an alternative in vitro model for human hepatocytes [10–12]. Due to their ubiquity, the metabolism of HepG2 cells may, however, differ depending on the source and culture conditions. Therefore, the characterization of these cells and their obtainment through a reliable source is recommended [13].

## 2 Materials

### 2.1 General Equipment

1. Incubator: 37 °C ± 1 °C, 90% ± 5% humidity, 5.0% ± 1% $CO_2$/air.
2. Laminar flow clean bench/cabinet (standard "biological hazard").
3. Water bath: 37 °C ± 1 °C.
4. Inverted phase-contrast microscope.
5. Analytical balance.
6. Microplate spectrophotometer (i.e., plate reader) equipped with 540 nm ± 10 nm filter.
7. Shaker for microtiter plates.
8. Cell counter or hemocytometer.
9. Pipettes, pipettors (multichannel and single channel; multichannel repeater pipette).
10. Vacuum aspiration system.
11. 96-well flat-bottom tissue culture microtiter plates.
12. Multichannel reagent reservoir.
13. Vortex mixer.
14. Sterile filters for filtration of media.
15. Sterile conical tubes (50 mL).

### 2.2 Reagents

1. HepG2 cells.
2. Routine culture medium: Dulbecco's Modified Eagle's Medium (DMEM) with L-glutamine (preferably) (*see* **Note 1**) and high glucose (4.5 g/L), 10% FBS (*see* **Note 2**), penicillin (100 IU/mL), streptomycin (100 μg/mL), and glutamine (4 mM) (if not already present in media) (*see* **Note 3**).
3. TripLE™ Express Enzyme.
4. Phosphate-buffered saline (PBS) without $Ca^{2+}$ and $Mg^{2+}$.
5. Neutral red (NR) solution: routine culture medium, NR (25 μg/mL).
6. Dimethyl sulfoxide (DMSO, cell culture grade).
7. Distilled water or any purified water suitable for cell culture (sterile).
8. Desorption solution: glacial acetic acid (1%), ethanol absolute (50%), $H_2O$ (49%).
9. Test compounds (acetaminophen and acetylsalicylic acid were used in this protocol).

## 3 Methods

All solutions, glassware, pipettes, etc., must be sterile and all procedures should be carried out under aseptic conditions and in the sterile environment of a laminar flow cabinet (biological hazard standard).

### 3.1 HepG2 Cell Culture

HepG2 cells are routinely grown as a monolayer in tissue culture grade flasks (e.g., 75 $cm^2$) at 37 °C ± 1 °C, 90% ± 5% humidity, and 5.0% ± 1% $CO_2$/air. When cells exceed 50% confluence (but do not reach 80% confluence), they should be passaged by removing them from the flask using TripLE™ Express Enzyme as follows:

1. Pre-warm the routine culture medium, TripLE™ Express Enzyme, and PBS in a water bath at 37 °C.
2. Aspirate the cell supernatants from the flask, briefly rinse cultures with 10 mL PBS, and add 5 mL TripLE™ Express Enzyme.
3. Incubate flask in the incubator for 2–5 min. Tap the flask on the side, to make sure all cells detach (check with a microscope).
4. Carefully resuspend the cells (*see* **Note 4**), and transfer the cell suspension into a 50 mL conical tube.
5. Rinse the culture flask two times with 10 mL pre-warmed routine culture medium, and add it to the 50 mL conical tube. Centrifuge the cell suspension at 385 × *g* for 5 min.
6. Remove supernatant. To avoid aspiration of cells, leave approximately 100 μL routine culture medium on top of the pellet. Resuspend the pellet in a 5 mL routine culture medium, and count the cells using a cell counter or a hemocytometer.
7. Prepare a cell suspension of 2.0–3.0 × $10^4$ cells/mL in a routine culture medium. Using a multichannel pipette, dispense 100 μL routine culture medium only into the peripheral wells (blanks) of a 96-well tissue culture microtiter plate. Seed between 7.5 and 10.0 × $10^3$ cells/well in the remaining wells (*see* **Note 5**).
8. Incubate cells for 24 ± 2 h (37 °C ± 1 °C, 90% ± 5% humidity, 5.0% ± 1% $CO_2$/air) (*see* **Note 6**).
9. Refresh routine culture medium every 2–3 days (*see* **Note 7**).
10. Examine each plate under a phase-contrast microscope to assure that cell growth is even across the microtiter plate and to identify potential systemic cell seeding errors.

**Table 1**
**Test concentration gradients of acetaminophen and acetylsalicylic acid (C1–C8: concentrations)**

| Concentration | C1 | C2 | C3 | C4 | C5 | C6 | C7 | C8 |
|---|---|---|---|---|---|---|---|---|
| Acetaminophen (μg/mL) | 5000 | 2325.6 | 1081.7 | 503.1 | 234.0 | 108.8 | 50.6 | 23.5 |
| Acetylsalicylic acid (μg/mL) | 2000 | 930.2 | 432.7 | 201.2 | 93.6 | 43.5 | 20.2 | 9.4 |

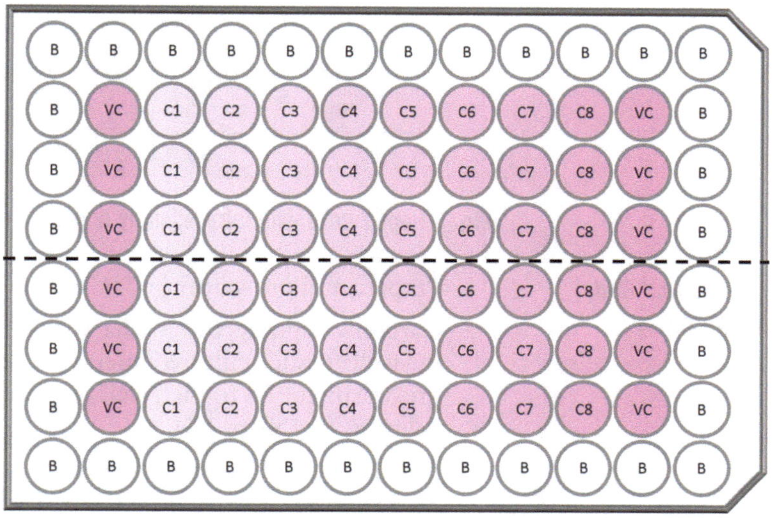


**Fig. 2** Plate layout for cell exposure (B blank; VC vehicle control; C1–C8 test concentration in descending order)

***3.2 Preparation of Test Chemicals***

1. Prepare stock solutions of test chemicals (*see* **Note 8**).
2. Prepare test chemical solutions by diluting stock solutions in a pre-warmed (37 °C) routine culture medium just before exposure to the cells (*see* **Notes 9** and **10**).
3. Table 1 shows an example of the concentration gradients for acetaminophen and acetylsalicylic acid. Hereby a dilution factor of 2.15 was used to prepare the serial dilutions.

***3.3 Chemical Exposure***

1. Aspirate the routine culture medium from the plates using a vacuum aspiration system.
2. Add 100 μL of each concentration of the test chemical according to a predefined plate layout. An example of a plate layout is given in Fig. 2 (*see* **Note 11**).
3. Incubate the plate under standard culture conditions, for 24 ± 1 h. Longer incubation times, e.g., 48 or 72 h, can also be used.

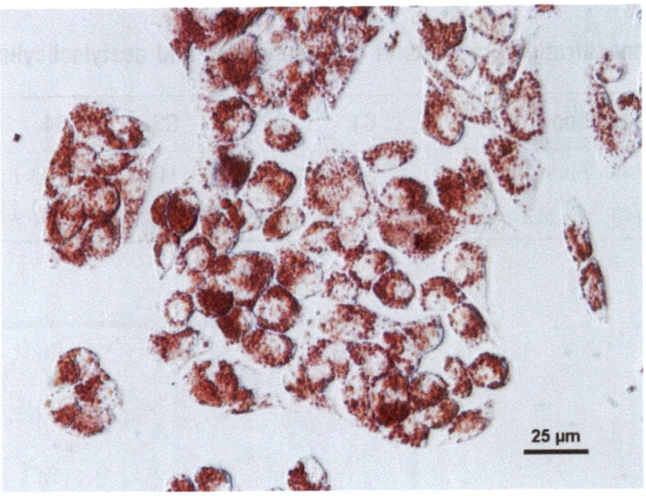


**Fig. 3** HepG2 cells with intracellularly bound neutral red

### 3.4 Neutral Red Uptake Assay

1. Prepare and pre-warm (37 °C) the NR solution (*see* **Note 12**).
2. Carefully aspirate the media with the test chemical from the 96-well tissue culture microtiter plate, and add 250 μL pre-warmed PBS to each well very carefully. Aspirate the rinsing solution (*see* **Note 13**).
3. Add 250 μL NR medium (to all wells including the blanks) and incubate (37 °C ± 1 °C, 90% ± 5% humidity, and 5.0% ± 1% $CO_2$/air) for 3 ± 0.1 h.
4. After incubation, remove the NR medium, and carefully rinse cells with 250 μL pre-warmed PBS. Figure 3 illustrates HepG2 cells with intracellular bound NR.
5. Add 100 μL desorption solution to all wells, including blanks.
6. Shake the microtiter plate on a microtiter plate shaker (e.g., 80 rpm) for 20–45 min to extract NR from the cells and form a homogeneous solution. Protect plates from light by covering them, e.g., with aluminum foil.
7. Remove plates from the plate shaker and wait 5 min before the measurement.
8. Measure the absorption (within 60 min of adding the desorption solution) at 540 nm ± 10 nm using a microplate spectrophotometer.

### 3.5 Data Analysis

The obtained spectrophotometric data is mostly presented as a concentration-response curve (often referred to as a dose-response curve) in which the effect caused by the xenobiotic can be visualized following a concentration gradient. Analysis of this curve provides information on the cytotoxic effect caused in the cells exposed to a xenobiotic for a determined period of time (*see* **Note

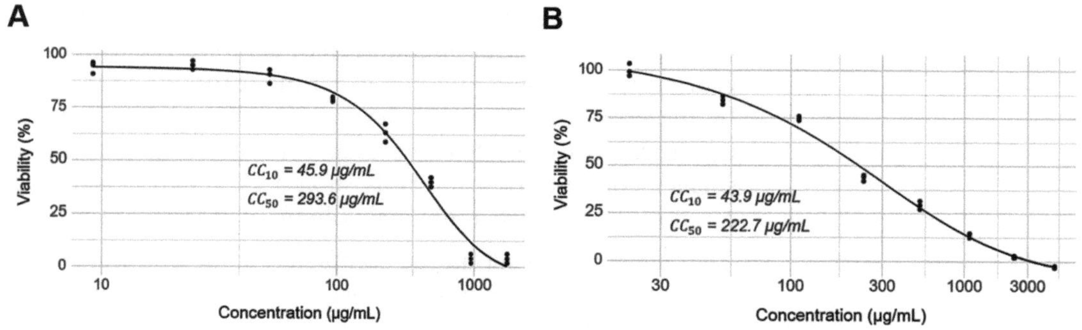


**Fig. 4** Concentration-response curves of HepG2 cells exposed for 24 h to (**a**) acetylsalicylic acid and (**b**) acetaminophen

**14**). This analysis is based on a four-parameter logistic nonlinear regression that can be conducted using several mathematical software packages such as RStudio (drc package), GraphPad Prism, or MasterPlex®. Figure 4 illustrates the concentration response of HepG2 cells exposed for 24 h to a concentration gradient of acetaminophen and acetylsalicylic acid.

## 4 Notes

1. In case DMEM does not contain L-glutamine, add 200 mM L-glutamine to the routine culture medium to a final concentration of 4 mM glutamine.
2. Due to the inconsistency of different sera, the cytotoxicity of different batches of FBS should be investigated. A sufficient amount of the same batch FBS should be reserved and used within the same experiments.
3. Completed media formulations should be kept at approximately 2–8 °C and stored for no longer than 1 month.
4. For this book chapter, the NRU assay is documented using the hepatic human cell line HepG2. This assay can, however, be applied using other animal or human cell systems.
5. Other plate formats than 96-well plates might also be used for the NRU assay. The incubation volumes should be adapted according to the cell culture recipient.
6. This incubation period assures cell recovery, adherence, and progression to the exponential growth phase.
7. At higher cell densities, slight acidification of the routine culture medium may occur (observed by a color shift of the phenol red indicator). In this case, the frequency of routine culture medium refreshment should be increased.

8. The stock solution for each test chemical should be prepared at the highest concentration found to be soluble. The lower concentrations in a range-finding experiment should be prepared by successive dilutions that decrease by, for example, one log unit each. Once the toxicity range for a compound is found, smaller concentration intervals should be tested. Each stock dilution should have a minimal volume of at least 1–2 mL.
9. The test chemical should be completely soluble in the culture medium, and the solutions must not be cloudy nor have a noticeable precipitate. Test chemical solutions in a culture medium should not be prepared in bulk for use in subsequent tests.
10. In case DMSO or EtOH is used to prepare stock solutions, the final DMSO or EtOH concentration in the test chemical solution to be applied to the cells must not exceed 0.5% (v/v). All test concentrations and vehicle controls should contain the same concentration of DMSO or EtOH.
11. In the presented plate layout, two different compounds are tested in the same plate (as illustrated by the horizontal dash line in Fig. 2). Each concentration of the chemicals is tested in threefold.
12. The NR medium can be filtered to reduce potentially formed NR crystals (e.g., Millipore filtering, 0.2–0.45 μm pore size).
13. To reduce the time of this step, the aspiration of NR medium and PBS can be replaced by "dumping" the content of the plate by a flip movement into a recipient with a large opening or the sink. Eventual liquid at the edges of the plates can be dried by pressing the plate onto a pile of paper tissues. This procedure can only be performed if no further culturing of the cells is envisaged.
14. The main endpoint readout of concentration-response curves is the determination of the concentration at which a particular percentage of the cells show a decrease in viability. Hereby, the $IC_{50}$ (inhibitory concentration 50), or the concentration of test substance at which 50% of cell death is observed, is conventionally used as a parameter of in vitro cytotoxicity. The $IC_{10}$, representing the concentration that induces 10% cell death, or lower concentrations are on the other hand often referred to as subcytotoxic concentrations. The use of inhibitory concentration (e.g., $IC_{10}$ and $IC_{50}$) as the parameter to indicate cytotoxicity of xenobiotics has become throughout the years a wide terminology. However, cytotoxicity concentration (CC) (e.g., $CC_{10}$ and $CC_{50}$), which is defined as the concentration required for the evaluation of a cytotoxic effect, might be a more appropriate representation of the endpoint determined by in vitro cytotoxicity assays.

## References


1. Borenfreund E, Puerner J (1984) A simple quantitative procedure using monolayer cultures for cytotoxicity assays (HTD/NR90). J Tissue Cult Methods 9(1):7–9
2. Repetto G, del Peso A, Zurita JL (2008) Neutral red uptake assay for the estimation of cell viability/cytotoxicity. Nat Protoc 3(7): 1125–1131
3. Aslantürk ÖS (2018) In vitro cytotoxicity and cell viability assays: principles, advantages, and disadvantages. In Genotoxicity - a predictable risk to our actual world
4. Zuang V (2001) The neutral red release assay: a review. Altern Lab Anim 29(5):575–599
5. Rodrigues RM, Bouhifd M, Bories G, Sacco M, Gribaldo L, Fabbri M, Coecke S, Whelan MP (2013) Assessment of an automated in vitro basal cytotoxicity test system based on metabolically-competent cells. Toxicol In Vitro 27(2):760–767
6. The National Toxicology Program Interagency Center for the Evaluation of Alternative Toxicological Methods (2003) Test method protocol for the BALB/c 3T3 neutral red uptake cytotoxicity test. A test for basal cytotoxicity for an in vitro validation study phase III. Available through: https://ntp.niehs.nih.gov/iccvam/methods/acutetox/invidocs/phiiiprot/3t3phiii.pdf. Consulted May 2022
7. Organisation for Economic Co-operation and Development (2004) Guideline 432: in vitro 3T3 NRU phototoxicity test. OECD guidelines for the testing of chemicals Available through: http://wwwoecd-ilibraryorg/environment/test-no-432-in-vitro-3t3-nru-phototoxicity-test_9789264071162-en. Consulted May 2022
8. European Commission Joint Research Center (2013) EURL ECVAM recommendation on the 3T3 neutral red uptake (3T3 NRU) cytotoxicity assay for the identification of substances not requiring classification for acute oral toxicity. Available through: https://eurl-ecvam.jrc.ec.europa.eu/eurl-ecvam-recommendations/3t3-nru-recommendation. Consulted May 2022
9. Bouhifd M, Bories G, Casado J, Coecke S, Norlén H, Parissis N, Rodrigues RM, Whelan MP (2012) Automation of an in vitro cytotoxicity assay used to estimate starting doses in acute oral systemic toxicity tests. Food Chem Toxicol 50(6):2084–2096
10. Knowles BB, Aden DP (1983) Human hepatoma derived cell line, process for preparation thereof, and uses therefor. US Patent 4,393,133
11. Schoonen WG, Westerink WM, de Roos JA, Débiton E (2005) Cytotoxic effects of 100 reference compounds on HepG2 and HeLa cells and of 60 compounds on ECC-1 and CHO cells. I mechanistic assays on ROS, glutathione depletion and calcein uptake. Toxicol In Vitro 19(4):505–516
12. Chiu JH, Hu CP, Lui WY, Lo SJ, Chang CM (1990) The formation of bile canaliculi in human hepatoma-cell lines. Hepatology 11: 834–842
13. Hewitt NJ, Hewitt P (2004) Phase I and II enzyme characterization of two sources of HepG2 cell lines. Xenobiotica 34(3):243–256

# Chapter 16

# Digital Holographic Microscopy to Assess Cell Behavior

**Brad Bazow, Van K. Lam, Thuc Phan, Byung Min Chung, George Nehmetallah, and Christopher B. Raub**


## Abstract

Digital holographic microscopy is an imaging technique particularly well suited to the study of living cells in culture, as no labeling is required and computed phase maps produce high contrast, quantitative pixel information. A full experiment involves instrument calibration, cell culture quality checks, selection and setup of imaging chambers, a sampling plan, image acquisition, phase and amplitude map reconstruction, and parameter map post-processing to extract information about cell morphology and/or motility. Each step is described below, focusing on results from imaging four human cell lines. Several post-processing approaches are detailed, with an aim of tracking individual cells and dynamics of cell populations.


**Key words** Cell motility, Holography, Microscopy, Quantitative phase, Cell lines, Cell morphology, Image processing, Machine learning

## 1 Introduction

Label-free imaging-based cell viability assays supplement biochemical and marker-based assays as imaging reveals additional spatial information. Single-shot micrographs of adherent cells yield information on cell morphology, from which mitotic status can be inferred [1]. Time-lapse image sequences further reveal and confirm cells undergoing division, motile cells, and nonmotile cells [2]. If images are acquired over enough time, individual cells can be tracked before, during, and after one or more divisions, elucidating each phase of the cell cycle [3]. However, the accuracy of cell shape, motility, and subcellular measurements depends highly on the image contrast modality and other imaging parameters, including lateral resolution and image acquisition interval [4, 5].

The extra information about cell shape and movement, beyond what is obtainable from standard viability assays, is biologically significant. Cell motility plays a critical role in many physiological and pathological processes such as wound healing, immune

Oliver Friedrich and Daniel F. Gilbert (eds.), *Cell Viability Assays: Methods and Protocols*, Methods in Molecular Biology, vol. 2644, https://doi.org/10.1007/978-1-0716-3052-5_16,

response, and cancer metastasis [6–8], with several excellent reviews [9, 10]. Cells coordinate interaction with the extracellular environment, regulation of signaling pathways, and remodeling of cytoskeleton to move [11–13]. In the process, most motile cells undergo changes in cell shape, and highly specialized structures such as lamellipodia, filopodia, and blebs develop at the leading edge to assist in cell motility [14]. In fact, to fully appreciate the contribution to cell motility, morphology of an entire cell must be considered. For example, carcinoma cells lose apicobasal polarity and become spindle shaped and more motile during an epithelial-to-mesenchymal transition [15, 16]. In cancer, recent studies have shown that cell morphology has a predictive value for tumor progression and metastatic potential [17, 18]. To accurately portray a representative cell, quantitation of cell morphology and motility are critical due to heterogeneity between individual cells in a population and variation of single cells across time [19].

Digital holographic microscopy (DHM) is one of a family of computational imaging techniques that build optical phase maps from interferograms [20–22]. In DHM, a digital hologram is generated by illuminating an object with a coherent optical source, such as a laser, and constructing an optical system to propagate the field to a digital optical intensity-capturing device [23]. The original field is then reconstructed digitally, and relevant quantitative information is extracted [24–26]. A common setup uses a Mach-Zender interferometer with a bi-telecentric lens system in the sample arm and an off-axis reference beam, with the interference pattern imaged at the Fourier plane [27]. Two microscope objectives with condensing and imaging functions cancel out phase and other aberrations from optical elements within the objectives, making phase reconstruction easier and more accurate. Once aligned and calibrated, the DHM is very stable within the limitations of the illumination source. State-of-the-art DHM systems are unique because they are capable of three-dimensional (3D) quantitative phase imaging (QPI) of objects with high vertical/lateral resolution (~2 nm/~300 nm) [28, 29] and high acquisition speed (~200 frames per second (fps)) [30, 31]. Other modifications of the DHM include the addition of a fluorescence modality for spatio-temporal mapping of fluorescently labeled cells and molecules [32], addition of a polarization modality to map birefringence and other polarization parameters [33–35], and index of refraction mapping [36].

Phase reconstruction, background correction, and biospecimen preparation are also critical to the success of DHM-based evaluation of cells. Phase reconstruction converts data from the hologram into a phase map, removing phase aberration and aliasing in the process [24–26]. Background correction removes non-uniform phase backgrounds [37]. In coherent interferometric techniques, the use of the reference beam allows for the recording of the phase of the object. However, drawbacks of the interferometric-based systems are their sensitivity to perturbations, tedious alignment, speckle noise, sophisticated aberration compensation, and phase unwrapping which may introduce artifacts in the reconstructed 3D image. Some of these drawbacks have been mitigated to a high degree by using partially coherent sources or different speckle-averaging techniques [38, 39]. Also, some DHM systems use a single path containing the object and the reference beams, making them more immune to vibrations [40–42].

Biospecimens to image with DHM include living cells or chemically fixed cells/tissue. Appropriate biospecimens include cells in suspension, cells adherent to glass or a flat transparent substrate, or cells adherent to or embedded in thin hydrogels. Specimen features with geometric or phase height exceeding the illumination wavelength (633 nm for a helium-neon laser) will result in poor phase unwrapping, the de-aliasing step that interprets $2\pi$ phase steps as continuous gradients [27]. This is true particularly for sharp edge variations in geometric and/or phase height. For even larger phase objects, the phase changes could be hundreds of multiples of $2\pi$, which are also problematic to assign accurate object phase values. Multi-wavelength DHM overcomes these challenges [27, 43, 44]. Finally, independently of the DHM technique, for live culture imaging, the temperature, humidity, and pH of the cell cultures must be tightly controlled, which is technically challenging depending on the space available around the specimen holder. Within these constraints, quantitative cytometry data from DHM images have numerous biological applications, including tracking of cell migration [45], morphology [46], proliferation [47], wound healing [48], and cytotoxicity [49]. Here, several use-cases are described.

## 2 Materials

Materials for a DHM imaging experiment include the imaging system, general equipment, solutions and reagents, biological specimens, and software. All materials are commercially available. Some materials are modified using common laboratory tools.

### 2.1 Imaging Systems

1. The major companies who sell DHM systems are Lyncée Tec SA, 4-Deep, Inc., Phi Optics, Inc., Nanolive SA, and Tomocube, Inc. Instrument selection depends on the range and kind of experiments and biospecimens to be analyzed (*see* **Note 1**).
2. Many DHMs are built from components on an optical bench (Fig. 1). To build a telecentric DHM system in a lab environment, you need the following: a gas laser or laser diode, a digital camera, two microscope objectives (MOs), two tube

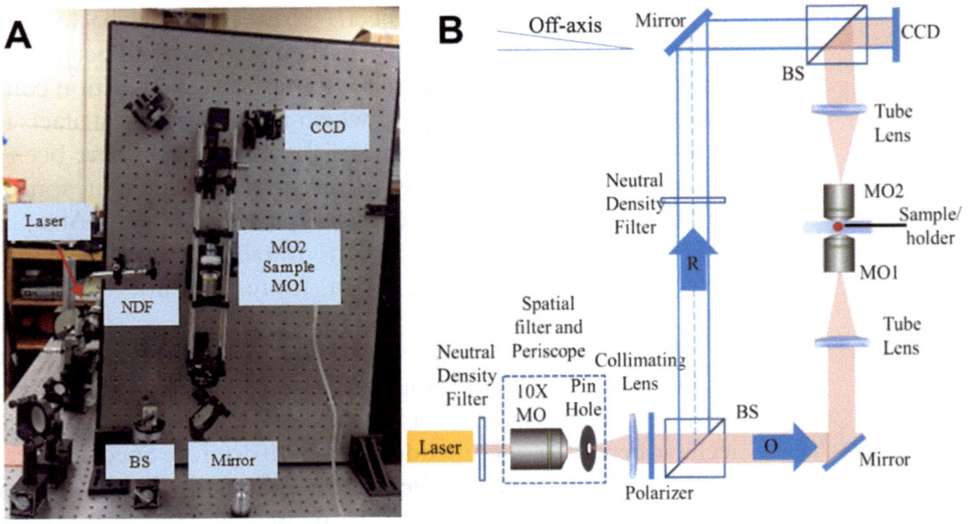


**Fig. 1** (**a**) Photo of a digital holographic microscope (DHM) setup on two optical benches, with horizontal and vertical working surfaces. (**b**) Schematic of the DHM from (a) showing the optical components and beam paths, and off-axis angle of the reference beam

lenses, a spatial filter (comprised of a 3-axis micrometer stage with a MO, a pinhole, and a collimating lens), two variable neutral density filters to control the intensity, two beam splitters, and two mirrors. Major optics companies including Thorlabs, Newport, and Edmund Optics sell these components. A side-by-side photograph and schematic of a typical DHM system are shown in Fig. 1 (*see* **Note 2**).

### *2.2 General Equipment*

1. An incubator with temperature and carbon dioxide control, e.g., HeraCell™ 150i (Thermo Fisher Scientific, Waltham, MA).
2. A transparent culture chamber for imaging: petri dish, chamber slide, or microfluidic chip (*see* **Note 3**).
3. An on-microscope incubation system: it must control heat (37 °C optimal), $CO_2$ (5% optimal), humidity (100% optimal), and eliminate fogging/condensation on the imaging chamber surfaces (*see* **Note 4**).
4. Pipettors, pipettes, micropipettes, and micropipette tips (sterile).
5. Large petri dishes (10 cm diameter or larger) and tissue culture flasks with vented caps and tilted necks, with surface area proportional to the number of cells you wish to collect and use for experiments—at least 25 $cm^2$, 75–150 $cm^2$ preferable.
6. A hemacytometer for counting cells.
7. An inverted phase contrast microscope for counting cells and checking cultures (e.g., the Olympus CK-2).

8. A molybdenum sputter-coated 1951 Air Force Resolution Standard pattern with measured phase and height, to calibrate the DHM. Reference data for the index of refraction of molybdenum at optical wavelengths is available [50]. At 633 nm, the index is $n = 3.70$.
9. A thermal evaporator (Vacuum Explorer 14, Denton LLC) for direct current magnetron sputtering.
10. A stylus profilometer (XP2, Ambios Technology, Inc.) to measure the phase standard feature height.
11. Any standard personal computer running Mac, Windows, or Linux platforms, with 8 GB RAM, at least 8 GB storage, and any Intel or AMD x86–64 processor.
12. A PC with Core i9-9940X CPU (3.3 GHz), 11 GB RAM, and NVIDIA GeForce RTX 2080Ti GPU, for rapid training of deep learning networks.

### 2.3 Solutions and Reagents

1. Culture media containing 10% fetal bovine serum and 1% penicillin-streptomycin (*see* **Note 5**). Selection of media type and supplements must be supported by an experimental rationale and best practices. For many cell lines, Dulbecco's Modified Eagle's Medium is sufficient, with the supplements listed above.
2. 10 mM *4-(2-hydroxyethyl)-1-piperazineethanesulfonic acid* (HEPES), diluted 100× into culture media from a 1 M, pH 7.5 stock solution. This provides extra pH buffering capacity to the media if 5% $CO_2$ is not maintained during the imaging session.
3. A sample collection kit for human cell line authentication (e.g., 135-XV™, ATCC®).
4. A mycoplasma detection kit (e.g., #30-1012 K, ATCC®; #LT07–418, Lonza Bioscience).
5. Hydrogel extracellular matrix proteins, for example, type 1 rat tail tendon collagen, acid-solubilized. This material is used in tumor spheroids, three-dimensional and biofabricated tissue constructs, and tissue chips (*see* Subheading 2.4).
6. Trypsin-EDTA solution, 0.25%, sterile-filtered and suitable for cell culture.
7. Phosphate-buffered saline to rinse cells.
8. Molybdenum pellets for thermal evaporation and sputter coating.

### 2.4 Biological Specimens

1. Cell lines or primary cells purchased from American Type Culture Collection (ATCC®) or another commercial source.
2. Tumor spheroids, three-dimensional and biofabricated tissue constructs, and tissue chips, prepared in the laboratory using cell lines or primary cells.
3. Tissue microarrays, formalin-fixed paraffin-embedded slide sections, and cytology slides for analysis of non-living cells. Purchase de-identified human specimens or obtain them with appropriate ethical oversight (e.g., Institutional Review Board approval in the United States) from a pathology core facility.

### 2.5 Software

1. MATLAB vR2006a or higher, preferably the most recent version (The Mathworks, Natick, MA). The installation needs to include the Statistics and Machine Learning Toolbox.
2. Python Environment Release 3.8.8. Also download the following libraries: imageio 2.9.0, keras 2.9.0, Matplotlib 3.3.4, NumPy 1.20.1, OpenCV-Python 4.5.4, scikit-image, scikit-learn, SciPy 1.6.2, and TensorFlow-GPU 2.9.0. For compatibility between libraries, it is recommended to install tensorflow first and check the compatibility of other libraries with tensorflow.
3. PyCharm, a Python-integrated development environment.
4. Spyder (a scientific Python development environment that is compatible with MATLAB).
5. Visual Studio 2019, for installing the C++ compiler to run the GPU.

## 3 Methods

### 3.1 Digital Holographic Microscopy of Live Cell Cultures

1. Turn the laser on and allow it to warm up for at least 30 min prior to the imaging session. Image the phase standard (molybdenum-coated 1951 Air Force Resolution standard) before beginning an experiment (Fig. 2).
2. Authenticate cell lines, and ensure all cell cultures are free of mycoplasma contamination (*see* **Note 6**).
3. Trypsinize and count cells, then plate an apportioned amount into glass-bottomed and glass-topped petri dishes, chamber slides, or another culture chamber suited to imaging. Cover glasses 150–170 μm thick or glass slides 1 mm thick are acceptable, but take care to match the top surface to the imaging objective requirements.
4. Record the fringe pattern with the digital camera at the Fourier plane, with no specimen in place in the sample arm. The sensor should capture regular, parallel interference fringes cutting obliquely across the imaged field of view (Fig. 3a). The fringe

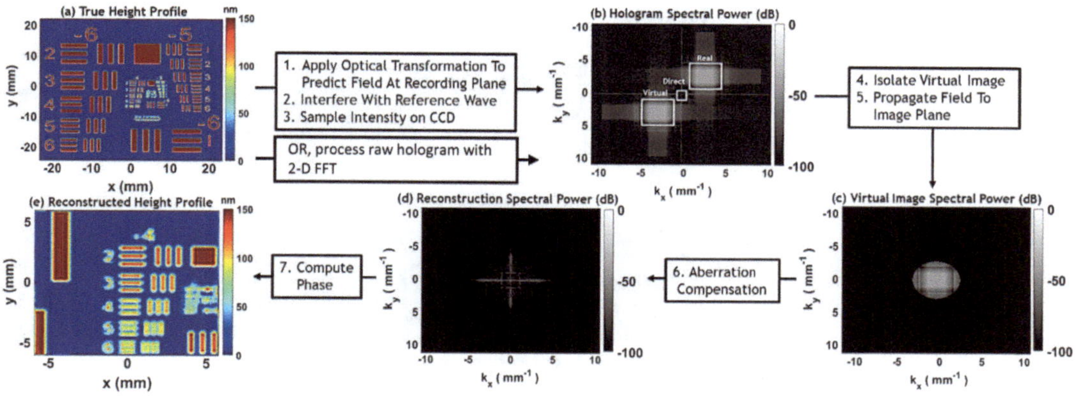


**Fig. 2** Seven steps to phase reconstruction using a 1951 Air Force Resolution slide sputter coated with molybdenum as a phase standard with a constant height profile. The specimen can be directly imaged, or the specimen and imaging process can be completely simulated, using optical transformation, interference, and imaging chip properties to predict the recorded intensity. A virtual image field containing light diffracted by the specimen is isolated in Fourier space, propagated to the image plane, corrected for aberrations, and converted by two-dimensional inverse Fourier transformation into spatial coordinates to create the reconstructed phase map

spatial separation relates to the highest spatial frequency/smallest phase resolution of the system (*see* **Note 7**).

5. Place the biospecimen in its culture chamber on the sample stage between the condensing and imaging objectives. Transparent phase objects appear in faint diffraction patterns superimposed on the background interference fringe pattern created by undeviated light in the specimen and reference arm interfering with each other (Fig. 3b). Focus these diffraction patterns, so the cells are at the focal plane of the imaging objective. This produces high-quality phase and amplitude maps.
6. Take special care for long-term live cell imaging with DHM (*see* **Note 8**). Collect time points at least twice the frequency of the fastest alteration you wish to observe. For lamellipodia dynamics and cytoplasmic streaming, acquire one image per 0.5–5 min. Breast cancer cell division from prophase to telophase lasts 2–3 h, so image at least once per hour to capture each cell division, or every 2–5 min to observe finer motions of chromosomes, cytoplasmic pinching, and de-adhesion/re-adhesion to the substrate. Without on-stage incubation, return cultures to the bench-top incubator after imaging over a period of 20 min. For time-lapse imaging, supplement culture media with 10 mM HEPES, and maintain cells in the culture dish on the microscope stage at 37 °C, 100% humidity, and 5% $CO_2$ (*see* **Note 4**).
7. Design an effective sampling plan for each DHM imaging experiment. Image adherent cell cultures in at least a 1 $cm^2$ circular culture area in evenly separated fields of view. For

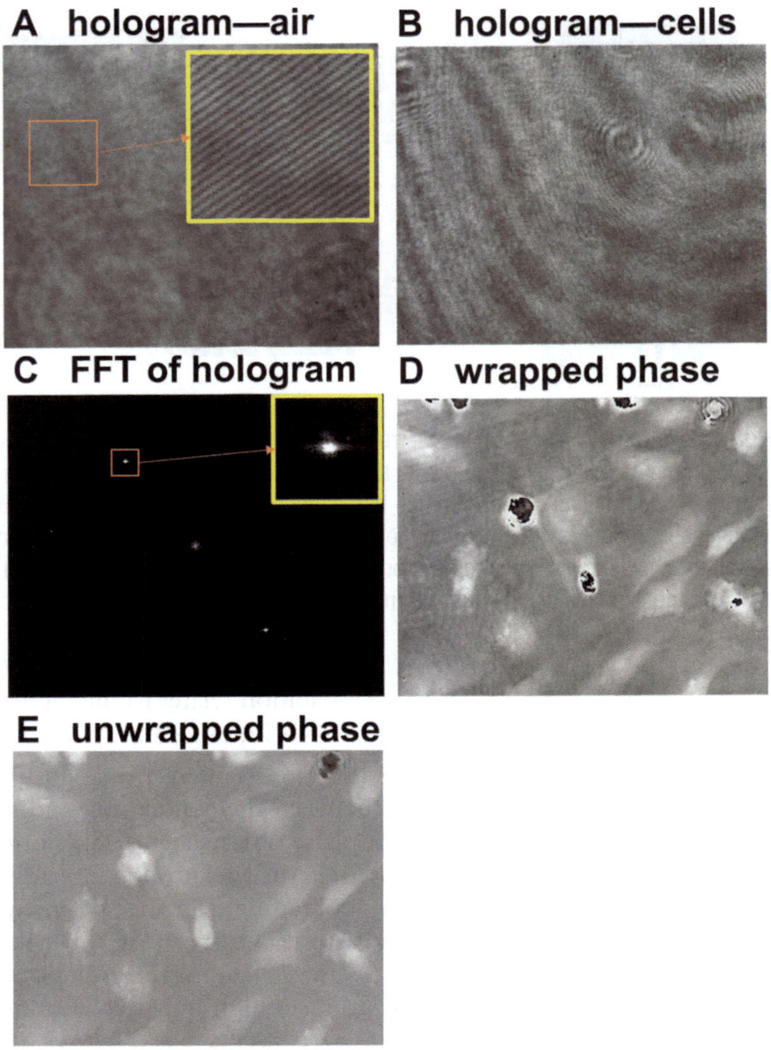


**Fig. 3** (**a**) A typical digital hologram recorded of free space, with no specimen in the sample holder. (**b**) A typical digital hologram recorded of cells adherent to a glass-bottomed petri dish. (**c**) A typical two-dimensional Fourier transform of the hologram with diffracted signal offset from the center in spatial frequency space. (**d**) A typical reconstructed phase map of adherent cells containing aliased phase signal termed "wrapped phase." (**e**) The same reconstructed phase map after phase unwrapping, which de-aliases phase maps

efficient spatial coverage, arrange sampling points in a rectangular or hexagonal lattice, with images taken from the vertices and the centroid, for 5–7 images/cm$^2$ (*see* **Note 9**).

### 3.2 Phase Map Reconstruction

1. Using MATLAB, digitally process raw holograms into uncorrected phase and amplitude maps (Figs. 2 and 3; *see* **Note 10**). Use previously published approaches [51] and publicly available codes (see links below). Follow these main steps for phase

and amplitude reconstruction of the biological samples (*see* Figs. 2 and 3), using the codes off_axis_DHM_example.m and off_axis_telecentric_example.m found at github.com/bazowbs1/Holographic-Simulation[52]. These codes apply Fourier transform to the optically captured digital hologram (Fig. 2a, b), isolate and center the virtual image through the middle of the spectrum by multiplication with digital replica of the recording reference wave (Fig. 2c), apply a low pass spatial filter, propagate the field to the image plane, perform aberration compensation (*see* Fig. 2c, d, and **step 2**, below), reconstruct an aberration-free spectrum (Fig. 2d), and compute the final corrected phase of the sample (Fig. 2e).

2. If the DHM images are at two wavelengths, decouple these to predict specimen refractive index maps (*see* **Note 11**).

### 3.3 Phase Aberration Correction

1. Additional processing corrects phase aberrations (Figs. 3, 4, and 5). Follow these additional steps for phase aberration compensation (*see* Fig. 4), using aberration_compensation_pca.m or aberration_compensation_zernike.m found at github.com/bazowbs1/Holographic-Simulation. These codes crop the spatial spectrum of the recorded hologram and apply inverse Fourier transform, compute the phase $\varphi$,

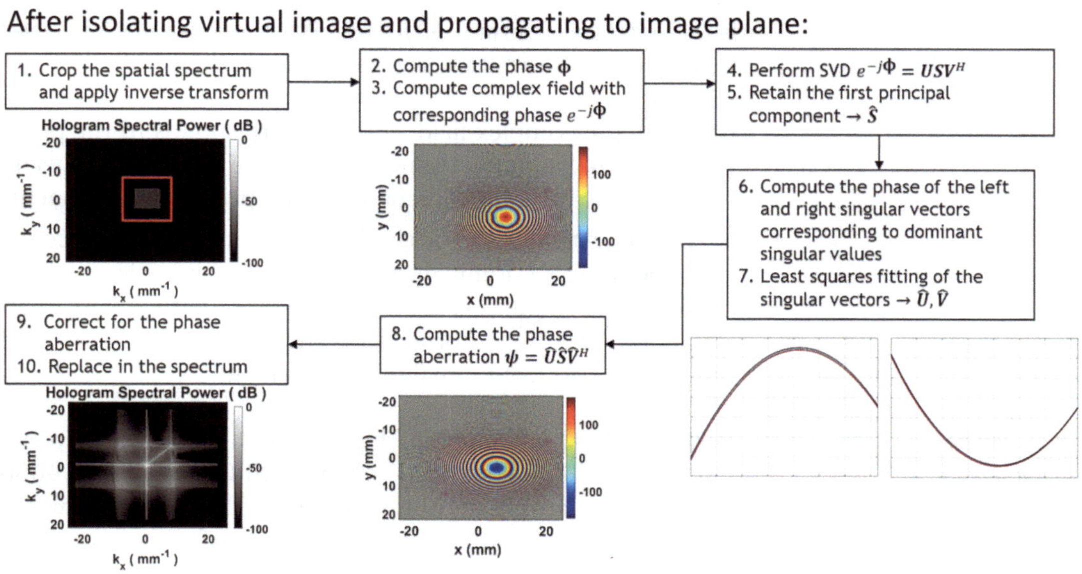


**Fig. 4** Ten steps to correct phase aberrations using the cropped power spectrum (two-dimensional Fourier transform) of the raw hologram. The complex field, containing phase and amplitude information, undergoes singular value decomposition. Least squares fitting of the left and right singular vectors allows for direct computation of the phase aberration, which is used to create a corrected power spectrum

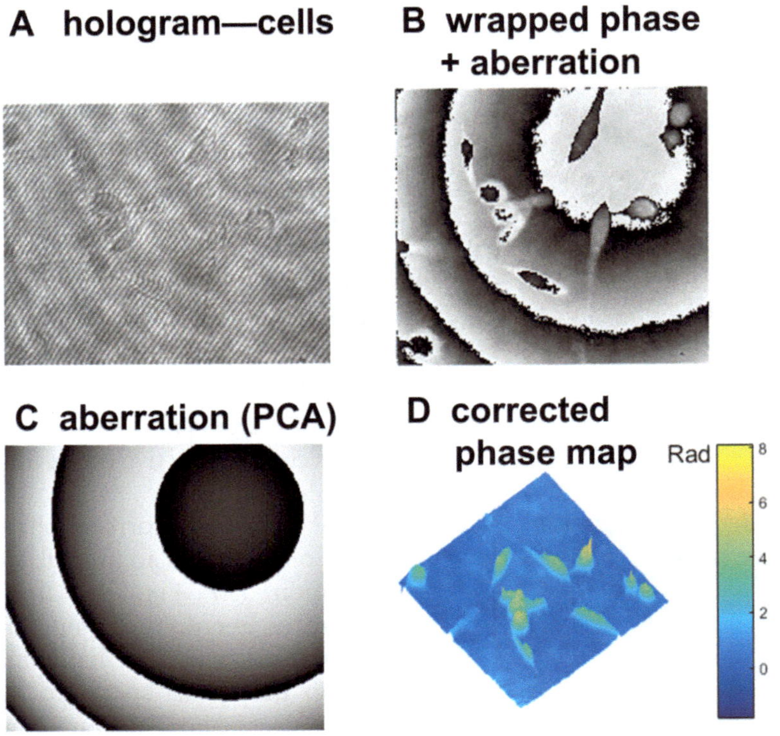


**Fig. 5** (**a**) A typical hologram containing adherent cells. (**b**) The wrapped phase map of the cells plus phase aberrations. (**c**) The calculated phase aberration to be used to correct the phase map. (**d**) A three-dimensional view of the corrected phase map with phase aberrations and aliasing removed

compute the complex field $e^{-j\varphi}$ that corresponds to the phase $\varphi$, perform singular value decomposition (SVD) using MATLAB: $e^{-j\varphi} = USV^H$, retain the first principal components $\widehat{S}$, compute the phase of the left and right singular vectors corresponding to dominant singular values, apply least square fitting of the singular vectors $\widehat{U}$, $\widehat{V}$, compute the phase aberration $\psi = \widehat{U}\widehat{S}\widehat{V}^H$, correct for the phase aberration, and replace the corrected signal in the spectrum.

### 3.4 Phase Unwrapping

1. Perform phase unwrapping (Figs. 3e and 5d). Use the phase unwrapping script contained at MATLAB Central, file #64630 [53], or alternatively at github.com/mfmullen/PhaseUnwrapping. Phase unwrapping removes aliasing of the phase signal with each pair of fringes, representing a full wavelength of $2\pi$, or 633 nm for a HeNe laser (Fig. 3d vs. e, Fig. 5b vs. d).
2. As an alternative to the codes in **steps 2** and **3**, use the graphical user interface (GUI) code linked at tuDHM GUI, found at oirl.github.io/tuDHM/ [54].

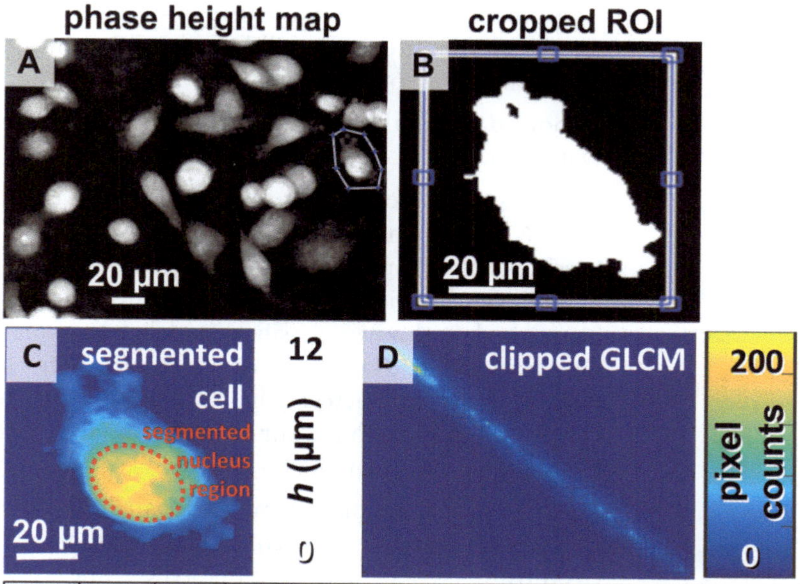


| µ (nm) | SD (nm) | kurt | sk | Ar (µm²) | Per (µm) | Ecc | Cn | Cr | En | Hm |
|---|---|---|---|---|---|---|---|---|---|---|
| 4778 | 3367 | 2.07 | 0.61 | 502 | 112 | 0.88 | 194 | 0.98 | 2E-4 | 0.22 |

**Fig. 6** A typical cell segmentation and parameter extraction workflow. (**a**) The corrected phase height map with initial outline of a single isolated cell. (**b**) The segmented cell outline after running a segmentation code. (**c**) The segmented cell phase height map, with the nucleus region, also indicated (red dashed outline). (**d**) The gray-level co-occurrence matrix derived from the cell in (c) with shape, phase, and texture parameters for the cell listed below

3. Extract quantitative parameters from corrected phase maps that are relevant to cell morphology and motility (*see* Subheading 3.3, Fig. 6, and **Note 12**) [46, 55].

### *3.5 Analysis of Cell Morphology and Motility*

1. Quantitative parameters from phase maps include shape descriptors, first-order pixel statistics, and second-order pixel statistics (texture parameters) [46, 56]. Segment cells in phase maps to extract single-cell morphology and texture parameters. Outline individual, isolated cells using the interactive region of interest (ROI) tool in MATLAB. A shrinking ROI algorithm detects the cell border by detecting a positive gradient in pixel phase values across 3 or more pixels.
2. Calculate shape descriptors and first-order pixel statistics from each cell using the regionprops() function in MATLAB.
3. Calculate the gray-level co-occurrence matrix for each cell, after converting the phase values to 8-bit depth (*see* **Note 12**). The function glcm() is used for this purpose.

4. Input parameters from phase map analysis to machine learning for cell classification—by cell line, motile phenotype, or other distinction (Fig. 6).
5. Import data to the MATLAB workspace using the Import Data tool.
6. Open the machine learning app by entering "classificationLearner" as a command in the Command Window. Select "New Session from Workspace." In the window that opens, select your data set variable from the drop-down box. Select your response data (known classification labels) as columns in your data set variable. Make numerical labels (0, 1) for binary classification. Select predictors (the calculated parameters for each cell) as column within your data set variable. Select fivefold cross-validation, then Start Session.
7. Perform principal component analysis (PCA), specifying 95% of explained variance for component reduction. This avoids overfitting errors. Select model type "linear SVM." Select "Train." Tune the model under "Tune Advanced Model Options," varying the box constraint level, and set the level that yields the highest classification performance in training. Keep the misclassification costs at default (0 and 1), and keep data standardized, to equally weight all morphology and texture parameters.
8. Import test data from the workspace. Divide your entire data set 3:1 training/testing by randomly sorting the single-cell measurements in Excel before importing the training, and test data sets as separate variables in the MATLAB workspace. View plots for validation and test results. Good test results have slightly worse performance than validation results.
9. Export the model to save and use it as a function on future data in the MATLAB command window. Extract support vector machine score outputs from the exported model, which identify cells as lying along a continuous spectrum between the two classification labels—a useful metric for cell shape phenotypes lying along a continuous spectrum, such as epithelial-mesenchymal (*see* **Note 13**) [57].
10. Identify unusual cells quantitatively in comparison to a normal population by defining a sum of squared deviations (SSD) score from standardized shape descriptors and pixel statistics (*see* **Note 14**) [55].

## 4 Notes

1. Commercial instruments are stand-alone or add-ons to existing imaging systems. Hologram processing and image analysis are typically built into the commercial software packages.
2. The dual horizontal and vertical optical bench is helpful for the interferometric DHM setup. The sample arm including the two MOs and specimen should be arranged vertically, so fluid in the specimen chamber does not spill. Without a vertical optical bench, additional mirrors in the sample arm placed at 45° are necessary to image cell and tissue culture chambers designed to lay flat. Common challenges of DHM are vibrations, tedious alignment, and speckle noise. Vibrations can be mitigated by pneumatically cushioned legs on the optical table. Speckle noise is partially mitigated by condensing the illuminating laser beam with the first MO. Once aligned, the DHM is fairly stable over many months, which can be checked by imaging a phase standard.
3. Borosilicate glass coverslips (140–180 μm thick) and slides (1 mm thick) are best for DHM imaging, superior to optically clear polycarbonate or polystyrene in many tissue culture chambers due to more uniformity of thickness. Several companies offer glass-bottomed and glass-topped imaging and culture chambers, for example, DS35-20-0-TOP from Cellvis. For endpoint imaging, the petri dish top can be removed and a standard circular coverslip placed over cells cultured within the inner well of the glass-bottomed petri dish.
4. In the telecentric DHM, space for the specimen chamber is limited by the combined working distances of the two (condensing and imaging) MOs. This makes on-microscope incubation difficult. A workaround description follows. Temperature can be controlled by a thin open disk Peltier device with temperature controller (e.g., WP-10 or WP-16, warmed platform, Warner Instruments), designed to hold a 35 mm petri dish at the set temperature. Carbon dioxide from a compressed 5% $CO_2$/balance air gas tank must be connected to the culture chamber by two gas regulators: a commercial regulator from high pressure to 5 mm Hg and an oxygen tank regulator that adjusts flow rate from 0 to 15 L/min. Medical tubing with appropriate adapters will connect the gas tank to these two regulators. Tubing from the second regulator sends gas through a water trap (an Erlenmeyer flask with rubber stopper) that provides humidity to the gas. Finally, microfluidics tubing (PTFE, 0.022″ inner diameter, 0.044″ outer diameter) sends humidified gas from the flask through the top of the culture chamber—either a microfluidics device

with metal stylet or else a petri dish lid. Use a diamond-coated lapidary drill to create the gas inlet hole in the dish lid. The microfluidics tubing should send one to two bubbles per second into the culture media. To avoid condensation on the petri dish lid, create a hollow disk-shaped trough from metal cylinders on top of the petri dish. Add the powder from hand warmer packets to the trough to avoid fogging of the central glass portion.

5. Keep track of FBS lot number as lot-to-lot variation may alter cell morphology, motility, and rate of division.
6. While mycoplasma contamination can be confirmed by sequencing, regular logging of cell culture data (time from seeding to confluency; cell appearance and morphology under phase contrast microscopy; and cell counts after splitting) will frequently flag potential problems when changes (sometimes subtle or gradual) are noted over time. Epifluorescence microscopy of cultured cells with fluorescent DNA stains may also reveal the presence of mycoplasma.
7. For off-axis DHM, manipulate mirrors in the reference and specimen beam paths to generate high contrast interference fringes with no biological specimen in the optical path (Fig. 3a). Then place the specimen in the beam path and focus and record the hologram (Fig. 3b). The specimen must be in the focal plane of the imaging objective. Finally, the CCD must be placed at a distance of one focal length from the tube lens. The lateral spatial resolutions ($\Delta\xi_{mag}$, $\Delta\eta_{mag}$) of a DHM are given by:

$\Delta\xi_{mag} = \Delta\xi/M = \lambda d/(N\Delta x M)$, $\Delta\eta_{mag} = \Delta\eta/M = \lambda d/(N\Delta y M)$, where $M$ is the magnification of the imaging objective, $\lambda$ is the imaging laser wavelength (e.g., 633 nm), $d$ is the distance of the specimen to the camera sensor, $N$ is the number of pixels in each dimension that record the hologram, and $\Delta x$ and $\Delta y$ are the two lateral dimensions of the pixels on the digital camera's sensor [27].

8. Stable culture temperature, media pH, media salinity, and limited phototoxicity are keys to successful time-lapse imaging experiments. To maintain temperature within 37 °C ± 0.5 °C, pH of 7.4 ± 0.2, media osmolarity of 260–320 mOsm, and 95–100% humidity, *see* **Note 4** or commercial on-stage incubators. To limit phototoxicity, reduce the incident optical power to what is needed for accurate phase reconstruction using the variable neutral density filters. In between images, a shutter can be placed in the beam path, manually removed or opened electronically, before triggering the camera acquisition.

9. Repeated imaging across time or independent microscopes requires co-registration of multiple fields of view, to identify and image the same cells. The best approach is to use multimodal DHM systems with on-stage incubation (*see* **Notes 4** and **8**) and to not move the specimen until the imaging session is complete. Otherwise, fiducial markers are required. The most inexpensive and effective way to find the same region across repeated independent imaging sessions is to etch a grid on the underside of the culture dish using a steel carbide awl or similar sharp instrument able to scratch glass. A simple grid pattern permits easy identification of regions to be repeatedly imaged.
10. For transmissive specimens, phase, geometric height, and pseudo-height parameters are related by: $h = \phi\lambda/2\pi\Delta n$, where $h$ is the specimen height in projection, $\phi$ is the optical phase delay introduced by the specimen, $\lambda$ is the wavelength (633 nm for a HeNe laser), and $\Delta n$ is the refractive index mismatch of specimen with surroundings. While $h$, $\phi$, and $\Delta n$ are mapped across the optical path projection through the specimen (e.g., collection of cells), an approximation for many cells is $\Delta n = 1.381 - 1.377 = 0.004$. In that case $h$ becomes the pseudo- or phase height. Several DHM techniques are available to map $\Delta n$ in two or three dimensions [36, 43].
11. Optionally, perform optical wavefront modeling [52, 58], and train deep learning neural networks to decouple multiple wavelengths and predict specimen refractive index. Utilize the codes data_loader.py and pix2pix.py found at github.com/bazowbs1/cgan-mw-decoupling. To test the decoupling code, use exemplar simulation data found at github.com/bazowbs1/cgan-mw-decoupling-data. Then, replace the simulation data with your own multi-wavelength DHM data.
12. Selected shape, histogram, and texture parameters for segmented cell phase height maps from the gray-level co-occurrence matrix are listed in Table 1.
13. For a binary classifier based on image parameters, the support vector machine score, $s_j$, is the Euclidean distance of observation $j$ to the decision boundary: $s_j = (x_j/s_k)'\boldsymbol{\beta} + b$, where $x_j$ is the predictor data for observation $j$, $s_k = 2.5196$ is the linear kernel scale, $\boldsymbol{\beta}$ is the vector of fitted linear coefficients, and $b$ is the intercept of the separating hyperplane [59]. Such a score treats all image parameters equally as they are standardized during training of the binary classifier. Therefore, $s_j$ is also a metric of cells, placing them on a spectrum across a chosen binary axis.

**Table 1**
**Phase, shape, and texture parameters from segmented cell phase (height) maps**

| | | |
|---|---|---|
| $\mu$ | Average phase height, segmented cell | 32-bit, in nm |
| sd | Phase height standard deviation, segmented cell | 32-bit, in nm |
| ku | Phase height kurtosis, segmented cell | 32-bit, unitless |
| sk | Phase height skewness, segmented cell | 32-bit, unitless |
| $A$ | Segmented cell area | Units of pixels→$\mu m^2$ |
| Per | Segmented cell perimeter | Units of pixels→μm |
| Ecc | Segmented cell fit ellipse eccentricity | 32-bit, unitless |
| Co | Contrast, from GLCM | 8-bit, unitless |
| Cor | Correlation, from GLCM | 8-bit, unitless |
| En | Entropy, from GLCM | 8-bit, unitless |
| Hm | Homogeneity, from GLCM | 8-bit, unitless |
| $\mu_n$ | Average nucleus region phase height | 32-bit, in nm→$\mu m^2$ |
| $\mu_{n,m}$ | Maximum nucleus region phase height | 32-bit, in nm |
| $A_n$ | Segmented nucleus region area | Units of pixels |
| $sd_n$ | Nucleus region phase height standard deviation | 32-bit, in nm |
| $ku_n$ | Phase height kurtosis, nuclear region | 32-bit, unitless |
| $sk_n$ | Phase height skewness, nuclear region | 32-bit, unitless |

14. Cells with SSD values outside the tolerance interval are significantly different in morphology and/or phase texture from the sampled population. Cells with SSD values closer to zero are characteristic of a population [55]. For 17 standardized phase parameters, $\varphi_i$, the segmented cell sum of squared deviations (SSD) is: $SSD_i = \sum_{i=1}^{17} (\varphi_i - \overline{\varphi}_i)^2$. The 99% tolerance interval (TI) of SSDs is: $99\% TI = \overline{SSD} \pm \sqrt{\upsilon(1+1/N) z^2_{\frac{1-p}{2}} / \chi^2_{1-\gamma,\upsilon}}\, \sigma_{SSD}$, where $\overline{SSD}$ and $\sigma_{SSD}$ are the mean and standard deviation of SSD values with $\upsilon$ degrees of freedom for $N$ cells, and with $z_{\frac{1-p}{2}}$ and $\chi^2_{1-\gamma,\upsilon}$ critical values from the $z$- and chi-squared distributions, respectively; $p = 0.99$ representing 99% tolerance; and $\gamma = 0.01$ representing $1 - \gamma$ representing 99% confidence.

## References


1. Grah JS, Harrington JA, Koh SB, Pike JA, Schreiner A, Burger M, Schonlieb CB, Reichelt S (2017) Mathematical imaging methods for mitosis analysis in live-cell phase contrast microscopy. Methods 115:91–99. https://doi.org/10.1016/j.ymeth.2017.02.001
2. Debeir O, Van Ham P, Kiss R, Decaestecker C (2005) Tracking of migrating cells under phase-contrast video microscopy with combined mean-shift processes. IEEE Trans Med Imaging 24(6):697–711. https://doi.org/10.1109/TMI.2005.846851
3. Padfield D, Rittscher J, Thomas N, Roysam B (2009) Spatio-temporal cell cycle phase analysis using level sets and fast marching methods. Med Image Anal 13(1):143–155. https://doi.org/10.1016/j.media.2008.06.018
4. Matula P, Maska M, Sorokin DV, Matula P, Ortiz-de-Solorzano C, Kozubek M (2015) Cell tracking accuracy measurement based on comparison of acyclic oriented graphs. PLoS One 10(12):e0144959. https://doi.org/10.1371/journal.pone.0144959
5. Hilsenbeck O, Schwarzfischer M, Skylaki S, Schauberger B, Hoppe PS, Loeffler D, Kokkaliaris KD, Hastreiter S, Skylaki E, Filipczyk A, Strasser M, Buggenthin F, Feigelman JS, Krumsiek J, van den Berg AJ, Endele M, Etzrodt M, Marr C, Theis FJ, Schroeder T (2016) Software tools for single-cell tracking and quantification of cellular and molecular properties. Nat Biotechnol 34(7):703–706. https://doi.org/10.1038/nbt.3626
6. Luster AD, Alon R, von Andrian UH (2005) Immune cell migration in inflammation: present and future therapeutic targets. Nat Immunol 6(12):1182–1190. https://doi.org/10.1038/ni1275
7. Stuelten CH, Parent CA, Montell DJ (2018) Cell motility in cancer invasion and metastasis: insights from simple model organisms. Nat Rev Cancer 18(5):296–312. https://doi.org/10.1038/nrc.2018.15
8. Rodrigues M, Kosaric N, Bonham CA, Gurtner GC (2019) Wound healing: a cellular perspective. Physiol Rev 99(1):665–706. https://doi.org/10.1152/physrev.00067.2017
9. Lauffenburger DA, Horwitz AF (1996) Cell migration: a physically integrated molecular process. Cell 84(3):359–369. https://doi.org/10.1016/s0092-8674(00)81280-5
10. Yamada KM, Sixt M (2019) Mechanisms of 3D cell migration. Nat Rev Mol Cell Biol 20(12):738–752. https://doi.org/10.1038/s41580-019-0172-9
11. Chung BM, Rotty JD, Coulombe PA (2013) Networking galore: intermediate filaments and cell migration. Curr Opin Cell Biol 25(5):600–612. https://doi.org/10.1016/j.ceb.2013.06.008
12. Devreotes P, Horwitz AR (2015) Signaling networks that regulate cell migration. Cold Spring Harb Perspect Biol 7(8):a005959. https://doi.org/10.1101/cshperspect.a005959
13. Te Boekhorst V, Preziosi L, Friedl P (2016) Plasticity of cell migration in vivo and in silico. Annu Rev Cell Dev Biol 32:491–526. https://doi.org/10.1146/annurev-cellbio-111315-125201
14. Bodor DL, Ponisch W, Endres RG, Paluch EK (2020) Of cell shapes and motion: the physical basis of animal cell migration. Dev Cell 52(5):550–562. https://doi.org/10.1016/j.devcel.2020.02.013
15. Pastushenko I, Blanpain C (2019) EMT transition states during tumor progression and metastasis. Trends Cell Biol 29(3):212–226. https://doi.org/10.1016/j.tcb.2018.12.001
16. Yang J, Antin P, Berx G, Blanpain C, Brabletz T, Bronner M, Campbell K, Cano A, Casanova J, Christofori G, Dedhar S, Derynck R, Ford HL, Fuxe J, Garcia de Herreros A, Goodall GJ, Hadjantonakis AK, Huang RYJ, Kalcheim C, Kalluri R, Kang Y, Khew-Goodall Y, Levine H, Liu J, Longmore GD, Mani SA, Massague J, Mayor R, McClay D, Mostov KE, Newgreen DF, Nieto MA, Puisieux A, Runyan R, Savagner P, Stanger B, Stemmler MP, Takahashi Y, Takeichi M, Theveneau E, Thiery JP, Thompson EW, Weinberg RA, Williams ED, Xing J, Zhou BP, Sheng G, Association EMTI (2020) Guidelines and definitions for research on epithelial-mesenchymal transition. Nat Rev Mol Cell Biol 21(6):341–352. https://doi.org/10.1038/s41580-020-0237-9
17. Wu PH, Phillip JM, Khatau SB, Chen WC, Stirman J, Rosseel S, Tschudi K, Van Patten J, Wong M, Gupta S, Baras AS, Leek JT, Maitra A, Wirtz D (2015) Evolution of cellular morpho-phenotypes in cancer metastasis. Sci Rep 5:18437. https://doi.org/10.1038/srep18437
18. Alizadeh E, Castle J, Quirk A, Taylor CDL, Xu W, Prasad A (2020) Cellular morphological features are predictive markers of cancer cell state. Comput Biol Med 126:104044. https://doi.org/10.1016/j.compbiomed.2020.104044

19. Gordonov S, Hwang MK, Wells A, Gertler FB, Lauffenburger DA, Bathe M (2016) Time series modeling of live-cell shape dynamics for image-based phenotypic profiling. Integr Biol (Camb) 8(1):73–90. https://doi.org/10.1039/c5ib00283d
20. Nguyen TL, Pradeep S, Judson-Torres RL, Reed J, Teitell MA, Zangle TA (2022) Quantitative phase imaging: recent advances and expanding potential in biomedicine. ACS Nano 16:11516. https://doi.org/10.1021/acsnano.1c11507
21. Popescu G, Park Y (2015) Quantitative phase imaging in biomedicine. J Biomed Opt 20(11):111201. https://doi.org/10.1117/1.JBO.20.11.111201
22. Popescu G (2011) Quantitative phase imaging of cells and tissues, 1st edn. McGraw-Hill, New York
23. Depeursinge C, Colomb T, Emery Y, Kuhn J, Charriere F, Rappaz B, Marquet P (2007) Digital holographic microscopy applied to life sciences. Annu Int Conf IEEE Eng Med Biol Soc 2007:6244–6247. https://doi.org/10.1109/IEMBS.2007.4353782
24. Kim M (2010) Principles and techniques of digital holographic microscopy. J SPIE Rev 1(1):018005
25. Javidi B, Carnicer A, Anand A, Barbastathis G, Chen W, Ferraro P, Goodman JW, Horisaki R, Khare K, Kujawinska M, Leitgeb RA, Marquet P, Nomura T, Ozcan A, Park Y, Pedrini G, Picart P, Rosen J, Saavedra G, Shaked NT, Stern A, Tajahuerce E, Tian L, Wetzstein G, Yamaguchi M (2021) Roadmap on digital holography [invited]. Opt Express 29(22):35078–35118. https://doi.org/10.1364/OE.435915
26. Guo CS, Wang BY, Sha B, Lu YJ, Xu MY (2014) Phase derivative method for reconstruction of slightly off-axis digital holograms. Opt Express 22(25):30553–30558. https://doi.org/10.1364/OE.22.030553
27. Nguyen T, Nehmetallah G, Raub C, Mathews S, Aylo R (2016) Accurate quantitative phase digital holographic microscopy with single- and multiple-wavelength telecentric and nontelecentric configurations. Appl Opt 55(21):5666–5683. https://doi.org/10.1364/AO.55.005666
28. Kühn J, Charrière F, Colomb T, Cuche E, Montfort F, Emery Y, Marquet P, Depeursinge C (2008) Axial sub-nanometer accuracy in digital holographic microscopy. Meas Sci Technol 19:074007. https://doi.org/10.1088/0957-0233/19/7/074007
29. Marquet P, Rappaz B, Magistretti PJ, Cuche E, Emery Y, Colomb T, Depeursinge C (2005) Digital holographic microscopy: a noninvasive contrast imaging technique allowing quantitative visualization of living cells with subwavelength axial accuracy. Opt Lett 30(5):468–470. https://doi.org/10.1364/ol.30.000468
30. Emery Y, Colomb T, Cuche E (2021) Metrology applications using off-axis digital holography microscopy. J Phys Photonics 3(3):034016. https://doi.org/10.1088/2515-7647/ac0957
31. Pedrini G, Froning P, Tiziani HJ, Gusev ME (1999) Pulsed digital holography for high-speed contouring that uses a two-wavelength method. Appl Opt 38(16):3460–3467. https://doi.org/10.1364/ao.38.003460
32. Kumar M, Quan X, Awatsuji Y, Tamada Y, Matoba O (2020) Digital holographic multimodal cross-sectional fluorescence and quantitative phase imaging system. Sci Rep 10(1):7580. https://doi.org/10.1038/s41598-020-64028-x
33. Yang TD, Park K, Kang YG, Lee KJ, Kim BM, Choi Y (2016) Single-shot digital holographic microscopy for quantifying a spatially-resolved Jones matrix of biological specimens. Opt Express 24(25):29302–29311. https://doi.org/10.1364/OE.24.029302
34. Colomb T, Durr F, Cuche E, Marquet P, Limberger HG, Salathe RP, Depeursinge C (2005) Polarization microscopy by use of digital holography: application to optical-fiber birefringence measurements. Appl Opt 44(21):4461–4469. https://doi.org/10.1364/ao.44.004461
35. Lam VK, Phan T, Ly K, Luo X, Nehmetallah G, Raub CB (2022) Dual-modality digital holographic and polarization microscope to quantify phase and birefringence signals in biospecimens with a complex microstructure. Biomed Opt Express 13(2):805–823. https://doi.org/10.1364/BOE.449125
36. Kemper B, Kosmeier S, Langehanenberg P, von Bally G, Bredebusch I, Domschke W, Schnekenburger J (2007) Integral refractive index determination of living suspension cells by multifocus digital holographic phase contrast microscopy. J Biomed Opt 12(5):054009. https://doi.org/10.1117/1.2798639
37. Nguyen T, Bui V, Lam V, Raub CB, Chang LC, Nehmetallah G (2017) Automatic phase aberration compensation for digital holographic microscopy based on deep learning background detection. Opt Express 25(13):

15043–15057. https://doi.org/10.1364/OE.25.015043

38. Pan F, Yang L, Xiao W (2017) Coherent noise reduction in digital holographic microscopy by averaging multiple holograms recorded with a multimode laser. Opt Express 25(18): 21815–21825. https://doi.org/10.1364/OE.25.021815
39. Dohet-Eraly J, Yourassowsky C, Mallahi AE, Dubois F (2016) Quantitative assessment of noise reduction with partial spatial coherence illumination in digital holographic microscopy. Opt Lett 41(1):111–114. https://doi.org/10.1364/OL.41.000111
40. Vora P, Trivedi V, Mahajan S, Patel N, Joglekar M, Chhaniwal V, Moradi AR, Javidi B, Anand A (2017) Wide field of view common-path lateral-shearing digital holographic interference microscope. J Biomed Opt 22(12):1–11. https://doi.org/10.1117/1.JBO.22.12.126001
41. Mico V, Garcia J (2010) Common-path phase-shifting lensless holographic microscopy. Opt Lett 35(23):3919–3921. https://doi.org/10.1364/OL.35.003919
42. Mico V, Zalevsky Z, Garcia J (2012) Super-resolved common-path phase-shifting digital inline holographic microscopy using a spatial light modulator. Opt Lett 37(23): 4988–4990. https://doi.org/10.1364/OL.37.004988
43. Rappaz B, Charriere F, Depeursinge C, Magistretti PJ, Marquet P (2008) Simultaneous cell morphometry and refractive index measurement with dual-wavelength digital holographic microscopy and dye-enhanced dispersion of perfusion medium. Opt Lett 33(7):744–746. https://doi.org/10.1364/ol.33.000744
44. Parshall D, Kim MK (2006) Digital holographic microscopy with dual-wavelength phase unwrapping. Appl Opt 45(3):451–459. https://doi.org/10.1364/ao.45.000451
45. Dubois F, Yourassowsky C, Monnom O, Legros JC, Debeir O, Van Ham P, Kiss R, Decaestecker C (2006) Digital holographic microscopy for the three-dimensional dynamic analysis of in vitro cancer cell migration. J Biomed Opt 11(5):054032. https://doi.org/10.1117/1.2357174
46. Lam VK, Nguyen TC, Chung BM, Nehmetallah G, Raub CB (2018) Quantitative assessment of cancer cell morphology and motility using telecentric digital holographic microscopy and machine learning. Cytometry A 93(3):334–345. https://doi.org/10.1002/cyto.a.23316
47. Janicke B, Karsnas A, Egelberg P, Alm K (2017) Label-free high temporal resolution assessment of cell proliferation using digital holographic microscopy. Cytometry A 91(5): 460–469. https://doi.org/10.1002/cyto.a.23108
48. Bettenworth D, Lenz P, Krausewitz P, Bruckner M, Ketelhut S, Domagk D, Kemper B (2014) Quantitative stain-free and continuous multimodal monitoring of wound healing in vitro with digital holographic microscopy. PLoS One 9(9):e107317. https://doi.org/10.1371/journal.pone.0107317
49. Kuhn J, Shaffer E, Mena J, Breton B, Parent J, Rappaz B, Chambon M, Emery Y, Magistretti P, Depeursinge C, Marquet P, Turcatti G (2013) Label-free cytotoxicity screening assay by digital holographic microscopy. Assay Drug Dev Technol 11(2):101–107. https://doi.org/10.1089/adt.2012.476
50. Werner WSM, Glantschnig K, Ambrosch-Draxl C (2009) Optical constants and inelastic electron-scattering data for 17 elemental metals. J Phys Chem Ref Data 38(4): 1013–1092. https://doi.org/10.1063/1.3243762
51. Nehmetallah G, Aylo R, Williams L (2015) Analog and digital holography with MATLAB. https://doi.org/10.1117/3.2190844
52. Bazow B, Phan T, Nguyen T, Raub C, Nehmetallah G (2021) Simulation of digital holographic recording and reconstruction using a generalized matrix method. Appl Opt 60(4): A21–A37. https://doi.org/10.1364/AO.404405
53. Mullen M (2022) 2D and 3D phase unwrapping using SRNCP. https://github.com/mfmullen/PhaseUnwrapping/releases/tag/v3.0.2. Accessed Retrieved August 22, 2022
54. Castaneda R, Doblas A (2021) Fast-iterative automatic reconstruction method for quantitative phase image with reduced phase perturbations in off-axis digital holographic microscopy. Appl Opt 60(32):10214–10220. https://doi.org/10.1364/AO.437640
55. Lam VK, Nguyen T, Phan T, Chung BM, Nehmetallah G, Raub CB (2019) Machine learning with optical phase signatures for phenotypic profiling of cell lines. Cytometry A 95(7):757–768. https://doi.org/10.1002/cyto.a.23774
56. Haralick RM, Shanmugam K, Dinstein IH (1973) Textural features for image classification. IEEE Trans Syst Man Cybern SMC-3 (6):610–621. https://doi.org/10.1109/TSMC.1973.4309314

57. Lam V, Nguyen T, Bui V, Chung BM, Chang LC, Nehmetallah G, Raub C (2020) Quantitative scoring of epithelial and mesenchymal qualities of cancer cells using machine learning and quantitative phase imaging. J Biomed Opt 25(2):1–17. https://doi.org/10.1117/1.JBO.25.2.026002
58. Bazow B, Phan T, Raub CB, Nehmetallah G (2022) Computational multi-wavelength phase synthesis using convolutional neural networks [invited]. Appl Opt 61(5):B132–B146. https://doi.org/10.1364/AO.439323
59. Cristianini N, Shawe-Taylor J (2000) An introduction to support vector machines and other kernel-based learning methods. Cambridge University Press, Cambridge. https://doi.org/10.1017/CBO9780511801389

# Chapter 17

# Second Harmonic Generation Morphometry of Muscle Cytoarchitecture in Living Cells

**Dominik Schneidereit, Stefanie Nübler, and Oliver Friedrich**

## Abstract

The architectural structure of cells is essential for the cells' function, which becomes especially apparent in the highly "structure functionally" tuned skeletal muscle cells. Here, structural changes in the microstructure can have a direct impact on performance parameters, such as isometric or tetanic force production. The microarchitecture of the actin-myosin lattice in muscle cells can be detected noninvasively in living cells and in 3D by using second harmonic generation (SHG) microscopy, forgoing the need to alter samples by introducing fluorescent probes into them. Here, we provide tools and step-by-step protocols to guide the processes of obtaining SHG microscopy image data from samples, as well as extracting characteristic values from the image data to quantify the cellular microarchitecture using characteristic patterns of myofibrillar lattice alignments.

**Key words** Multiphoton imaging, Second harmonic generation, Skeletal muscle, Quantitative morphometry, Sarcomere architecture

## 1 Introduction

Cytoarchitecture is inherently designed to support the specific cellular functions of respective tissues. Whether neuronal cell architecture is reflected by its main purposes of connectivity and signal integration through its numerous dendritic and axonal branching, red blood cell architecture serving optimized flow characteristics and gas binding and exchange properties through their discoid shape or epithelial cell architecture to serve barrier function by their brick-like shapes with tightly regulated junctions among them, structure-function relationships are inherently connected to an optimized cell function by its structure. An archetype of such a highly ordered structure-function relationship is represented in

**Supplementary Information** The online version contains supplementary material available at https://doi.org/10.1007/978-1-0716-3052-5_17.

Oliver Friedrich and Daniel F. Gilbert (eds.), *Cell Viability Assays: Methods and Protocols*, Methods in Molecular Biology, vol. 2644, https://doi.org/10.1007/978-1-0716-3052-5_17,

skeletal muscle cells where the highly ordered arrangement of parallel bundles of contractile myofibrils is tuned to provide maximum mechanical force production in a linear biopolymer activation scenario [10]. Although a complex cascade of electrochemical coupling of electrical membrane excitation to intracellular $Ca^{2+}$ signaling contributes to a large extent to mechanical force output [1], deviations from the parallel arrangement of neighboring myofibrillar motor-protein tubes have been shown to structurally reduce maximum force output if the integrated force vectors deviate from the main contractile power axis [19]. As such, contractile function crucially relies on a hierarchical degree of myofibrillar order that may limit power output pure "beyond signaling." Thus, cell viability in contractile tissues, like skeletal or cardiac muscle, also involves a holistic estimate of contractile function predictable from the contractile cytoarchitecture in 3D. In particular, the last requirement still poses some technical and methodological challenges to the assessment of the myofibrillar structure. Although electron microscopy undoubtedly provides unparalleled structural resolution, sample preparation is often tedious and irreversibly destroys the sample by embedding and cutting, and 3D aspects are difficult to obtain, although substantial progress has been made in the last decades, e.g., using tilt stages [6]. Visible light confocal microscopy, on the other hand, has the advantage of enabling studying living cells and their cytoskeletal architecture and dynamics. This has been successfully applied to skeletal muscle fibers in the context of disease-related remodeling of the contractile cytoarchitecture [11, 13, 14]. However, two of the main disadvantages of confocal imaging over its noninvasive use are (i) the requirement for external labeling with fluorescent probes (that can potentially interfere with a cell's homeostasis) and (ii) the limited penetration depths of visible light in thick tissues due to marked scattering. In regard to the latter, even a single muscle cell with diameters of ~80–120 μm must be considered thick [9]. Multiphoton imaging using infrared pulsed laser light already can overcome both limitations due to increased penetration depths, less tissue scattering [5] and also exploiting non-linear effects of higher order harmonic generation capabilities of biomolecules, in particular collagen and muscle myosin [15, 18]. Those two biomolecules, therefore, act as an endogenous second harmonic generation (SHG) susceptible biomarker for the structure that is accessible to noninvasive optical assessment by use of a multiphoton imaging system employing high-energy pulsed laser excitation. When attempting to translate SHG images of ultrastructural networks of extracellular matrix collagens or intracellular myosin sarcomere patterns to a firm statement of "normal" or "deranged" architecture, for instance, seen in degenerative myopathies [4, 7], it is imperative to define structural patterns for image processing and analysis that can also be applied in an automated fashion to a large number of library images [8, 12]. Since muscle myofibrils are highly structurally ordered with repetitive *Z*- and *M*-line elements being aligned in normal muscle and in parallel, we have introduced and refined the specific pattern variables of "cosine angle sums" (CAS) and "Vernier densities" (VD) into the muscle community and validated the concepts in several disease models over the years in an approach termed "quantitative morphometry" [3, 4]. These parameters provide a sensitive pair of continuous variables that correlate well with the degree of ultrastructural myofibrillar order in muscle fibers (e.g., normal state, CAS close to 1, VD close to 0; less ordered state, for instance, in chronic disease, CAS ↓, VD ↑) and have recently been shown to be a predictor of muscle force output [19]. This concept is shown in Fig. 3b, c (*see* later). Thus, in principle, the structurally related aspect of active force performance of skeletal muscle fibers could be predicted from image pattern analysis alone. However, the variety of pattern recognition algorithms and often their lack of standardization across labs complicates comparability between studies and weakens scientific rigor. Thus, in the present chapter, we provide a step-by-step guide for performing a quantitative morphometry analysis for CAS and VD in 2D SHG images of muscle fibers, either obtained from single fibers or of thick sections from whole muscle samples. We also provide a ready-to-use GUI as an executable analysis file in the Electronic Supplementary Materials.

## 2 Materials

### 2.1 Optics Equipment

1. A benchtop reflected light stereomicroscope for sample preparation and manipulation.
2. An upright or inverse laser scanning microscope (e.g., TriM Scope II, Miltenyi Biotec, Germany).
3. A fs-pulsed far-red to NIR wavelength-range LASER source capable of at least 5 nJ pulse energy (e.g., Coherent Chameleon Vision II Ti/Sa femtosecond pulse laser, tuned to a wavelength of 810 nm).
4. A fully opaque optical shielding for the microscope, detectors and LASER beam path.
5. A high-resolution objective for excitation purposes that offers a diffraction-limited resolution of 1 μm or smaller, with sufficient working distance to access the sample (e.g., Leica 25× HC FLUOTAR L or Zeiss 40× LD C-Apochromat).
6. A low-resolution objective or condenser lens for signal collection purposes in transmission direction (e.g., Zeiss 20× W Plan-Apochromat or Olympus U-AAC condenser).

7. Non-descanned ultrafast single-pixel detector placed in transmission direction (e.g., Hamamatsu Photomultiplier tube H 7422–40 LV 5 M).
8. A short-pass filter with an edge wavelength just below your excitation laser wavelength, capable of blocking high-intensity laser light (e.g., Chroma Technology 750 nm ET750sp-2p8).
9. A long-pass dichroic mirror with a cutoff between excitation wavelength and SHG emission wavelength to selectively direct signal toward the detector (e.g., Chroma Technology 685 nm T685lpxr).
10. A bandpass filter with a center wavelength of the SHG emission wavelength (*see* **Note 1**) and bandwidth of about 20 nm (e.g., Chroma Technology 405 nm ET405/20×).
11. A bright-field illumination light source.

### 2.2 Software

1. A system control and image generation software, suited to your optics equipment (e.g., Miltenyi Biotec ImSpector).
2. "Fiji is just ImageJ" (FIJI) [17] as a tool for viewing and handling image data and image metadata.
3. A programming language that is capable of handling multidimensional matrixes well (e.g., MATLAB) or a specialized software package for CAS and VD detection (e.g., CASVD2d, provided here as Electronic Supplementary Material).

### 2.3 General Laboratory Equipment

1. Culture dishes with 35 mm and 60 mm diameters that are coated with a silicone elastomer (*see* Subheading 3.1).
2. Insect pins, size 000 to 0.
3. One pair of spring scissors, 2 mm cutting edge, non-serrated, straight edge.
4. Two forceps, Dumont #5, super fine.
5. A rocking platform shaker.
6. Coverslip or specialized container for single fiber imaging (*see* **Note 2**).

### 2.4 Solutions and Reagents

1. Ringer's solution: 145 mM NaCl, 5 mM KCl, 2.5 mM $CaCl_2$, 1 mM $MgCl_2$, 10 mM HEPES and 10 mM glucose. Adjust pH to 7.4 using KOH.
2. HKS (high $K^+$ solution): 140 mM monopotassium glutamate, 10 mM HEPES, 10 mM $MgCl_2$, 10 mM glucose, 1 mM EGTA, pH 7.0.
3. A silicone elastomer kit that forms a preferably transparent elastomer (e.g., Dow *SYLGARD 184* silicone elastomer kit).
4. Petroleum jelly, pure.

5. Fixing solution: 4% weight of solid paraformaldehyde in Ringer's solution (*see* **Note 3** for details).
6. TDE 2,2′-thiodiethanol, purity ≥99%.
7. PBS phosphate-buffered saline, pH 7.4.

# 3 Methods

### *3.1 Coating Culture Dishes with Silicone Elastomer*

1. Mix the base and curing agent of the silicone elastomer kit according to manufacturer's specifications (in the case of SYLGARD 184, 1:10). Prepare a volume that fills a 50 mL centrifuge tube to about ¾ (33 mL).
2. Mix the components by either stirring with a spatula or sealing the tube and placing it on a roller mixer for 30 min.
3. Remove excess air inclusions (bubbles) by centrifuging the mix at 9 g for 2 min.
4. Add the silicone elastomer mix to culture dishes with 35 mm and 60 mm diameters. The optimal thickness range of the elastomer layer is 2–4 mm.
5. Cure the elastomer. This step may take up to 2 days, depending on the elastomer kit and curing temperature. In the case of *SYLGARD 184*, the curing time at room temperature is 48 h, and at 60 °C, it is 8 h. Refer to Fig. 1 to estimate the relation of the curing temperature to the curing time of *SYLGARD 184* in more detail.

### *3.2 Single Fiber Extraction*

1. Remove relevant limbs from a sacrificed laboratory animal post-mortem.
2. Store limbs in Ringer's solution at 4 °C for short periods only (e.g., for transport between labs).
3. Remove skin and hair from the limb (e.g., dissection protocol [2]).
4. Rinse and replace Ringer's solution to remove any stray hair.
5. Remove the relevant muscle by opening its fascicle with a scalpel and severing its tendons. Retain as much of the length of the tendons as possible. Specifics on the muscle dissection are inherent to each desired muscle and its localization.
6. Place the muscle in an elastomer-coated petri dish (*see* Subheading 3.1) that is filled with Ringer's solution.
7. Immobilize muscle under slight tension by pulling with forceps and affixing the tendons to the petri dish using insect needles.
8. Replace the Ringer's solution in the culture dish with HKS solution to depolarize the membrane potential of the muscle cells and inhibit any accidental stimulation of excitation-contraction coupling (*see* **Note 4**).

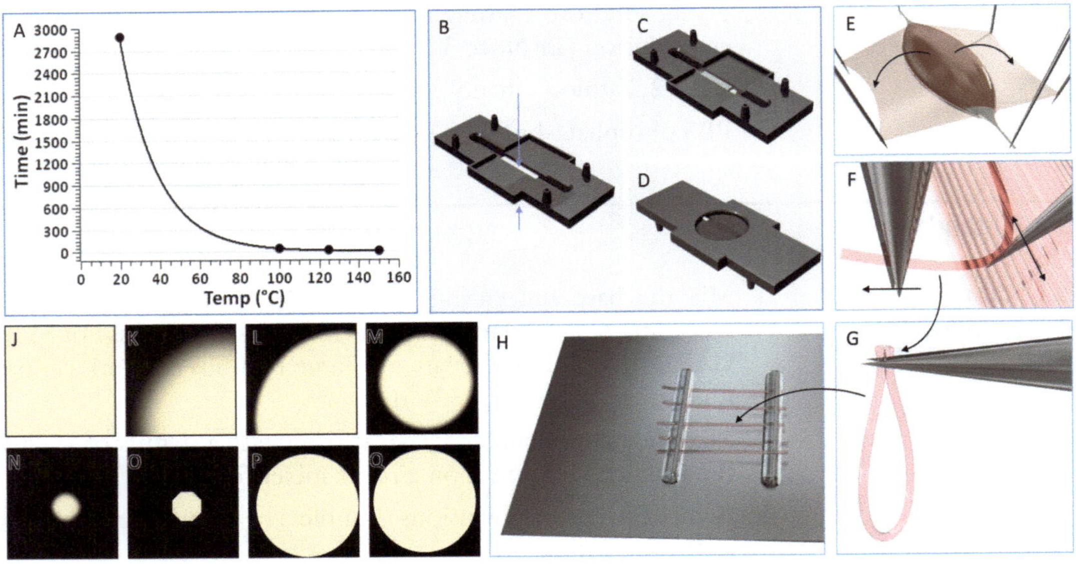


**Fig. 1** Equipment and sample preparation. (**a**) Interpolated correlation of curing temperature to curing time of *SYLGARD 184* silicone elastomer kit based on the Arrhenius correlation. (**b**) An example of a 3D-printed single fiber imaging chamber with coverslip insets and hose connections for medium perfusion in a microscopy slide form factor. (**c**) Assembled top view and (**d**) bottom view of the chamber with a water droplet suspended between two glass coverslips. An STL file for reprinting the chamber design is provided in the Electronic Supplementary Material. Steps of preparing a single muscle fiber sample are illustrated in the panels (**e–h**). (**e**) Spreading of the cut fascicle and pinning it to the elastomer substrate layer. (**f**) Grabbing and dislodging a cut single muscle fiber from the fiber bulk. (**g**) Forming a fiber loop for transferring the fiber short distances through air. (**h**) Placement of fibers on a glass coverslip with strips of petroleum jelly to keep them attached and under slight tension. The panels (**j–q**) represent the shapes that are visible through the ocular while aligning the excitation and collection objectives (*see* Subheading 3.4)

9. Incubate the muscle in HKS for 20 min at 4 °C before proceeding with manipulating the muscle.
10. Place the culture dish under a stereomicroscope, and manipulate the samples in the upcoming steps while observing them through the microscope.
11. Open the fascicle of the muscle longitudinally by cutting with spring scissors.
12. Spread the fascicle with forceps, and affix it to the elastomer bottom of the petri dish under tension using insect needles. If possible, remove the fascicle completely (*see* Fig. 1e).
13. If the muscle is composed of multiple ligaments, spread the ligaments out, and affix the individual ligament tendons.
14. Open the fascicles of ligaments longitudinally by cutting with spring scissors.
15. Spread the fascicles open with forceps, and remove them or fix them to the elastomer bottom with insect needles.

16. Grip a single fiber near its tendon attachment point, and sever it between the tendon and the griping location with spring scissors.
17. Keep the grip on the single fiber, and dislodge it from the bulk by slightly pulling orthogonally to the fiber direction. Aid the dislodging process by gently brushing closed forceps in between the dislodged fiber and the bulk (Fig. 1f). Sever any interconnecting filaments along the fiber (nerve connections, blood vessels) with spring scissors.
18. Once sufficient length (2–3 mm) of a fiber has been dislodged or the opposing tendon connection is reached, re-grip the fiber just before its end, or the tendon connection, and cut the fiber at the far end of the new grip site.
19. Transfer dislodged fibers over short distances through the air into the imaging chamber by forming a fiber loop. To form a loop, grip both ends of a dislodged single fiber in the same forceps (*see* Fig. 1g and **Note 5**).
20. Immobilize the fibers to the bottom of the imaging chamber by applying droplets of petroleum jelly to each end of the fiber as an adhesive. Alternatively, each end can be pushed into a pre-positioned petroleum jelly drop on the container surface (*see* Fig. 1h and **Note 6**).

### 3.3 TDE Clearing

1. Start by performing **steps 1–7** of the previous Subheading 3.2.
2. Replace the Ringer's solution with fixing solution, and incubate for at least 2 h at 4 °C (*see* **Note 7** details).
3. Wash the fixed sample once with 30% TDE solution.
4. Incubate the sample in 30% TDE solution for 4 h at RT and while shaking at 30 rpm.
5. Replace the solution with 50% TDE solution, and incubate again for 4 h, RT, 30 rpm.
6. Replace the solution with 70% TDE solution, and incubate again for 4 h, RT, 30 rpm.
7. Wash the sample once with 80% TDE solution.
8. Incubate the sample with 80% TDE solution overnight at 4 °C while shaking at 30 rpm.
9. The sample is now cleared and ready to be imaged or stored at 4 °C until imaging (*see* **Note 8**).

### 3.4 Aligning Excitation and Collection Objectives

1. Turn on the bright-field illumination source.
2. Ensure an unblocked beam path to the sample area by checking for light emission from the collection objective or condenser lens.
3. Ensure an unblocked light path to the ocular by checking for bright-field light in the ocular (*see* Fig. 1j).

4. Adjust the objectives to be roughly positioned coaxially on the optical axis and with a distance of the sum of their working distances from each other along the optical axis (usually the *Z*-axis in sample coordinates; *see* **Note 10**).
5. Slowly close the iris diaphragm that is located in the intermediate image plane between the collection objective and the bright-field illumination while observing through the ocular.
6. A dark circle will encroach on the field of view from the fringes (*see* Fig. 1k). Stop closing the iris diaphragm when the border between bright and dark areas has come into good view.
7. Adjust the distance between the objectives until the border between bright and dark areas comes into focus (*see* Fig. 1l).
8. Continue slowly closing the iris diaphragm until it is fully closed; adjust the radial position of the collection objective (usually *Y*- and *X*-axes in sample coordinates) to ensure the bright zone remains in the field of view (*see* Fig. 1m).
9. When the iris diaphragm is fully closed, adjust the radial position of the collection objective until the bright zone is roughly in the center of the field of view (*see* Fig. 1n).
10. Adjust the distance between the objectives again until the bright spot is in full focus (usually depicting an octagonal shape; *see* Fig. 1o).
11. Open the iris diaphragm until the bright zone nearly fills the whole field of view, and leave a small dark border at the fringe of the field of view (*see* Fig. 1p).
12. Adjust the radial position of the collection objective until the dark border at the fringe of the field of view has a constant thickness along its circumference (*see* Fig. 1q).
13. Fully open the iris diaphragm, and the objectives are now aligned.
14. Turn off the bright-field illumination.

### 3.5 Pre-Focusing a Sample Using Bright-Field Illumination

1. Space the microscope objectives sufficiently to allow the sample chamber to pass between them.
2. Turn on the bright-field illumination source.
3. Place the sample container in the microscope, and roughly center the bright-field illumination spot on the sample.
4. Adjust the position of the excitation objective to be roughly its working distance away from the sample.
5. Apply immersion liquid to the excitation objective, and ensure that the liquid and all optical surfaces are free of microbubbles. When using a dipping objective, confirm a bubble-free immersion of the objective into the sample medium.

6. Move the focal plane of the collection objective 1–10 mm away from the focal plane of the excitation objective to increase the contrast of the bright-field image (*see* **Note 11**).
7. Find the outline of your sample by moving the sample container or the *XY*-stage while observing through the microscope ocular.
8. Bring the sample into focus by adjusting the excitation objective along the optical axis (*Z*-axis).
9. Return the collection objective to its original position, realigning the focal planes of the two objectives (*see* **Note 12**).
10. Turn off bright-field illumination.

### 3.6 Placement of Filters and Dichroic Mirrors

1. Place the short-pass laser filter with an edge wavelength just below the excitation LASER wavelength on the main optical axis after the collection objective in order to block laser light from reaching the detector (*see* Fig. 2a, b).

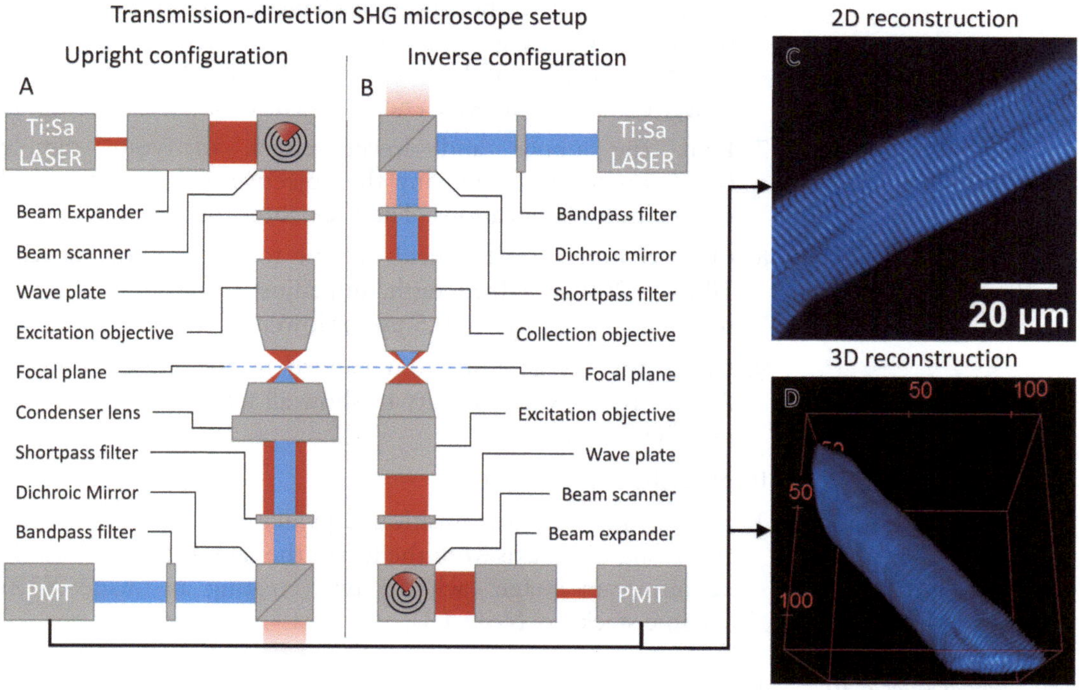


**Fig. 2** Multiphoton second harmonic generation (SHG) microscope setup sketches and resulting images of muscle fibers. (**a**, **b**) Upright and inverse configuration of an SHG microscopy setup. Both consist of a femtosecond laser source, a beam expander, a beam scanner, a wave plate (optional, *see* **Note 9**), a high-resolution excitation objective, a collection objective or condenser lens, a short-pass laser filter, a long-pass dichroic mirror, a narrow bandpass emission filter and an ultrafast single-pixel detector with a wide aperture (here, a photomultiplier tube, PMT). (**c**) Reconstructed SHG image of a 2D plane scan of a section of single mouse EDL muscle fiber (pixel size $0.1 \times 0.1$ µm$^2$). (d) Reconstructed SHG image of a 3D volume scan of a section of single mouse EDL fiber (voxel size $0.5 \times 0.5 \times 0.5$ µm$^3$)

2. Place the long-pass dichroic mirror with a cutoff wavelength between excitation LASER and the SHG emission on the main optical axis of the microscope to redirect the shorter wavelength SHG emission toward the detector setup (*see* Fig. 2a, b).
3. Place the bandpass filter with a center wavelength at the SGH emission directly in front of the detector (*see* Fig. 2a, b).

### 3.7 Single Fiber 3D SHG Image Acquisition

1. Ensure that the microscope system control software is started and initialized.
2. Ensure that the LASER is on and has had sufficient time to stabilize (e.g., in the case of Chameleon Vision II roughly 30 min).
3. After roughly positioning your fiber sample in the focal plane (*see* Subheading 3.5), close the optical shielding around the microscope.
4. Open laser safety shutters if applicable.
5. Adjust the LASER power to roughly 300 mW.
6. Start imaging with the largest available field of view and a low resolution (e.g., 4 μm pixel size) in a live-view mode. Micro-adjust the position of the sample and the objectives to properly center the sample in *X*, *Y* and *Z* dimensions.
7. Find the upper and lower edges of your sample (in *Z* dimension) by moving the objective up and down, and save or note the locations. Then re-center the sample.
8. Once centered, set the field of view to a square of between 100 and 200 μm edge length, and adjust the number of pixels to obtain a pixel size of 0.5 μm or lower.
9. Check for a visible striation pattern in the muscle sample. Adjust the LASER intensity to show all features but not saturate the detector.
10. Stop the live-view mode.
11. Start an automated acquisition sequence that moves the objective along the *Z*-axis from the top of the sample to the bottom, acquiring an image each 0.5 μm (creating an image *XYZ*-volume stack; *see* **Note 13**).

### 3.8 Whole Muscle 3D SHG Image Acquisition

1. Follow **steps 1–7** of Subheading 3.7, but using a whole cleared muscle, keep it immersed in the clearing solution for imaging (*see* Subheading 3.3).
2. Set the field of view as large as possible to still achieve a pixel size of 0.5 μm or lower with the maximum available pixel count in the image.
3. Check for a visible striation pattern in the muscle sample. Adjust the LASER intensity to show all features but not

saturate the detector. If possible in your system, attenuate the LASER power depending on the *Z*-axis position, in order to allow features to be discernible throughout the whole depth of the sample.

4. Start an automated acquisition sequence that moves the objective along the *Z*-axis from the top of the sample to the bottom, acquiring an image each 0.5 μm (creating an image *XYZ*-volume stack; *see* **Note 14**).
5. The maximally available field of view is usually smaller than a whole muscle sample, so a mosaic-style acquisition of multiple stacks across the *X* and *Y* dimensions of the sample is required.
6. If your system supports an automated acquisition of a mosaic of stacks, start by locating your sample borders in *X* and *Y* dimensions and setting up a mosaic field that encompasses them. If your system does not support an automated mosaic acquisition, you manually move the field of view in *X* and *Y* and acquire a new image stack. In both cases, the stacks need to overlap in the range of 10% for later image reconstruction (stitching).

### 3.9 Exporting an Image Slice from a 3D Stack While Maintaining the Size Information Metadata

1. Open an SHG 3D image stack of a muscle fiber in FIJI using the Bioformats importer to assure that size information metadata is taken into account. If more than one color channel is present in the stack, open the data as hyperstack in composite color mode.
2. Adjust the brightness and contrast of each channel, either manually ("Image > Ajust > Brichtness/Contrast") or automatically ("Process > Enhance Contrast").
3. Find a slice to export and duplicate it with "Image > Duplicate", and uncheck the duplicate stack/hyperstack option.
4. Saving the resulting image slice as a TIFF file ("File > Save As > Tiff...") will retain its size information metadata (*see* **Note 15**).

### 3.10 3D Reconstruction of an Image Stack

1. Open an SHG 3D image stack of a muscle fiber in FIJI using the Bioformats importer to assure that size information metadata is taken into account. If more than one color channel is present in the stack, open the data as hyperstack in composite color mode.
2. Adjust the brightness and contrast of each channel, either manually ("Image > Ajust > Brichtness/Contrast") or automatically ("Process > Enhance Contrast").
3. Once the desired features are properly visible, convert the image stack to the RGB color data format by selecting "Image > Type > RGB Color" (if using a Hyperstack, select

"Image > Color > Stack to RGB"). This step allows properly transferring the brightness/contrast settings to the 3D reconstruction.

4. To create a volumetric image (*see* Fig. 2d) select "Plugins > 3D Viewer" (*see* **Note 16**).

#### *3.11 3D Reconstruction of a Mosaic of Image Stacks*

1. Open Fiji and navigate to "Plugins > Stitching > Grid/Collection stitching" [16].
2. Select the order in which the stacks were acquired or which identifiers indicate the stack position in the mosaic. If you acquired the mosaic manually, compute the exact overlap from the image data here.
3. After stitching is complete, the resulting file can be treated as an image stack and any other protocols aimed at stacks (such as Subheadings 3.9 and 3.10) can be applied.

#### *3.12 Determining Mean Sarcomere Length in an Image*

1. Open an SHG image of a muscle fiber in FIJI using the Bio-formats importer to ensure that size information metadata is taken into account.
2. Perform a 2D fast Fourier transform on the image using the "FFT" function from the Process/FFT submenu of FIJI.
3. In the Fourier-transformed image, mouse over the centroid of the first high-intensity area from the image center (*see* Fig. 3a and **Notes 17** and **18**).
4. Read out the frequency of the Fourier-space bin from the info section of the FIJI main window (*see* Fig. 3a upper right). The bin frequency represents the mean sarcomere length of all structures in the image (*see* **Notes 19** and **20**).

#### *3.13 Determining the Major Orientation Axis in an Image*

1. Convolve the pixel matrix of the image $I_m$ with the *x*-axis Sobel kernel $S_x$ and the *y*-axis Sobel kernel $S_y$ producing the approximate derivatives of image $I_m$ in vertical horizontal direction $G_x$ and horizontal direction $G_y$ (*see* **Note 21**).
2. Form a two-component vector of each pixel $g_x^{i,j}$ of the matrix $G_x$, and corresponding pixel $g_y^{i,j}$ of $G_y$. The vector represents the orientation and magnitude of an intensity gradient through each pixel.
3. Determine the orientation $\theta_{i,\ j}$ of each vector $(g_x^{i,j}, g_y^{i,j})$ by applying the four-quadrant inverse tangent function $\text{atan2}\left(g_x^{i,j}, g_y^{i,j}\right)$ on the vector.

$$\theta_{i,j} = \text{atan2}\left(g_y^{i,j}, g_x^{i,j}\right).$$

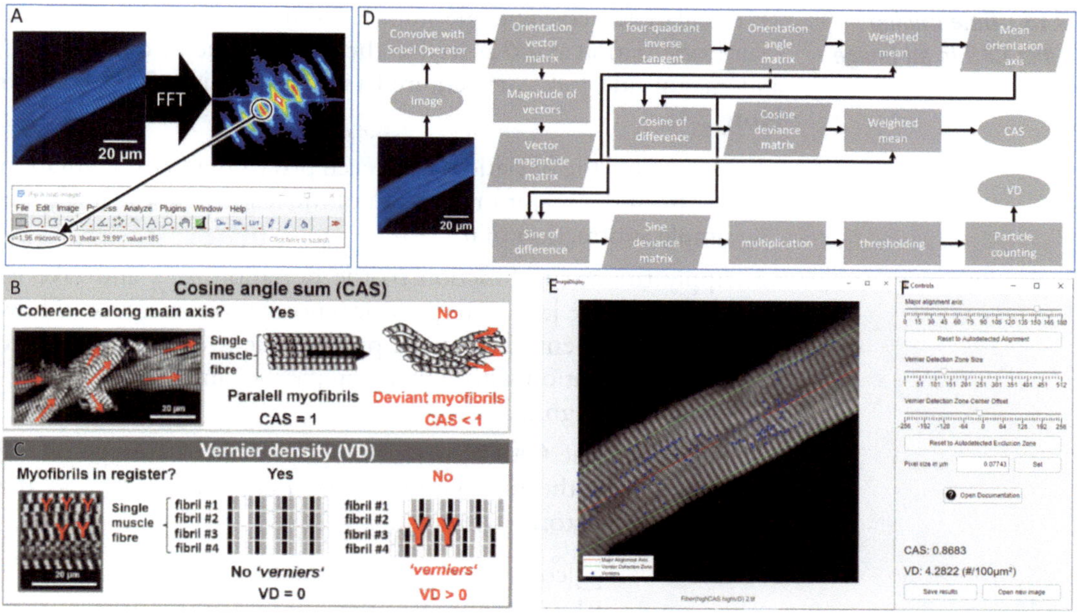


**Fig. 3** Morphometric pattern extraction strategies to produce characteristic numbers from SHG images of muscle fibers to quantify cytoarchitectural morphology. (**a**) A manual process to quickly identify the mean sarcomere length in a muscle fiber image by using the 2D fast Fourier transform in *FIJI*. (**b**) The concept of CAS as a measure of angular misalignment of the myofibrils and their actin-myosin filament lattice within muscle fibers. (**c**) The concept of the VD as a measure for the axial misalignment of myofibrils within a muscle fiber. (**d**) A process flowchart for extracting CAS and VD from a fiber image. The start and end points are shown as oval shapes, the process steps as rectangles and the intermediate input/output stages as rhomboids. (**e**) A fiber image processed by the CASVD2d software package shows identified Verniers marked with blue asterisks. The control and output window of the software (**f**) displays the CAS and VD resulting from the current size and alignment of the detection zone. Detailed information on the software is provided in the software documentation file in the installation folder or by clicking the "Open Documentation" button in the control window of the software

4. Determine the vector magnitude $|m|_{i,\ j}$ is determined via the Pythagorean theorem from the vector components.

$$|m|_{i,j} = \sqrt{\left(g_{\mathrm{y}}^{i,j}\right)^2 + \left(g_{\mathrm{x}}^{i,j}\right)^2}$$

5. Determine the major orientation axis $\theta$ by the weighted average of all $\theta_{i,\ j}$ with $|m|_{i,\ j}$ being the weighting factor (*see* **Note 22**).

$$\theta = \frac{\sum\limits_{i,j} \theta_{i,j} \cdot |m|_{i,j}}{\sum\limits_{i,j} |m|_{i,j}}$$

### 3.14 Determining Fiber CAS in an Image

To calculate the cosine angle sum (CAS) in an image, knowing the major orientation axis $\theta$ as well as the gradient vector direction $\theta_{i,\,j}$ and magnitude $|m|_{i,\,j}$ is required (*see* Subheading 3.13; also *see* **Note 23**).

1. Determine the deviation $\delta_{\mathrm{i,j}}$ of each pixel's gradient orientation $\theta_{i,\,j}$ from the major orientation of the features in the image via simple subtraction of $\theta_{i,\,j} - \theta$.
2. Apply the cosine function to the deviation $\delta_{i,\,j}$, and take its absolute value $|\cos|_{i,\,j}$ to provide a normalized measure of how much the orientation of the pixel gradient deviates from the major orientation with 1 meaning parallel and 0 meaning perpendicular alignment.
3. Determine the CAS by performing a weighted average of all $|\cos|_{i,\,j}$ with the orientation vector magnitude $|m|_{i,\,j}$ as the weighting factor.

$$\mathrm{CAS} = \frac{\sum_{i,j} |\cos|_{i,j} \cdot |m|_{i,j}}{\sum_{i,j} |m|_{i,j}} = \frac{\sum_{i,j} |\cos(\theta_{i,j} - \theta)| \cdot |m|_{i,j}}{\sum_{i,j} |m|_{i,j}}$$

### 3.15 Determining Fiber VD in an Image

To calculate the cosine angle sum (CAS) in an image, knowing the major orientation axis $\theta$ as well as the gradient vector direction $\theta_{i,\,j}$ and magnitude $|m|_{i,\,j}$ is required (*see* Subheading 3.13; also *see* **Note 23**).

1. Determine the deviation $\delta_{i,\,j}$ of each pixel's gradient orientation $\theta_{i,\,j}$ from the major orientation of the features in the image via simple subtraction of $\theta_{i,\,j} - \theta$.
2. Apply the sine function to the deviation $\delta_{i,\,j}$, and take its absolute value $|\sin|_{i,\,j}$ to provide a normalized measure of how much the orientation of the pixel gradient deviates from the major orientation with 1 meaning perpendicular and 0 meaning parallel alignment.
3. Multiply the $|\sin|_{i,\,j}$ with the orientation vector magnitude $|m|_{i,\,j}$ to form the weighted Vernier image matrix $V$.
4. Apply a region of interest (ROI) to $V$ that encompasses only the inside of the fiber, excluding the background and the fiber edges.
5. Consider regions of pixels within the ROI of $V$ whose value is over the threshold value $t$ ($t = 0.025$) to be Vernier regions $V_{\mathrm{r}}$.
6. Count the number of Vernier regions $n(V_{\mathrm{r}})$.

7. Determine the Vernier area density VD by normalizing the number of Vernier regions over the area of the ROI $A_{\mathrm{ROI}}$.

$$\mathrm{VD} = \frac{n(V_r)}{A_{\mathrm{ROI}}} = \frac{n\left(\sin\left(\theta_{i,j} - \theta\right) > t, \ \in \mathrm{ROI}\right)}{A_{\mathrm{ROI}}}$$

## 4 Notes

1. Differentiating between SHG and two-photon (2p) fluorescence emission; the SHG emission wavelength is always half of the wavelength of the excitation LASER. 2p fluorescence emission wavelength is only dependent on the emitting fluorochrome properties. The fluorescence emission intensity may be influenced by the choice of LASER excitation wavelength; however, the emission wavelength is not influenced.
2. The most simple sample chamber suited for both upright and inverse multiphoton imaging is a coverslip, covered with a drop of HKS. Coverslips with elongated form factor (e.g., 22 × 100 mm) are preferable for better handling. Microscopy slides provide even better handling due to their rigidity; however, they are exclusively suited for upright microscopy, due to their thickness. Specialized imaging chambers provide maximized handling comfort and streamline imaging; these may be 3D printed [19] or milled from polymethyl methacrylate (PMMA) according to custom designs. A design file for a 3D-printed chamber is provided in the Electronic Supplementary Material.
3. Regarding preparing the fixing solution, adding PFA to Ringer's solution acidifies the solution. Raise the pH back to 7.4 by adding NaOH solution (5 M NaOH) while stirring vigorously. PFA does not dissolve well in Ringer's solution (or water, for that matter) even when the pH is adjusted. Heating the solution to 60 °C after or during pH adjustment will aid in dissolving the PFA completely.
4. The transfer from Ringer's solution into HKS may induce muscle contraction, depending on sample type and condition. Ensure a thorough immobilization of the tendons to prevent excessive muscle shortening or contracture. Gradually add HKS to the Ringer's solution to reduce the chance of muscle contracture during the exchange of the solution.
5. When re-gripping a single fiber after having it dislodged, just before cutting the second end of the fiber, try retaining the initial grip site in the forceps when gripping the new site,

forming a fiber loop. Forming a fiber loop is advised for transferring the fiber through the surface tension of liquid surfaces, and forming a loop during this step simplifies the procedure by retaining a solution drop within the loop.

6. When fibers are immobilized in the imaging chamber with petroleum jelly, the fibers should be as straight as possible, and the petroleum jelly should be positioned as far toward the ends of the fibers as possible. The best case scenario is that the fibers are held under slight tension that is upheld by the adhesive force of the petroleum jelly.
7. The sample fixation time is dependent on sample size; if the sample is larger than 3 mm at its smallest diameter through its center, consider fixing overnight to ensure thorough fixation. Gentle shaking is recommended during an extended fixing period.
8. TDE clearing is best suited for small extracted muscle samples, such as mouse *extensor digitorum longus* (EDL) or *Soleus* muscle. It is cheap, simple and quick. It retains the macro- and microarchitecture and preserves the myosin SHG signal [18]. However, the technique is not suited to deal with large and/or high-density tissue such as whole limbs or bone inclusions.
9. Wave plates may be placed in the excitation LASER beam path to modify the light polarity. The orientation of light polarization relative to the sample structure influences SHG intensity. A lambda/4 wave plate will convert linearly polarized light to circularly polarized light, removing all influence of sample orientation on the SHG intensity. A rotationally mounted lambda/2 plate will allow rotating the orientation of linearly polarized light around the optical axis, effectively allowing for tuning the setup to excite maximum SHG intensity in the sample [20].
10. The two objectives in the setup are moved relative to each other in this step. Keep in mind that this also implies that the tips of the objectives might collide. Avoid the collision of the objectives with utmost care.
11. The sample will only provide very little contrast in bright-field illumination if the objectives are used in the configuration with aligned focal planes that is essential for SHG imaging. Make sure to move the collection objective or condenser lens out of focus to be able to locate the sample in bright field with ease. Some intracellular structure of muscle fibers may already be discernible in the bright-field image once the fiber is well focused.

12. To align the focal planes, the aligning procedure (*see* Subheading 3.4) may be performed when the sample is already in focus. This helps compensating for focus distance modulations by the immersion liquid or, in case of larger samples, by the sample itself.
13. A cubic voxel size of (e.g., of $0.5 \times 0.5 \times 0.5\ \mu m^3$) is desirable for 3D reconstruction and 3D analysis. However, if imaging time is a crucial factor, the step size along the *Z*-axis may be increased by one to two orders of magnitude. Image stacks, acquired in this fashion, will still allow for 2D and 2.5D analysis.
14. Like with single fibers, a cubic voxel size of (e.g., of $0.5 \times 0.5 \times 0.5\ \mu m^3$) is desirable. But as the sample volume increases, the imaging time increases proportionally with a cubic correlation, as does the file size. Consider increasing resolution or reducing stack count if your application is not heavily affected by the changes.
15. If you are exporting a slice for use with CASVD2d, the software currently only works with images with a single color channel (for further details, refer to the software documentation that can be accessed in your software installation folder or from the control interface; available in the Electronic Supplementary Material). In this case, split the composite image into its single channels ("Image > Color > Split Channels"), then select and save the channel with the relevant SHG information as a TIFF file.
16. When your data contain a low amount of color channels, consider switching to high contrast lookup tables (such as "Cyan Hot" or "16_colors") before converting to the RGB color data type, to visualize the image features more clearly in 3D.
17. The high-intensity areas may be hard to discern in grayscale; consider applying a high-contrast lookup table such as "16_colors" (as seen in Fig. 3a upper right) to the Fourier-transformed image as a visual aid.
18. The Fourier-transformed image may be grainy due to high-frequency noise; consider applying a Gaussian blur filter with a sigma value of 2 on the Fourier-transformed image to help in properly discerning the centroids of the high-intensity regions.
19. Determining the mean sarcomere length of a fiber can be tedious, especially in fibers with high inhomogeneity. The presented Fourier-space method represents a quick and easy way for an unbiased determination of the mean sarcomere length of a whole image and is resilient to image noise.

20. An estimation of the mean orientation axis $\theta$ can be obtained by reading out the angle theta of the frequency bin that represents the mean sarcomere length. The value is also found in the info section of the main FIJI window, next to the frequency value.
21. Keep in mind that the derivates $G_x$ and $G_y$ will contain negative values. Assure that your image data structure can accommodate negative values.
22. Weighting the mean pixel orientation with the Sobel vector magnitude ensures that well-pronounced gradients in the image have a large influence on the determined major orientation, effectively eliminating the influence of background noise without the need for thresholding.
23. Albeit it is possible to determine CAS and VD in FIJI, the IJ1 Macro language is ill-suited for the task. Especially, the step of determining the direction and magnitude of the vectors ($g_x^{i,j}$, $g_y^{i,j}$) is time-consuming (minute range). Other programming languages that are suited for handling multidimensional matrixes can perform the operations to determine CAS and VD very quickly (millisecond range). As such, a MATLAB-based implementation of CAS and VD detection, including a minimalistic intuitive GUI, is provided as a windows installer package. The package includes a small selection of sample data.

## References


1. Berchtold MW, Brinkmeier H, Müntener M (2000) Calcium ion in skeletal muscle: its crucial role for muscle function, plasticity, and disease. Physiol Rev 80:1215–1265. https://doi.org/10.1152/physrev.2000.80.3.1215
2. Bröllochs A (2018) Dissection of mouse EDL and soleus muscles. Protocols.Io. https://doi.org/10.17504/protocols.io.jcrciv6
3. Buttgereit A, Weber C, Friedrich O (2014) A novel quantitative morphometry approach to assess regeneration in dystrophic skeletal muscle. Neuromuscul Disord 24:596–603. https://doi.org/10.1016/j.nmd.2014.04.011
4. Buttgereit A, Weber C, Garbe CS, Friedrich O (2013) From chaos to split-ups – SHG microscopy reveals a specific remodelling mechanism in ageing dystrophic muscle: Remodelling of dystrophic muscle. J Pathol 229:477–485. https://doi.org/10.1002/path.4136
5. Centonze VE, White JG (1998) Multiphoton excitation provides optical sections from deeper within scattering specimens than confocal imaging. Biophys J 75:2015–2024. https://doi.org/10.1016/S0006-3495(98)77643-X
6. Franzini-Armstrong C (2015) Electron microscopy: from 2D to 3D images with special reference to muscle. Eur J Transl Myol 25:4836. https://doi.org/10.4081/ejtm.2015.4836
7. Friedrich O, Both M, Weber C, Schürmann S, Teichmann MDH, von Wegner F, Fink RHA, Vogel M, Chamberlain JS, Garbe C (2010) Microarchitecture is severely compromised but motor protein function is preserved in dystrophic mdx skeletal muscle. Biophys J 98:606–616. https://doi.org/10.1016/j.bpj.2009.11.005
8. Garbe CS, Buttgereit A, Schürmann S, Friedrich O (2012) Automated multiscale morphometry of muscle disease from second harmonic generation microscopy using tensor-based image processing. IEEE Trans Biomed Eng 59:39–44. https://doi.org/10.1109/TBME.2011.2167325
9. Gu M, Gan X, Kisteman A, Xu MG (2000) Comparison of penetration depth between two-photon excitation and single-photon excitation in imaging through turbid tissue media.

Appl Phys Lett 77:1551–1553. https://doi.org/10.1063/1.1308059

10. Haug M, Ritter P, Michael M, Reischl B, Schürmann S, Prölß G, Friedrich O (2022) Structure-function relationships in muscle Fibres: MyoRobot online assessment of muscle fibre elasticity and sarcomere length distributions. IEEE Trans Biomed Eng 69:148–155. https://doi.org/10.1109/TBME.2021.3089739
11. Kiriaev L, Kueh SLL, Morley JW, North KN, Houweling PJ, Head SI (2018) Branched fibers from old fast-twitch dystrophic muscles are the sites of terminal damage in muscular dystrophy. Am J Physiol: Cell Physiol:C662–C674. https://doi.org/10.1152/ajpcell.00161.2017
12. Liu W, Ralston E, Raben N (2013) Quantitative evaluation of skeletal muscle defects in second harmonic generation images. J Biomed Opt 18:026005. https://doi.org/10.1117/1.JBO.18.2.026005
13. Lovering RM, Michaelson L, Ward CW (2009) Malformed mdx myofibers have normal cytoskeletal architecture yet altered EC coupling and stress-induced $Ca^{2+}$ signaling. Am J Phys Cell Phys 297:C571–C580. https://doi.org/10.1152/ajpcell.00087.2009
14. Lovering RM, O'Neill A, Muriel JM, Prosser BL, Strong J, Bloch RJ (2011) Physiology, structure, and susceptibility to injury of skeletal muscle in mice lacking keratin 19-based and desmin-based intermediate filaments. Am J Phys Cell Phys 300:C803–C813. https://doi.org/10.1152/ajpcell.00394.2010
15. Mohler W, Millard AC, Campagnola PJ (2003) Second harmonic generation imaging of endogenous structural proteins. Methods 29:97–109. https://doi.org/10.1016/S1046-2023(02)00292-X
16. Preibisch S, Saalfeld S, Tomancak P (2009) Globally optimal stitching of tiled 3D microscopic image acquisitions. Bioinformatics 25:1463–1465. https://doi.org/10.1093/bioinformatics/btp184
17. Schindelin J, Arganda-Carreras I, Frise E, Kaynig V, Longair M, Pietzsch T, Preibisch S, Rueden C, Saalfeld S, Schmid B, Tinevez J-Y, White DJ, Hartenstein V, Eliceiri K, Tomancak P, Cardona A (2012) Fiji: an open-source platform for biological-image analysis. Nat Methods 9:676–682. https://doi.org/10.1038/nmeth.2019
18. Schneidereit D, Bröllochs A, Ritter P, Kreiß L, Mokhtari Z, Beilhack A, Krönke G, Ackermann JA, Faas M, Grüneboom A (2021) An advanced optical clearing protocol allows label-free detection of tissue necrosis via multiphoton microscopy in injured whole muscle. Theranostics 11:2876. https://doi.org/10.7150/thno.51558
19. Schneidereit D, Nübler S, Prölß G, Reischl B, Schürmann S, Müller OJ, Friedrich O (2018) Optical prediction of single muscle fiber force production using a combined biomechatronics and second harmonic generation imaging approach. Light: Sci Appl 7:79. https://doi.org/10.1038/s41377-018-0080-3
20. Schürmann S, von Wegner F, Fink RHA, Friedrich O, Vogel M (2010) Second harmonic generation microscopy probes different states of motor protein interaction in myofibrils. Biophys J 99:1842–1851. https://doi.org/10.1016/j.bpj.2010.07.005

# Chapter 18

# Optimization of Cell Viability Assays for Drug Sensitivity Screens

**Peter Larsson and Toshima Z. Parris**


## Abstract

During the preclinical stages of the drug discovery process, cell viability assays are fundamental tools for studying the phenotypic properties and overall health of cells following in vitro drug sensitivity screens. Therefore, it is important to optimize your viability assay of choice to obtain reproducible and replicable results, as well as use relevant drug response metrics (e.g., IC50, AUC, GR50, and GRmax) to identify candidate drugs for further evaluation in vivo. Herein, we used the resazurin reduction assay which is a quick, cost-effective, simple-to-use, and sensitive method for examining the phenotypic properties of cells. Using the MCF7 breast cancer cell line, we provide a detailed step-by-step protocol for optimizing drug sensitivity screens using the resazurin assay.


**Key words** High-throughput screen, Optimal cell number, Doubling time, Resazurin reduction assay, Colorimetric assay, Protocol, Phenotypic screen, Drug discovery, MCF7 cell line

## 1 Introduction

Cell viability assays are used to estimate the proportion of healthy, living cells at a given time point [1]. Although viability assays are routinely performed in biomedical research to study optimal cell culture conditions, toxicity to pharmaceutical compounds, receptor binding, and cellular protease activity, they are commonly used for high-throughput drug sensitivity screens on multi-well plates (e.g., 96- or 384-well plates) to determine their efficiency and potency at different concentrations [1, 2]. Drug sensitivity screens are, therefore, fundamental for the de novo drug development process, as well as for identifying new clinical uses for existing drugs [3]. Most viability assays exploit the metabolic or enzymatic activity of a cell and are broadly classified depending on the metabolic activity marker used and the equipment needed (e.g., spectrophotometer or flow cytometer), including dye exclusion assays (e.g., Congo red and trypan blue), colorimetric assays (e.g., MTT, MTS, and

Oliver Friedrich and Daniel F. Gilbert (eds.), *Cell Viability Assays: Methods and Protocols*, Methods in Molecular Biology, vol. 2644,
https://doi.org/10.1007/978-1-0716-3052-5_18,

resazurin reduction assay), fluorometric assays (e.g., resazurin assay), luminometric assays (e.g., adenosine triphosphate [ATP]), and flow cytometric assays (e.g., membrane permeability or asymmetry) [1, 4].

Since cell viability assays are normally performed over several days, high demands are placed on finding optimal cell culturing conditions to retain the natural properties of the cells (e.g., nutrition supply and cell confluency) while at the same time being reliable, reproducible, cost-effective, and quick [1, 5–7]. However, some viability assays (e.g., dye exclusion assays and flow cytometric assays) are too time-consuming and/or expensive for use in high-throughput screens. Herein, we chose to use the resazurin assay, which is relatively safe, simple to use, cost-effective, and highly sensitive. Resazurin (7-hydroxy-10-oxidophenoxazin-10-ium-3-one) is a blue, non-fluorescent molecule that is reduced by metabolically active cells to resorufin (7-hydroxyphenoxazin-3-one), which is a pink, fluorescent molecule that can be measured using either absorbance or fluorescence [1, 4, 8].

To ensure the successful interpretation of a drug sensitivity screen, it is important that the cell viability assay is optimized to determine optimal experimental parameters. First, experimental factors influencing the performance of the cell viability assay need to be identified using Quality Control Metrics (QCM), e.g., one-way ANOVA. Optimization of any confounding factors should minimize the influence of technical (e.g., evaporation and drug storage) and biological factors (e.g., cell type and culture medium composition) that can potentially affect data precision, and thereby the replicability and reproducibility of a drug sensitivity screen [8]. Thereafter, the performance of the optimized assay should then be assessed using QCMs (e.g., *Z*-factor [*Z*], signal window [SW], coefficient of variation [CV]) that can establish the assay signal dynamic range [9]. Ideally, the optimized assay should demonstrate robustness between replicate experiments and different labs, despite minor variations in the experimental setup. The assay should also be able to identify biologically active drugs (*Z*-factor close to or equal to 1), have a large signal dynamic range (high SW), and have little to no signal variation (low CV). Lastly, appropriate drug response metrics (e.g., IC50 [half maximal inhibitory concentration], AUC [area under the dose-response curve], GR50 [growth rate inhibition concentration at GR(c) = 0.5], and GRmax [maximal measured GR value]) should be chosen to evaluate drug potency and efficiency. In this work, we present an optimization strategy for drug sensitivity screening with the resazurin reduction cell viability assay and provide a detailed description of experimental factors that can potentially affect cell viability and how to deal with them. Moreover, we have included an optimized protocol for the MCF-7 breast cancer cell line.

## 2 Materials

### 2.1 Pharmaceutical Compounds

1. Chemical stock solutions (1 mM): Compound, drug solvent (e.g., dimethyl sulfoxide [DMSO], dimethylformamide [DMF], ethanol, water, and sodium chloride [NaCl]; *see* **Note 1**).
2. DMSO.
3. 15 mL Falcon tubes.
4. 2 mL Eppendorf tubes.
5. 96-well PCR plates.
6. Aluminum sealing tape.
7. Phosphate-buffered saline, pH 7.4 (PBS): Sterile $Ca^{2+}$- and $Mg^{2+}$-free phosphate-buffered saline.

### 2.2 Cell Culture

1. MCF7 human breast carcinoma cell line (e.g., HTB-22 from ATCC).
2. T75 (75 $cm^2$ growth area) flask with filter cap.
3. 15 mL Falcon tubes.
4. Cell culture medium: Dulbecco's modified Eagle's medium (DMEM) with phenol red, 10% fetal bovine serum (FBS), 2 mM L-glutamine (*see* **Note 2**).
5. PBS.
6. Trypsin-EDTA (0.25%), phenol red.
7. 96-well clear, flat-bottom microplates.
8. Hemocytometer (e.g. Invitrogen Countess Automated Cell Counter).
9. Trypan blue solution (0.4%).

### 2.3 Resazurin Reduction Cell Viability Assay

1. Resazurin sodium salt (high-purity biological stain quality). Resazurin stock solution (0.2 mg/mL solution, 0.02% (w/v)). Dissolve 10 mg resazurin sodium salt in 50 mL 1XPBS (pH 7.4). Vortex vigorously and filter sterilize the solution using a 0.22 μm filter. Since resazurin solution is light sensitive, it should be stored protected from light (covered with aluminum foil) at 2–8 °C for up to 12 months.
2. 50 mL Falcon tubes.
3. Aluminum foil.
4. Microplate reader to measure absorbance with a 560 nm excitation filter and a 615 nm emission filter, or fluorescence with an F560 excitation filter (540–570 nm) and an F590 emission filter (580–610 nm).

### 2.4 Data Analysis

1. Microsoft Excel.
2. R/Bioconductor version 4.0 or higher and RStudio.
3. R package "GRmetrics."
4. Web-based GR Calculator tool (http://www.grcalculator.org/grcalculator/).

## 3 Methods

Unless otherwise stated, all experimental steps are performed at room temperature (18–25 °C) and in sterile conditions (in a laminar flow cabinet sterilized with 70% ethanol before use), while incubation steps are performed at 37 °C and 5% $CO_2$ in a humidified incubator.

### 3.1 Dilute Pharmaceutical Compounds

1. In 2 mL Eppendorf tubes, dilute the drug stock solutions (e.g., 1 mM in DMSO) with PBS to create a nine-concentration 1: 2 and 1:5 dilution series (Fig. 1a, *see* **Note 3**).
2. Dilute nine matched drug solvent concentration vehicle controls (e.g., DMSO) with PBS in 2 mL Eppendorf tubes (Fig. 1b, *see* **Note 4**).
3. Plate the diluted drugs and matched drug solvent vehicle controls in a 96-well PCR plate (Fig. 1c).
4. Cover the PCR plate with aluminum sealing tape and store at −20 °C for up to 4 weeks (*see* **Note 5**). Avoid more than four freeze-thaw cycles.

### 3.2 Cell Culture

Cell culture procedures are cell line specific. The following description includes standard culturing procedures for adherent MCF7 breast cancer cells.

1. Expansion: Seed 1 × $10^6$ cells suspended in 20 mL of culture medium in a T75 flask, and incubate the cells at 37 °C and 5% $CO_2$ in a humidified incubator. Replace the culture medium every 2–3 days. Passage the culture (or use for drug screening) while the cells are still in the log phase (around 80% confluent) before they reach 100% confluence (*see* **Note 6**).
2. Harvesting: Remove the culture medium from the flask and wash the cells gently with 5 mL of PBS. Remove the PBS, and add 2 mL of Trypsin-EDTA. Incubate the cells at 37 °C for 4–5 min, until most of the cells have detached from the flask. Add 5 mL culture medium (containing 10% FBS) to the flask, and transfer the cell suspension to a 15 mL Falcon tube. Centrifuge at 220 × *g* for 5 min and discard the supernatant. Resuspend the cell pellet in 5 mL of culture medium.

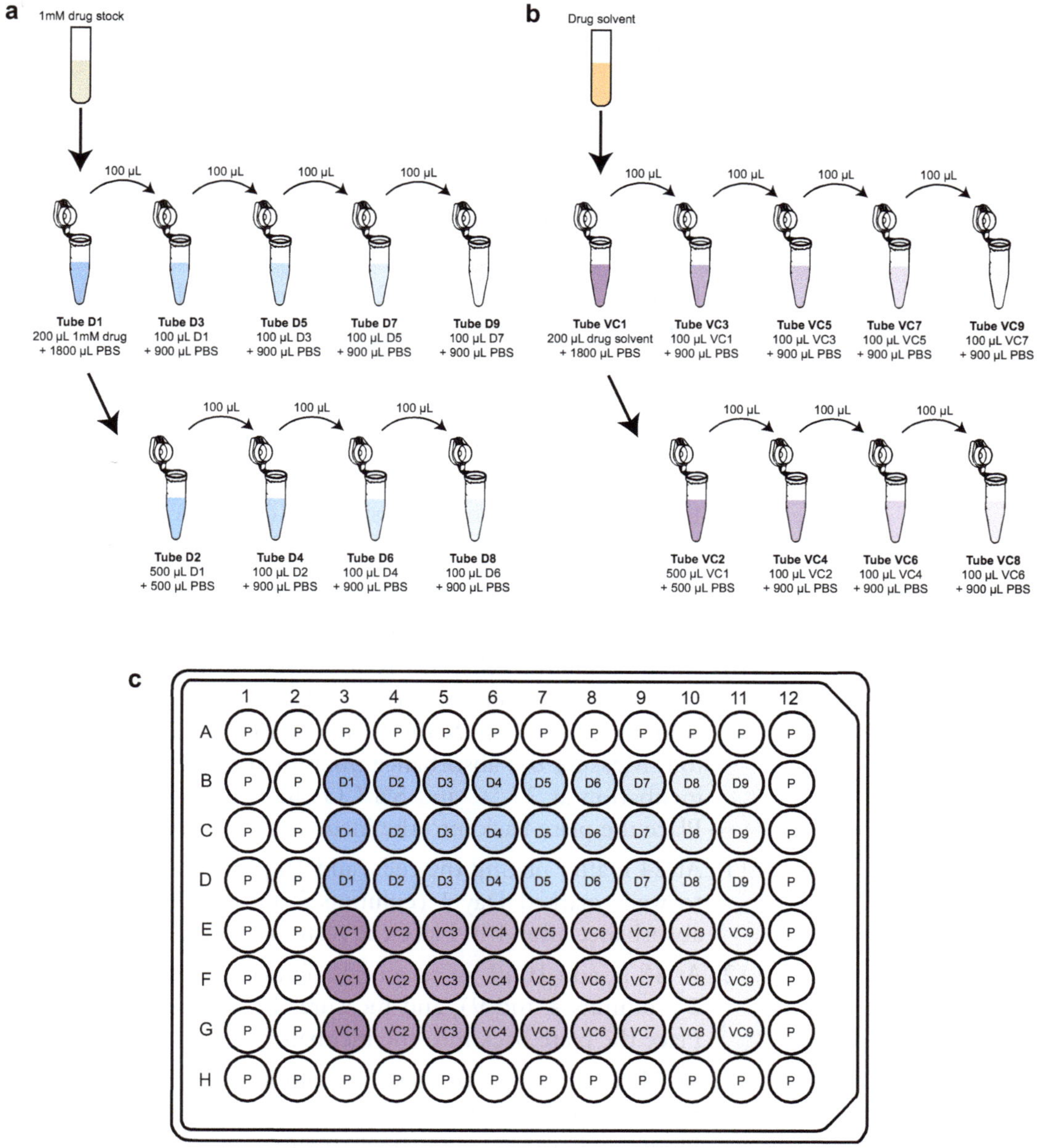


**Fig. 1** Serial dilution protocol for the pharmaceutical compounds. (**a**) Preparation of nine drug concentrations from a 1 mM stock solution. (**b**) Preparation of matched drug solvent (e.g., DMSO) concentration vehicle controls. (**c**) Plate layout for diluted pharmaceutical compounds and matched drug solvent concentration vehicle controls. P PBS, D1–D9 drug concentration, and VC1–VC9 vehicle control concentrations in descending order

3. Counting: Mix 10 µL of the cell suspension with 10 µL of trypan blue solution in a 1.5 mL Eppendorf tube and pipette 10 µL of the mixture into a Countess chamber slide (*see* **Note** 7). Record the concentration of cells per mL, and adjust the cell concentration to $1 \times 10^6$ cells per mL with culture medium.

### 3.3 Determine the Optimal Seeding Density

1. Prepare a serial dilution of the cell suspension in five 15 mL Falcon tubes. Add 2 mL of culture medium to tubes 2–5 (Fig. 2a).
2. In tube 1, add 4 mL of culture medium and $6.4 \times 10^5$ cells (final concentration: $1.6 \times 10^4/100$ μL medium). Mix thoroughly, but carefully.
3. Transfer 2 mL from tube 1 to tube 2 (final concentration: $8.0 \times 10^3/100$ μL medium) and mix thoroughly, but carefully.
4. Transfer 2 mL from tube 2 to tube 3 (final concentration: $4.0 \times 10^3/100$ μL medium) and mix thoroughly, but carefully.
5. Transfer 2 mL from tube 3 to tube 4 (final concentration: $2.0 \times 10^3/100$ μL medium) and mix thoroughly, but carefully.
6. Transfer 2 mL from tube 4 to tube 5 (final concentration: $1.0 \times 10^3/100$ μL medium) and mix thoroughly, but carefully.
7. Transfer 100 μL of the cell suspension (in duplicate) from tubes 1–5 to a clear 96-well microplate (Fig. 2b).
8. Add 100 μL of PBS in the outer perimeter wells (*see* **Note 8**).
9. Add 100 μL of medium (BLANK) in wells B2-C6.
10. After plating the cells, add 10 μL of resazurin solution (10% of cell culture volume) to wells marked as Day 0 (time $t = 0$, columns 2 and 7; *see* **Note 9**), and wrap the plate in aluminum foil. Measure the absorbance on a microplate reader after 4 h incubation at 37 °C and 5% $CO_2$ in a humidified incubator. Place the plate back into the incubator (*see* **Note 10**).
11. At 24 h after plating the cells, add 10 μL of resazurin solution to wells marked Day 1 (columns 3 and 8), and measure the absorbance after 4 h. Place the plate back into the incubator.
12. At 48 h, add 10 μL of resazurin solution to wells marked Day 2 (columns 4 and 9), and measure the absorbance after 4 h. Place the plate back into the incubator.
13. At 72 h, add 10 μL of resazurin solution to wells marked Day 3 (columns 5 and 10), and measure the absorbance after 4 h. Place the plate back into the incubator.
14. At 96 h, add 10 μL of resazurin solution to wells marked Day 4 (columns 6 and 11), and measure the absorbance after 4 h. The 96-well plate can now be discarded.
15. Export the raw absorbance data from the microplate reader to Excel. For each day (Day 0, 1, 2, 3, or 4), calculate the absorbance for each seeding density (seeding densities $1.0 \times 10^3$ to $1.6 \times 10^4$) by subtracting the average absorbance for the BLANK from the corresponding wells containing cells (seeding densities $1.0 \times 10^3$ to $1.6 \times 10^4$).

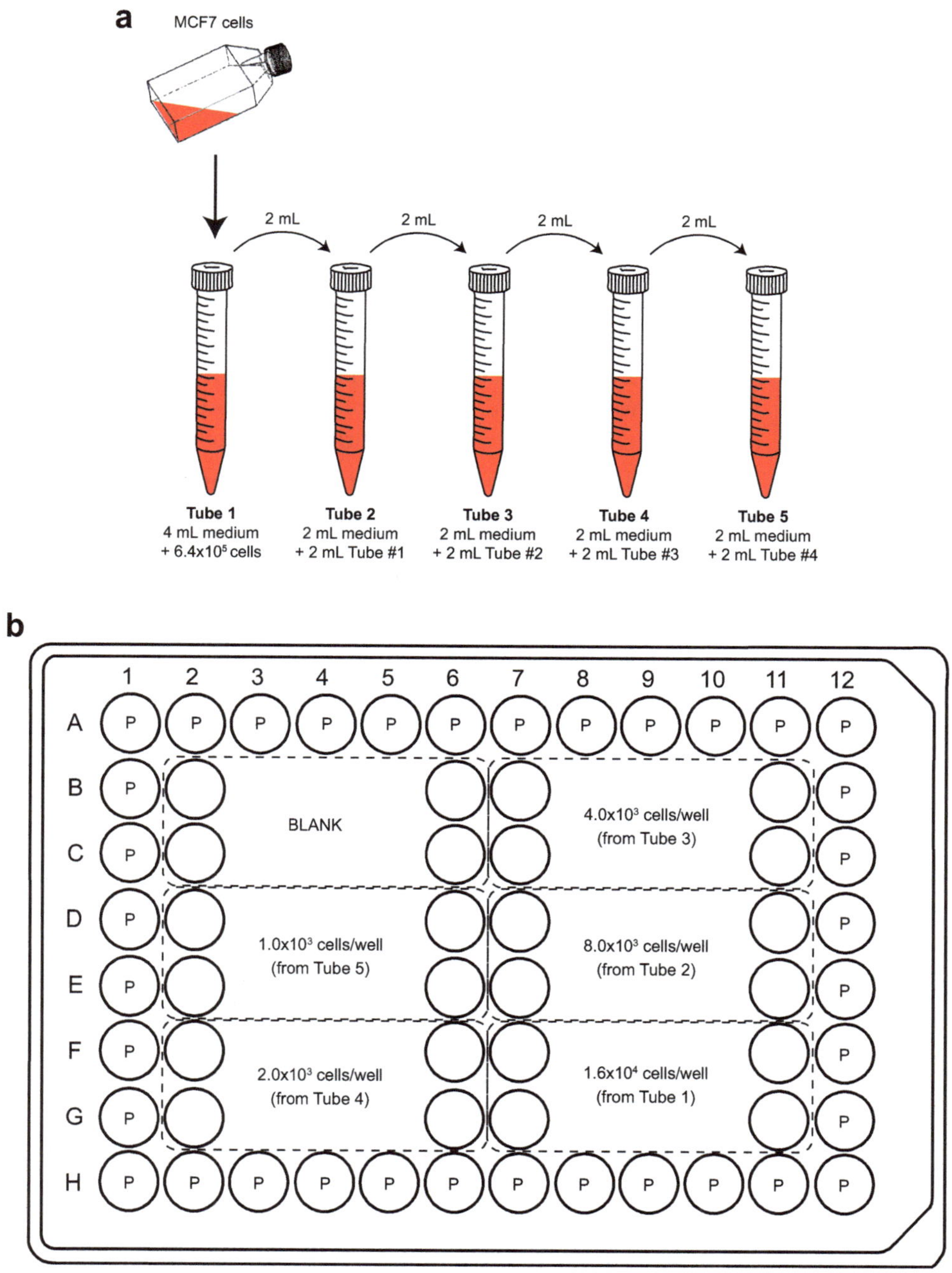


**Fig. 2** Determining the optimal seeding density. (**a**) Preparation of a serial dilution of MCF7 cell suspension. (**b**) Plate layout for determining the optimal seeding density. P PBS, BLANK (medium)

16. Construct a line graph with "absorbance" on the *x*-axis and "days after plating" on the *y*-axis. The optimal seeding density is the highest seeding density with linear growth between Days 1 and 3 (Fig. 2b; *see* **Note 11**). For MCF7 cells, the optimal seeding density was determined to be around $8.0 \times 10^3$ cells.

### 3.4 Estimate the Cell Doubling Time ($T_d$)

1. Transfer $8.0 \times 10^3$ cells (the optimal seeding density for MCF7 cells) in 100 μL of culture medium to wells B2-D11 in a clear 96-well microplate (Fig. 3a).
2. Add 100 μL of PBS in the outer perimeter wells.
3. Add 100 μL of medium (BLANK) to wells E2-G11.
4. After plating the cells, add 10 μL of resazurin solution (10% of cell culture volume) to wells marked as Day 0 (time $t = 0$, columns 2 and 3). Measure the absorbance on a microplate reader after 4 h incubation at 37 °C and 5% $CO_2$ in a humidified incubator. Place the plate back into the incubator.
5. At 24 h after plating the cells, add 10 μL of resazurin solution to wells marked Day 1 (columns 4 and 5), and measure the absorbance after 4 h. Place the plate back into the incubator.
6. At 48 h, add 10 μL of resazurin solution to wells marked Day 2 (columns 6 and 7), and measure the absorbance after 4 h. Place the plate back into the incubator.
7. At 72 h, add 10 μL of resazurin solution to wells marked Day 3 (columns 8 and 9), and measure the absorbance after 4 h. Place the plate back into the incubator.
8. At 96 h, add 10 μL of resazurin solution to wells marked Day 4 (columns 10 and 11) and measure the absorbance after 4 h. The 96-well plate can now be discarded.
9. Export the raw absorbance data from the microplate reader to Excel. For each day (Day 0, 1, 2, 3, or 4), calculate the absorbance (for the optimal seeding density) by subtracting the average absorbance for the BLANK from the corresponding wells containing cells.
10. Construct a line graph with "absorbance" on the *x*-axis and "days after plating" on the *y*-axis for the optimal seeding density ($8.0 \times 10^3$ cells/well for MCF7 cells), excluding the acclimatization phase (Day 0) and subsequent stationary phase (Fig. 3b, *see* **Note 12**).
11. Insert an exponential trend line into the graph, and display the equation and R-squared value (*see* **Note 13**).
12. To calculate the cell doubling time, use the trend line equation ($y = 86971e^{0.3808x}$).
    (a) Cell doubling time $(T_\text{d}) = \frac{\ln 2}{\text{Exponential slope}} \times 24$
    (b) For MCF7 cells: $\frac{0.693}{0.3808} \times 24 = 1.819 \text{ days} \times 24 \text{ h} = 43.7 \text{ h}$

### 3.5 Drug Sensitivity Screening

To determine the replicability of your assay, at least three independent experiments should be performed per drug screen. Cell viability should then be calculated using the arithmetic average background absorbance values for the three experiments. For comparison purposes, cell viability should however be calculated for each experiment separately.

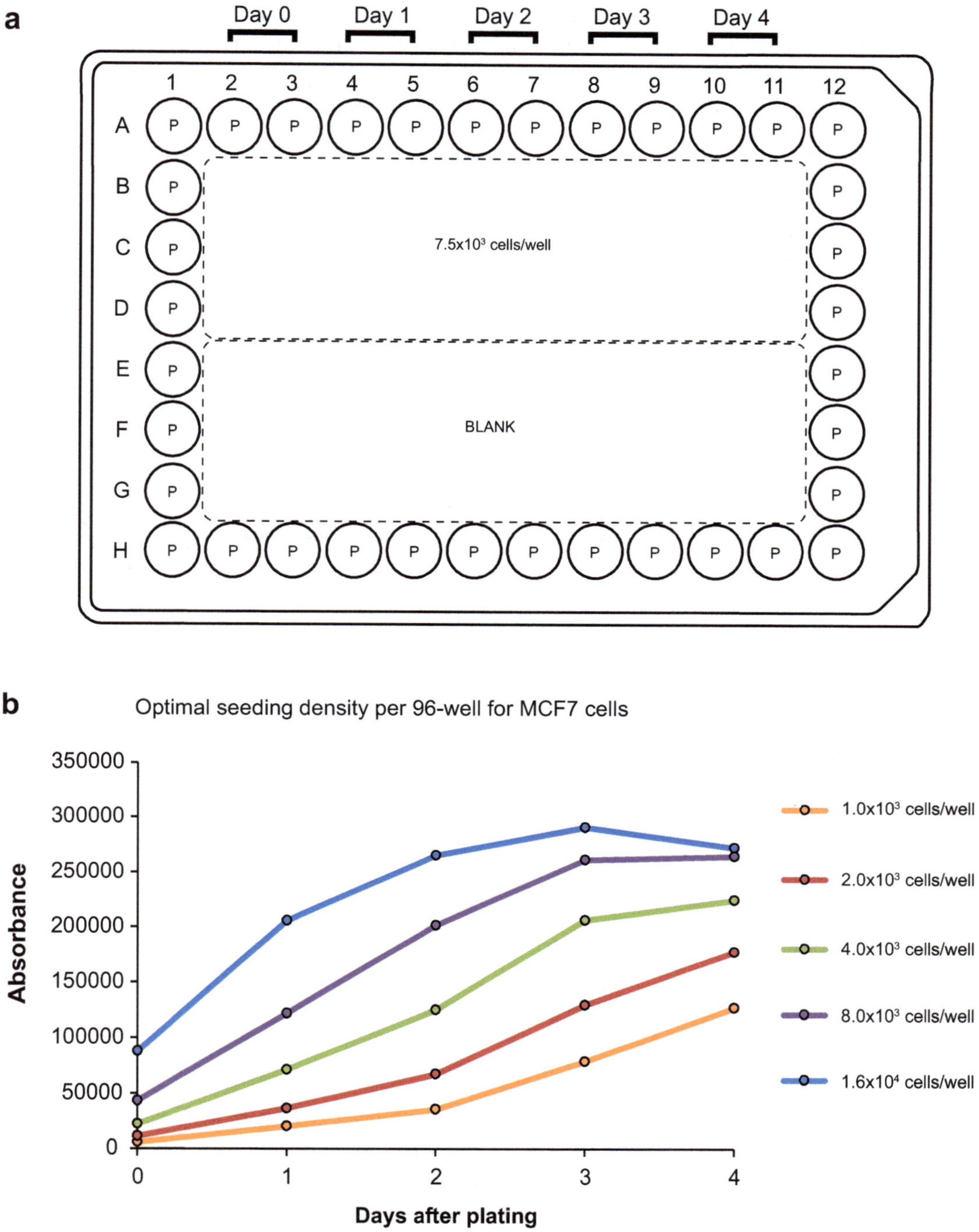


**Fig. 3** Estimating the cell doubling time. (**a**) Plate layout to estimate the cell doubling time using the optimal seeding density for MCF7 cells ($8.0 \times 10^3$ cells/100 μL medium). P PBS, BLANK (medium). (**b**) Line plot depicting the growth curves for MCF7 cells seeded at different densities. The optimal seeding density was determined to be around $8.0 \times 10^3$ cells/well since the line plot was linear between Days 1 and 3

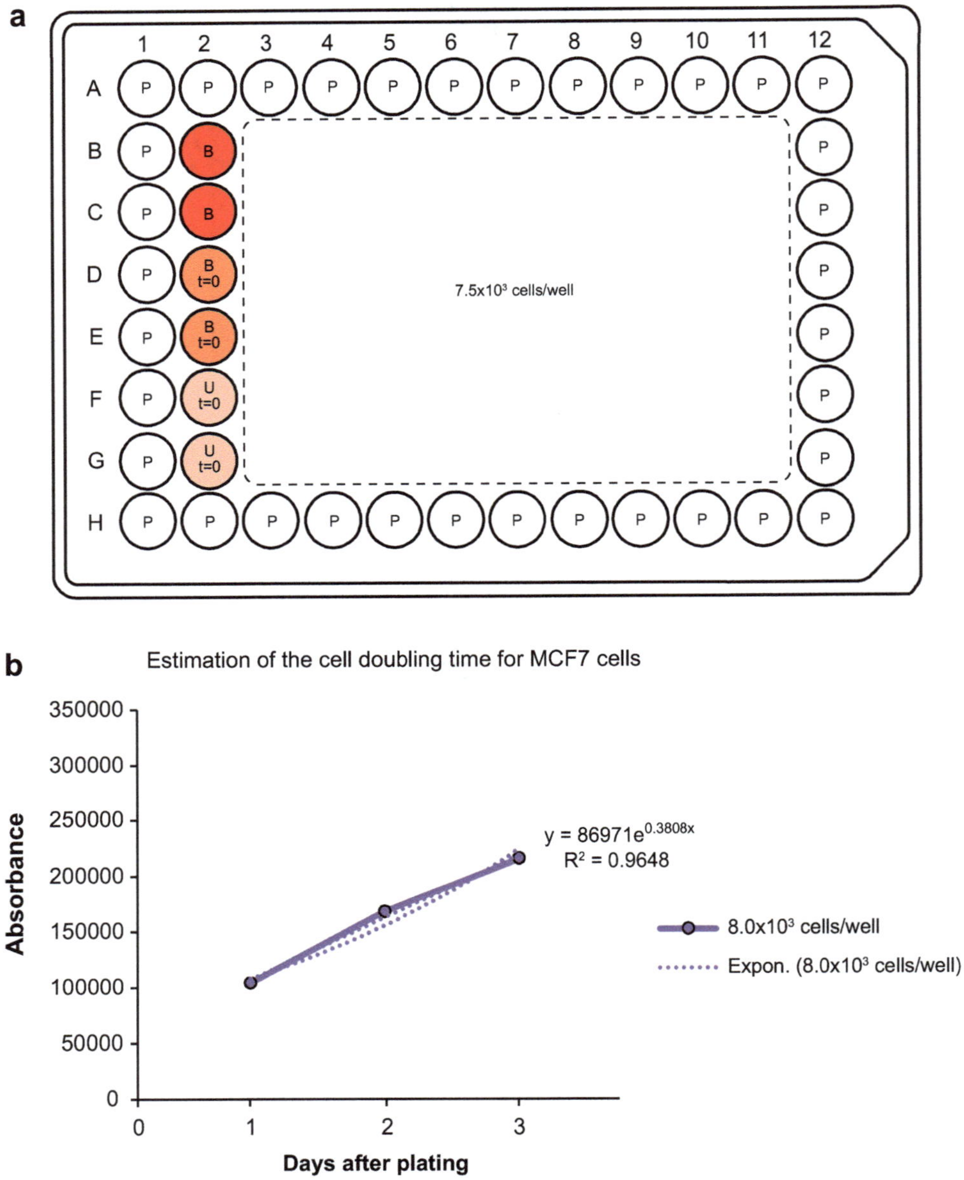


**Fig. 4** Drug sensitivity screen. (**a**) Plate layout to test the sensitivity of MCF7 cells (optimal seeding density: $8.0 \times 10^3$ cells/100 μL medium) to one or more drugs. P PBS, B BLANK (medium), B $t = 0$ BLANK (medium) at time 0, U $t = 0$ untreated cells at time 0. (**b**) The cell doubling time for MCF7 cells was calculated using the equation of the exponential trend line for the determined optimal seeding density (around $8.0 \times 10^3$ cells/well)

1. Transfer $8.0 \times 10^3$ cells (the optimal seeding density for MCF7 cells) in 100 μL of medium to wells F2-G2 and B3-G11 in a clear 96-well microplate (Fig. 4a, *see* **Note 14**).
2. Add 100 μL of PBS in the outer perimeter wells.

3. Add 100 μL of medium (BLANK) to wells B2-E2.
4. After plating the cells, add 10 μL of resazurin solution (10% of cell culture volume) to wells marked as $t = 0$ (wells D2-G2). Measure the absorbance on a microplate reader after 4 h incubation at 37 °C and 5% $CO_2$ in a humidified incubator. Place the plate back into the incubator for an additional 20 h.
5. At 24 h after plating the cells, defrost the diluted drugs prepared in **step 3.1** and briefly centrifuge the 96-well PCR plate at 1300 × *g* before use.
6. Transfer 10 μL of diluted drugs to the 96-well plate-containing cells (from wells B2-G11 in the plate (Fig. 1c) containing diluted drugs to wells B2-G11 in the plate-containing cells (Fig. 4a). Mix gently and place the plate back into the incubator.
7. At 24 h after drug treatment, add 10 μL of resazurin solution to all wells on the plate and measure the absorbance after 4 h. Discard the 96-well plate.

### 3.6 Determine Signal Dynamic Range Using Quality Control Metrics

Using the absorbance values from the drug sensitivity screen, calculate the *Z*-factor (*Z*), signal window (SW), and coefficient of variation (CV) in Excel. Cutoffs should be set to $Z > 0.4$ (*Z*-factor close to 1), SW > 2 (large signal dynamic range above 2 standard deviations), and CV < 20% (low CV with little or no signal variation) [9]. If the Z, SW, and/or CV values are not within the recommended cutoff, this could indicate that one or more biological (medium composition, medium volume, or seeding density) and/or technical (edge effect, drug storage, viability assay duration time, or dose-response metric) factors still need to be optimized. In this case, calculate the one-way ANOVA to determine which experimental parameters have a negative impact on cell viability, re-run your drug sensitivity screen using the adjusted protocol, and calculate the *Z*, SW, and CV values again.

1. Calculate the *Z*-factor using the following equation: $Z = 1 - \frac{3 \times \left(\mathrm{SD_{Sample}} + \mathrm{SD_{Control}}\right)}{\mathrm{Mean_{Sample}} - \mathrm{Mean_{Control}}}$.
2. Calculate the signal window using the following equation: $\mathrm{SW} = \frac{\mathrm{Mean_{Sample}} - \mathrm{Mean_{Control}} - 3 \times \left(\mathrm{SD_{Sample}} + \mathrm{SD_{Control}}\right)}{\mathrm{SD_{Sample}}}$
3. Calculate the coefficient of variation using the following equation: $\mathrm{CV} = \frac{\mathrm{SD}}{\mathrm{Mean}}$ x 100

### 3.7 Calculate Cell Viability (Microsoft Excel)

1. Export the raw absorbance data from the microplate reader to Excel.
2. Calculate the arithmetic average background absorbance value of the triplicate wells for the BLANK (Fig. 4a, wells B2-C2).

3. Subtract the average background value for the BLANK from all non-perimeter wells, i.e., BLANK $t = 0$ (wells D2-E2), untreated cells $t = 0$ (wells F2-G2), and treated cells (wells B3-G11).
4. Calculate the arithmetic average of the triplicate wells for the matched drug solvent vehicle controls (wells E3-G11).
5. For treated cells (wells B3-D11), calculate the % viability for each replicate using the following formula: %viability = $\frac{\text{Treated cells} - \text{BLANK}}{\text{Mean}_{(\text{DMSO control} - \text{BLANK})}} \times 100$.
6. Calculate the arithmetic average and standard deviation of the triplicate % viability values for each drug dose.

### 3.8 IC50 and GR50 Value Calculation (R/Bioconductor)

Half-maximum inhibitory concentration (IC50) values can be calculated using, e.g., GraphPad Prism software, but we also recommend that growth rate inhibition metrics (GR50 and GRmax) should be calculated, which will normalize the viability data according to cellular division rate. The GRmetrics package in R or the web-based GR Calculator tool can be used to calculate multiple growth rate inhibition metrics, including IC50, GR50, and GRmax.

1. In Excel, prepare an input file (.txt format) for GRmetics/GR Calculator containing columns for *cell_line*, *agent*, *perturbation*, *replicate*, *time*, *concentration*, *cell_count*, *cell_count__ctrl*, and *cell_count__time0*. A description of the column variables in the input file is shown in Table 1. Transfer the absorbance values from the Excel spreadsheet described in Subheading 3.7.
2. Open RStudio and load the GRmetrics package:

```
library(GRmetrics)
```

First-time users will need to install GRmetrics package:

```
BiocManager::install("GRmetrics")
```

3. Load your input file into R and save it into a variable *Cells* using the following command:

```
Cells <- read.table("File_name.txt", header=T, sep="\t")
```

4. Calculate the growth rate inhibition metrics using appropriate “groupingVariables” for your particular experiment:

```
drc_output = GRfit(Cells, groupingVariables = c('cell_
line','agent','time','perturbation'))
```

**Table 1**
**Description of the column variables in the input file for GRmetrics**

| Variable name | Description |
|---|---|
| cell_line | Name of the cell line |
| agent | Drug name |
| perturbation | Notes regarding the experiment, e.g., medium composition, seeding density, experiment number |
| replicate | Technical replicates 1–3 for each drug treatment |
| time | Drug treatment time |
| concentration | Drug dose in μM |
| cell_count | Absorbance values (minus BLANK) for treated cells (wells B3-D11) from **step 3** in Subheading 3.7 |
| cell_count__ctrl | Absorbance values (minus BLANK) for Vehicle controls (wells E3-G11) from **step 3** in Subheading 3.7 |
| cell_count__time0 | Absorbance values (minus average BLANK) for untreated cells $t = 0$ (wells F2-G2) from **step 3** in Subheading 3.7 |

5. Use the *head* function to review the output table for GRmetrics:

```
head(GRgetMetrics(drc_output))
```

6. Use the *head* function to review the output table for individual GR values:

```
head(GRgetValues(drc_output))
```

7. Export the output tables for GRmetrics and the individual GR values using the following commands:

```
write.table(GRgetMetrics(drc_output), file = "Cells_GRmetrics_output.txt", quote = FALSE, sep = "\t", row.names = FALSE)

write.table(GRgetValues(drc_output), file = "Cells_GRvalues_output.txt", quote = FALSE, sep = "\t", row.names = FALSE)
```

8. Plot dose-response curves using the *GRdrawDRC(drc_output)* function or ggplot2 package in *R*.

## 4 Notes

1. Chemical solutions should be aliquoted in 2 mL Eppendorf tubes and stored at −80 °C. Avoid more than 5 freeze-thaw cycles. Solubility of a drug at 25 °C is dependent on the solvent. Use the solubility information provided by the drug manufacturer to determine a suitable drug solvent. For example, cisplatin is insoluble in ethanol at 25 °C but is soluble in DMF and warm water (no higher than 50 °C). More importantly, DMSO has been shown to be inappropriate for platinum complexes like cisplatin since it inhibits the cytotoxicity of the drug by changing the structure of the complex. The concentration of the solvent should also be kept low (e.g., <0.3%) to avoid having a negative impact on cell viability.
2. Antibiotics are commonly used in culture medium to avoid bacterial contamination in the cell culture, but this could affect cellular gene expression patterns. With appropriate sterile lab work, no additives like antibiotics are necessary [10].
3. To fit dose-response curves with sufficient quality, at least five drug concentrations and a large concentration range should be tested [11, 12]. The drug concentration range is especially critical if different cell types with varying sensitivity to a particular drug are tested. Drug concentrations are frequently diluted according to a logarithmic scale of the base *e*, 2 or 10. In this book chapter, we demonstrate drug sensitivity screening using a nine-concentration 1:2 and 1:5 dilution series (e.g., 1, 5, 10, 50, 100, 500, 1000, 5000, and 10,000 nM).
4. Using a single vehicle control (e.g., DMSO) in drug screening experiments can not only lead to an overestimation and underestimation of cell viability, but can also result in dose-response curves starting at viability levels higher than 100%. We, therefore, recommend using matched drug solvent concentration vehicle controls.
5. Storage of diluted drugs in PCR plates sealed with aluminum sealing tape (rather than 96-well flat-bottom microplates with a loose lid) results in less evaporation during storage. Storage in the freezer prolongs the drug shelf life.
6. Cells will often stop growing at 100% confluence due to contact inhibition. Therefore, expansion of cells and cell viability assays should be performed using cells maintained in the log phase. As drug screening will frequently be done in 96- or 384-well microplates from one up to several days, it is important that the initial seeding density is optimized in order to prevent cellular confluency during the course of the experiment that in turn can have an effect on drug sensitivity.

7. Cell viability and total cell counts can also be determined using a manual hemocytometer.
8. Evaporation of the culture medium during the course of a drug screening experiment can lead to inconsistent resazurin absorbance values. Therefore, all wells on the microplate should either be filled with either medium or PBS. To minimize the edge effect, cells cultured in medium should not be plated in the perimeter wells where evaporation is more likely to occur. Instead, it is recommended that the perimeter wells be filled with PBS or medium only.
9. To compute GR metrics (i.e., GR50 and GRmax), the cell number should also be measured prior to drug exposure, i.e., $t = 0$ measurement [13, 14].
10. In our experimental setup, 4 h resazurin incubation time was optimal for all cell lines tested, but this should always be optimized for each cell line. Be aware of the risk of contamination when the plates are moved between sterile and non-sterile working environments.
11. In the graph, choose the line for the seeding density with linear growth between Days 1 and 3. Linear growth between these time points indicates that the cells have not reached 100% confluence (stationary phase). If the optimal seeding density is determined to be somewhere between $8.0 \times 10^3$ and $1.6 \times 10^4$ cells per 96-well, a new serial dilution of the cell suspension can be created containing $8.0 \times 10^3$, $1.0 \times 10^4$, $1.2 \times 10^4$, $1.4 \times 10^4$, and $1.6 \times 10^4$ cells/100 μL medium.
12. Acclimatization (the cells adapt to their new environment) and the stationary phase (the cells reach a plateau in their growth curve when no further growth occurs) can both affect cellular doubling time. Therefore, cells should always be in the exponential phase during the course of an experiment.
13. The R-squared value is used to determine the fit of the trend line to the exponential curve. The R-squared value should be 1 or close to 1.
14. Herein, we use the optimal seeding density, medium composition, medium volume, and drug solvent for MCF7 cells. These experimental parameters should be optimized for each cell line and viability assay.

## References


1. Riss TL, Moravec RA, Niles AL, Duellman S, Benink HA, Worzella TJ, Minor L (2004) Cell viability assays. In: Sittampalam GS, Grossman A, Brimacombe K et al (eds) Assay guidance manual. Eli Lilly & Company and the National Center for Advancing Translational Sciences, Bethesda, pp 353–377
2. Stoddart MJ (2011) Cell viability assays: introduction. Methods Mol Biol 740:1–6. https://doi.org/10.1007/978-1-61779-108-6_1

3. Adan A, Kiraz Y, Baran Y (2016) Cell proliferation and cytotoxicity assays. Curr Pharm Biotechnol 17(14):1213–1221. https://doi.org/10.2174/1389201017666160808160513
4. Kamiloglu S, Sari G, Ozdal T, Capanoglu E (2020) Guidelines for cell viability assays. Food Front 1(3):332–349. https://doi.org/10.1002/fft2.44
5. Aslantürk ÖS (2018) In vitro cytotoxicity and cell viability assays: principles, advantages, and disadvantages. In: Genotoxicity – a predictable risk to our actual world. IntechOpen, London, pp 13–30. https://doi.org/10.5772/intechopen.71923
6. Präbst K, Engelhardt H, Ringgeler S, Hübner H (2017) Basic colorimetric proliferation assays: MTT, WST, and Resazurin. Methods Mol Biol 1601:1–17. https://doi.org/10.1007/978-1-4939-6960-9_1
7. Shokrzadeh M, Modanloo M (2017) An overview of the most common methods for assessing cell viability. J Res Med Dent Sci 5(2):33. https://doi.org/10.5455/jrmds.2017526
8. Larsson P, Engqvist H, Biermann J, Werner Rönnerman E, Forssell-Aronsson E, Kovács A, Karlsson P, Helou K, Parris TZ (2020) Optimization of cell viability assays to improve replicability and reproducibility of cancer drug sensitivity screens. Sci Rep 10(1):5798. https://doi.org/10.1038/s41598-020-62848-5
9. Ivanov DP, Parker TL, Walker DA, Alexander C, Ashford MB, Gellert PR, Garnett MC (2014) Multiplexing spheroid volume, resazurin and acid phosphatase viability assays for high-throughput screening of tumour spheroids and stem cell neurospheres. PLoS One 9(8):e103817. https://doi.org/10.1371/journal.pone.0103817
10. Ryu AH, Eckalbar WL, Kreimer A, Yosef N, Ahituv N (2017) Use antibiotics in cell culture with caution: genome-wide identification of antibiotic-induced changes in gene expression and regulation. Sci Rep 7(1):7533. https://doi.org/10.1038/s41598-017-07757-w
11. Quast J-P (2022) Dose-response data analysis workflow. https://cran.r-project.org/web/packages/protti/vignettes/data_analysis_dose_response_workflow.html
12. Hafner M, Niepel M, Subramanian K, Sorger PK (2017) Designing drug-response experiments and quantifying their results. Curr Protoc Chem Biol 9(2):96–116. https://doi.org/10.1002/cpch.19
13. Hafner M, Niepel M, Chung M, Sorger PK (2016) Growth rate inhibition metrics correct for confounders in measuring sensitivity to cancer drugs. Nat Methods 13(6):521–527. https://doi.org/10.1038/nmeth.3853
14. Niepel M, Hafner M, Chung M, Sorger PK (2017) Measuring cancer drug sensitivity and resistance in cultured cells. Curr Protoc Chem Biol 9(2):55–74. https://doi.org/10.1002/cpch.21

# Chapter 19

# Cellasys #8: A Microphysiometric Assay for Real-Time Cell Analysis Within 24 Hours

**Sebastian Eggert, Svitlana Stetsenko, and Joachim Wiest**

## Abstract

Traditional biochemical assays present a vital toolbox to identify the effects of a test substance on cells. However, current assays are single-point measurements, only reveal one parameter at a time, and introduce potential interferences with labels and fluorescent lights. We have addressed these limitations by introducing the cellasys #8 test which is a microphysiometric assay for real-time cell analysis. Within 24 h, the cellasys #8 test is able to identify not only the effect of a test substance but also measure recovery effects. Due to the multi-parametric read-out, the test provides insights into metabolic as well as morphological changes in real-time. The following protocol provides a detailed introduction to the materials as well as a step-by-step description to support scientists with protocol adoption. The automated and standardized assay opens up manifold new application areas for scientists to study biological mechanisms, develop new therapeutic approaches, and validate serum-free media formulation.

**Key words** Microphysiometry, Metabolism, Morphology, Recovery effect, Sensors, Cell-based assays, Reproducibility, Automation, Real-time

## 1 Introduction

An in vitro assay is defined as an analytical measurement procedure with the ability to quantify a biological process by measuring a detectable signal [1]. It is an indispensable tool in various areas such as life science, drug discovery and development and biomedical research [2–6]. While biochemical assays are widely accepted and applied to identify the effects of a test substance on cells, they exhibit three pronounced limitations. Firstly, since biochemical assays are based on single-point measurement, they fail to provide real-time data, also referred to as time-resolved data [7]. Real-time data is, in particular, very important to identify recovery effects after removing the test substance [8]. However, single-point measurements are either not possible to detect recovery effects or a high

Oliver Friedrich and Daniel F. Gilbert (eds.), *Cell Viability Assays: Methods and Protocols*, Methods in Molecular Biology, vol. 2644, https://doi.org/10.1007/978-1-0716-3052-5_19,

number of assays, and hence, samples are required to increase the measurements. Secondly, most biochemical assays are only able to measure one parameter with the performed assay [1]. If additional parameters are required for a detailed understanding, further assays have to be performed, and hence, additional samples have to be prepared. This is not only very cumbersome but also increases the time and money spent on the experiment. Finally, biochemical assays also introduce the risk of interferences with fluorescent light, chemical reactivity or labels in general [1]. Failure to identify potential interferences can result in increased time and resources analyzing substances that are rather false positives and, in the case of a drug or chemical, may damage human health [9].

To address these limitations and provide scientists with reduced assay time, we have established a fully automated and standardized microphysiometric assay, referred to as cellasys #8, on the Intelligent Mobile Lab for In Vitro Diagnostics (IMOLA-IVD, cellasys GmbH, Germany). The cellasys #8 test builds upon the principle of microphysiometry which is defined as the in vitro measurement of the functions and activities of life or of living matter (as organs, tissues, or cells) and of the physical and chemical phenomena involved on a very small (micrometer) scale [10]. By converging microphysiometry with the fundamentals of a cell-based assay [11], the cellasys #8 test allows to overcome the aforementioned limitations. By culturing cells on biocompatible microsensors, a label-free multi-parametric read-out can be generated automatically without the need for human interference [12]. In combination with a unique protocol design, not only the impact of the test substance is investigated, but also potential recovery effects are monitored to give further insights. This is, in particular, important for investigations of the reversibility of toxic effects [13, 14]. The presented cellasys #8 test is an automated assay that monitors the metabolism and the morphology of a cellular model [15]. The automated assay runs for 24 h and includes four replicates and positive, negative and blank control (Fig. 1). The following paragraphs provide an overview of the required materials, including reagents and consumables, step-by-step protocol descriptions, and further important notes to ensure easy protocol adoption in the future.

## 2 Materials

Prepare all solutions using double-distilled water ($ddH_2O$) at a sensitivity of >10 MΩ/cm at room temperature and analytical grade reagents. If not otherwise noted, prepare and store all reagents at room temperature. Diligently follow all waste disposal regulations when disposing waste materials.

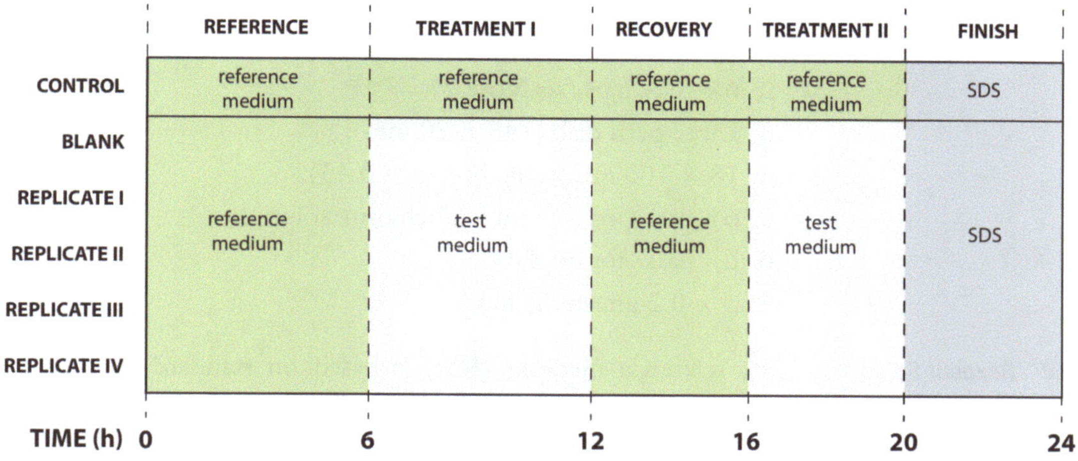


**Fig. 1** Setup of a cellasys #8 assay: The automated and standardized assay includes a control that is fed with the reference medium (negative control) for 20 h (green) and then with the reference medium supplemented with sodium dodecyl sulfate (SDS) as a positive control (dark gray). The blank is an empty control (without cells) to determine possible interactions between the sensors and the media. The assay enables the analysis of four replicates which are each treated with the treatment medium (light gray) for a further 6 h after 6 h of reference medium. After another 4 h with the reference medium, another treatment for 4 h with the treatment medium takes place. Finally, media supplemented with SDS is applied as a positive control

### 2.1 Reagents

1. Double-distilled water ($ddH_2O$).
2. Buffer-free cell culture medium (e.g., DMEM (Dulbecco's Modified Eagle's Medium) with 4500 mg/L glucose and L-glutamine, without sodium bicarbonate).
3. Buffered cell culture medium (e.g., DMEM with 4500 mg/L glucose and L-glutamine, with sodium bicarbonate).
4. 2% (w/v) sodium hypochlorite solution (diluted in $ddH_2O$).
5. 70% denatured ethanol solution.
6. 10% (w/v) sodium dodecyl sulfate (SDS) (diluted in $ddH_2O$).
7. 3 M KCl solution.
8. Test substance (powder or stock dilution).

### 2.2 Equipment and Tools

1. Analytical balance.
2. pH meter.
3. 6× IMOLA-IVD [12].
4. 6× fluidic heads.
5. 12× O-rings.
6. 18× distribution caps with Luer-steel needles.
7. 6× reference electrodes with barrels.
8. Osmometer.

### 2.3 Consumables

1. Cells under investigation (approx. 5× 50,000 cells) (*see* **Note 1**).
2. 6× BioChips (*see* **Note 2**).
3. Large petri dish (145 × 20 mm).
4. 18 × 100 mL sterile bottles (GL45).
5. 6× flasks for sodium hypochlorite solution.
6. 6× flasks for dd$H_2O$.
7. 2 × 0.2 μm sterile filters.

### 2.4 Reagent Setup

#### 2.4.1 Reference Medium

The reference medium (RM) is based on standard cell culture media compositions (e.g., DMEM) with the following variations. As any buffer, such as $NaHCO_3$ and HEPES, will impact the extracellular pH, the media is prepared without any buffers to allow the measurement of extracellular acidification. It is recommended to prepare the media according to the manufacturer's formulation (*see* **Note 3**) and, subsequently, adjust the pH to 7.4 ± 0.07 at 37 °C and the osmolarity to 340 mOsmol/kg ± 17. After preparation, sterile filter the reference medium and store at 4 °C for a maximum of 6 months. A buffer-based version of the reference medium is used to seed the cells in the BioChips and culture the cells in a $CO_2$ incubator. When the reference medium is supplemented with SDS to act as the positive control, the solution is referred to as "RM+" in the following. When the reference medium is not supplemented with SDS, it is termed "RM−."

#### 2.4.2 Treatment Medium

The treatment medium (TM) is the reference medium with the addition of the test substance (*see* **Note 4**). When the treatment medium is supplemented with SDS to act as the positive control, the solution is referred to as "TM+" in the following. When the treatment medium is not supplemented with SDS, it is termed "TM−."

## 3 Methods

The SOP cellasys #8 test is to be performed within 3 days in a cell culture laboratory, and the required tasks are divided in the following into day prior to assay, assay day, and day after assay.

### 3.1 Day Prior to Assay

#### 3.1.1 Print and Fill Quality Control Document

1. Print and fill "QC S001-002_cellasys_8" (document is available via homepage; *see* **Note 5**).

*3.1.2 Setup 6× IMOLA-IVD*

2. Turn on the IMOLA-IVD platform.
3. Close pump cartridges, which are part of the IMOLA-IVD platform and connected to the fluidic system.
4. Assemble and mount reference electrodes to each IMOLA system (6×). The correct order of the assembly in flow direction is (i) inflow tube, (ii) fluidic head, (iii) barrel, and (iv) outflow tube (*see* **Note 6**).

*3.1.3 Sterilize 6× IMOLA-IVD*

5. Fill six flasks with sodium hypochlorite solution, and put three Luer-steel connectors per IMOLA-IVD into it.
6. Open the software "DALiA client," and load the application "cellasys #8 prefi-steri" (*see* **Notes** 7 and **8**).
7. Press "START" to start the sterilization protocol which takes approx. 25 min.

*3.1.4 Prepare Medium Flasks with Reference Medium*

8. Prepare 7 × 40 mL with reference medium and label with **RM−**. For the positive control, prepare 1 × 25 mL with reference medium, add 500 μL of 10% (w/v) SDS solution, and label with **RM+** (*see* **Note 9**).

*3.1.5 Prepare Medium Flasks with Treatment Medium*

9. Prepare 5 × 40 mL with treatment medium and label with **TM−**. For the positive control, prepare 5 × 25 mL with treatment medium, add 500 μL of 10% (w/v) SDS solution, and label with **TM+** (*see* **Note 10**).

*3.1.6 Setup 6× IMOLA-IVD with Medium*

10. Press "STOP" to finish the sterilization procedure.
11. Set up the M2 position (labeled at the fluidic modules) according to the following configuration (*see* also Fig. 1 for a detailed description of control, blank, and replicates): Control (IMOLA 1) with **RM+** and blank (IMOLA 2) with **TM+**, and replicate I–IV (IMOLA 3, 4, 5, and 6) with **TM+**.
12. Set up the M3 position according to the following configuration: Control (IMOLA 1) with **RM−** and blank (IMOLA 2) with **TM−**, and replicate I–IV (IMOLA 3, 4, 5, and 6) with **TM−**.
13. Set up the M4 position according to the following configuration: Control (IMOLA 1) and blank (IMOLA 2), and replicate I–IV (IMOLA 3, 4, 5, and 6) with **RM−**.

*3.1.7 First Prefilling with Medium*

14. Press "START" via the DALiA client to start the prefilling protocol, which prefills the fluidic system with media.
15. Switch on the 6× IMOLA-IVD incubator to reach 37 °C for the experimentation (*see* **Note 11**).

*3.1.8 Seed Cells onto BioChips*

16. Harvest and re-suspend cells and subsequently seed 50,000 cells in 300 μL growth medium (total volume with cells) onto BioChips for the control and replicate I–IV.
17. Add 300 μL growth medium only (no cells) to the blank.
    Place 6 BioChips in a large petri dish and transfer to a humidified 37 °C $CO_2$ incubator and incubate overnight.

### 3.2 Assay Day

*3.2.1 Second Prefilling with Medium (see **Note 12**)*

1. Open DALiA client and press "STOP."
2. Press "START" to initialize the second prefilling.
3. Press "STOP" after approx. 25 min.

*3.2.2 Assemble BioChips with Fluidic Heads*

4. Sterilize fluidic heads in sodium hypochlorite solution.
5. After 20 min, work in a sterile environment to flush six fluidic heads equipped with six O-rings and six O-rings in $ddH_2O$.
6. Get six BioChips from the incubator and gently press O-rings into BioChip housing (one O-ring per BioChip). Start with the BioChip for the blank (without cells) to prevent any cell take-over from the other BioChips seeded with cells.
7. Gently press fluidic heads into BioChips.
8. Close Luer connectors of fluidic heads (*see* **Note 13**).

*3.2.3 Assemble BioChips into 6× IMOLA-IVD*

9. Connect BioChips with 6× IMOLA-IVD in the following configuration: Position BioChip without cells to blank (IMOLA 2) and BioChips with cells to control (IMOLA 1) and replicates (IMOLA 3, 4, 5, and 6).
10. Connect Luer connectors of the BioChips with the fluidic system.

*3.2.4 Start Assay*

11. In DALiA client, load application "cellasys #8."
12. Press "START" and ensure that all air bubbles are flushed out from the BioChip.

### 3.3 Day After Assay

*3.3.1 Stop the Experiment and Export the Data*

1. Open DALiA client and press "STOP."
2. From DALiA client, export data via "Database" and "Export" as exp.file format.
3. Load the data file into DALiA analytics to automatically calculate gradient data for detailed analysis.

*3.3.2 Dismount and Clean*

4. Dismount medium flasks with distribution caps and put Luer-steel tubes in flasks with $ddH_2O$ (*see* **Note 14**).
5. Flush distribution caps with $ddH_2O$, and prepare for sterilization.

6. Flush medium flasks with water, and prepare for the dishwasher.
7. Remove BioChips/fluidic head/reference electrode setup from 6× IMOLA-IVD.
8. Disassemble BioChip/fluidic head/reference electrode/barrel setup.
9. Flush BioChip, fluidic head, and barrel with $ddH_2O$.
10. Store barrels of reference electrodes in 3 M KCl.
11. In DALiA client, load the application "cellasys #8 prefi-steri," and press "START" with $ddH_2O$ filled into six flasks.
12. After approx. 25 min, press "STOP," and release cartridges of 6× IMOLA-IVDs pump.
13. Empty waste flasks.
14. Shut down 6× IMOLA-IVD platform.

## 4 Notes

1. The optimal cell seeding number and, hence, required cell number varies between cell types but are generally between 50,000 and 250,000 cells per BioChip for adherent cells. A confluency of 70–80% after 24 h culture generates signals in the desirable range.
2. As indicated in the quality control document (*see* **Note 5**), perform a chip test before and after the cellasys #8 test. The chip test ensures that the BioChip is working correctly.
3. Media can be also prepared without phenol red, as no color indicator is required for this assay.
4. It is recommended to prepare a stock solution of the test substance and then dilute the test substance into the treatment medium to the defined concentration.
5. A detailed standard operation procedure, quality control documents, and a training video are available at https://www.cellasys.com/downloads/cellasys-8/.
6. Make sure that the barrel is completely filled with 3 M KCl solution and that the reference electrode is tightly screwed into it. It is important that the barrel stands upright and no air bubbles interrupt the conductive connection of the reference electrode's wire to the fluid system.
7. DALiA client software analyses the connection status of the hardware interfaces and provides information about issues (e.g., power failure during the assay). The status is indicated with three colors (green, yellow, and red):

7.1. Green: The hardware is connected correctly to the software.

7.2. Yellow: The hardware is connected correctly to the software, but there had been a problem in the past.

7.3. Red: The hardware is not connected correctly to the software. Data exchange is not working.

8. The comment feature of the DALiA client software allows commenting on the experiment. The commenting function can be used, for example, to write statements on the occurrence of special events while an experiment is running. Comments are automatically assigned to a timestamp and are saved together with the measurement data in the measurement protocol.

9. The usage of an unbuffered assay medium is required for the cellasys #8 test to ensure meaningful and accurate functional measurements of metabolic activity.

10. Depending on the study objective, any test substance (e.g., drug molecule, chemical, etc.) can be added to the treatment medium. A separate blank control ensures that there is no interference between the microsensors and the test substance.

11. It is essential to switch on the incubator 24 h before the start of the measurement to assure homogeneous temperature distribution throughout all components.

12. A second prefilling is performed to wash out any air bubbles which may have been formed overnight.

13. Inspect the cells using a microscope before loading the BioChips into the system. The confluency should be between 70% and 80%, and the cells should be nicely distributed on the chip surface. Check the blank sample to ensure that no cells are attached to the chip.

14. When the medium flasks are dismounted after the experiment, it is recommended to check the remaining volume for consistency. Variable volumes may indicate issues with the fluidic system.

## Acknowledgments

The authors like to thank for the support of the Deutscher Tierschutzbund – Akademie für Tierschutz (German Animal Welfare Federation—Animal Welfare Academy) for the ongoing cooperation.

## References


1. Sittampalam G, Coussens N, Arkin M et al (2016) Assay guidance manual. Eli Lilly & Company and the National Center for Advancing Translational Sciences, Bethesda
2. Horvath P, Aulner N, Bickle M et al (2016) Screening out irrelevant cell-based models of disease. Nat Rev Drug Discov 15:751–769. https://doi.org/10.1038/nrd.2016.175
3. Broach JR, Thorner J (1996) High-throughput screening for drug discovery. Nature 384S:14–16. https://doi.org/10.1038/384014a0
4. Gjorevski N, Sachs N, Manfrin A et al (2016) Designer matrices for intestinal stem cell and organoid culture. Nature 539:560–564. https://doi.org/10.1038/nature20168
5. von Erlach T, Saxton S, Shi Y et al (2020) Robotically handled whole-tissue culture system for the screening of oral drug formulations. Nat Biomed Eng 4:544–559. https://doi.org/10.1038/s41551-020-0545-6
6. Khalili AA, Ahmad MR (2015) A review of cell adhesion studies for biomedical and biological applications. Int J Mol Sci 16:18149
7. Brandenberg N, Hoehnel S, Kuttler F et al (2020) High-throughput automated organoid culture via stem-cell aggregation in microcavity arrays. Nat Biomed Eng 4:863. https://doi.org/10.1038/s41551-020-0565-2
8. Wiest J (2022) Systems engineering of microphysiometry. Organs-on-a-Chip 4:100016. https://doi.org/10.1016/J.OOC.2022.100016
9. Mullard A (2014) New drugs cost US$2.6 billion to develop. Nat Rev Drug Discov 13: 877–877. https://doi.org/10.1038/nrd4507
10. Brischwein M, Wiest J (2018) Microphysiometry. In: Matysik F, Wegener J (eds) Bioanalytical reviews. Springer, Berlin/Heidelberg, pp 163–188
11. Alexander F, Eggert S, Wiest J (2018) Skin-on-a-chip: transepithelial electrical resistance and extracellular acidification measurements through an automated air-liquid interface. Genes (Basel) 9:114. https://doi.org/10.3390/genes9020114
12. Weiss D, Brischwein M, Grothe H, et al (2013) Label-free monitoring of whole cell vitality. In: Conference proceedings: IEEE engineering in medicine and biology society 2013, pp 1607–1610. https://doi.org/10.1109/EMBC.2013.6609823
13. Mahadevan MS, Yadava RS, Yu Q et al (2006) Reversible model of RNA toxicity and cardiac conduction defects in myotonic dystrophy. Nat Genet 38:1066. https://doi.org/10.1038/ng1857
14. Hall AM, Gollan B, Helaine S (2017) Toxin–antitoxin systems: reversible toxicity. Curr Opin Microbiol 36:102
15. Eggert S, Stetsenko S, Wiest J (2022) Mikrophysiometrischer Test zur Evaluierung serumfreier Zellkulturmedien. BIOspektrum 28(1): 60–61. https://doi.org/10.1007/S12268-022-1686-9

# Chapter 20

# Live Cell Adhesion, Migration, and Invasion Assays

Jordi Pijuan, Anna Macià, and Anaïs Panosa


## Abstract

Cell migration is a fundamental procedure involved in many physiological processes such as embryological development, tissue formation, immune defense or inflammation, and cancer progression. Here, we provide four in vitro assays that describe step-by-step cell adhesion, migration and invasion strategies, and their corresponding image data quantification. These methods include the following: two-dimensional wound healing assays, two-dimensional individual cell-tracking experiments by live cell imaging, and three-dimensional spreading and transwell assays. These optimized assays will facilitate physiological and cellular characterization of cell adhesion and motility, which may be used for fast screening of specific therapeutic drugs for adhesion function, novel strategies in pathophysiological diagnosis, and assaying new molecules involved in migration and invasion metastatic properties of cancer cells.


**Key words** Cell migration/invasion, Wound healing assay, Transwell assay, Spreading assay, Live cell imaging, Data analysis, ImageJ/Fiji

## 1 Introduction

Cell motility is a key process where individual cells and cell groups must be able to change and reach their proper position in a given environment to execute their function [1]. In multicellular organisms, this phenomenon plays an important role in several biologic processes such as gastrulation, embryonic morphogenesis, development of the nervous system, tissue homeostasis, and immune cell trafficking. However, moving cells can be deregulated and contribute to many pathological processes such as inflammation and cancer metastasis [2–4]. Cell adhesion, migration, and invasion are pivotal steps during cancer development and progression; hence, their study and understanding are crucial in order to fight against cancer pathological processes.

Cells can strongly adhere to surfaces coated with extracellular matrix (ECM) proteins via focal adhesion complexes and associated stress fibers [5]. Cell-substrate adhesion and cell-cell adhesion are

Oliver Friedrich and Daniel F. Gilbert (eds.), *Cell Viability Assays: Methods and Protocols*, Methods in Molecular Biology, vol. 2644, https://doi.org/10.1007/978-1-0716-3052-5_20,

mechanisms involved in cancer cell movement that should be analyzed in different treatment conditions.

Cancer cells use different movement strategies. They can migrate collectively as cohesive cell groups (wound healing assay) that retain cell-cell junctions and coordinate cytoskeletal activity between neighboring cells as well as with the surrounding tissue. Further, they can migrate individually mediated by cytoskeletal activity (single-cell tracking) and have the ability to migrate and invade (transwell) with directionality responding to various chemoattractants and treatments (Fig. 1).

In this chapter, we aim to explain four different in vitro assays for assessing live cell migratory behavior and emphasize the importance of accurate and quantitative data analysis [6].

## 2 Materials

When handling the chemical and biological materials used in these assays, always wear suitable personal protective equipment, including a lab coat, gloves, safety goggles, and, where indicated, a face shield. For any reagent listed in these methods, appropriate institutional and governmental safety guidelines must be followed. Please refer to the appropriate materials safety data sheets.

All glass material, buffers, solutions, and pipettes have to be sterile, and all procedures should be carried out under sterile conditions of a laminar flow hood to maintain cell lines alive.

### *2.1 General Equipment and Consumables*

1. Cell culture incubator ($CO_2$ at 5%, humidified at 37 °C).
2. Laminar flow hood.
3. Water bath set to 37 °C.
4. Adjustable pipettes: P-2, P-20, P-200, and P-1000.
5. Pipette tips: 10 μL, 20 μL, 200 μL, and 1000 μL.
6. Pipette controller.
7. Tissue culture pipettes: 5 mL and 10 mL.
8. Sterile conical (micro) centrifuge tubes: 1.5 mL, 15 mL, and 50 mL.
9. Hemocytometer.
10. Cotton swabs.
11. Forceps.
12. Sterile 100 mm culture dishes (or other sizes).
13. Sterile 24-well plates for cell culture.
14. Transwell inserts for 24-well plates (membrane 8.0 μm pores).
15. Confocal microscope or fluorescence microscope with automated stage and cage incubator and $CO_2$/temperature controllers.

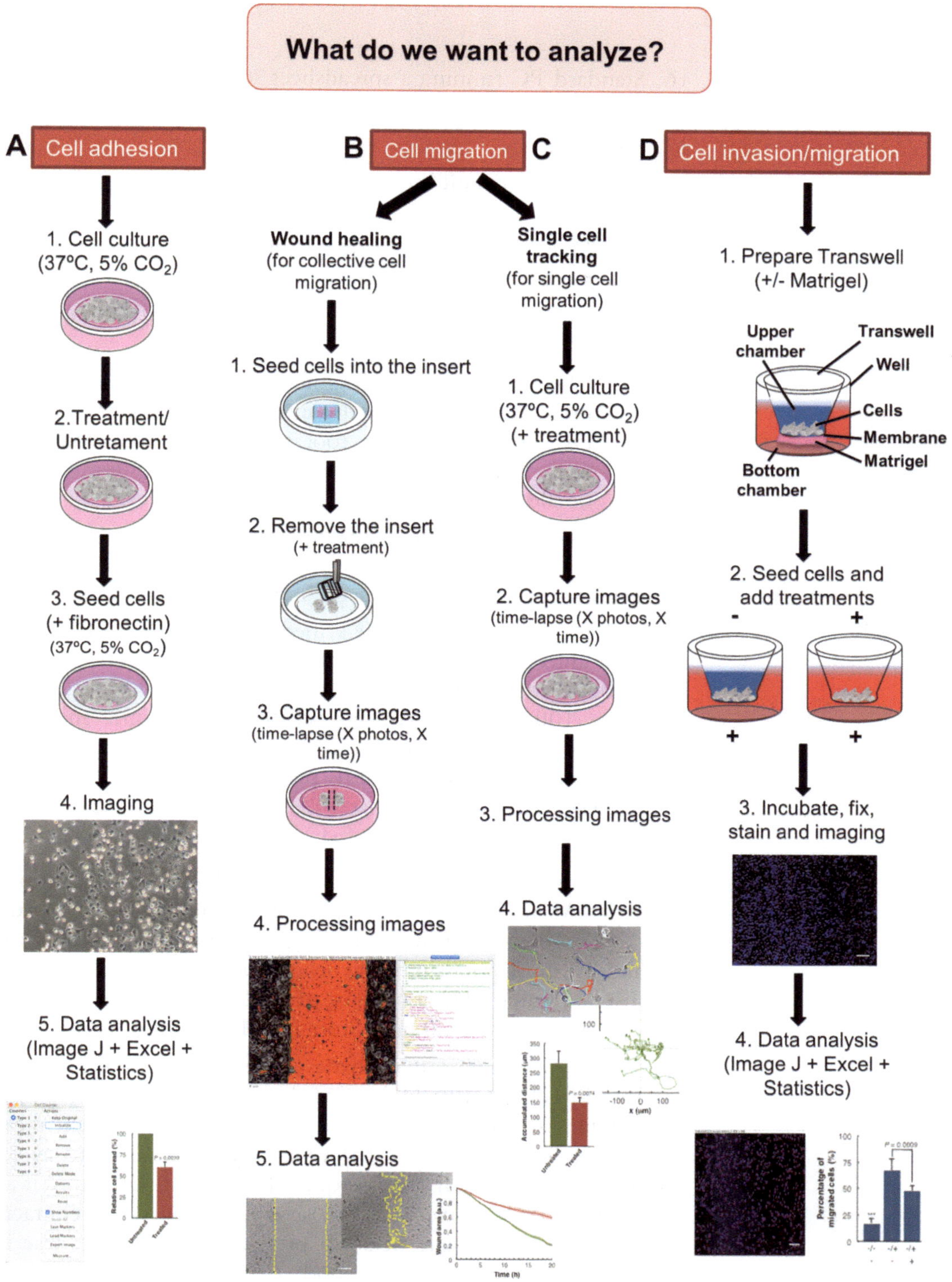


**Fig. 1** Overview of the different approaches in cell adhesion, migration, and invasion assays. (**a**) Cell spreading assay: seed treated cells in a fibronectin-coated plate, and count the percentage of attached and spreading cells. (**b**) Wound healing assay: in a 100% confluent cell monolayer, make a wound in the middle of the well to quantify the velocity and percentage of wound closure after specific treatment. (**c**) Single-cell migration assay: analyze the trajectory and velocity of single-cell motility after specific treatment. (**d**) Transwell assay: analyze single-cell migration and invasion through a porous membrane due to chemoattractant or treatment stimulation

16. Standard PC running a spreadsheet software (e.g., Microsoft Excel).
17. Software:
    (a) Fiji or ImageJ: http://fiji.sc/ or https://imagej.nih.gov/ij/h.
    (b) Manual Tracking: https://imagej.nih.gov/ij/plugins/track/track.html.
    (c) Cell Counter: http://rsbweb.nih.gov/ij/plugins/cell-counter.html.
    (d) Chemotaxis and migration tool or Chemotaxis tool: http://ibidi.com/fileadmin/products/software/chemotaxis_tool/chemotaxis_tool.jar or http://ibidi.com/software/chemotaxis_and_migration_tool/?x504f5=f74e46e10dfbebe758db0683f9de1303.
    (e) Macros [6]: Wound_healing.ijm; NotSpread&All_.ijm.

### 2.2 Reagents and Solutions

1. Target cells. Cancer cell line or another cell line of interest.
2. Dulbecco's modified Eagle's medium (DMEM), (4.5 g/L) high glucose, 2 mM L-glutamine. Media cell culture: (4.5 g/L) high glucose DMEM supplemented with 100 U/mL penicillin, 100 μg/mL streptomycin, 0.1% fungizone, and 10% (vol/vol) FBS heat-inactivated. Store it at 4 °C penicillin-streptomycin (10,000 U/mL penicillin and 10,000 μg/mL streptomycin).
3. Trypsin-EDTA, 0.25% (wt/vol).
4. 1× phosphate-buffered saline (PBS), pH 7.40. Composition: 137 mM sodium chloride (NaCl), 2.7 mM potassium chloride (KCl), 10 mM disodium hydrogen phosphate ($Na_2HPO_4$), and 1.8 mM potassium dihydrogen phosphate ($KH_2PO_4$) in distilled water ($_dH2O$). Adjust pH to 7.40, and autoclave it at 121 °C, 1 atm for 20 min. Store it at 4 °C.
5. 4% (wt/vol) paraformaldehyde (PFA) fixation solution. 4 g PFA in 100 mL PBS by heating the solution to ~70 °C under constant stirring with a magnetic stirrer in a safety chemical fume hood. Cool down the PFA solution, filter it to avoid precipitates in the fixative, aliquot the solution, and store it at −20 °C. PFA is toxic and corrosive. Avoid any direct contact, and wear appropriate personal protective equipment. Use it only inside the chemical hood. Collect and discard waste appropriately.
6. 10 μg/mL Fibronectin bovine plasma: 100 μL of fibronectin (1 mg/mL) with 900 μL of PBS. Freshly prepare the solution and use it immediately.

7. Heat-denatured bovine serum albumin (BSA): 10 mg of BSA in 1 mL of PBS. Filter the solution through a 0.22 μm filter to remove undissolved protein and incubate it at 90 °C for 15 min. Use after cooling.
8. Hoechst 33342 nuclear stain.
9. Matrigel™ basement membrane matrix (produced by Engelbreth-Holm-Swarm (EHS) mouse sarcoma cells): Thaw aliquots of Matrigel slowly on ice at 4 °C overnight. Add 100 μL of Matrigel to a microcentrifuge tube containing 300 μL of cold DMEM (without FBS) (ratio 1:3) and mix thoroughly. Matrigel solution is liquid at 4 °C, but it gels very quickly at room temperature. Remember to work in cold conditions. The mixture of Matrigel and DMEM should be prepared fresh every time.

## 3 Methods

### 3.1 Spreading Assay

1. Coat the wells of a 24-well plate with 250 μL of fibronectin (10 μg/mL) and incubate for 1 h at 37 °C or overnight at 4 °C. Other extracellular matrix (ECM) substrates can be used, e.g., gelatin or collagen, depending on your cell line.
2. Remove the unbound fibronectin by gentle aspiration, and wash twice with sterile 1× PBS.
3. (Optional step to block chemoattractant proteins present in the extracellular matrix coating). Add 200 μL of heat-denatured BSA solution (10 mg/mL) into each well, incubate for 30 min at room temperature or overnight at 4 °C, and wash twice with sterile 1× PBS.
4. Seed $2 \times 10^4$ cells in 500 μL of proper medium (according to cell line used) onto the fibronectin-coated 24-well plate (approximately 60–70% confluence).
5. Incubate at 37 °C and 5% $CO_2$ for 1 h (*see* **Note 1**).
6. Fix cells. Add PFA in the medium (final concentration 2%) of each well for 15 min at room temperature (*see* **Note 2**).
7. Remove 2% PFA culture medium by gentle aspiration.
8. Wash the wells twice with sterile 1× PBS to remove debris, non-attached cells, and fixation solution excess (*see* **Note 3**).
9. Acquire several images of fixed cells in a phase-contrast microscope using a 10× objective and avoid overlapping fields (Fig. 2a).
10. Quantitative data analysis can be performed with open-source software (ImageJ/Fiji). In ImageJ/Fiji, acquired images can be opened by clicking *File* → *Open* (*see* **Note 4**).

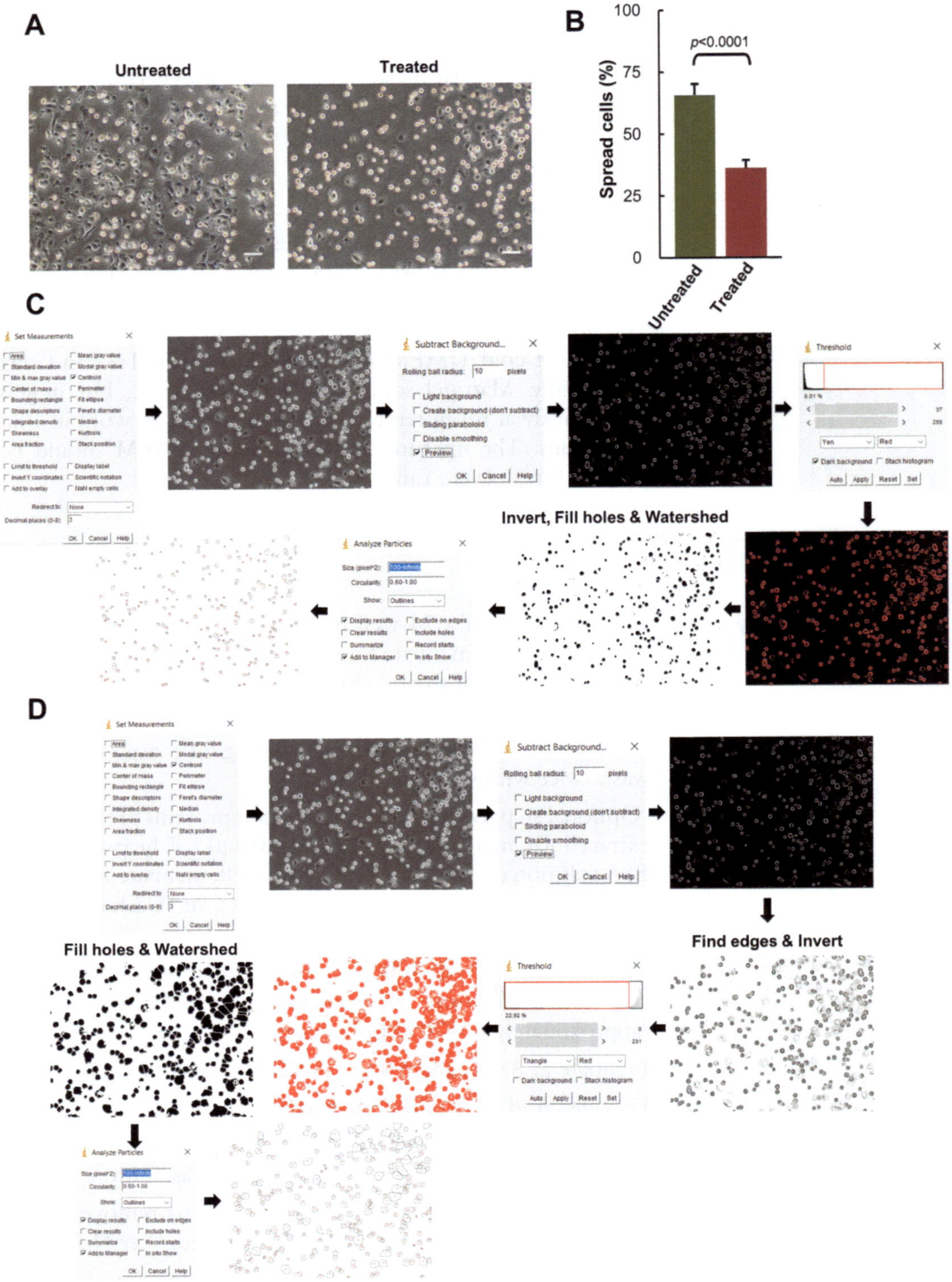


**Fig. 2** Cell spreading assay in melanoma cells. (**a**) Representative phase-contrast images of untreated and treated melanoma cells. Round bright cells are considered unspread. Scale bars, 50 μm. (**b**) Plot showing the percentage of spreading cells. (**c**) Brightfield images of the fixed cells are background-subtracted and thresholded (using the *Yen method*), and non-attached cell number is counted after a few binary operations, including fill holes and watershed. (**d**) Brightfield images of the fixed cells are background-subtracted, edges are found, and the edged image is thresholded (using the *Triangle method*), and the total cell number is counted after a few binary operations, including fill holes and watershed. (Reproduced and adapted from [6])

11. Subtract image background using a rolling ball of 10px radius by clicking *Process* → *Subtract background*. Afterward, the image is duplicated by clicking *Image* → *Duplicate*. One image will be used to quantify the unspread cells, while in the other one we will quantify the total cells.
12. To quantify unspread cells (Fig. 2c), the image is thresholded by the Yen method, just leaving the brightest sections of the image (corresponding to unspread cells) by clicking *Image* → *Adjust* → *Threshold* and selecting the Yen method. This creates a binary image. Afterward, the image is inverted by clicking on *Edit* → *Invert*.
13. Some binary operations, like "Erode," "Dilate," and "Fill Holes," are performed in order to polish the thresholding; you can find them in *Process* → *Binary* → ... Later, the watershed algorithm is used to segment cells that are touching each other, by clicking *Process* → *Binary* → *Watershed*.
14. Particle number is counted automatically by clicking *Analyze* → *Analyze particles*, setting a minimal size of 100 $px^2$ (as smaller particles are just little remains of other features coming from the thresholding, not real unspread cells).
15. To quantify total cells (Fig. 2d), the duplicated image created in **step 11** is used. First, to better detect the spread cells, highlight all the cell edges by clicking *Process* → *Find Edges*. Afterward, the image is inverted by clicking on *Edit* → *Invert*.
16. The image is thresholded using the Triangle method, which is less restrictive than the Yen method but still selecting where the cells are, by clicking *Image* → *Adjust* → *Threshold* and selecting the Triangle method, creating a binary image.
17. Some binary operations, like "Erode," "Dilate," and "Fill Holes," are performed in order to polish the thresholding; you can find them in *Process* → *Binary* → ... Later, the watershed algorithm is used to segment cells that are touching each other, by clicking *Process* → *Binary* → *Watershed*.
18. Particle number is counted automatically by clicking *Analyze* → *Analyze particles*, and setting a minimal size of 100 $px^2$ (as smaller particles are just little remains of other features coming from the thresholding, not real cells).
19. We have written the **steps** between **11** and **18** in a macro for ImageJ/Fiji to automatically analyze and determine the number of unspread and total cells in the image. The code "Unspread&All_.ijm" is provided in [6]. Execute this code at ImageJ/Fiji, and your images will be automatically quantified (*see* **Notes 5** and **6**).
20. Import data to the spreadsheet software, and determine the percentage of spread cells to total cells (Fig. 2a, b). Statistical significance of the differences between samples can be

estimated by unpaired two-tailed Student's *t*-test or multi-comparison by one-way ANOVA followed by Tukey-Kramer post hoc test.

### 3.2 Wound Healing Assay

1. Seed the cell suspension into each well with the silicone culture-insert on 35 mm culture-insert μ-dish to perform wound healing assay (Fig. 3a), or alternatively seed cells onto all surfaces of 35 mm culture μ-dish to perform scratch assay. Incubate the dishes for 6 h or overnight at 37 °C and 5% $CO_2$, allowing cells to adhere and spread on the substrate (100% confluence). The number of cells to create a confluent monolayer depends on your cell type and the size of dishes (*see* **Note 7**).
2. Check the cells under the microscope to ensure that you have confluent cell monolayer (95–100%).
3. Remove the culture-insert with sterile forceps. Alternatively, make a wound by scratching the cell monolayer in a straight line in the middle of the well with a sterile P-200 pipette tip (try to move the tip in a continuous way and always with a similar size in each well to avoid variation between conditions).
4. Remove the debris and non-attached cells by washing the cell layer twice with 1 mL of sterile 1× PBS, and then replace it with 2 mL of appropriate medium for the assay (if it is necessary add a treatment to the medium) (*see* **Note 8**).
5. Place the dish under a phase-contrast scanning confocal microscope with a $CO_2$ microscope cage incubator (*see* **Note 9**); image acquisition is performed at 37 °C and 5% $CO_2$. Define exact positions and focal planes. Start image acquisition by taking images several times throughout 20–24 h (usually with 10× objective). Select three or four fields from each wound and at least three wounds per sample (Fig. 3b).
6. After acquiring the time-lapse images, review the frames in a sequence format (oif., zvi., .avi, or similar format).
7. Quantitative data analysis can be performed with open-source software (ImageJ/Fiji). In ImageJ/Fiji, time-lapse sequences can be opened by clicking *File* → *Open.*
8. Calibrate the images sequence by using *Analyze* → *Set scale*, and introduce the correct scale (pixels/μm) depending on the objective and microscope used (Fig. 3d), if this information is not stored automatically by your microscope's software in the meta-data file.
9. Select the measurement parameters *Analyze* → *Set measurements* → *Area* (*see* **Note 10**) (Fig. 3d).
10. Click on duplicate the sequence images *Image* → *Duplicate.*
11. Detect the change intensity on the edge of the duplicate sequence *Process* → *Find edges* (Fig. 3d).

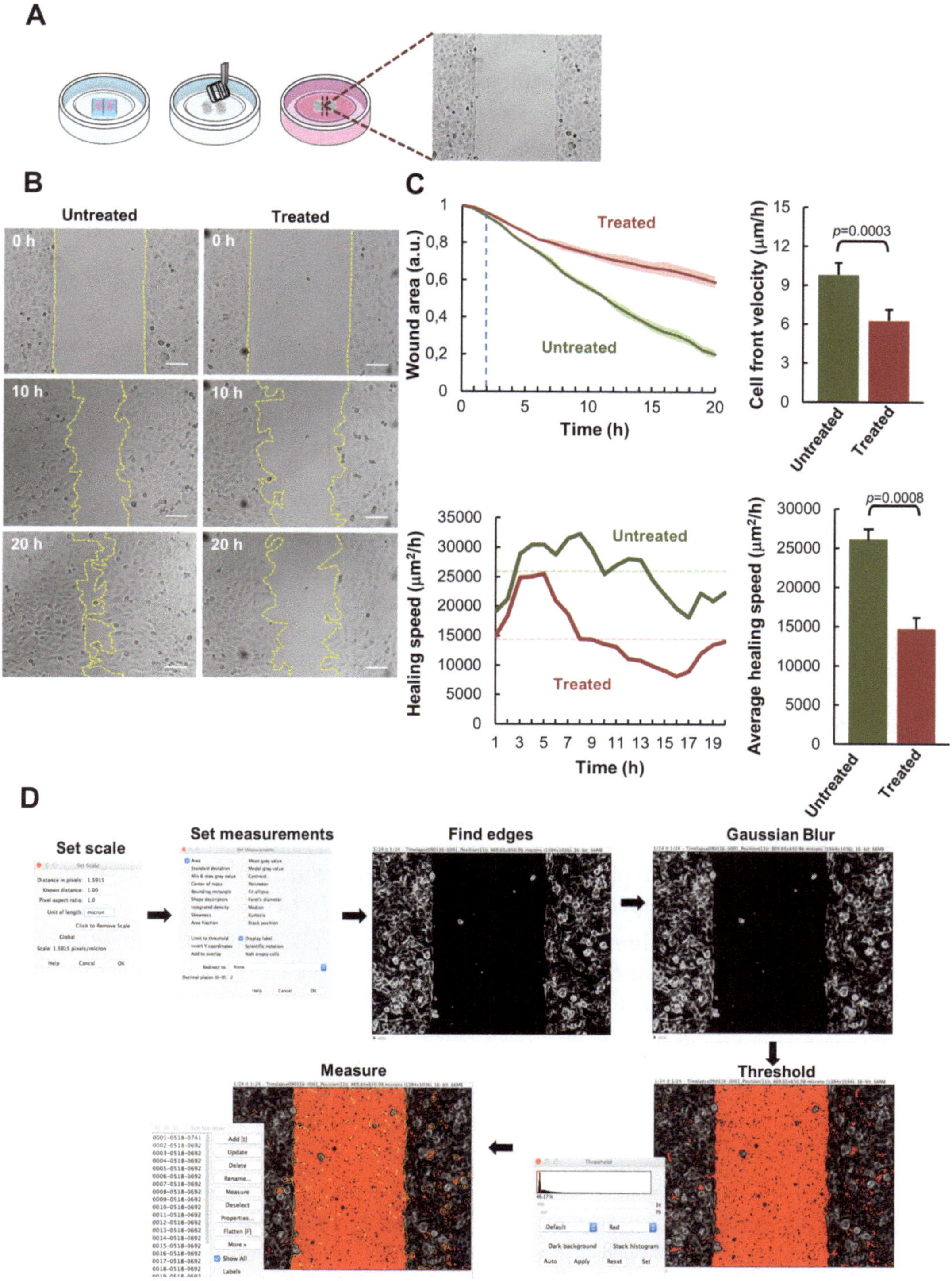


**Fig. 3** Analysis of collective migration melanoma cells by wound healing assay. (**a**) Schematic representation shows the different steps in the wound healing assay using a culture-insert. Cells are seeded onto the two wells of the culture-insert and allowed to adhere for 6 h or overnight. The culture-insert is removed with sterile forceps, and fresh culture medium is added to the dish. Phase-contrast microscopy shows the appearance of the gap and both cell fronts just before starting the time-lapse, at the initial time (0 h). (**b**) Wound closure in

12. Apply a blurred edge filter *Process → Filters → Gaussian Blur* (e.g., Sigma (Radius) = 2) (Fig. 3d).
13. Adjust the intensity threshold to detect the wound area *Image → Adjust → Threshold* (Fig. 3d).
14. Create a selection around the threshold area *Edit → Selection → Create selection.*
15. The area selection is enlarged (e.g., 8 pixels) *Edit → Selection → Enlarge.*
16. Add all time points into ROI manager *Analyze → Tool → ROI manager.*
17. Obtain the area value in the different time points *Analyze → Measure* (Fig. 3d).
18. We have written the **steps** (**10–17**) into a macro for ImageJ/Fiji to automatically analyze the wound healing area in each image sequence. The code "Wound_healing.ijm" is provided in [6]. Execute this code at ImageJ/Fiji and automatically your time-lapse images will be quantified.
19. Import the data to a spreadsheet software and determine the relative wound area closure, velocity, and other parameters (Fig. 3c). Estimate statistical significance between samples by any statistics software.

### 3.3 Individual Cell Tracking Assay

1. Seed $1.2 \times 10^4$ cells in 500 μL of proper complete medium onto each well on 24-well plates, and allow to attach them overnight at 37 °C and 5% $CO_2$ (*see* **Note 11**).
2. Carefully remove the medium with the debris and dead cells by washing the wells once with 500 μL of sterile 1× PBS, and then replace it with 500 μL of proper medium for the assay and add specific treatment, if necessary (*see* **Note 8**).
3. Immediately place the 24-well plate under the phase-contrast or fluorescence microscope with a $CO_2$ microscope cage incubator (at 37 °C and 5% $CO_2$). Perform time-lapse imaging (e.g., taking an image every 20 min for 20 h using a 20× objective) (Fig. 4a). You should acquire about 10–15 fields for each of the three biological replicates in order to obtain data from at least 300 cells (*see* **Notes 9**, **12**, and **13**).

**Fig. 3** (continued) untreated and treated melanoma cells after 0, 10, and 20 h of imaging. Scale bars, 100 μm. (**c**) Quantification of the wounded area invaded during 20 h, cell front velocity (μm/h), healing spread (μm$^2$/h), and average healing spread (μm$^2$/h) in untreated and treated melanoma cells. (**d**) Step-by-step analysis of wound healing data in ImageJ/Fiji software. *Set scale* and *Set measurements* dialog boxes allow adjusting the scale (pixel/μm) and the parameters to measure. Different settings to determine the measurable area by applying *Find edges*, *Gaussian blur* (of radius 2 pixels), and *Threshold adjust.* Final step and results (area values) of analysis. (Reproduced and adapted from [6])

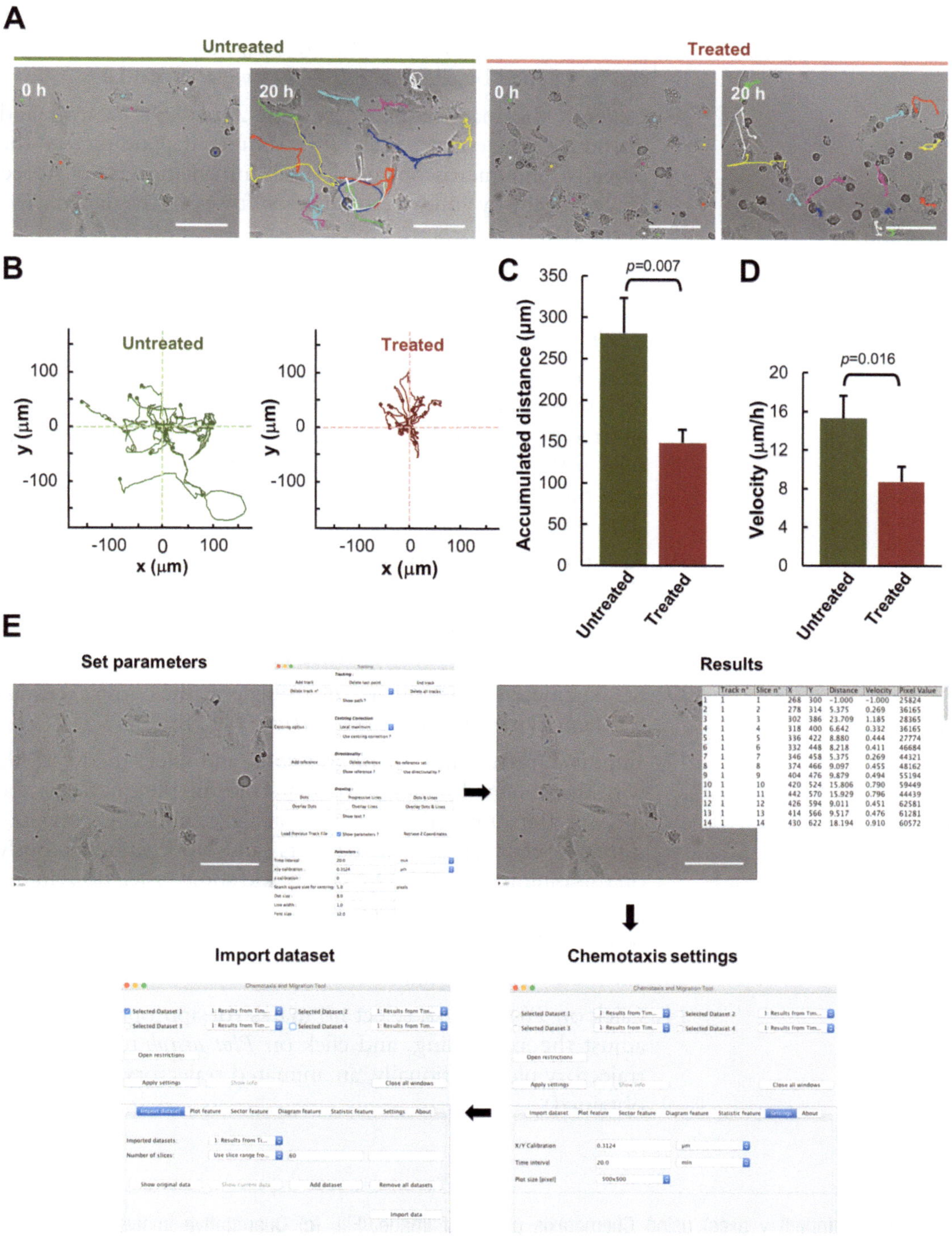


**Fig. 4** Analysis of single-cell migration assay. (**a**) Representative phase-contrast images of untreated and treated melanoma cells (captured by time-lapse microscopy at 20 min intervals) at initial (0 h) and end (20 h) time. The Manual Tracking plugin of ImageJ/Fiji was used to manually trace 14 representative cell trajectory tracks, marked in colors. Scale bars, 100 μm. (**b**) Trajectory plots showing melanoma cells trajectory during 20 h in untreated (green) and treated (red) cells. All tracks are set to a common origin (intersection of *x* and

4. Quantitative data analysis can be performed with open-source software (ImageJ/Fiji). In ImageJ/Fiji, time-lapse sequences can be opened by clicking *File* → *Open* (*see* **Note 14**).
5. Calibrate the image sequence by using *Analyze* → *Set scale*, and introduce the correct scale (pixels/μm) depending on the objective and microscope used, if this information is not stored automatically by your microscope's software in the data file (*see* **Note 10**).
6. Open *Plugins* → *Tracking* → *Manual Tracking*.
7. Set the parameters *Time interval*, *x/y calibration*, and *z calibration* in the *Manual Tracking* (Fig. 4e).
8. Start individual cell-tracking analysis, click on *Add track*, and then select one cell in the first time point and follow it through all time points, finally click on *End track*. Repeat it for all the cells in the frame through all the time points. A new window with the results (*track*, *slice*, *distance*, and *velocity*) will appear (Fig. 4e).
9. Save the values in (.txt) format and click on *Overlay dots and line* to save the sequence with the track.
10. Download Chemotaxis plugin and open it in ImageJ/Fiji *Plugins* → *Chemotaxis tool* (*see* **Note 15**).
11. Set parameters. Click on *Settings*, and apply the correct settings for *Time interval* and *x/y calibration* (Fig. 4e).
12. Import dataset. Click on *Import data*, and select the (.txt) file with the obtained values in Manual Tracking plugin. Select the file, set the number of slices to analyze, and click on *Add dataset*. Select up to four datasets, and work simultaneously on this plugin (Fig. 4e). Optionally add some restrictions in the parameters. Click on *Open restrictions*, and set *split dataset*, *threshold distance*, and *threshold velocity*.
13. Click on *Apply settings*.
14. Click on *Plot feature*, select no marks (or appropriate marks), adjust the axis scaling, and click on *Plot graph* to obtain the trajectory plot (optionally, an animated trajectory plot can be obtained).

**Fig. 4** (continued) y axes) using Chemotaxis plugin of ImageJ/Fiji. (**c**) Quantitative analysis of average accumulated distance (μm) and (**d**) velocity (μm/h) of untreated and treated melanoma cells. (**e**) Time-lapse sequence and main window of Manual Tracking with the different parameters to set *time interval*, *x/y calibration*, and *track* and results window after tracking cells at different time points by Manual Tracking, *track nº*, *slice nº*, *X*, *Y*, *distance*, *velocity*, and *pixel value* are shown. Main window of Chemotaxis tool plugin with *time interval*, *x/y calibration*, and *plot size* and import dataset settings. (Reproduced and adapted from [6])

15. (Optional step, depending on the analysis). Click on *Diagram feature* and select *Plot histogram*, *Plot rose diagram*, or *Circular plot* to obtain several plots that show an orientation/distribution of all cell migration directions.
16. Click on *Statistics feature* and select *velocity*, *distance* (Euclidean and accumulated), *FMI* (Forward Migration Index), and directionality. Save or export these data to spreadsheet software.
17. Determine the mean and standard deviation (SD) of velocity (μm/h), accumulated distance (μm), Euclidean distance (μm), and other parameters (Fig. 4a–c). Statistical significance of the differences between samples can be estimated, for example, by unpaired two-tailed Student's *t*-test or multi-comparison by one-way ANOVA followed by Tukey-Kramer post hoc test.

### 3.4 Transwell Cell Migration/Invasion Assay

1. Thaw the Matrigel solution at 4 °C overnight. Coat the downside surface of the transwell membrane with 25 μL of Matrigel solution (ratio 1:3 with DMEM) (Fig. 5a).
2. Incubate the transwell with Matrigel at 37 °C for 30 min for gelation.
3. Add 500 μL of culture medium (with or without chemoattractant (10% FBS was used in melanoma cells) into each well on a 24-well plate. The use of chemoattractant is to enforce the direction for cancer cell migration/invasion (Fig. 5b).
4. (Optional step) Add a specific treatment in the culture medium if necessary.
5. Use sterile forceps to transfer the transwell insert (with or without Matrigel) into each well of the 24-well plate that is already filled with culture medium (with or without chemoattractant).
6. Add 400 μL of cell suspension (at final confluence 50–60%) onto the transwell upper chamber (Fig. 5b) (*see* **Note 16**).
7. Incubate the plate at 37 °C and 5% $CO_2$ for 20–24 h.
8. Fix cells by adding paraformaldehyde (PFA) at a final concentration of 4% in culture medium into each transwell insert for 15 min at room temperature.
9. Remove the culture medium with 4% PFA by gentle aspiration.
10. Wash transwell insert at both sides of the membrane, twice with sterile 1× PBS to remove debris, non-attached cells, and the excess of fixation solution (*see* **Note 3**).
11. Stain the cells by incubating the transwell insert with Hoechst (final concentration 10 μg/mL in 400 μL of 1× PBS) for 15 min at room temperature protected from light (Crystal violet or Hematoxylin could be also used to stain the cells).

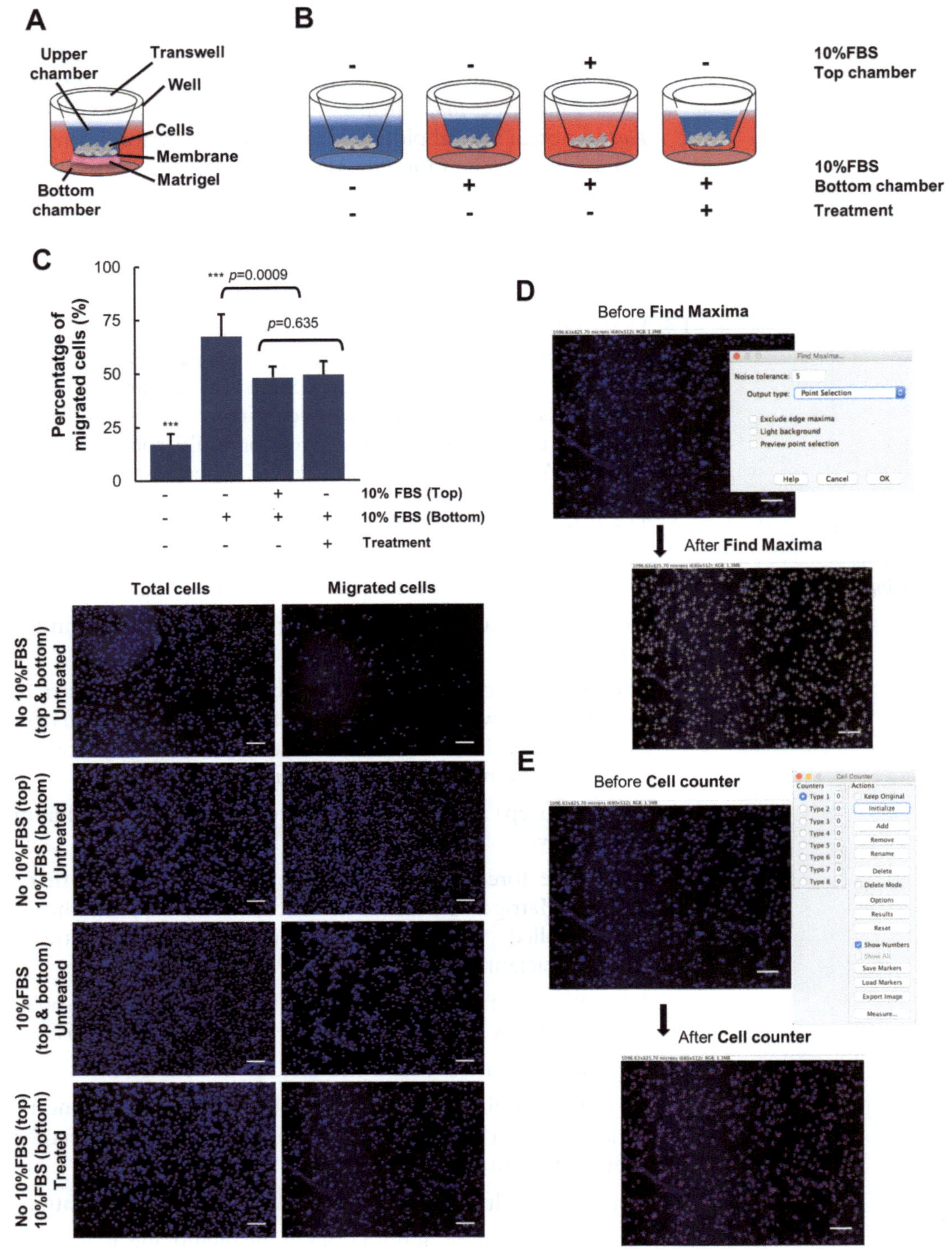


**Fig. 5** Quantitative and qualitative analysis of melanoma cell migration/invasion assessed by in vitro transwell assay. (**a**) Schematic illustration of transwell system. (**b**) Experimental design scheme of transwell migration/invasion assay. (**c**) Percentage of migrated cells after 20–24 h, with (+) or without (−) treatment and chemoattractant (10% FBS) in upper and lower compartments. Representative fluorescence images of nuclear Hoechst staining (10 μg/mL) of total cells (left panel) and migrated cells (right panel). Scale bars, 100 μm. (**d**) Setting to count the numbers of cells in ImageJ/Fiji after applying the Find maxima parameter (e.g., noise tolerance 5) (automatic recount of cells) or (**e**) after using Cell Counter application plugin (manual recount of cells). (Reproduced and adapted from [6])

12. Wash the transwell insert twice with sterile 1× PBS to remove Hoechst solution excess.
13. Maintain the inserts in the 24-well plate with sterile 1× PBS to avoid drying the membrane.
14. Capture several images in order to cover up all the surfaces of the transwell insert membrane, using 10× or 4× objectives on a fluorescence microscope. The number of cells counted using these images will be the total number of cells in migration and/or invasion assay (left panel in Fig. 5c).
15. Immediately, remove the non-invaded/migrated cells from the upper surface of the transwell membrane by gently scrapping with a cotton swab. If crystal violet or hematoxylin is used, the membrane turns blue. Scrape the membrane surface until the last swab used remains white (or clean) (*see* **Note 17**).
16. Repeat **step 14**, but in this case the acquired images represent the invaded or migrated cells (right panel in Fig. 5c).
17. Quantitative data analysis can be performed with open-source software (Fiji/ImageJ). In ImageJ, acquired images can be opened by clicking *File* → *Open*.
18. Select *Process* → *Find Maxima* to automatically measure the total number of cells in each image. Adjust the parameter *Noise tolerance* to avoid background noise, detect, and count all cells (nuclei stained by Hoechst) (Fig. 5d).
19. (Alternatively, to **step 17**), a manual counting can be obtained by opening *Plugins* → *Analyze* → *Cell counter*. Select *Initialize* and *Type I*, and click on the cells (nuclei stained by Hoechst) manually to obtain the total number of cells (Fig. 5e).
20. Import data to the spreadsheet software, and determine the percentage of migrated and/or invaded cells (cells after the cotton swap/total cells × 100) (Fig. 5c). Statistical significance of the differences between samples can be, for instance, estimated by unpaired two-tailed Student's *t*-test or multi-comparison by one-way ANOVA followed by Tukey-Kramer post hoc test.

## 4 Notes

1. The incubation period depends on the cell type and the treatment used. This cell type-specific binding time must be determined previously by the researchers.
2. Perform the cell fixation in a safety chemical fume hood, and wear appropriate personal protective equipment.

3. Be gentle with all washes to avoid detaching the cells. In this step, the samples can be stored in 1× PBS at 4 °C for up to 1 month.
4. In order to be analyzed, images must be 8-bit. If they are in a different bit depth, modify them by clicking *Image* → *Type* → *8-bit*.
5. The macro may have trouble in identifying the total number of cells if the cells are seeded too confluent; it is better to seed at lower density and take images of a few more fields in order to ensure a reliable quantification. Otherwise, the analysis may be done manually (*Plugins* → *Analyze* → *Cell counter*), but this will slow down the quantification a lot, and still, it is hard to identify the boundary between cells if the culture is too confluent.
6. Round and refringent cells are considered unspread, while dark cells with cytoplasm surrounding the entire circumference of the nucleus are considered spread cells. The sum of spread and unspread cells is the total number of cells. We can infer the number of spread cells from the difference between total number of cells and unspread cells.
7. The number of cells to create a confluent monolayer depends on your cell type and the size of dishes. It is very important to create a confluent cell monolayer (95–100% confluence) and maintain a constant seeding number specific to each cell line, which will ensure reproducibility of the results. Depending on the cell type used, coating dishes with extracellular matrix (ECM) substrate (e.g., gelatin, collagen, or fibronectin) will help cells to adhere. Coating must be done at 4 °C overnight.
8. Be gentle with all washes to avoid detaching the cells.
9. Prior to the setup of time-lapse microcopy, switch on the $CO_2$ and temperature to equilibrate the system at least 30 min before starting the acquisition. Ensure that the parameters are set at 5% $CO_2$ and 37 °C, respectively.
10. It is important to set the correct scale for the cell migration analysis.
11. The number of seeded cells depends on the characteristics of the cell type and the size of dishes. To analyze individual cell tracking, cells are seeded at low density (30–40%). Perform at least three biological replicates for each different experimental condition.
12. Define the image acquisition setup depending on the previous knowledge about your cell line (e.g., higher temporal resolution if your cell line is known to have high motility). Avoid overlapping fields.

13. Fluorescently tagged cells can be used in this assay. You must also be careful with light intensity during time-lapse acquisition, as it may cause phototoxicity.
14. After acquiring the time-lapse images, review the frames in a sequence format (*.tif, *.oif, *.zvi, or equivalent).
15. Alternatively, the *Chemotaxis and migration tool* software can be downloaded and used with the same function.
16. The number of seeded cells depends on the characteristics of the cell type and the size of dishes. To analyze migration/invasion, cells are seeded at medium density (50–60%). Perform at least three biological replicates for each different experimental condition.
17. Be careful with scraping. Scraping must be gentle with little pressure to remove all the cells on the upper surface of the membrane but not affecting the migrated/invaded cells of the bottom. If it is necessary, repeat the scraping procedure with a second cotton swab.

## References


1. te Boekhorst V, Preziosi L, Friedl P (2016) Plasticity of cell migration in vivo and in silico. Annu Rev Cell Dev Biol 32:491–526. https://doi.org/10.1146/annurev-cellbio-111315-125201
2. Charras G, Sahai E (2014) Physical influences of the extracellular environment on cell migration. Nat Rev Mol Cell Biol 15:813–824. https://doi.org/10.1038/nrm3897
3. Mayor R, Etienne-Manneville S (2016) The front and rear of collective cell migration. Nat Rev Mol Cell Biol 17:97–109. https://doi.org/10.1038/nrm.2015.14
4. van Helvert S, Storm C, Friedl P (2018) Mechanoreciprocity in cell migration. Nat Cell Biol 20:8–20. https://doi.org/10.1038/s41556-017-0012-0
5. Stuelten CH, Parent CA, Montell DJ (2018) Cell motility in cancer invasion and metastasis: insights from simple model organisms. Nat Rev Cancer 18:296–312. https://doi.org/10.1038/nrc.2018.15
6. Pijuan J, Barceló C, Moreno DF et al (2019) In vitro cell migration, invasion, and adhesion assays: from cell imaging to data analysis. Front Cell Dev Biol 7:107. https://doi.org/10.3389/FCELL.2019.00107

# Chapter 21

# Cell Viability and Immunogenic Function of T Cells Loaded with Nanoparticles for Spatial Guidance in Magnetic Fields

Felix Pfister, Christoph Alexiou, and Christina Janko

## Abstract

Immune cell therapies, such as adoptive T cell therapies, are an innovative and powerful treatment option for previously non-treatable diseases. Although immune cell therapies are thought to be very specific, there is still the danger of developing severe to life-threatening side effects due to the unspecific distribution of the cells throughout the body (on-target/off-tumor effects). A possible solution for the reduction of these side effects and the improvement of tumor infiltration is the specific targeting of the effector cells (e.g., T cells) to the desired destination (e.g., tumor region). This can be achieved by the magnetization of cells with superparamagnetic iron oxide nanoparticles (SPIONs) for spatial guidance via external magnetic fields. A prerequisite for the use of SPION-loaded T cells in adoptive T cell therapies is that cell viability and functionality after nanoparticle loading are preserved. Here, we demonstrate a protocol to analyze cell viability and functionality such as activation, proliferation, cytokine release, and differentiation at a single cell level using flow cytometry.

**Key words** T cells, Cell viability, Intracellular cytokine staining, Flow cytometry, T cell activation, T cell proliferation

## 1 Introduction

Cancer is one of the most impactful and deadly diseases worldwide, with an increasing number of cases in the future expected [1]. Nevertheless, the treatment of cancer has been improved in recent years with the development of personalized immunotherapies such as T cell therapies, which allow the treatment of previously hard-to-manage cancers [2]. Although these advancements have increased the long-term survival of patients with hematological cancers, T cell therapy still has limited effects against solid tumors due to tumor heterogeneity, the immunosuppressive microenvironment, and the need to penetrate tissues [3, 4]. Additionally, T cell therapy may induce severe systemic side effects, such as cytokine release syndrome, that can result in death [5, 6]. Therefore, improvements in the safety and efficacy of T cell therapy are needed. A possible

Oliver Friedrich and Daniel F. Gilbert (eds.), *Cell Viability Assays: Methods and Protocols*, Methods in Molecular Biology, vol. 2644, https://doi.org/10.1007/978-1-0716-3052-5_21,

solution for the reduction of these side effects and the improvement of tumor infiltration is the specific targeting of the effector cells (e.g., T cells) to the desired destination (e.g., tumor region). This can be achieved by the introduction of spatial controllability of T cells by external magnetic fields via loading of magnetic iron oxide nanoparticles, which allows the local enrichment of T cells [7, 8].

A prerequisite for using "magnetized" T cells is that they maintain their viability and effector functions after nanoparticle loading. To analyze viability, cells are stained with a mix containing Annexin A5 (AxV)-Fitc, propidium iodide (PI), MitoProbe™ DiIC1(5), and Hoechst 33342 [9, 10]. Hoechst 33342 is a plasma membrane permeable dye which we use to discriminate cells from possible nanoparticle agglomerations. AxV calcium dependently binds to phosphatidylserine exposed on the outer layer of the plasma membrane during apoptosis [11]. As a plasma membrane impermeable DNA dye, propidium iodide only stains cells with disrupted plasma membranes and is therefore used to discriminate apoptotic from necrotic cells [11]. MitoProbe™ DiIC1(5) provides information about the mitochondrial membrane potential, a marker for metabolic activity of the cell [9].

The functionality of the T cells must be investigated after their activation, which can be achieved by co-incubation with antigen-presenting cells that have received an antigen for presentation. However, due to the broad antigen recognition capabilities of T cells, this only activates a small fraction of the T cells. Therefore, a different approach can be used by antigen-independent activation to achieve a broad T cell activation. This can be accomplished by stimulation with antibodies targeting T cell costimulatory receptors such as CD3, CD28, and CD2, resulting in an unspecific but higher frequency of cytokine-producing T cells. Additionally, antigen-specific transgenic T cell receptors can also be introduced to increase the frequency of antigen recognition.

T cell stimulation results in the expression of activation-associated membrane receptors like CD25 and CD69. CD25, also known as the α-chain of the IL-2 receptor, is able to bind IL-2 together with the β- and γ-chain with high affinity, promoting cell survival and proliferation [12]. CD69 is expressed after activation by tissue-resident T cells and functions as a signal-transmitting receptor [13]. The expression of both allows us to analyze the activation status of effector T cells [14]. The T cell effector subclass can be identified by their expression of CD45RO and/or CCR-7 [15, 16]. Naïve T cells only express CCR-7; however, after activation, they express both CCR-7 and CD45RO as central memory T cells, while effector memory T cells only express CD45RO [17]. A further sign of T cell activation is the production of cytokines. Important T cell cytokines include interleukin (IL)-2, interferon (IFN)-γ, and tumor necrosis factor (TNF)-α. IL-2 is secreted by activated T cells in an autocrine manner and promotes further activation, proliferation, and CD8+ T cell differentiation into effector and memory cells [18]. IFNγ, on the other hand, is a pro-inflammatory cytokine that promotes activation and chemotaxis of T helper 1 cells, monocytes, and macrophages and has an antiviral and anti-tumorigenic effect [19]. TNFα is a member of the TNF superfamily and is able to induce broad activation and survival via NFκB but also apoptosis of effector CD8+ T cells after prolonged exposure [20].

After activation, T cells start to proliferate. A method for measuring the frequency of cell proliferation is by staining T cells with the membrane-permeable nonfluorescent carboxyfluorescein diacetate succinimidyl ester (CFDA-SE). Inside the cells, CFDA-SE is processed by intracellular esterase cleavage [21]. The resulting fluorescent carboxyfluorescein succinimidyl ester (CFSE) is membrane impermeable and therefore retained inside the cell. During cell division, CFSE is diluted to the daughter cells, enabling us to investigate up to eight T cell divisions [22]. Furthermore, the current state of proliferation can also be examined by propidium iodide staining of the cell nuclei. Absolute cellular DNA content increases during cell division until mitosis, where it reaches its highest fluorescence intensity after propidium iodide staining. This allows the identification of the different cell cycle phases G1 (low intensity), S (medium intensity), and G2/M (high intensity) [23, 24].

In this chapter, we will describe a step-by-step guide on how to analyze the viability and cell cycle of T cells, as well as cytokine production, activation, and subset marker expression of T cells loaded with magnetic iron oxide nanoparticles using flow cytometry (Fig. 1).

## 2 Materials

### 2.1 General Equipment and Consumables

1. Blood from healthy human volunteers.
2. Kit for nonmagnetic isolation of human T cells (e.g., IBA CD3 Fab-TACS Gravity Kit, IBA Lifesciences) (*see* **Note 1**).
3. ImmunoCult™ human CD3/CD28/CD2 T cell activator (e.g., STEMCELL Technologies).
4. Magnetic iron oxide nanoparticles (obtained by commercial suppliers or synthesized in-house, e.g., according to [7]).
5. 50 mL sterile polypropylene tubes.
6. F-bottom 6-well plate.
7. F-bottom 96-well plate.
8. FACS tubes.

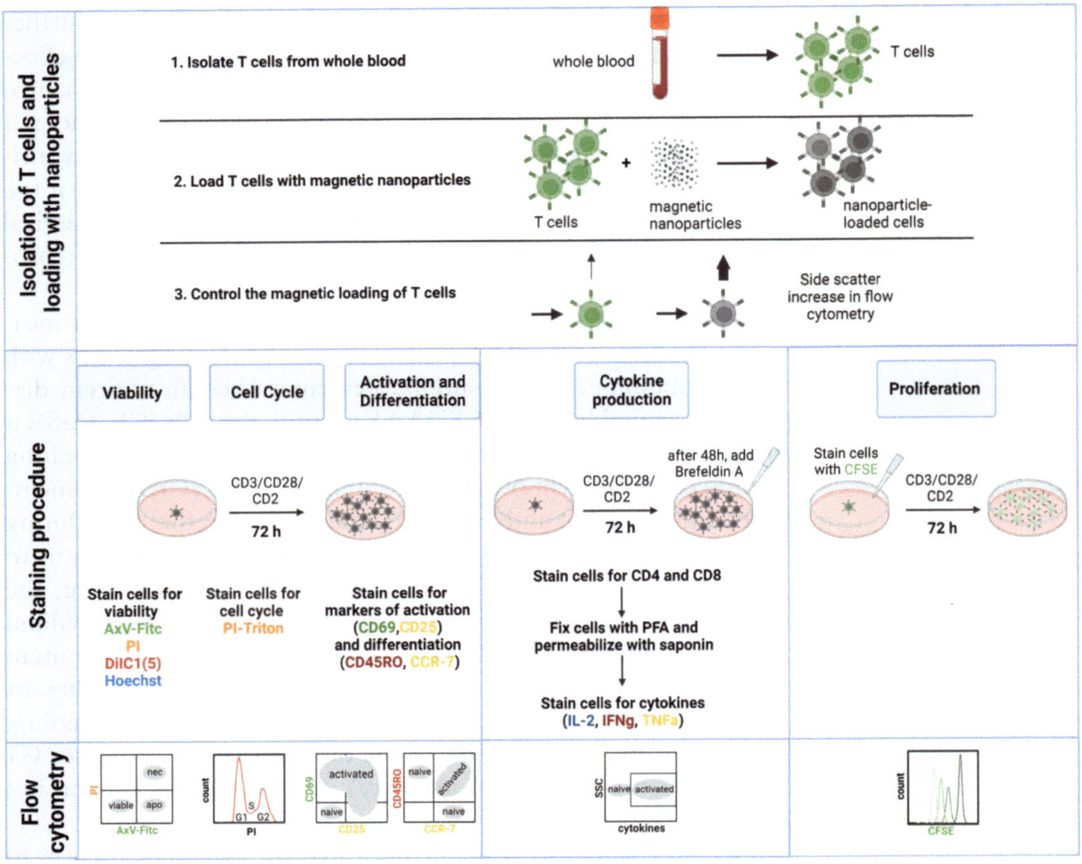


**Fig. 1** Overview of T cell isolation, nanoparticle loading, staining procedures, and flow cytometry analysis

9. Cell incubator.
10. Flow cytometer with measurement and analysis software.
11. 1000× brefeldin A (Thermo Fisher).
12. 4% w/v paraformaldehyde (PFA) diluted in PBS, pH 7.4.
13. 0.5% w/v saponin diluted in PBS, pH 7.4.
14. Hemocytometer.

### 2.2 Antibodies

The final fluorochrome-conjugated antibody concentration has to be determined beforehand (*see* **Notes 2** and **3**). The selection of fluorochrome conjugation to the antibodies has to be adapted to the hardware of your flow cytometer. Indicated fluorescence labeling is an example. Use appropriate controls (*see* **Note 4**). In case fluorophores produce spectral overlap, use fluorescence compensation to eliminate fluorescence-bleed through (*see* **Note 5**).

1. Anti-human CD3-FITC (co-receptor for the T cell receptor on T cells, BD Biosciences, catalogue Nr. 561806).

2. Anti-human CD4-PerCP (co-receptor for the T cell receptor on T cells, subclass identification, BioLegend™, catalogue Nr. 317428).
3. Anti-human CD8a-Pacific Blue™ (co-receptor for the T cell receptor on T cells, subclass identification, BioLegend™, catalogue Nr. 300928).
4. Anti-human CD8a-APC (co-receptor for the T cell receptor on T cells, subclass identification, BioLegend™, catalogue Nr. 300912).
5. Anti-human CD25-PE (activation marker, BioLegend™, catalogue Nr. 302606).
6. Anti-human CD69-FITC (activation marker, BioLegend™, catalogue Nr. 310904).
7. Anti-human CD45RO-APC (differentiation marker, BioLegend™, catalogue Nr. 304210).
8. Anti-human CCR-7-PE (differentiation marker, BioLegend™, catalogue Nr. 353204).
9. Anti-human IFNγ-APC-Vio® 770 (cytokine, Miltenyi Biotec, catalogue Nr. 130-113-491).
10. Anti-human TNFα-PE (cytokine, Miltenyi Biotec, catalogue Nr. 130-118-974).
11. Anti-human IL-2-eFluor™ 450 (cytokine, Thermo Fisher Scientific, catalogue Nr. 48-7021-82).

### 2.3 Buffers and Cell Culture Media

1. $Ca^{2+}$ and $Mg^{2+}$ free phosphate-buffered saline (PBS), cell culture grade.
2. Ringer's solution, pH 7.4.
3. Roswell Park Memorial Institute (RPMI) 1640 medium.
4. T cell medium: RPMI 1640 supplemented with 10% heat-inactivated fetal calf serum (FCS), 2 mM L-glutamine, 100 IU/mL penicillin, 100 μg/mL streptomycin, 5 μg/mL amphotericin B; 10 ng/mL recombinant human IL-7, pH 7.4.
5. FACS buffer: PBS containing 2% FCS, pH 7.4.

### 2.4 Staining Solutions

1. Viability staining: Ringer's solution supplemented with 20 nM Hoechst 33342 (Hoe), 2 μL/mL AxV with fluorescein-5-isothiocyanate conjugation (AxV-FITC), 4 nM MitoProbe™ DiI-$C_1(5)$ and 66.6 ng/mL propidium iodide (PI).
2. Proliferation staining: T cell medium with 5 μM CFSE.
3. Cell cycle staining: $dH_2O$ supplemented with 1 g/L trisodium citrate dihydrate, 1 mL/L Triton X-100, and 1 mg/L PI.
4. Antibody staining: FACS buffer with antibodies.

## 3 Methods

***3.1 Isolation of CD3+ T Cells via IBA CD3 Fab-TACS Gravity Kit***

1. Isolate T cells from whole blood with IBA CD3 Fab-TACS Gravity Kit, and wash with T cell medium (*see* **Note 1**).
2. Count T cells via a hemocytometer, and seed cells at a concentration of $1 \times 10^6$ T cells/mL in a 6-well plate. Use at least 2 mL per well to prevent drying out.
3. Determine the purity of T cells via staining with anti-CD3, anti-CD4, and anti-CD8 and a subsequent analysis by flow cytometry.

***3.2 Loading of T Cells with Magnetic Nanoparticles***

1. Incubate T cells with the desired concentration of nanoparticles overnight in an incubator at 37 °C and 5% $CO_2$. Use control cells without nanoparticle loading (*see* **Notes 6** and **7**).
2. Wash T cells twice with T cell medium (*see* **Note 8**). The cell pellet should have a brown color.
3. Resuspend cells in PBS, and count them via a hemocytometer. Blue discoloration indicates dead cells. Adjust the cells to a density of $1 \times 10^6$ T cells/mL in T cell medium and seed in triplicates in a 12-well cell culture plate.
4. Successful loading is controlled by an increase in cellular sideward scatter via flow cytometry due to the nanoparticle uptake/attachment in/onto the cells (*see* **Note 9**, and Fig. 2).

***3.3 Analysis of Cell Viability and Cell Cycle***

1. Adjust loaded T cell concentration to $1 \times 10^6$ cells/mL.
2. Add 25 μL/mL of ImmunoCult™ human CD3/CD28/CD2 T cell activator. Incubate in a 37 °C and 5% $CO_2$ environment for 72 h. Use unstimulated cells as controls (*see* **Note 10**).
3. Directly after in vitro activation, wash the cells twice with PBS with $Ca^{2+}$ or Ringer's solution (*see* **Note 11**).
4. Resuspend cells in 50–100 μL $Ca^{2+}$ or Ringer's solution and add 200 μL of viability staining solution. Incubate on ice or 4 °C in the dark for 20 min and analyze in flow cytometry (Fig. 3).
5. For the analysis of the cell cycle, take another 100 μL of sample, and add 500 μL of cell cycle staining. Incubate overnight at 4 °C for optimal DNA extraction, then analyze by flow cytometry (Fig. 4).

***3.4 Intracellular Cytokine Staining***

1. Perform all steps on ice or 4 °C if not written else.
2. Adjust cell concentration to $1 \times 10^6$ cells/mL.
3. Add 25 μL/mL of ImmunoCult™ human CD3/CD28/CD2 T cell activator. Incubate in a 37 °C and 5% $CO_2$ environment for 72 h. Use unstimulated cells as controls (*see* **Note 10**).

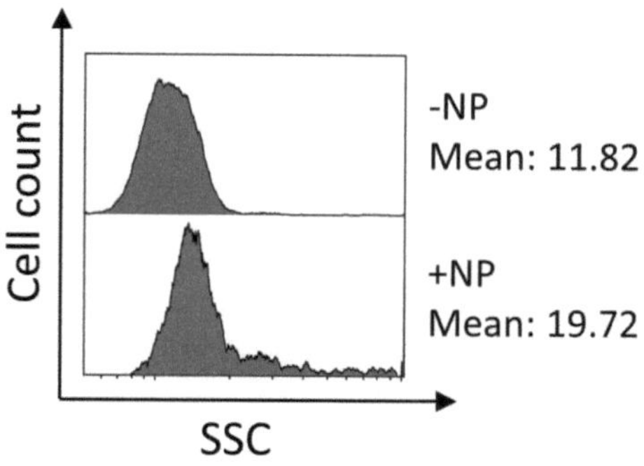


**Fig. 2** Nanoparticle uptake/attachment increases cell granularity. Isolated T cells are incubated overnight with (+NP) or without magnetic nanoparticles (−NP) and analyzed in flow cytometry. Cells which take up nanoparticles have an increased light scattering; thus, flow cytometry allows us to estimate if nanoparticles are associated with the cells. Since cell morphology is altered during cell death as well, it is important to add viability staining (e.g., AxV-Fitc or $DilC_1(5)$) into the measurement and gate on viable cells for analysis of the side scatter. Note that with this method it is not possible to discriminate between nanoparticle uptake and attachment to the plasma membrane of the cells. If you need information about absolute iron amounts per cell, an alternative readout such as atomic emission spectroscopy in combination with cell counting is necessary. The depicted loaded cells have an average iron content of 1.0 pg Fe/cell, while T cells without magnetic particles only have an average iron content of 0.1 pg Fe/cell

4. After 48 h, add 0.1% v/v of 1000× brefeldin A to the cells. Brefeldin A inhibits the protein transport to the Golgi complex, resulting in cytokine accumulation inside of the T cell, increasing detectability.
5. On the next day, wash cells twice with PBS.
6. Resuspend cells at a concentration of $1 \times 10^6$ T cells/mL in FACS buffer supplemented with anti-CD8 and/or anti-CD4 antibodies. Incubate cells in the dark for 30 min.
7. Wash cells twice with cold PBS.
8. Resuspend T cells in 4% PFA, and incubate for 20 min at room temperature in the dark.
9. After fixation, wash cells twice with PBS.
10. Resuspend T cells in 0.5% saponin supplemented with anti-IFNγ, anti-TNFα, and anti-IL-2 antibodies (*see* **Note 2**), and incubate in the dark for 30 min (*see* **Note 12**).
11. Wash cells twice with FACS buffer and resuspend in 100 μL FACS buffer for analysis by flow cytometry (Fig. 5).

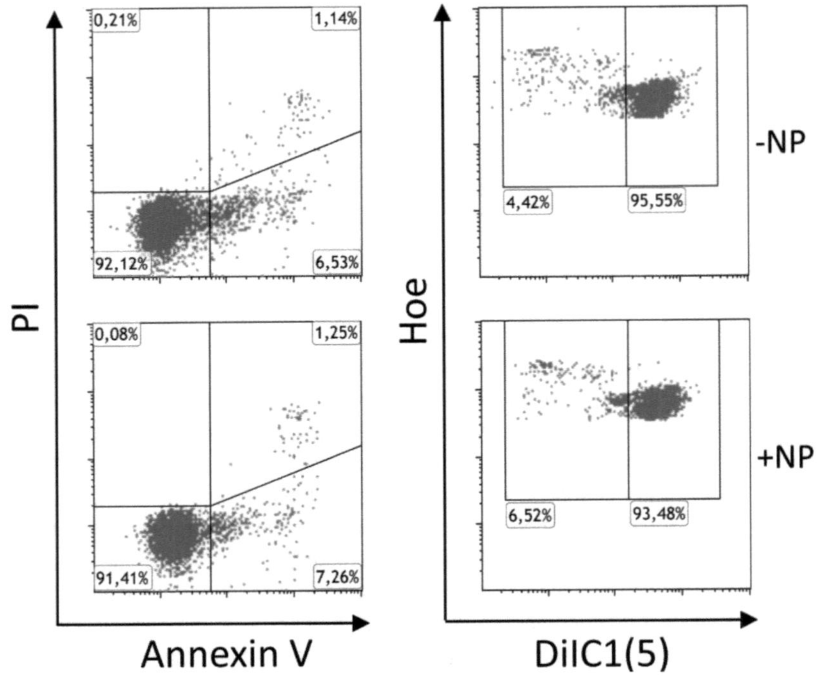


**Fig. 3** Analysis of cell viability after nanoparticle loading and stimulation. Isolated T cells are incubated overnight with (+NP) or without magnetic nanoparticles (−NP). Then, T cells are stained with AxV-Fitc, PI, Hoe, and DiIC$_1$(5) to identify cell viability in flow cytometry. We gate on T cell size and mononuclear cells. The plasma membrane permeable dye Hoe helps us to discriminate between cells (Hoe+) and possible nanoparticle agglomerations (Hoe−). During cell death, cells lose their energy supply, and phosphatidylserine is exposed on the outer layer of the plasma membrane (apoptosis), which can be detected by AxV. In later cell death phases or immediately after severe insults, the plasma membrane ruptures (necrosis), which can be measured by the plasma membrane impermeable dye PI. Viable T cells are considered AxV-Fitc and PI double negative, while apoptotic T cells are AxV-Fitc single positive, and necrotic cells are AxV-Fitc and PI double positive. Viable cells having an active mitochondrial membrane potential are DiIC$_1$(5) positive, while disruption of the mitochondrial membrane potential is associated with apoptosis. Note that AxV needs calcium for binding; thus, staining must be performed in an appropriate buffer (e.g., Ringer's solution)

### *3.5 Analysis of T Activation Markers and Cell Subsets*

1. Adjust loaded T cell concentration to $1 \times 10^6$ cells/mL.
2. Add 25 μL/mL of ImmunoCult™ human CD3/CD28/CD2 T cell activator. Incubate in a 37 °C and 5% $CO_2$ environment for 72 h. Use unstimulated cells as controls (*see* **Note 10**).
3. Resuspend cells at a concentration of $1 \times 10^6$ T cells/mL in FACS buffer supplemented with anti-CD8 and/or anti-CD4.
4. If you want to analyze T cell subsets, additionally stain cells with anti-CCR-7 and anti-CD45RO antibodies (*see* **Note 13** and Fig. 6a).

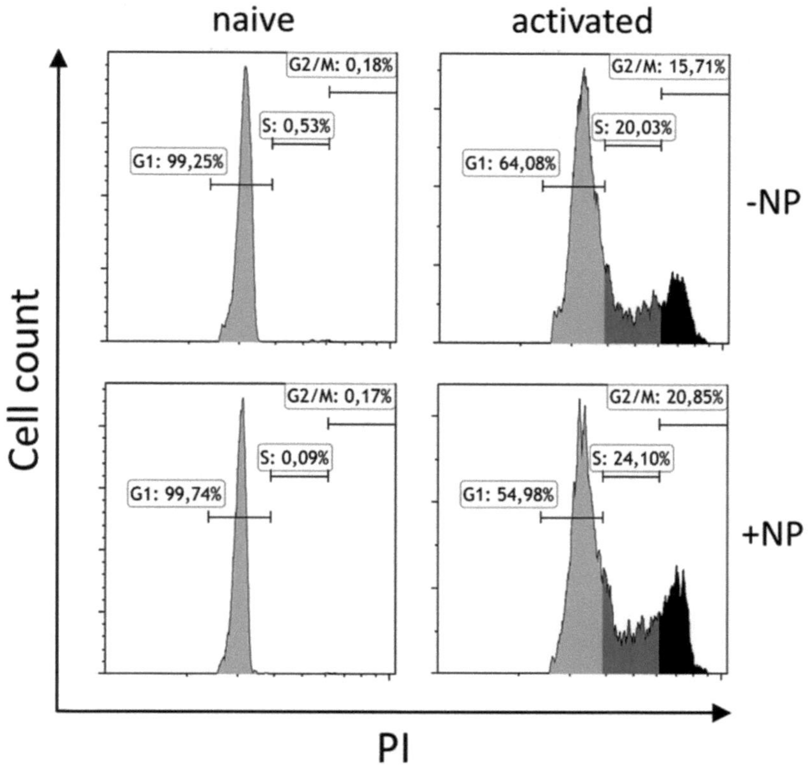


**Fig. 4** Analysis of cell cycle after nanoparticle loading and stimulation. Isolated T cells are incubated overnight with (+NP) or without magnetic nanoparticles (+NP) and stimulated (activated) or not (naïve) for 72 h. Then, cells are incubated in a Triton X-100, sodium citrate, and PI-containing buffer overnight, which permeabilizes the cells and stains the nuclear DNA before flow cytometry. The fluorescence intensity of the stained cells correlates with the amount of containing DNA and enables the discrimination of cell in different cell cycle phases (G1-phase: diploid DNA, S-phase: DNA-replication; G2-phase: double-diploid DNA). Since cells in G2-phase and agglomerations of two cells in G1-phase show the same PI fluorescence, it is mandatory to exclude doublets by plotting PI height versus PI area and exclude cells with higher or disproportional area. Note: If cell cycle cannot be discriminated, it is often helpful to reduce the number of cells and increase the amount of staining buffer. Also, the treatment with RNase A can improve results since PI also stains RNA

5. If you want to analyze T cell activation markers, additionally stain cells with anti-CD25 and anti-CD69 antibodies (*see* **Note 13** and Fig. 6b).
6. Incubate cells in the dark for 30 min at 4 °C.
7. Wash cells twice with cold FACS buffer and resuspend cells in 100 μL cold FACS buffer and immediately analyze by flow cytometry.

### *3.6 Analysis of Cell Proliferation*

1. Adjust loaded T cell concentration to $1 \times 10^6$ cells/mL.
2. Before activation, stain T cells with 1 mL proliferation staining per $1 \times 10^6$ of T cells. Incubate cells for 5 min at room temperature in the dark.

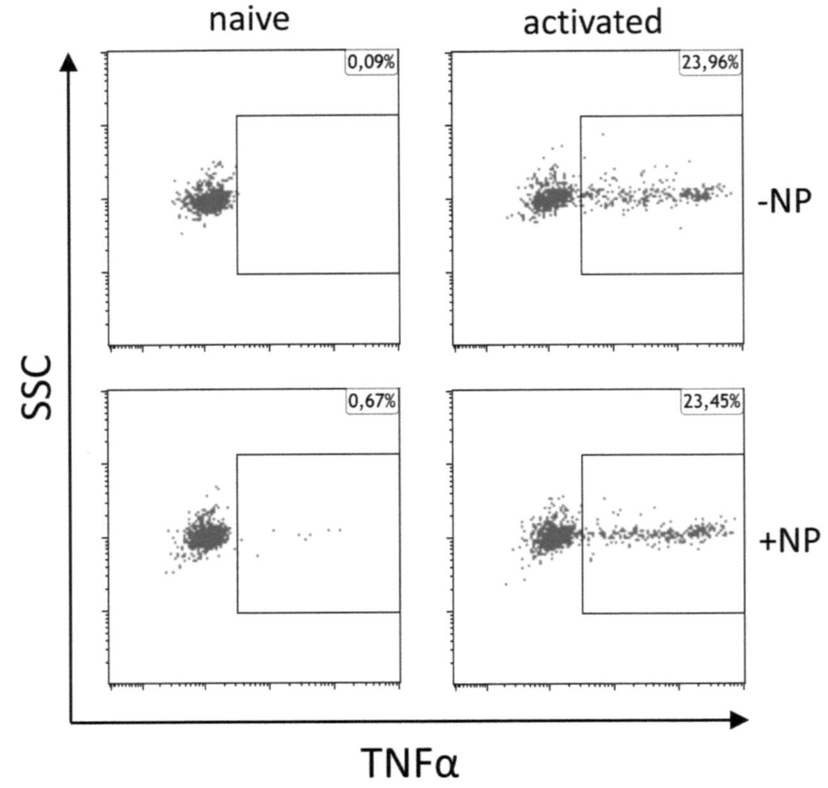


**Fig. 5** Analysis of TNFα-producing activated CD8+ T cells by intracellular cytokine staining after nanoparticle loading and stimulation. Isolated T cells are incubated overnight with (+NP) or without magnetic nanoparticles (−NP). The cells are then stimulated (activated) or not (naïve). 24 h before flow cytometry, the protein transport inhibitor brefeldin A is given to the cells to increase intracellular accumulation of cytokines. To discriminate between T helper and cytotoxic T cells, we stain extracellularly for CD4 and CD8. After fixation and permeabilization of the plasma membrane using saponin, intracellular cytokines are stained by antibodies against TNFα, IFNγ, and IL-2. The depicted dot plots are gated on T cell-sized events and CD8 expression and show the percentage of CD8+ T cells expressing TNFα. SSC: side scatter

3. Stop the staining by the addition of 20 mL T cell medium and wash twice with PBS.
4. Add 25 μL/mL of ImmunoCult™ human CD3/CD28/CD2 T cell activator. Incubate in a 37 °C and 5% $CO_2$ environment for 72 h. Use unstimulated cells as controls (*see* **Note 10**).
5. After the activation, wash twice with PBS. For the analysis of proliferation, stain 100 μL of the sample with 66.6 ng/mL PI for 20 min at 4 °C in the dark. Wash cells once with PBS, and resuspend in FACS buffer. Analyze the mean fluorescence intensity of CFSE using flow cytometry (Fig. 7 and *see* **Note 14**).

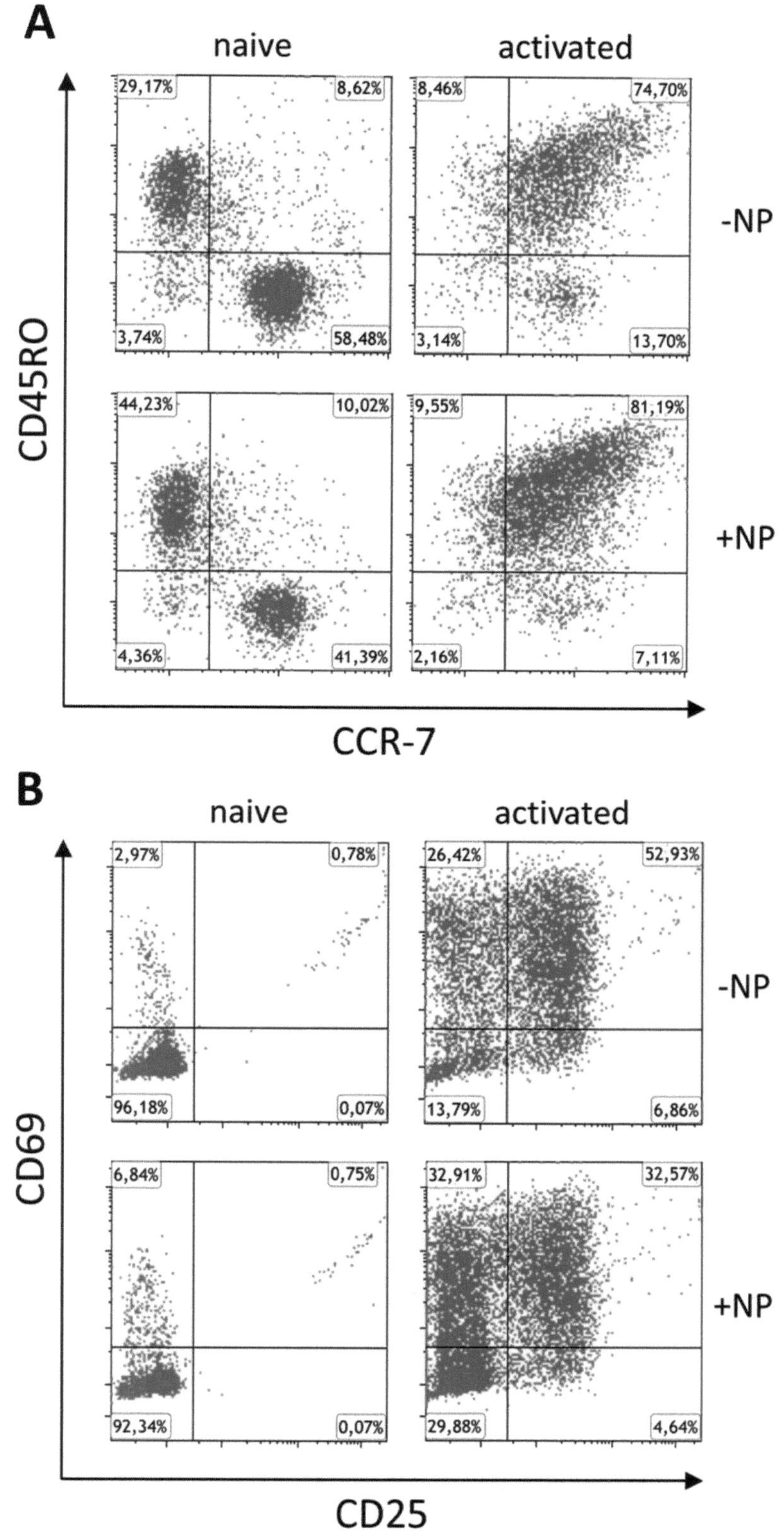


**Fig. 6** Analysis of T cell subsets and activation markers after nanoparticle loading and stimulation. Isolated T cells are incubated overnight with (+NP) or without magnetic nanoparticles (+NP). The cells are then

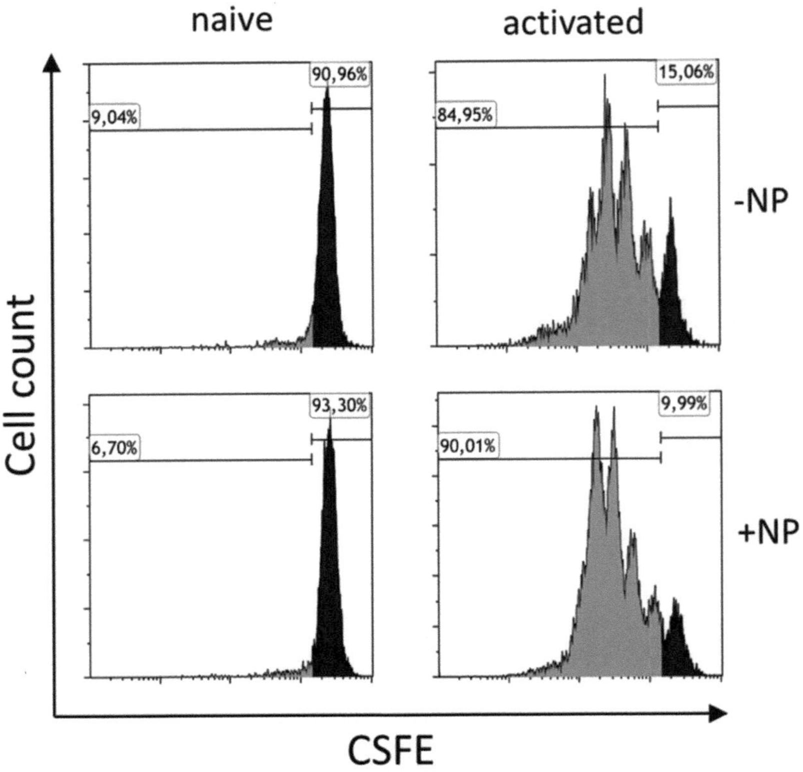


**Fig. 7** Analysis of T cell proliferation and cell cycle after nanoparticle loading and stimulation. Isolated T cells are first stained with cell-permeable CFDA-SE, which becomes impermeable after cleavage into CFSE. Stained cells are incubated overnight with (+NP) or without magnetic nanoparticles (−NP). The depicted histograms are gated on T cell-sized events to analyze CFSE intensity. CFSE is partitioned approximately equally between dividing cells, so that cell division can be analyzed by halving the fluorescence intensity through multiple generations

**Fig. 6** (continued) stimulated (activated) or not (naïve) for 72 h. (**a**) T cell subsets can be identified via the expression of CD45RO and CCR-7. Naïve T cells express only CCR-7 but not CD45RO (CD45RO$^{-}$/CCR-7$^{+}$). Central memory T cells are positive for both CD45RO and CCR7 (CD45RO$^{+}$/CCR-7$^{+}$), while effector memory T cells express only CD45RO (CD45RO$^{+}$/CCR-7$^{-}$). The depicted dot plots are gated on viable T cell-sized events and CD8 expression and show the percentage of CD8+ T cells expressing CD45RO and/or CCR-7. (**b**) Activated T cells express CD25 and CD69. The depicted dot plots are gated on viable T cell-sized events and CD8 expression and show the percentage of CD8+ T cells expressing CD25 and/or CD69

## 4 Notes

1. Ensure that you use an isolation method with which the wanted cells are not magnetically labeled. We isolate T cells using a kit for nonmagnetic cell isolation (e.g., IBA CD3 Fab-TACS Gravity Kit, IBA Lifesciences). Untouched isolation by magnetic labeling of unwanted cells would be also possible.
2. Protect fluorescence-labeled proteins, antibodies, etc. from light during staining and washing steps.
3. Optimal antibody concentration has to be determined by titration. Use cells which have the desired antigen expressed and reduce the concentration of antibodies generally starting at 10 μg/mL and do several two-fold dilutions. Perform the titration under the same conditions as in the experiment (cell number, temperature, incubation time, staining protocol). Use the lowest concentration which enables the best differentiation between negative and positive populations with low background.
4. Use appropriate isotype controls to determine unspecific binding of the antibody or fluorophore to the cells. Isotype antibodies should be stained under the same conditions (isotype, fluorophore, concentration). Cut-offs between background fluorescence and positive populations are determined using a fluorescence-minus one (FMO) control, which is a tube of cells stained with all fluorochromes used in the experiment except one. A multi-color immunofluorescent experiment has one fluorescent minus one control for each fluorochrome.
5. If the used fluorophores show a spectral overlap, fluorescence bleed-through must be eliminated by compensation. Use single-stained samples to determine the levels of compensation for each dye. Avoid channels which have a large amount of fluorescence bleed-through.
6. Nanoparticles can impact the viability and functionality of T cells; therefore, biocompatibility at utilized concentration has to be investigated beforehand. Dilute nanoparticles in sterile $dH_2O$ to achieve preferred concentration, and add them to the T cells. Do not surpass $dH_2O$ volume of more than 10% of the final volume, since this can influence the osmotic pressure of the medium in a negative way.
7. We have observed optimal T cell viability and nanoparticle stability with the described T cell medium. However, if your particle system is not stable at these conditions, we found it helpful to vary FCS concentrations. Additionally, T cell survival can be increased by higher concentrations of IL-7 or the addition of IL-2, although this can induce unwanted T cell activation.

8. Washing of cells is performed by centrifugation (5 min, 400 g) and subsequent resuspension in the described medium. Depending on the size of your nanoparticles, it may also be helpful to reduce the centrifugation speed to avoid nanoparticle precipitation. If nanoparticles are not colloidally stable and thus cannot be removed by washing, choose another method. We have good experience with nonmagnetic chromatographic separation [25].
9. Absolute nanoparticle concentration per cell can be measured by their amount of iron oxide via atomic emission spectroscopy, while relative SPION concentration can be identified via the increased sideward scatter in flow cytometry due to the higher granularity [8, 26].
10. The duration of activation can be adjusted to match your experiment; however, at least 48 h are recommended to observe significant T cell activation.
11. We found it helpful to add 2% v/v FCS to PBS during washing to increase overall T cell viability. This should however be avoided during the viability staining, since albumin in the FCS can bind free $Ca^{2+}$, which is necessary for AxV binding.
12. If your flow cytometer does not allow the simultaneous analysis of five parameters, we recommend splitting the staining of intracellular cytokines into multiple analyses to match your parameters. For example, you can analyze the expression of cytokines separately.
13. If you analyze the viability and have used an additional cell type for stimulation, stain T cells with antibodies against CD3, CD4, or CD8 for absolute and subclass identification.
14. Cell division results in the dilution of intracellular CFSE. Therefore, each division lowers the CFSE intensity measurable per flow cytometry, which is visible as distinct peak. This allows the quantification of divisions each cell has undergone. Analyze unlabeled cells as a control to distinguish cells which have undergone high proliferation from unlabeled cells in the same specimen.

## References


1. Mattiuzzi C, Lippi G (2019) Current cancer epidemiology. J Epidemiol Glob Health 9(4): 217–222. https://doi.org/10.2991/jegh.k.191008.001
2. van den Bulk J, Verdegaal EM, de Miranda NF (2018) Cancer immunotherapy: broadening the scope of targetable tumours. Open Biol 8(6). https://doi.org/10.1098/rsob.180037
3. D'Aloia MM, Zizzari IG, Sacchetti B, Pierelli L, Alimandi M (2018) CAR-T cells: the long and winding road to solid tumors. Cell Death Dis 9(3):282. https://doi.org/10.1038/s41419-018-0278-6
4. Marofi F, Motavalli R, Safonov VA, Thangavelu L, Yumashev AV, Alexander M, Shomali N, Chartrand MS, Pathak Y,

Jarahian M, Izadi S, Hassanzadeh A, Shirafkan N, Tahmasebi S, Khiavi FM (2021) CAR T cells in solid tumors: challenges and opportunities. Stem Cell Res Ther 12(1):81. https://doi.org/10.1186/s13287-020-02128-1

5. Schubert ML, Schmitt M, Wang L, Ramos CA, Jordan K, Muller-Tidow C, Dreger P (2021) Side-effect management of chimeric antigen receptor (CAR) T-cell therapy. Ann Oncol 32(1):34–48. https://doi.org/10.1016/j.annonc.2020.10.478
6. Lee S, Margolin K (2011) Cytokines in cancer immunotherapy. Cancers (Basel) 3(4): 3856–3893. https://doi.org/10.3390/cancers3043856
7. Muhlberger M, Janko C, Unterweger H, Friedrich RP, Friedrich B, Band J, Cebulla N, Alexiou C, Dudziak D, Lee G, Tietze R (2019) Functionalization of T lymphocytes with citrate-coated superparamagnetic iron oxide nanoparticles for magnetically controlled immune therapy. Int J Nanomedicine 14: 8421–8432. https://doi.org/10.2147/IJN.S218488
8. Boosz P, Pfister F, Stein R, Friedrich B, Fester L, Band J, Muhlberger M, Schreiber E, Lyer S, Dudziak D, Alexiou C, Janko C (2021) Citrate-coated superparamagnetic iron oxide nanoparticles enable a stable non-spilling loading of T cells and their magnetic accumulation. Cancers (Basel) 13(16). https://doi.org/10.3390/cancers13164143
9. Munoz LE, Maueroder C, Chaurio R, Berens C, Herrmann M, Janko C (2013) Colourful death: six-parameter classification of cell death by flow cytometry – dead cells tell tales. Autoimmunity 46(5):336–341. https://doi.org/10.3109/08916934.2012.755960
10. Vermes I, Haanen C, Reutelingsperger C (2000) Flow cytometry of apoptotic cell death. J Immunol Methods 243(1–2): 167–190. https://doi.org/10.1016/s0022-1759(00)00233-7
11. Vermes I, Haanen C, Steffens-Nakken H, Reutelingsperger C (1995) A novel assay for apoptosis. Flow cytometric detection of phosphatidylserine expression on early apoptotic cells using fluorescein labelled Annexin V. J Immunol Methods 184(1):39–51. https://doi.org/10.1016/0022-1759(95)00072-i
12. Arenas-Ramirez N, Woytschak J, Boyman O (2015) Interleukin-2: biology, design and application. Trends Immunol 36(12): 763–777. https://doi.org/10.1016/j.it.2015.10.003
13. Cibrian D, Sanchez-Madrid F (2017) CD69: from activation marker to metabolic gatekeeper. Eur J Immunol 47(6):946–953. https://doi.org/10.1002/eji.201646837
14. Wieland E, Shipkova M (2016) Lymphocyte surface molecules as immune activation biomarkers. Clin Biochem 49(4–5):347–354. https://doi.org/10.1016/j.clinbiochem.2015.07.099
15. Picker LJ, Treer JR, Ferguson-Darnell B, Collins PA, Buck D, Terstappen LW (1993) Control of lymphocyte recirculation in man. I Differential regulation of the peripheral lymph node homing receptor L-selectin on T cells during the virgin to memory cell transition. J Immunol 150(3):1105–1121
16. Mahnke YD, Brodie TM, Sallusto F, Roederer M, Lugli E (2013) The who's who of T-cell differentiation: human memory T-cell subsets. Eur J Immunol 43(11):2797–2809. https://doi.org/10.1002/eji.201343751
17. Mangare C, Tischer-Zimmermann S, Riese SB, Dragon AC, Prinz I, Blasczyk R, Maecker-Kolhoff B, Eiz-Vesper B (2019) Robust identification of suitable T-cell subsets for personalized CMV-specific T-cell immunotherapy using CD45RA and CD62L microbeads. Int J Mol Sci 20(6). https://doi.org/10.3390/ijms20061415
18. Ross SH, Cantrell DA (2018) Signaling and function of Interleukin-2 in T lymphocytes. Annu Rev Immunol 36:411–433. https://doi.org/10.1146/annurev-immunol-042617-053352
19. Kak G, Raza M, Tiwari BK (2018) Interferon-gamma (IFN-gamma): exploring its implications in infectious diseases. Biomol Concepts 9(1):64–79. https://doi.org/10.1515/bmc-2018-0007
20. Mehta AK, Gracias DT, Croft M (2018) TNF activity and T cells. Cytokine 101:14–18. https://doi.org/10.1016/j.cyto.2016.08.003
21. Quah BJ, Parish CR (2010) The use of carboxyfluorescein diacetate succinimidyl ester (CFSE) to monitor lymphocyte proliferation. J Vis Exp 44. https://doi.org/10.3791/2259
22. Parish CR (1999) Fluorescent dyes for lymphocyte migration and proliferation studies. Immunol Cell Biol 77(6):499–508. https://doi.org/10.1046/j.1440-1711.1999.00877.x
23. Shen Y, Vignali P, Wang R (2017) Rapid profiling cell cycle by flow cytometry using concurrent staining of DNA and mitotic markers. Bio Protoc 7(16). https://doi.org/10.21769/BioProtoc.2517
24. Crowley LC, Chojnowski G, Waterhouse NJ (2016) Measuring the DNA content of cells in apoptosis and at different cell-cycle stages

by propidium iodide staining and flow cytometry. Cold Spring Harb Protoc 2016(10):pdb.prot087247. https://doi.org/10.1101/pdb.prot087247

25. Mühlberger M, Janko C, Unterweger H, Band J, Schreiber E, Lehmann C, Dudziak D, Lee G, Alexiou C, Tietze R (2019) Non-magnetic chromatographic separation of colloidally metastable superparamagnetic iron oxide nanoparticles and suspension cells. J Chromatogr B 1122-1123:83–89. https://doi.org/10.1016/j.jchromb.2019.05.033

26. Friedrich RP, Janko C, Poettler M, Tripal P, Zaloga J, Cicha I, Durr S, Nowak J, Odenbach S, Slabu I, Liebl M, Trahms L, Stapf M, Hilger I, Lyer S, Alexiou C (2015) Flow cytometry for intracellular SPION quantification: specificity and sensitivity in comparison with spectroscopic methods. Int J Nanomedicine 10:4185–4201. https://doi.org/10.2147/IJN.S82714

# Part II

## Single Cell Assays in 3D Environments

# Chapter 22

# Transwell In Vitro Cell Migration and Invasion Assays

**Calvin R. Justus, Mona A. Marie, Edward J. Sanderlin, and Li V. Yang**

## Abstract

Cell migration and invasion have essential roles in both normal physiology and disease. As such, methodologies to assess cell migratory and invasive capacities are necessary to elucidate normal cell processes and underlying mechanisms of disease. Here, we describe commonly used transwell in vitro methods for the study of cell migration and invasion. The transwell migration assay involves the chemotaxis of cells through a porous membrane after the establishment of a chemoattractant gradient using two medium-filled compartments. The transwell invasion assay involves the addition of an extracellular matrix on top of the porous membrane which only permits chemotaxis of cells which possess invasive properties such as tumor cells.

**Key words** Cell migration, Cell invasion, Transwell assay, Matrigel, Extracellular matrix

## 1 Introduction

Migration of cells, such as leukocytes, stem cells, fibroblasts, and tumor cells, is involved in a variety of physiological and pathophysiological processes [1–10]. As such, cellular migration and invasion assays have been widely employed and made a significant impact in the fields of biology and biomedicine [1, 5, 7, 11, 12].

Numerous assays have been developed to evaluate spontaneous migration or chemotactic response toward a chemoattractant such as chemokines, bioactive lipids, or growth factors [4, 13–21]. Here, we outline the widely utilized transwell migration and invasion assays which involve two medium-filled chambers that are divided by a porous membrane [13–15, 17, 18]. Medium containing the chemotactic agent is added to the bottom chamber followed by the seeding of cells in the top chamber to establish a chemoattractant gradient for directional cellular migration. Following incubation, the extent of cell chemotaxis can be assessed by quantification of adherent cells that have migrated to the basal side of the porous membrane. Additionally, migration of nonadherent cells to the bottom chamber can be quantified. For transwell invasion assays,

Oliver Friedrich and Daniel F. Gilbert (eds.), *Cell Viability Assays: Methods and Protocols*, Methods in Molecular Biology, vol. 2644, https://doi.org/10.1007/978-1-0716-3052-5_22,

addition of an extracellular matrix layer on top of the porous transwell membrane allows for the investigation of cell invasion and migration [13, 14].

## 2 Materials

### 2.1 Cells

1. J774A.1 mouse macrophage/monocyte cell line (TIB-67).
2. Mouse bone marrow-derived macrophages (BMDMs).
3. B16-F10 mouse melanoma cell line (CRL-6475) or any suitable cell line.

### 2.2 Media and Solutions

1. Cell culture medium for J774A.1 macrophage cell line and B16-F10 melanoma cell line: Dulbecco's Modified Eagle's Medium (DMEM) + 10% fetal bovine serum (FBS) + 1% penicillin-streptomycin (5000 U/mL and 5000 μg/mL).
2. Cell culture medium for bone marrow-derived macrophages (BMDMs): DMEM/F12 + 15% L929-conditioned medium (*see* **Note 1**) + 10% FBS + 1% penicillin-streptomycin.
3. Migration buffer: DMEM + 10 mM HEPES (N-[2-hydroxyethyl]piperazine-N′-[2-ethanesulfonic acid]) + 0.1% bovine serum albumin (BSA), pH 7.4.
4. Dulbecco's phosphate-buffered saline (DPBS), $Ca^{2+}$/$Mg^{2+}$ free.
5. 0.25% (w/v) Trypsin-EDTA.
6. Crystal violet solution: 0.2% (w/v) crystal violet acetate in deionized water.
7. 1 μg/mL 4′,6-diamidino-2′-phenylindole, dihydrochloride (DAPI) in phosphate-buffered saline (PBS).
8. 0.4% trypan blue solution.
9. Matrigel extracellular matrix.

### 2.3 Transwell Inserts

1. 6.5 mm diameter inserts, 8.0 μm pore size, tissue culture treated, polycarbonate membrane, sterile polystyrene plates.
2. 6.5 mm diameter insets, 5.0 μm pore size, tissue culture treated, polycarbonate membrane, sterile polystyrene plates (*see* **Note 2**).

### 2.4 Chemoattractant

1. Recombinant complement component 5a (C5a).
2. Stromal cell-derived factor 1α (SDF-1α).
3. NIH3T3 fibroblast-conditioned medium (*see* **Note 3**).

### 2.5 Equipment

1. EVOS Fl-inverted microscope or similar model, used for fluorescence detection.
2. DISCOVER ECHO Revolve 3 microscope or similar model, used for colorimetric detection.
3. Beckman Coulter Allegra X-12 centrifuge or similar model.

## 3 Methods

### 3.1 Cell Culture

1. Prepare cells for transwell cell migration assay by culturing up to 80–90% confluence in a tissue culture plate or a flask prior to initiating the experimental protocol (*see* **Note 4**).
2. For nonadherent cells, pipette 10 mL cell solution into a 15 mL conical tube and proceed with **step** 7.
3. For adherent cells, aspirate culture media from a 10 cm cell culture plate, and gently add 5 mL room temperature (R.T.) phosphate-buffered saline (PBS) without calcium and magnesium to the inside wall of the cell culture plate, swirl gently to cover entire plate, and aspirate and add another 5 mL PBS for an additional wash (*see* **Note 5**).
4. Aspirate PBS solution from the plate, add 2 mL 0.25% Trypsin-EDTA, and swirl gently allowing trypsin to cover the entire plate (*see* **Note 6**).
5. Insert the cell culture plate into an incubator set at 37 °C and 5% $CO_2$, and incubate for 3 min or until cells are detached from the plate.
6. Remove cell culture plate from the incubator, add 3 mL warmed medium supplemented with 10% fetal bovine serum (FBS) to the plate, mix gently with a pipette to further detach cells from the culture plate, and make single-cell suspension. Transfer the 5 mL cell solution into a 15 mL conical centrifuge tube.
7. Centrifuge the conical tube at 335 × *g* for 5 min to pellet the cells.
8. Aspirate the solution leaving the cell pellet undisturbed and subsequently flick the conical tube to loosen the cell pellet.
9. Add 5 mL PBS to the conical tube and pipette gently to mix cells. Repeat **steps** 7 and **8**.
10. Add the migration buffer and pipette gently to mix cell solution.
11. Pipette 15 μL of cell solution into 15 μL of trypan blue in the well of a 96-well plate, pipette up and down four to five times to mix solution, and carefully transfer 10 μL to a hemocytometer. Count live and dead cells. Resuspend viable cells to a concentration of $1 \times 10^6$ cells/mL in the migration buffer (*see* **Note 7**).

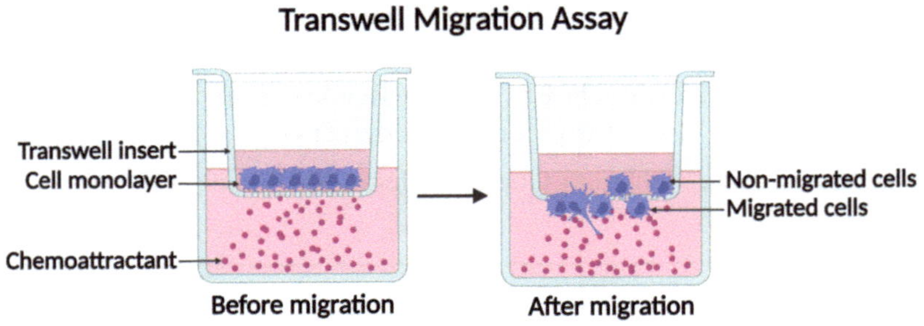


**Fig. 1** Diagram of the transwell cell migration assay. Cells in migration buffer are added into a transwell insert with a porous membrane. A chemoattractant is added into the bottom chamber to form a chemotactic gradient. Cells sense the chemotactic gradient and migrate through the pores of the transwell membrane. (Created with BioRender.com)

### *3.2 Transwell Cell Migration Assay (see Fig. 1)*

1. Carefully pipette 100 μL cell solution containing $1 \times 10^5$ cells onto the membrane of the 24-well transwell insert (5 or 8 μm pore size), and incubate for 10 min at 37 °C and 5% $CO_2$ (*see* **Notes 8** and **9**).
2. Carefully add 600 μL of migration buffer containing chemoattractant (e.g., 5 ng/mL C5a or 10% NIH3T3 fibroblast-conditioned medium) to the well directly below the transwell insert (*see* **Notes 10**, **11**, and **12**).
3. Incubate the plate at 37 °C and 5% $CO_2$ for 2–5 h (*see* **Note 13**).
4. For nonadherent cells, migrated cells will drop into the medium in the bottom chamber. The transwell insert should be carefully removed with forceps, and the migrated cells in the medium of the bottom chamber can be quantified as described in Subheading 3.7.
5. For adherent cells, follow **steps 6–11**.
6. Remove the plate from the incubator (*see* **Note 14**).
7. Using a cotton-tipped applicator, gently remove any non-migrated cells from the apical side of the transwell insert membrane. Repeat to ensure there are no remaining non-migrated cells.
8. Pipette 1 mL of 70% ethanol into a well of a 24-well plate.
9. Place the transwell insert into a well-containing 70% ethanol. Incubate at R.T. for 10–15 min to fix the migrated cells on the basal side of the transwell insert membrane.
10. Place the insert into an empty well to allow the transwell membrane to dry.
11. Proceed with the preferred staining method in Subheadings 3.4 and 3.5.

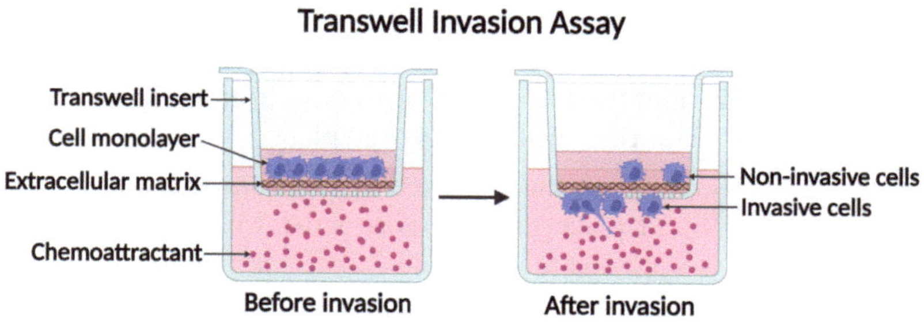


**Fig. 2** Diagram of the transwell cell invasion assay. An extracellular matrix such as Matrigel is added to the apical side of the transwell membrane to form a uniform thin layer of gel. Cells in migration buffer are added into the transwell insert. A chemoattractant is added into the bottom chamber to form a chemotactic gradient. Cells with invasive capability can sense the chemotactic gradient, invade through the extracellular matrix, and migrate through the pores of the transwell membrane. (Created with BioRender.com)

### 3.3 Transwell Cell Invasion Assay (see Fig. 2)

1. Remove Matrigel solution from −20 °C freezer and store in 4 °C refrigerator overnight the day prior to starting the experimental protocol.
2. Keep Matrigel solution on ice while completing the experimental protocol (*see* **Note 15**).
3. Prepare the Matrigel solution by diluting the stock solution with sterile ice-cold deionized water (1:1–1:2), then mix well (*see* **Note 16**).
4. In a biosafety hood, gently add 50–100 μL of diluted Matrigel solution to the membrane of a transwell insert (*see* **Note 17**).
5. Incubate the 24-well plate containing the transwell inserts with Matrigel for 30 min to 1 h at 37 °C and 5% $CO_2$ to allow Matrigel to solidify.
6. Carefully pipette 100 μL cell suspension, such as B16-F10 melanoma cells, in migration buffer onto the Matrigel and incubate for 10 min at 37 °C and 5% $CO_2$ (*see* **Note 18**).
7. Carefully add 600 μL of migration buffer containing chemoattractant (e.g., 10% NIH3T3 fibroblast-conditioned medium or 5 ng/mL SDF-1α) to the well directly below the 24-well transwell insert (*see* **Notes 10**, **11**, and **12**).
8. Incubate the plate at 37 °C and 5% $CO_2$ for 16–24 h (*see* **Note 13**).
9. For nonadherent cells, the transwell insert should be carefully removed with forceps and the migrated cells in the medium of the bottom chamber can be quantified as described in Subheading 3.7.
10. For adherent cells, follow **steps 11–16**.
11. Remove the plate from the incubator. Gently but firmly remove Matrigel from the inside of the transwell insert by use of a cotton-tipped applicator. Repeat with a fresh cotton-

tipped applicator to ensure all Matrigel and remaining non-migratory cells are removed (*see* **Note 14**).

12. Pipette 1 mL PBS into a well and 1 mL 70% ethanol into another well of a 24-well plate.
13. Remove the transwell insert from the well-containing media, and place the insert into the well-containing PBS to briefly wash the membrane.
14. Remove the transwell insert from the well-containing PBS, and place the insert into the well-containing 70% ethanol.
15. Incubate at R.T. for 10–15 min to fix the migrated cells, and subsequently place the insert into an empty well to allow the transwell membrane to dry.
16. Proceed with preferred staining methods in Subheadings 3.4 and 3.5.

### 3.4 Colorimetric Staining for Migrated Adherent Cells: Crystal Violet

1. Place the transwell insert into the well with 0.2% (w/v) crystal violet and incubate for 3–5 min at R.T. (*see* **Note 19**).
2. Remove the transwell insert, and place it into a well-containing PBS or water to wash any remaining crystal violet from the membrane.
3. Repeat the wash with PBS or water to remove excess crystal violet.
4. Allow the membrane to dry by placing the transwell insert into an empty well.
5. Using a microscope, multiple images of the transwell membrane can be taken with a 10× or 20× objective to ensure a representative field of view (FOV) is captured (*see* Fig. 3; *see* **Note 20**).

### 3.5 Fluorescent Staining for Migrated Adherent Cells—DAPI

1. Dilute the stock solution of DAPI (1 mg/mL) by adding 1 μL into 1000 μL PBS, and add 600 μL of diluted DAPI solution (1 μg/mL) into a well of a 24-well plate (*see* **Note 21**).
2. Place the transwell insert into the well with diluted DAPI, and incubate for 10 min while protected from light.
3. Pipette 1 mL of PBS into three separate wells in the 24-well plate.
4. Place the transwell insert into one of the wells containing PBS to wash the transwell insert, and repeat with the other two wells for a total of three washes.
5. Using a fluorescence microscope, images of the transwell membrane can be taken with a 10× or 20× objective to ensure a representative FOV is captured (*see* Fig. 4; *see* **Note 20**).

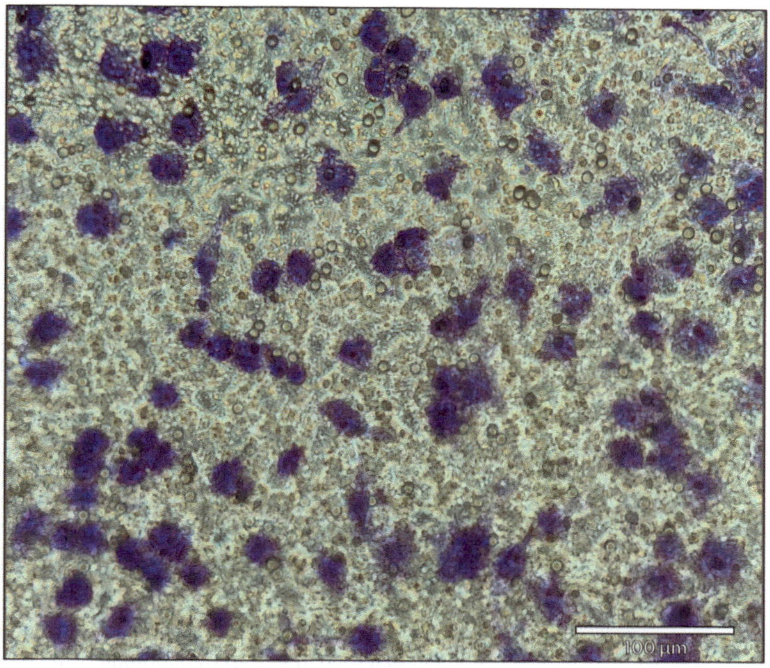


**Fig. 3** Representative image of migrated cells stained with crystal violet. J774A.1 macrophage/monocyte migration using 100 ng/mL recombinant mouse C5a. Migrated cells are stained with 0.2% crystal violet solution. Images of the transwell membrane are taken with a 20× objective using the ECHO Revolve 3 microscope. Migrated J774A.1 cells are stained purple. Scale bar = 100 μm

### 3.6 Quantification of Migrated Adherent Cells

1. As described above, adherent cells migrate through the pores of the transwell membrane and adhere to the basal side of the membrane. The migrated cells can be fixed and stained. Multiple representative images can be taken for quantification of cell migration.
2. Adobe Photoshop's counting tool can be used to enumerate the number of crystal violet- or DAPI-positive cells per FOV.
3. Open each image, and click to label each cell in the FOV.
4. The average number of migrated cells per FOV can be calculated.
5. Alternatively, the multipoint tool of the ImageJ software can be used to count cells per FOV (*see* **Note 22**).

### 3.7 Quantification of Migrated Nonadherent Cells

1. Nonadherent cells migrating through the pores of the transwell membrane will drop into the medium in the bottom chamber. The 600 μL of cell-containing medium in the bottom chamber can be collected for quantification.

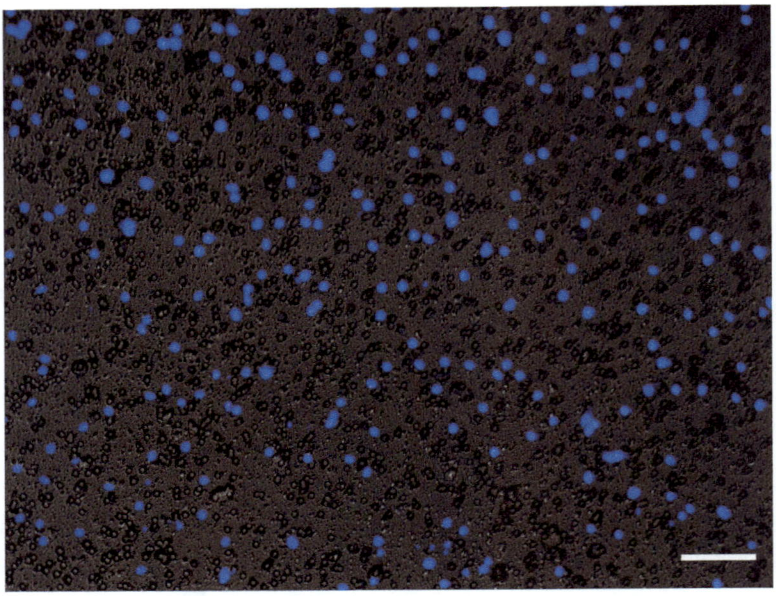

**Fig. 4** Representative image of migrated cells stained with DAPI. Bone marrow-derived macrophage (BMDM) migration using 5 ng/mL recombinant human C5a. Migrated cells are stained with 1 μg/mL DAPI solution. Images of the transwell membrane are taken with a 10× objective using the EVOS Fl fluorescence microscope. The nuclei of migrated BMDMs are stained with DAPI and visualized using a fluorescence microscope. Scale bar = 100 μm

2. Migrated cells in the medium can be directly counted using a hemocytometer. The total number of migrated cells = cells/mL × collected volume. If the cell concentration is low, the samples may be centrifuged to concentrate the cells, then resuspended in a smaller volume of medium and counted using a hemocytometer.
3. Alternatively, a flow cytometer or cell counting instrument (e.g., automated cell counter) can be used to count the migrated cells in the medium.

## 4 Notes

1. L929-conditioned medium is obtained from the mouse fibroblast L929 cell line. L929 cells secrete macrophage colony-stimulating factor (M-CSF) which is required for the differentiation of BMDMs [22, 23].
2. Transwell pore size should be chosen based on cell type and size. Generally, the selected membrane pore size should be smaller than the cell diameter when in suspension. Pilot experiments may be needed to determine the appropriate membrane

pore size, which allows sufficient cell migration toward a chemoattractant while minimizing random cell migration without chemoattractant.

3. NIH3T3 conditioned medium is generated by growing the mouse NIH3T3 fibroblasts in DMEM + 10% FBS to ~95% confluency. The medium is collected and centrifuged to remove cell debris. The conditioned medium contains soluble chemotactic factors secreted by NIH3T3 fibroblasts.
4. J774A.1 cells and B16-F10 cells are cultured using the tissue culture plate. BMDMs are cultured using a non-tissue culture petri dish.
5. PBS without calcium and magnesium may be used to disassociate specific cell types without strong adherence to substrates. J774A.1 cells are semi-adherent and can be mechanically dissociated into single-cell suspension by direct pipetting.
6. Percentage of Trypsin-EDTA solution required to detach cells is cell type specific and should be empirically determined prior to completing the transwell cell migration assay. Prolonged exposure to Trypsin-EDTA can impact cell migration due to cleavage of cell surface receptors. As such, Trypsin-EDTA exposure time should be limited. If a cell surface receptor under investigation is sensitive to trypsin cleavage, a nonenzymatic cell dissociation buffer can be used.
7. Percentage of live and dead cells should be considered prior to continuing with the experimental protocol. It is recommended that the experimental protocol below should only be initiated if cells are more than 95% viable. If cell clumps are observed, a cell strainer can be used to remove the clumps.
8. Be careful not to disturb the transwell membrane with the pipette tip. The transwell membrane pore size and cell number used are dependent on the cell type being investigated. Generally, a 5–8 μm pore size and cell concentration of $1 \times 10^6$ cells/mL are a good starting point.
9. This incubation step is to allow the cells time to settle on the transwell insert membrane.
10. Be careful not to disturb the transwell insert. Slowly add the chemoattractant to the lower chamber to form a chemoattractant gradient. Avoid air bubbles.
11. Ensure that the chemoattractant is in direct contact with the transwell membrane and that there are no bubbles present under the transwell membrane to ensure a proper chemoattractant gradient is present. This is critical for the success of the cell chemotaxis assay.
12. The chemoattractant used and the optimal concentration depend on cell type and should be empirically determined.

13. The migration time depends on cell type and should be empirically determined.
14. A sterile environment is no longer required from this step.
15. Matrigel liquefies at 4 °C or on ice and solidifies at 37 °C.
16. Avoid creating bubbles, and keep all reagents chilled on ice.
17. Carefully pipette Matrigel solution onto the transwell membrane to avoid damaging the membrane. Make sure that Matrigel forms a uniform thin layer of gel on top of the transwell membrane.
18. Be careful not to disturb the transwell membrane or Matrigel with the pipette tip.
19. Other colorimetric dyes, such as hematoxylin, can also be used to stain the migrated cells.
20. Remaining non-migrated cells may be observed at the edge of the apical side of the transwell membrane. Avoid taking images of these areas.
21. Other fluorescent dyes, such as Hoechst 33358, can also be used to stain the migrated cells.
22. ImageJ is a software that can be freely downloaded from https://imagej.nih.gov/ij/.

## Acknowledgments

The work was supported by research grants from the National Institutes of Health (R03CA249473 and R15DK109484, to L.V. Y.). C.R.J., M.A.M., and E.J.S. contributed equally to this work.

### References

1. Eksteen B, Liaskou E, Adams DH (2008) Lymphocyte homing and its role in the pathogenesis of IBD. Inflamm Bowel Dis 14(9): 1298–1312. https://doi.org/10.1002/ibd.20453
2. Bauer A, Tatliadim H, Halin C (2022) Leukocyte trafficking in lymphatic vessels. Cold Spring Harb Perspect Med 12. https://doi.org/10.1101/cshperspect.a041186
3. Liesveld JL, Sharma N, Aljitawi OS (2020) Stem cell homing: from physiology to therapeutics. Stem Cells 38(10):1241–1253. https://doi.org/10.1002/stem.3242
4. Maddaluno L, Urwyler C, Werner S (2017) Fibroblast growth factors: key players in regeneration and tissue repair. Development 144(22):4047–4060. https://doi.org/10.1242/dev.152587
5. Hall A (2009) The cytoskeleton and cancer. Cancer Metastasis Rev 28(1–2):5–14. https://doi.org/10.1007/s10555-008-9166-3
6. Nagano M, Hoshino D, Koshikawa N, Akizawa T, Seiki M (2012) Turnover of focal adhesions and cancer cell migration. Int J Cell Biol 2012:310616. https://doi.org/10.1155/2012/310616
7. Yilmaz M, Christofori G (2010) Mechanisms of motility in metastasizing cells. Mol Cancer Res 8(5):629–642. https://doi.org/10.1158/1541-7786.MCR-10-0139
8. Sanderlin EJ, Leffler NR, Lertpiriyapong K, Cai Q, Hong H, Bakthavatchalu V, Fox JG, Oswald JZ, Justus CR, Krewson EA, O'Rourke D, Yang LV (2017) GPR4 deficiency alleviates intestinal inflammation in a mouse model of acute experimental colitis.

Biochim Biophys Acta 1863(2):569–584. https://doi.org/10.1016/j.bbadis.2016.12.005

9. Krewson EA, Sanderlin EJ, Marie MA, Akhtar SN, Velcicky J, Loetscher P, Yang LV (2020) The proton-sensing GPR4 receptor regulates paracellular gap formation and permeability of vascular endothelial cells. iScience 23(2):100848. https://doi.org/10.1016/j.isci.2020.100848
10. Marie MA, Sanderlin EJ, Satturwar S, Hong H, Lertpiriyapong K, Donthi D, Yang LV (1868) GPR65 (TDAG8) inhibits intestinal inflammation and colitis-associated colorectal cancer development in experimental mouse models. Biochim Biophys Acta Mol basis Dis 2022(1):166288. https://doi.org/10.1016/j.bbadis.2021.166288
11. Amantea D, Nappi G, Bernardi G, Bagetta G, Corasaniti MT (2009) Post-ischemic brain damage: pathophysiology and role of inflammatory mediators. FEBS J 276(1):13–26. https://doi.org/10.1111/j.1742-4658.2008.06766.x
12. Itatani Y, Kawada K, Inamoto S, Yamamoto T, Ogawa R, Taketo MM, Sakai Y (2016) The role of chemokines in promoting colorectal cancer invasion/metastasis. Int J Mol Sci 17(5). https://doi.org/10.3390/ijms17050643
13. Castellone RD, Leffler NR, Dong L, Yang LV (2011) Inhibition of tumor cell migration and metastasis by the proton-sensing GPR4 receptor. Cancer Lett 312(2):197–208. https://doi.org/10.1016/j.canlet.2011.08.013
14. Justus CR, Leffler N, Ruiz-Echevarria M, Yang LV (2014) In vitro cell migration and invasion assays. J Vis Exp 88:e51046. https://doi.org/10.3791/51046
15. Radu CG, Yang LV, Riedinger M, Au M, Witte ON (2004) T cell chemotaxis to lysophosphatidylcholine through the G2A receptor. Proc Natl Acad Sci U S A 101(1):245–250. https://doi.org/10.1073/pnas.2536801100
16. Rodriguez LG, Wu X, Guan JL (2005) Wound-healing assay. Methods Mol Biol 294:23–29. https://doi.org/10.1385/1-59259-860-9:023
17. Justus CR, Sanderlin EJ, Dong L, Sun T, Chi JT, Lertpiriyapong K, Yang LV (2017) Contextual tumor suppressor function of T cell death-associated gene 8 (TDAG8) in hematological malignancies. J Transl Med 15(1):204. https://doi.org/10.1186/s12967-017-1305-6
18. Yang LV, Radu CG, Wang L, Riedinger M, Witte ON (2005) Gi-independent macrophage chemotaxis to lysophosphatidylcholine via the immunoregulatory GPCR G2A. Blood 105(3):1127–1134. https://doi.org/10.1182/blood-2004-05-1916
19. Junger WG (2008) Purinergic regulation of neutrophil chemotaxis. Cell Mol Life Sci 65(16):2528–2540. https://doi.org/10.1007/s00018-008-8095-1
20. Peter C, Waibel M, Radu CG, Yang LV, Witte ON, Schulze-Osthoff K, Wesselborg S, Lauber K (2008) Migration to apoptotic "find-me" signals is mediated via the phagocyte receptor G2A. J Biol Chem 283(9):5296–5305. https://doi.org/10.1074/jbc.M706586200
21. Roussos ET, Condeelis JS, Patsialou A (2011) Chemotaxis in cancer. Nat Rev Cancer 11(8):573–587. https://doi.org/10.1038/nrc3078
22. Zhang X, Goncalves R, Mosser DM (2008) The isolation and characterization of murine macrophages. Curr Protoc Immunol Chapter 14:Unit 14 11 https://doi.org/10.1002/0471142735.im1401s83
23. Ying W, Cheruku PS, Bazer FW, Safe SH, Zhou B (2013) Investigation of macrophage polarization using bone marrow derived macrophages. J Vis Exp 76:e50323. https://doi.org/10.3791/50323

# Chapter 23

# Calcium Imaging of Non-adherent Cells

**Lydia J. Bye, Rocio K. Finol-Urdaneta, and David J. Adams**

## Abstract

Live-cell imaging can reveal dynamic and multimodal cell signaling by monitoring calcium flux. Spatiotemporal changes in $Ca^{2+}$ concentrations instigate specific downstream processes and by categorizing these events, we can examine the language cells use to communicate both to themselves and with each other. Thus, calcium imaging is an understandably popular and versatile technique that relies on high-resolution optical data as measured by fluorescence intensity. This is executed with relative ease on adherent cells, as changes in fluorescence intensity can be monitored over time in fixed regions of interest. However, perfusion of non-adherent or mildly adherent cells leads to their mechanical displacement thereby hindering the spatial resolution of fluorescence intensity changes through time. Here we provide details of a simple and cost-effective protocol using gelatin to prevent cell dislodgement during the solution exchanges that occur during recording.

**Key words** Calcium imaging, THP-1 monocytes, Gelatin, Suspension cells, Fura-2 AM, Fluo-4 AM

## 1 Introduction

Non-adherent cells represent just a fraction of cell types, but they are crucial for immune system function [1]. Circulating blood cells such as lymphocytes, monocytes, erythrocytes, granulocytes, and dendritic cells predominate, but non-adherent cells also comprise lymph nodes, bone marrow, and cord blood cells, gametes, adipocytes, urinary cells, and circulating tumor cells. Monocyte-derived macrophages are largely heterogeneous and can differentiate into both adherent and non-adherent populations. Non-mammalian non-adherent cells include plant protoplasts and yeast cells. Non-adherent cells are such because they lack cadherins that bind to adhesion receptors, or the number is so low they cannot maintain cell adhesion [2]. This poses a particular challenge for techniques that rely on adherence as not all procedures and equipment can be adapted for non-adherent cell purposes. Here we have adapted conventional live-cell imaging for use with non-adherent cells [3].

Oliver Friedrich and Daniel F. Gilbert (eds.), *Cell Viability Assays: Methods and Protocols*, Methods in Molecular Biology, vol. 2644, https://doi.org/10.1007/978-1-0716-3052-5_23,

Live-cell imaging has a wide range of applications from investigating how plants respond to stimuli to measuring action potentials generated within individual neurons. Its many advantages include the ability to observe cell function in real-time and upon stimulation at the single-cell and population levels. However, imaging live non-adherent cells can become challenging due to their propensity to mechanical perturbation.

$Ca^{2+}$ acts as a near-universal second messenger, responding to both external and internal mechanisms through both extracellular influx and mobilization from intracellular stores. $Ca^{2+}$ is highly sequestered because it can be toxic if left unregulated. Upon activation, it flows down its electrochemical gradient from either (or both) extracellular space ($10^{-3}$ M) or intracellular stores ($10^{-4}$ M) into the cytoplasm ($10^{-7}$ M) before it is actively pumped back across the same gradients [4]. Activation of single calcium channels increases intracellular $Ca^{2+}$ concentration locally that in turn triggers neighboring calcium channels and may activate more widespread signaling in various compartments thereby amplifying the originating signal [5].

Imaging $Ca^{2+}$ signals relies on the ability of fluorescent $Ca^{2+}$ indicators to intensify at a given wavelength corresponding to $Ca^{2+}$ concentration. Fluorescent molecules are not limited to $Ca^{2+}$, but due to the nature of $Ca^{2+}$ as a second messenger as well as its distinct amplification effect, $Ca^{2+}$ seems a highly attractive target. Fluorescent indicators were originally modeled off the $Ca^{2+}$ chelating properties of EGTA [6] and have since undergone various iterations. They can be either ratiometric or non-ratiometric, each with its own appeal. Fura-2 is a model ratiometric indicator. Unbound, its excitation spectrum is at 380 nm with emission at 500 nm, which when bound shifts its excitation to 340 nm. By measuring both wavelengths (380 and 340 nm), cytosolic $Ca^{2+}$ can be quantified based on the equation [7]:

$$\left[Ca^{2+}\right]_i = K_d(R - R_{min})/R_{max} - R)(R_{380\ nm}/R_{340\ nm})$$

$R_{max}$ can be calculated by permeabilizing the cell membrane with ionomycin when studying extracellular $Ca^{2+}$ influx, or by discharging intracellular stores with thapsigargin when observing intracellular $Ca^{2+}$ flux. $R_{min}$ can be obtained by flooding the cell with EGTA to mop up excess $Ca^{2+}$ and obtain a near-zero $Ca^{2+}$ fluorescence [8]. Fluo-4 is a single-wavelength fluorescent indicator with a large dynamic range and has a relatively high intensity at its target emission wavelength of 525 nm, with excitation at 480 nm [9]. Unlike Fura-2, Fluo-4 is essentially non-fluorescent until bound to $Ca^{2+}$, and while not as accurate for quantitative measurement of maximum intensity as is Fura-2, Fluo-4 is more time sensitive as there is only one wavelength being recorded as opposed to switching back and forth between wavelengths in the case of Fura-2.

Suspension or non-adherent cells make up only a fraction of cells in culture as reflected in situ. Most cells prefer to be attached to others like them and/or some type of surface to anchor onto. Because of this, adherent cells are ideally suited for imaging and other types of manipulation as they grow and cling fiercely to plastic or glass cover slips. However, suspension cells like red blood cells and lymphocytes are notoriously difficult to image. They can sediment to cover slips to allow for treatments such as immunostaining [10], but this involves cell fixation and is not suitable for live cell imaging. Gelatin is on nearly every laboratory shelf, cost-effective, and not unduly time-consuming to prepare. Here we have aimed at establishing a method for cell immobilization using gelatin as a matrix that allows for the solution perfusion and exchange required for live cell calcium imaging.

## 2 Materials

Use personal protective equipment, and follow handling and waste disposal regulations diligently. Prepare all solutions and reagent stocks using analytical grade reagents.

### 2.1 Materials

1. 12 mm glass coverslips.
2. 24-well plate.
3. 1.5 mL reaction tubes.
4. T75 culture flasks.

### 2.2 Reagent Stocks

1. Fura-2 AM: 10 mM Fura-2 AM in sterile DMSO. Store at −20 °C (*see* **Note 1**).
2. Fluo-4 AM: 10 mM Fluo-4 in sterile DMSO. Store at −20 °C.
3. Pluronic® F-127: 20% (w/v). Store at room temperature.
4. Ionomycin: 10 mM in DMSO. Store at −20 °C.
5. Thapsigargin: 10 mM in DMSO. Store at −20 °C.
6. $K_2$ATP (optional, *see* **Note 2**): 100 mM in $H_2O$. Adjust pH to 7.4, and store at −20°C.
7. Probenecid: 2 mM in 1 M NaOH. Prepare as needed.
8. Trypan blue: 0.4%

### 2.3 Imaging Solutions

All imaging solutions should be sterile filtered using a 0.22 μm nitrocellulose filter and stored at 4°C before use.

1. $Ca^{2+}$-loading solution: 130 mM NaCl, 5.4 mM KCl, 2 mM $MgCl_2$, 10 mM glucose, 10 mM HEPES, 2 mM $CaCl_2$, in 1 L ultrapure water. Add 1 mL of the probenecid stock after all dry components are dissolved, and adjust pH to 7.4 (NaOH).

2. $Ca^{2+}$-buffered imaging solution: 130 mM NaCl, 5.4 mM KCl, 2 mM $MgCl_2$, 10 mM glucose, 10 mM HEPES, and 2 mM $CaCl_2$ in 1 L ultrapure water, pH 7.4 (NaOH).
3. $Ca^{2+}$-free buffered imaging solution: 130 mM NaCl, 5.4 mM KCl, 3 mM $MgCl_2$, 10 mM glucose, 10 mM HEPES, 1 mM EGTA in 1 L ultrapure water, pH 7.4 (NaOH).
4. Ionomycin working solution: 25 μM Ionomycin in $Ca^{2+}$-buffered imaging solution. Store at 4 °C. Keep on ice during experimentation.
5. Thapsigargin working solution: 10 μM thapsigargin in $Ca^{2+}$-free buffered imaging solution. Aliquot and store at −20 °C. Keep aliquots on ice during experimentation.

### 2.4 Cell Culture Medium

RPMI culture medium: 10% heat-inactivated FBS, 1% Pen/Strep, 4 mM L-glutamine in 500 mL RPMI. Store at 4 °C.

### 2.5 5% (w/v) Gelatin Solution

Gelatin solution: 5% porcine gelatin in 1 mL $Ca^{2+}$-buffered imaging solution (*see* **Note 1**). Keep at 37 °C (water bath) until use. Store at 4 °C for up to a week.

## 3 Methods

### 3.1 THP-1 Cell Culture

1. Maintain suspension cells in 15 mL cell culture media on a T75 flask at 37 ± 1 °C, 90 ± 5% humidity, 5 ± 1% $CO_2$. Monitor cells frequently. When cells reach 80% confluence, remove 5–10 mL and replace with fresh media.

### 3.2 Loading Cells with Calcium Indicator

1. On the day of experimentation, remove 100 μL of cell suspension, stain with trypan blue, and count under the microscope using a hemocytometer.
2. Remove the appropriate volume of culture to contain ~$10^6$ cells. Centrifuge cell suspension for 2 min, 300 g at room temperature (*see* **Note 2**). Discard the supernatant and resuspend cell pellet into 1 mL $Ca^{2+}$-loading solution.
3. Add 5 μM of calcium indicator stock and incubate at 37 °C, 5% $CO_2$) for either 30 min (Fura-2 AM) or 45 min (Fluo-4 AM).

### 3.3 Cell Immobilization

1. Remove calcium indicator loading solution by centrifugation (2 min, at 300 g, RT). Aspirate as much supernatant as possible, and gently resuspend cells by pipetting up and down into 50 μL pre-warmed (37 °C) 5% gelatin (*see* **Note 3**).
2. Place one 12 mm glass coverslip per well of a 24-well plate, and pipet one 10 μL droplet of cell-laden gelatin at the center of each coverslip. Place 24-well plate into humidified incubator (37 °C, 5% $CO_2$), and allow cells to sediment to the bottom of the coverslip for 5–10 min.

3. Remove 24-well plate from incubator, and add 1 mL room temperature $Ca^{2+}$-buffered imaging solution to each well. Place 24-well plate in the refrigerator (4–8 °C) for 10 min to speed gelatin polymerization (*see* **Note 4**).

### 3.4 $Ca^{2+}$ Fluorescence Acquisition Experiments

1. Mount the coverslip on the imaging chamber of a fluorescence microscope equipped with a perfusion system (*see* **Note 5**). Carefully flood the chamber by manually pipetting 0.5–1 mL imaging-buffered solution (*see* **Note 6**).
2. Start perfusion at 1.5 mL/min, and verify coverslip is fixed in place. Gelatin should remain stable during constant perfusion flow and solution exchanges (*see* **Note 7**).
3. Program imaging software to the appropriate acquisition settings (*see* **Note 8**):

   Fura-2: Alternate excitation wavelengths at 340 and 387 nm, 1 Hz.

   Fluo-4: Excitation wavelength at 488 nM, 1 Hz, with peak emission at 521 nm.
4. Focus on cells of interest.
5. Commence imaging and acquire fluorescence data until a stable baseline is reached, typically 5–10 min (Fig. 1b(a) and *see* **Note 9**).
6. Add modulator (e.g., $K_2ATP$) while acquiring fluorescence images (*see* **Note 10**). At 300 s (Fig. 1b(b)), we added 1 mM $K_2ATP$ and observed a sharp rise in fluorescence intensity.
7. Elicit maximum calcium signal if using $Ca^{2+}$ imaging solution: permeabilize cell membranes with 2.5 μM ionomycin and allow the signal to reach a maximum plateau (typically 5–10 min). *See* Fig. 1b (Iono).
8. Elicit maximum calcium signal if using $Ca^{2+}$-free imaging solution: empty intracellular $Ca^{2+}$ stores by treatment with 1 μM thapsigargin, and wait 2 min before ending the experiment.
9. Data Analysis.
   1. Draw regions of interest around cells imaged in automated image analysis software (*see* **Note 11**). If imaging with Fluo-4 AM, data will consist of a single measurement at 525 nm per time point; if using Fura-2 AM, data will consist of paired fluorescence measurements at 340 and 380 nm per time point.
   2. If using Fura-2 AM, fluorescence traces acquired at 340 and 380 nm need to be divided (F340/F380) at this ratio conveys the change in $Ca^{2+}$ concentration.

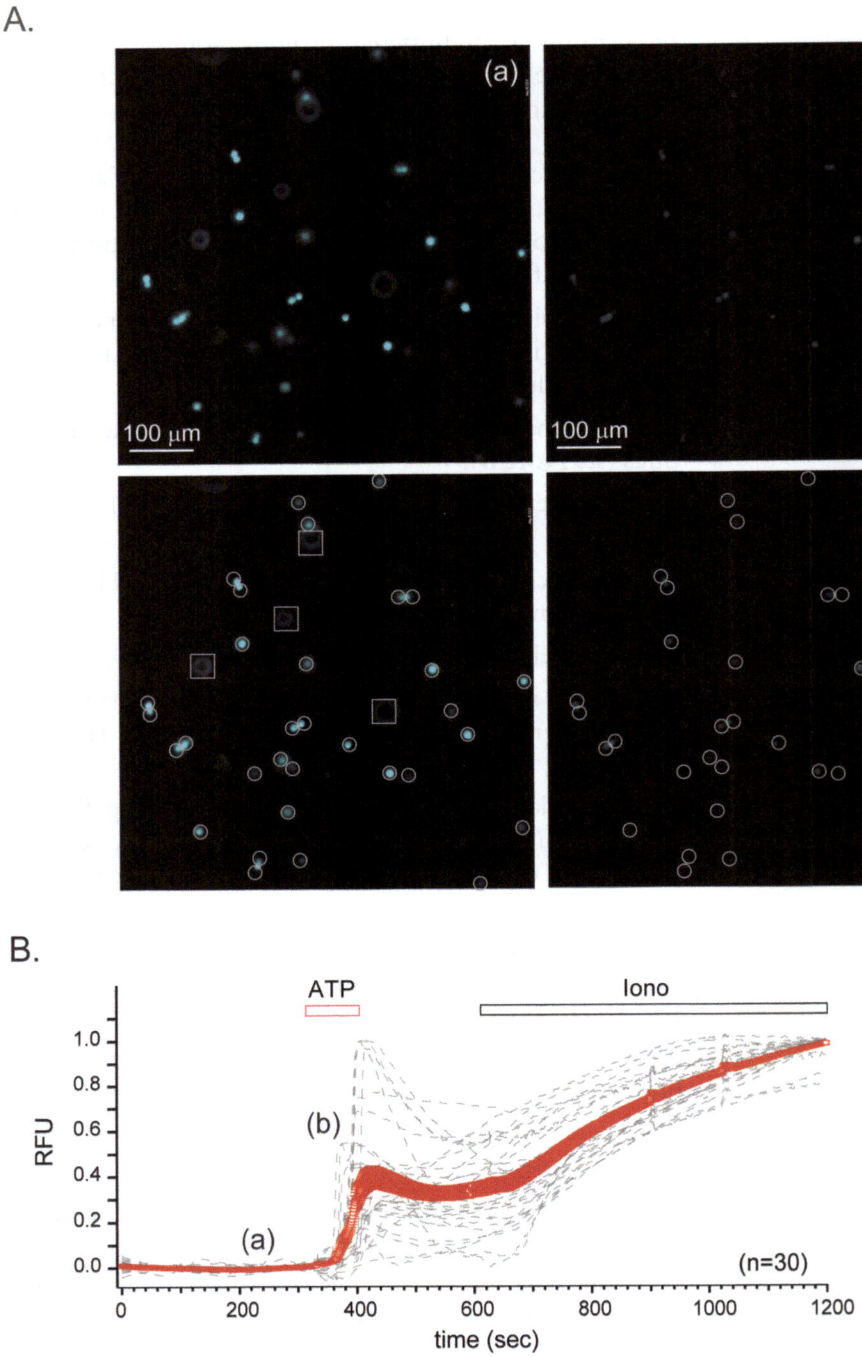


**Fig. 1** Representative imaging experiment from gelatin immobilized THP-1 cells exposed to ATP and Ionomycin. (**a**) Left: Fura-2 fluorescence image of THP-1 cells upon application of 1 mM ATP, at time (a). Right: Fura-2 fluorescence image in $Ca^{2+}$ imaging solution at time (b). Bottom panels show regions where fluorescence intensity was measured within regions of interest (ROIs) selected (circles). Rejected ROIs from air bubbles or blurred cells laying at different focal depths are shown inside squares. (**b**) Time course of fluorescence intensity changes in control (a) and in the presence of ATP (b). 2.5 μM ionomycin (Iono) application elicits maximal fluorescence response and is used for normalization

3. Baseline recording. Measure the fluorescence intensity within a region where it remained stable (somewhere during the first 5–10 min of recording) and use it to substract the whole trace, effectively making this level of intensity = 0.
4. Normalize to $R_{max}$. Divide each trace by the maximal intensity ($R_{max}$) recorded in the presence of Ionomycin or thapsigargin (*see* **Note 12**).

## 4 Notes

1. Or microwave at 3 s intervals pipetting up and down after each interval, until fully dissolved.
2. Fluorescent indicators are light sensitive, aliquot in amber vials and use in a darkened room away from direct light sources.
3. Use any compound of interest. We used $K_2ATP$ because it activates P2X receptor channels present in THP-1 cells [11].
4. It is important to pipette carefully to avoid creating bubbles. Bubbles in the gelatin are visible under the microscope and interfere with image acquisition.
5. Cells can be stored in the fridge for up to 1.5 h. Remove cells 15 min prior to imaging to bring up to room temperature.
6. Observe that the chamber does not overflow to avoid saline spills into the microscope.
7. We used a gravity-fed perfusion system for inlet and a peristaltic pump-driven outlet flow.
8. A tiny daub of vacuum grease can be used to hold the coverslip in place.
9. Fluctuations in the fluorescence signal are pronounced at the beginning of the recordings and tend to stabilize in a few minutes. Provide sufficient time to allow the signal to reach a steady state.
10. It is preferable to apply modulators through the perfusion system. Alternatively, the volume of modulator needed can be minimized by direct addition to the perfusion chamber. If direct addition is required, be careful not to introduce artifacts by interfering with the light sources or mechanically dislodge the gelatin.
11. We used LASX software (Leica Microsystems) for data acquisition.
12. Relative maximums can provide useful data to compare efficacies of certain drugs as to what percentage of calcium has been mobilized.

To quantify the $Ca^{2+}$ concentration from:

1. Fura-2 AM data use:

$$[\mathrm{Ca}] = K_\mathrm{d} \times \frac{(R - R_\mathrm{min})}{(R_\mathrm{max} - R)} \times \frac{\mathrm{Sf2}}{\mathrm{Sb2}} \quad [6]$$

Under standard experimental conditions $K_d$ can be assumed to be 225 nM [7]. Alternatively, $K_d$ can be determined through calcium calibration by measuring excitation/emission at known calcium concentrations. Sf2 corresponds to Fura-2 AM excitation/emission when $Ca^{2+}$ is not bound ($[Ca^{2+}] \sim 10^{-8}$ M), and Sb2 is Fura-2 AM excitation/emission when fully bound with $Ca^{2+}$ ($[Ca^{2+}] \sim 10^{-3}$ M). $R_{max}$ is the experimental maximum of interest, and $R_{min}$ is the experimental minimum or baseline.

2. Fluo-4 AM data use:

$$\frac{\Delta[\mathrm{Ca}^{2+}]}{K_\mathrm{D}} = \frac{F_\mathrm{max}}{F_0}\left(1 - R_\mathrm{f}^{-1}\right)\frac{\delta F}{(\delta F_\mathrm{max} - \delta F)\delta F_\mathrm{max}}$$
$$\frac{[\mathrm{Ca}^{2}]_\mathrm{rest}}{K_\mathrm{D}} = \frac{\left(1 - R_\mathrm{f}^{-1}\right)}{\delta F_\mathrm{max}} - R_\mathrm{f}^{-1} \quad [13]$$

Under normal experimental conditions the $K_d$ for Fluo-4 AM can be assumed to be ~300 nM [12]. Alternatively, $K_d$ can be determined through calcium calibration by measuring excitation/emission at known $Ca^{2+}$ concentrations. The calcium concentration at rest must first be determined before the change in calcium concentration can be used to calculate calcium concentration at point of interest. $F_{max}$ is fluorescence intensity at saturation while $F_0$ is fluorescence intensity at minimum $[Ca^{2+}]$. $R_f$ is the dynamic range of the indicator or $F_{max}/F_{min}$ while $\delta F$ is the change in fluorescence from baseline.

## References


1. Beutler B (2004) Innate immunity: an overview. Mol Immunol 40(12):845–859. https://doi.org/10.1016/j.molimm.2003.10.005
2. French JJ, Cresswell J, Wong WK, Seymour K, Charnley RM, Kirby JA (2002) T cell adhesion and cytolysis of pancreatic cancer cells: a role for E-cadherin in immunotherapy?. Br J Cancer 87(9):1034–1041. https://doi.org/10.1038/sj.bjc.6600597
3. Ritter P, Bye LJ, Finol-Urdaneta RK, Lesko C, Adams DJ, Friedrich O, Gilbert DF (2020) A method for high-content functional imaging of intracellular calcium responses in gelatin-immobilized non-adherent cells. Exp Cell Res 395(2):112210. https://doi.org/10.1016/j.yexcr.2020.112210
4. Bagur R, Hajnóczky G (2017) Intracellular $Ca^{2+}$ sensing: its role in calcium homeostasis and signaling. Mol Cell 66(6):780–788. https://doi.org/10.1016/j.molcel.2017.05.028
5. McCarron JG, Saunter C, Wilson C, Girkin JM, Chalmers S (2018) Chapter 9 – Mitochondria structure and position in the local control of calcium signals in smooth muscle cells. In: Trebak M, Earley S (eds) Signal transduction and smooth muscle. CRC Press/Taylor & Francis© LLC., Oxon, pp 173–190. https://doi.org/10.1201/b20699

6. Russell JT (2011) Imaging calcium signals in vivo: a powerful tool in physiology and pharmacology. Br J Pharmacol 163(8):1605–1625. https://doi.org/10.1111/j.1476-5381.2010.00988.x
7. Grynkiewicz G, Poenie M, Tsien RY (1985) A new generation of $Ca^{2+}$ indicators with greatly improved fluorescence properties. J Biol Chem 260(6):3440–3450. https://doi.org/10.1016/S0021-9258(19)83641-4
8. Tanaka M, Matsui T (2021) Chapter 11 – Methodologies for investigating the vasorelaxation action of peptides. In: Toldrá F, Wu J (eds) Biologically active peptides. Academic Press, pp 255–274. https://doi.org/10.1016/B978-0-12-821389-6.00011-X
9. Gee KR, Brown KA, Chen WN, Bishop-Stewart J, Gray D, Johnson I (2000) Chemical and physiological characterization of fluo-4 $Ca^{2+}$-indicator dyes. Cell Calcium 27(2):97–106. https://doi.org/10.1054/ceca.1999.0095
10. Tsang M, Gantchev J, Ghazawi FM, Litvinov IV (2017) Protocol for adhesion and immunostaining of lymphocytes and other non-adherent cells in culture. BioTechniques 63(5):230–233. https://doi.org/10.2144/000114610
11. Syed NH, Kennedy C (2012) Pharmacology of P2X receptors. Wiley Interdiscip Rev Membr Transp Signal 1(1):16–30. https://doi.org/10.1002/wmts.1
12. Maravall M, Mainen ZF, Sabatini BL, Svoboda K (2000) Estimating intracellular calcium concentrations and buffering without wavelength ratioing. Biophys J 78(5):2655–2667. https://doi.org/10.1016/s/0006-3495(00)76809-3

# Chapter 24

# FUCCI Reporter Gene-Based Cell Cycle Analysis


## Lena Fischer and Ingo Thievessen



## Abstract

One of the most commonly assessed parameters in cellular analyses is the proliferative activity of a cell population. The fluorescence ubiquitin cell cycle indicator (FUCCI)-based system allows live and in vivo observation of cell cycle progression. Based on the mutually exclusive activity of two fluorescently labeled proteins cdt1 and geminin during the G0/1 and S/G2/M phases of the cell cycle, individual cells can be assigned to their respective cell cycle phase by fluorescence imaging of the nucleus. Here, we describe the generation of NIH/3T3 cells containing the FUCCI reporter system by lentiviral transduction and their use in 3D culture assays. The protocol can be adapted to other cell lines.



**Key words** Cell cycle analysis, Cell proliferation, FUCCI, Geminin, Cdt1, Lentivirus


## 1 Introduction

Traditionally, the study of cell cycle progression uses immunofluorescence or biochemical analyses of cell populations for proliferation markers, such as Ki67, or DNA incorporation of fluorescent dyes, such as bromodeoxyuridine (BrdU), with subsequent fluorescence imaging. While these methods generally do not require extensive pretreatment of the samples prior to harvesting, they involve time-consuming and often complex biochemical sample preparation or fixation procedures during harvesting [1]. In contrast, the FUCCI cell cycle reporter system applies the simultaneous expression of markers for different cell-cycle phases, labeled with different fluorescent proteins. It allows cells to be observed live and in vivo. Expression can be achieved either transiently via lipofection or adenoviral infection of a target cell population, or stably via lentiviral transduction and genomic integration of the cDNAs encoding for the fluorescent marker proteins. Straightforward fluorescence imaging of live samples without pretreatment or sample preparation/fixation allows for efficient assessment of cell proliferation, which, in addition, can be repeatedly performed on the same sample.


Oliver Friedrich and Daniel F. Gilbert (eds.), *Cell Viability Assays: Methods and Protocols*, Methods in Molecular Biology, vol. 2644, https://doi.org/10.1007/978-1-0716-3052-5_24,

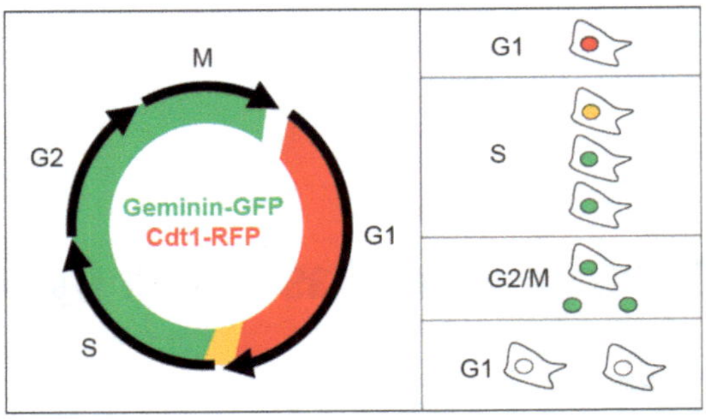


**Fig. 1** Schematic depiction of the FastFUCCI reporter system. Fluorescently labelled geminin (GFP) and cdt1 (RFP) levels alternate through the cell cycle phases. High cdt1 levels during G1 phase are associated with a red colour of the nucleus, while high geminin levels during S, G2 and M phases lead to a green colour of the cellular nucleus. Between G1 and S phases, the abundance of both cdt1 and Geminin can result in a yellow-coloured nucleus

The FUCCI system uses the reciprocal activity of cdt1 and geminin during the cell cycle to visualize the G1/S transition [1] (Fig. 1). Cdt1 is a licensing factor for DNA replication and therefore most abundant from the G0/G1 to S phase. Following DNA synthesis in the S phase, its levels start to drop [2]. This decrease in cdt1 levels is mediated by its inhibitor geminin which is primarily present from S to G2 phase to prevent repeated replication of DNA [3]. In addition to geminin-mediated inhibition of cdt1, the levels of both proteins are controlled by sequential activation of the E3 ubiquitin ligases APC/C$^{Cdh1}$ and SCF$^{Skp2}$ during G1 and S and G2 phases, respectively [2, 4], which in turn control each other and thereby oscillate during the cell cycle as well [5]. In the FUCCI reporter system, cdt1 and geminin are labelled with different fluorescent proteins (e.g., kusabira-orange and azami-green), resulting in either red fluorescence of the cells in G0/1 phase or green fluorescence in S1/G2/M phase. This allows for the assignment of each individual cell to its respective cell cycle stage based on the colour of its nucleus. In addition, the fraction of S/G2/M phase cells (geminin-positive) versus G0/1 phase cells (cdt1-positive) can serve as a measure for the proliferative activity of the whole cell population and can easily be assessed in live samples by the ratio of green to red cells.

The FUCCI system can be applied to various research questions, such as the relationship between other cellular processes and cell cycle stages, or the simple distinction between actively proliferating and quiescent cells, for instance in the context of pharmaceutical treatment [6]. The FastFUCCI reporter system allows simultaneous delivery of both fluorescent transgenes, cdt1-kusabira orange and geminin-azami green, into cells by lentiviral transduction [7]. This allows for a stable expression of the reporter system in many cell lines as well as primary cell types, facilitating straightforward assessment of the proliferative activity of live cells, which can be performed repeatedly on the same sample, without time-consuming sample fixation and preparation procedures. In this chapter, we provide a step-by-step guide to demonstrate the generation of NIH/3T3 reporter cells stably expressing the FastFUCCI reporter system by lentiviral transduction.

## 2 Materials

### 2.1 General Equipment

1. Fluorescence microscope-system.
2. Fluorescence imaging filter: RFP/TRITC/cy3/Alexa532 or similar; GFP/FITC/cy2/Alexa488 or similar.
3. Sterile hood.
4. Centrifuge.
5. Micropipettes with pipette tips.
6. Cell culture incubator (37 °C, 90% humidity, 5% $CO_2$).
7. Hemocytometer or cell counter.
8. Cell culture-treated flasks (75 $cm^2$ surface area).
9. 12-well plate.
10. 6-well plate.
11. 100 × 20 mm cell culture dish.
12. 15 mL conical centrifuge tube (15 mL Falcon tube).
13. 50 mL conical centrifuge tube (50 mL Falcon tube).
14. Parafilm.

### 2.2 Cell Lines

1. NIH/3T3 cells (CRL 1658, ATCC).
2. Lenti-X 293 T cells (Takara) or other HEK-based cell lines.

### 2.3 Cell Culture Media

1. Dulbecco's Modified Eagle's Medium (DMEM): 4500 mg/L glucose, 4.0 mM L-glutamine, 1.0 mM sodium pyruvate.
2. Bovine calf serum (BCS).
3. Fetal bovine serum (FBS/FCS).
4. Penicillin/streptomycin (100×).

### 2.4 Solutions and Reagents

1. Phosphate-buffered saline ($Ca^{2+}$ and $Mg^{2+}$ free), sterile solution (PBS).
2. 0.25% trypsin/EDTA.
3. Puromycin.
4. Lipofectamine 2000: 3:1 mixture of DOSPA (2,3-dioleoyloxy-N- [2(sperminecarboxamido)ethyl]-N,N-dimethyl-1-propaniminium trifluoroacetate and DOPE (1,2-dioleoyl-*sn*-glycerophosphoethanolamine).

5. OptiMEM.
6. Lenti-X concentrator (Takara): Polyethylene glycol 400 (PEG 400)-based.
7. Hexadimethrine bromide (polybrene): 1 mg/mL polybrene in 0.9% NaCl.

### 2.5 Lentiviral Plasmids

1. Transfer plasmid: pBOB-EF1-FastFUCCI-Puro (Addgene), amplified and isolated.
2. Packaging plasmid: psPAX.2 (Addgene), amplified and isolated.
3. Envelope plasmid: pCMV-VSV-G (Addgene), amplified and isolated.

### 2.6 Analysis Software

1. ImageJ [8].

## 3 Methods

Preparative work includes the ordering, amplification and isolation of lentiviral plasmid DNA (transfer plasmid, packaging plasmid, envelope plasmid), as well as thawing and cultivation of all cell lines needed (Lenti-X 293 T, NIH/3T3). All procedures should be carried out at room temperature and in the sterile environment of a laminar flow cabinet unless stated otherwise. Furthermore, since lentivirus are infectious and regarded as biosafety level-2-material, all steps must be performed under safety precautions and in biosafety level-2 BSL-2 approved or similarly categorized facilities. The protocol can be adapted to different cells as well.

### 3.1 Cell Culture, Harvesting, and Counting

1. Prepare complete NIH/3T3 culture medium by supplementing DMEM with 10% BCS and 1% penicillin/streptomycin. Prepare complete Lenti-X 293 T culture medium by supplementing DMEM with 10% FBS/FCS and 1% penicillin/streptomycin.
2. Culture cells in 8 mL complete cell culture medium per T75 flask at 37 °C, 5% $CO_2$ in a humidified incubator. Replace medium every second day. Subculture cells at 80% confluency.
3. Place complete cell culture medium and PBS in water bath prior to pre-warm to 37 °C.
4. Remove complete cell culture medium from flask, and wash cells with 8 mL PBS to remove residual cell culture medium which might block trypsin/EDTA activity.
5. Aspirate PBS, and add 1 mL trypsin/EDTA to detach cells from the flask surface.

6. Incubate the flask with trypsin/EDTA for 2–5 min in the incubator until all cells have detached.
7. Rinse flask with 5 mL pre-warmed complete cell culture medium, and resuspend cells by gently pipetting up and down for approx. five times. Transfer cell suspension into Falcon tube (*see* **Note 1**).
8. Count cells using a hemocytometer or cell counter. To subculture cells, seed appropriate aliquots of the cell suspension into fresh T75 flasks, and fill up to 8 mL with complete cell culture medium.

### 3.2 Determination of Puromycin Concentration

1. Harvest, count and seed 7.5 × $10^4$ NIH/3T3 cells per well into 11 wells of a 12-well cell culture plate. Fill up with complete NIH/3T3 culture medium to 1 mL per well. Incubate cells for 24 h at 37 °C, 5% $CO_2$.
2. After 24 h add puromycin at increasing concentrations (1–10 μg/mL) to 10 wells, containing NIH/3T3 cells. Spare one well to have a non-treated control. Incubate cells for at least 48–72 h in cell culture medium containing the selective antibiotic. The goal is to find the lowest concentration which kills all seeded cells. This concentration will be used after lentivirus transduction to kill all non-transduced cells (*see* **Note 2**).

### 3.3 Generation and Concentration of Lentivirus (Transfection of LentiX 293 T Cells)

Here, we describe the transfection of Lenti-X 293 T cells to produce lentivirus carrying the FastFUCCI reporter system (Fig. 2a). This protocol can be applied to other HEK-based cell lines.

1. One day before transfection, harvest, count and seed (*see* Subheading 3.1) 4–5 × $10^6$ Lenti-X 293 T cells on a 100 × 20 mm cell culture dish. Add 8 mL complete LentiX 293 T culture medium, and incubate cells for 24 h at 37 °C, 5% $CO_2$.
2. On the day of transfection, prepare antibiotic-free Lenti-X 293 T culture medium by supplementing DMEM with 10% FBS/FCS.
3. Prepare the DNA transfection mixture by adding 9.9 μg pBOB-EF1-FastFUCCI-Puro (transfer plasmid), 8.6 μg psPAX.2 (packaging plasmid) and 5.2 μg pCMV-VSV-g (envelope plasmid) (*see* **Note 3**) to 1.5 mL OptiMEM-reduced serum medium and mix gently. Incubate for approx. 5 min.
4. Prepare the transfection reagent mixture: Add 45 μL of lipofectamine to 1.5 mL OptiMEM-reduced serum medium, and mix gently (lipofectamine mixture). Incubate for approx. 5 min. Lipofectamine is a cationic lipid which forms a transfection complex with nucleic acids and enables their cellular uptake through the interaction of its positively charged lipid head groups and the negatively charged cell membrane.

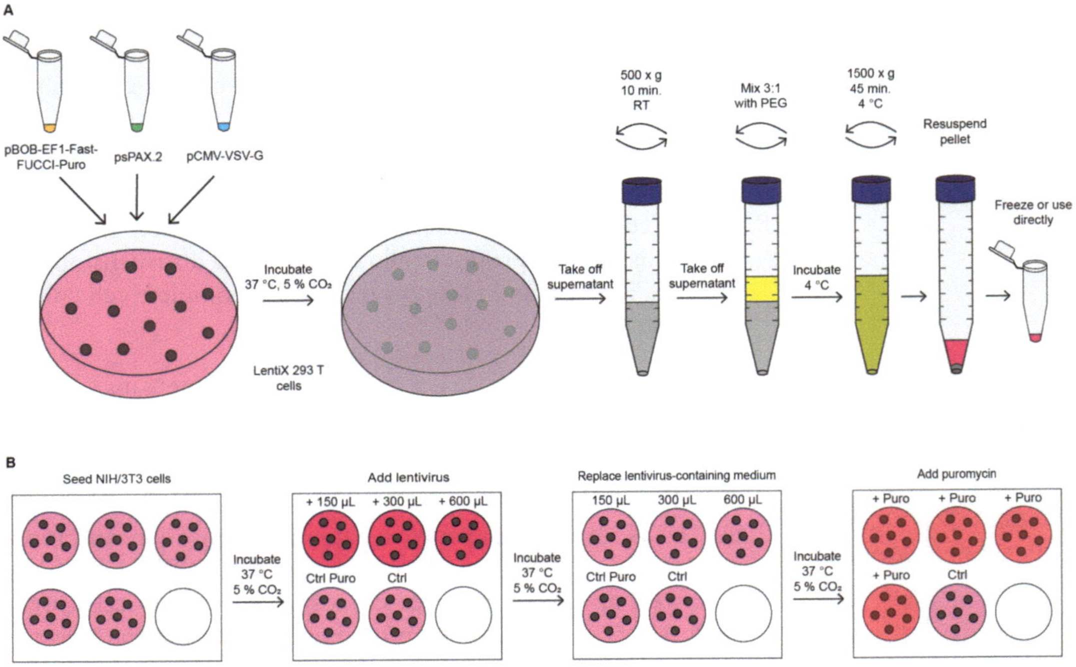


**Fig. 2** Generation of lentiviral particles containing the FastFUCCI reporter system, and transduction of NIH/3T3 target cells. (**a**) Transfection of Lenti-X 293 T cells seeded on a 100 × 20 mm cell culture dish with lentiviral plasmids. After 48 h of incubation, the lentivirus-containing supernatant is briefly centrifuged and mixed with the Lenti-X concentrator (PEG-based). After incubation at 4 °C overnight, the precipitated lentivirus is centrifuged and the resulting pellet resuspended in cell culture medium. (**b**) NIH/3T3 cells are pre-seeded into five wells of a 6-well plate. After 24 h of incubation, cells are transduced with different amounts of lentivirus. Following incubation, the lentivirus-containing medium is replaced, and after a further incubation step, the puromycin-based selection of infected is started

5. Mix 1.5 mL DNA transfection mixture and 1.5 mL transfection reagent mixture and incubate for at least 15–20 min.
6. Replace the complete cell culture medium from pre-seeded Lenti-X 293 T cells with 5 mL of antibiotic-free cell culture medium.
7. When the incubation time of the lipofectamine/DNA transfection mixture is over, add 3 mL lipofectamine/DNA transfection mixture to the cell culture dish containing Lenti-X 293 T cells.
8. Incubate cells with transfection mixture for 48 h at 37 °C, 5% $CO_2$. At this time, Lenti-X 293 T cells will generate lentivirus based on the genetic information encoded by the transfected plasmids.
9. Transfer lentivirus-containing supernatant from transfected Lenti-X 293 T cells after 48 h of incubation into a 15 mL Falcon tube, and briefly centrifuge at 500 × *g* for 10 min or filter through a 0.45 μm filter to remove cell debris.

10. Following centrifugation, transfer clear supernatant to a fresh 50 mL Falcon tube, and add Lenti-X concentrator in a 1:3 (Lenti-X concentrator/supernatant) ratio. Mix gently by inverting the tube and seal the lid with parafilm.
11. Incubate the Lenti-X concentrator/Lentivirus mixture at 4 °C overnight. The PEG-containing Lenti-X concentrator will precipitate the lentivirus which can then be concentrated without the need for ultracentrifugation.
12. Centrifuge precipitated lentivirus at 1500 × *g* speed at 4 °C for 45 min.
13. Discard the supernatant, and resuspend the lentivirus-containing pellet in complete NIH/3T3 culture medium (1/10 of the original volume). This will result in a 10-times concentration. The concentration factor can be increased or decreased accordingly.
14. Directly use the lentivirus for the transduction of NIH/3T3 cells, or freeze in aliquots at −80 °C until further use.

### 3.4 Infection of Target Cells (Transduction of NIH/3T3 Cells)

The following protocol describes the infection of NIH/3T3 fibroblasts only (Fig. 2b). It can be applied to different adherent and suspension cell lines, as well as various types of primary cells which may require protocol adaptations not outlined in detail here (*see* **Note 4**).

1. Harvest, count cells and seed (*see* Subheading 3.1) approx. $1 \times 10^5$ NIH/3T3 cells per well into five wells of a 6-well cell culture plate, and incubate for 24 h at 37 °C, 5% $CO_2$.
2. Prepare NIH/3T3 transduction medium by supplementing DMEM with 2% BCS, 1% penicillin/streptomycin and 10 μg/mL polybrene.
3. After 24 h of cell incubation, thaw lentivirus (*see* Subheading 3.3) on ice.
4. To perform the transduction, the lentivirus is added onto the attached cells. To determine the amount of lentivirus that will result in an effective transduction without affecting cell survival, it is recommended to test different amounts of virus (*see* **Note 5**). To do so, remove cell culture medium from cell-containing wells, and replace with different amounts of virus (e.g., 150 μL, 300 μL, 600 μL). Keep two wells free of lentivirus to serve as non-transduced controls. During antibiotic selection after 48 h, only one of them will receive the selective antibiotic while the other one will serve as a negative control, to ensure proper selection.
5. Fill up all wells to 1.5 mL with NIH/3T3 transduction medium, containing polybrene. Polybrene increases the

transduction efficiency by neutralizing the charge repulsion between the virus and the cell surface. Incubate cells for 24 h at 37 °C, 5% $CO_2$.

6. Replace the virus-containing or control NIH/3T3 transduction medium from each well with 2 mL complete NIH/3T3 culture medium per well. Before adding puromycin, cells are incubated for another 24 h at 37 °C, 5% $CO_2$, to enable expression of the transgenes.
7. Add puromycin to four wells of the cell culture plate (three transduced wells + one non-transduced control well) at the lowest efficient concentration determined before (*see* Subheading 3.2). This eventually results in the death of all non-transduced cells.
8. Monitor the cell culture plate over the following days until all non-transduced cells in the antibiotic-treated control well are killed. At this point, the amount of the selective antibiotic to the cell culture medium of the reporter cells can be reduced or left out completely.
9. Expand selected, polyclonal reporter cells and freeze down aliquots of low passage number cells.

### 3.5 Imaging

1. Mount sample on the microscope stage (*see* **Note 6**). Use an environmental chamber when possible, and keep sample exposure time to room air to a minimum (*see* **Note 7**). Choose an appropriate objective lens for imaging (*see* **Note 8**). When using a Dip-In objective, ethanol-clean it immediately prior to imaging.
2. Choose 10–15 fields of view per sample from different sample areas for imaging, each field of view containing about 20–40 cells depending on the optical magnification/size of the field of view. This should result in a total number of 200–500 cells per sample gathered from different sample regions (*see* **Note 9**). When imaging three-dimensional (3D) samples, capturing image stacks and maximum-projecting them for image analysis (*see* **step 1** in Subheading 3.6) (Fig. 3), instead of capturing single-plane images, can be used to image cells from different sample depths, dependent on the cell density in the sample. However, similar to single-plane imaging of 2D samples, 10–15 fields of view per sample from different sample regions and sample depths should be imaged.
3. Adjust exposure time, laser/fluorescent light intensity and/or electronic gain in order to generate images with high signal intensity and good signal/noise ratio (*see* **Note 10**).

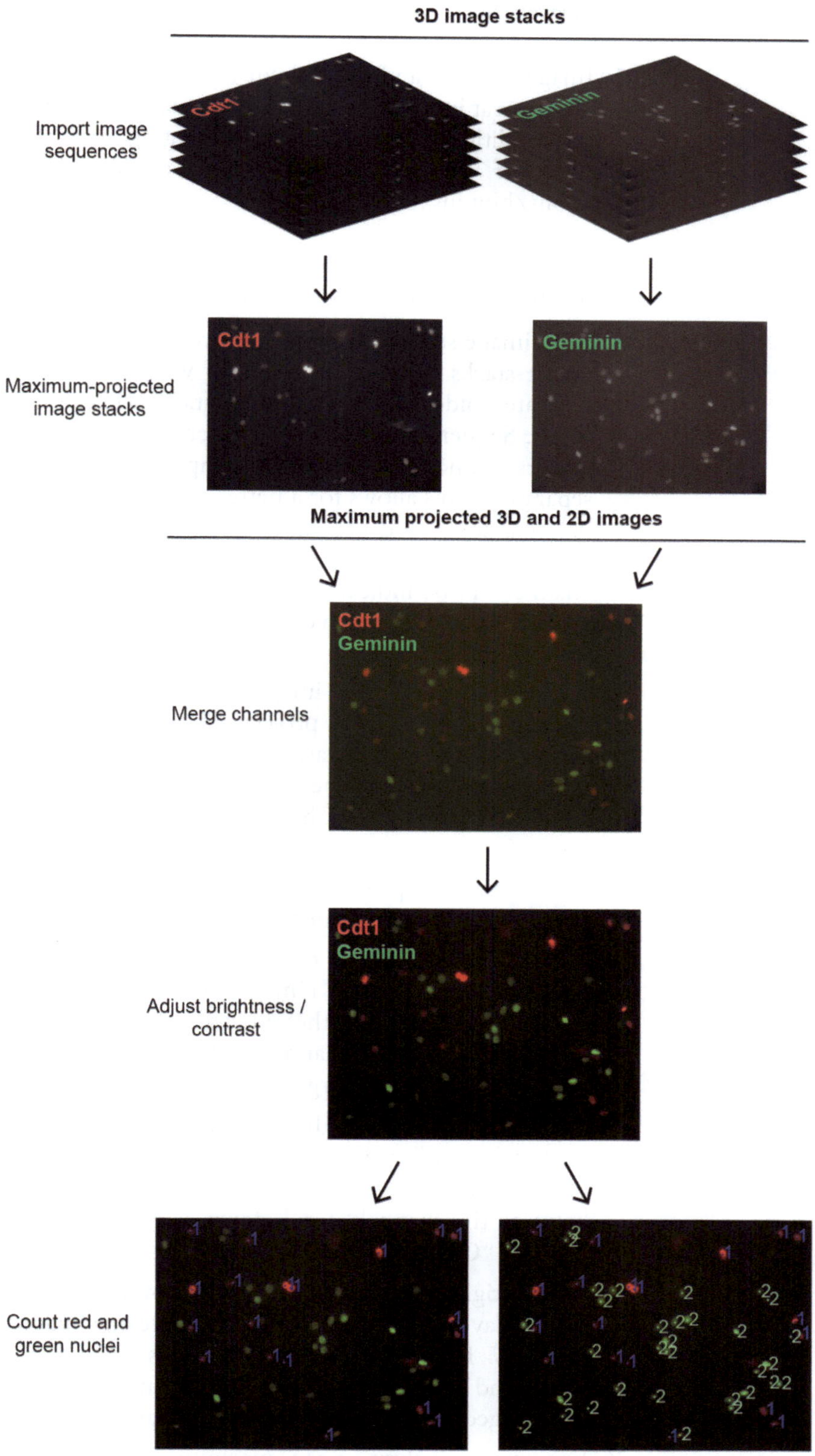


**Fig. 3** Processing of fluorescence 2D images or 3D image stacks and subsequent evaluation. Image stacks should be maximum-projected before channels are merged, and the resulting image can be optimized for counting

4. Image samples and place them back into incubator. If necessary, repeat imaging of the samples at later stages under identical conditions (keep track of all imaging conditions such as objective lens, stack size, exposure time/detector gain, zoom/binning, etc.).

### 3.6 Image Evaluation

In this protocol, ImageJ is used to evaluate cell proliferation, based on the ratio of green/red nuclei [8].

1. If no image stacks were taken, proceed directly to **step 4**. The image stacks of different fields of view should be stored in separate folders. Import image stacks (File → Import → Image Sequence). Choose the folder containing the individual images of one stack. If possible, import fluorescence channels separately. This allows for a better adjustment of both individual channels. To do so, distinguish channels using the "File name contains" field. Typically, the red and green fluorescent channel will be indicated differently in the image name (e.g., ch_00 and ch_01) which can be entered here. After that, repeat with the other channel.
2. Maximum-project the images in both channels separately (Image → Stacks → Z project). A window will open which allows you to set the start slice, stop slice and the projection type. As per default, the slice range will automatically cover your whole stack. Choose "Max. Intensity" for the projection type.
3. Store maximum-projected images separately by clicking File → Save As . . . → Tiff. Do not close the projections yet.
4. Merge the projected or single-plane images from both corresponding channels (Image → Color → Merge channel). Assign the image from the red channel to "C1 (red)" and the image from the green channel to "C2 (green)." The images will now be displayed as merged in one window.
5. If necessary, adjust brightness and contrast in both channels separately, until red and green nuclei can be clearly distinguished (Image → Adjust → Brightness/Contrast). To switch between the channels, scroll down or move the bar below the picture ("C") from left to right.
6. When brightness and contrast are set, overlay channels (Image → Overlay → Flatten). This will generate one image from both channels. From this point, all changes will be made to this image and cannot be made to both channels individually. Save the merged, overlayed image (File → Save As . . . → Tiff).
7. For image analysis, open maximum-projected image (File → Open).

8. Download and install the Cell Counter Plugin (Install → Plugins) (*see* **Note 11**).
9. Open the Cell Counter Plugin (Plugins → Cell counter), and initialize the opened image by clicking "Initialize." If there is more than one image opened, make sure to click on the image you want to analyse, before the initialization.
10. Assign counters (e.g., "Type 1" for red nuclei). Select one counter (e.g., "Type 1"), and click on the nuclei with the assigned colour (red) on the initialized image. Every click will be counted. By selecting "Delete" the last count can be deleted.
11. When you are finished counting one colour, proceed to the next counter type (e.g., "Type 2"), and count green nuclei.
12. After counting red and green nuclei, click on "Results" to have your counts displayed in a tabular format. This file can be directly saved (File → Save As ...) or copied to an excel file. You can save markers (Save Markers), and re-load them at a later point (Load Markers), when the image is initialized again.

## 4 Notes

1. Some cell lines are more sensitive towards trypsin and therefore should be centrifuged before being seeded or experimentally used. After centrifugation, the supernatant containing trypsin can be discarded, and the cell pellet can be resuspended in fresh cell culture medium.
2. If a different lentiviral construct is used, the selection antibiotic might be different.
3. DNA concentrations in the transfection mixture are optimized to have a 1:1:1 molar ratio of packaging plasmid (psPAX.2), envelope plasmid (pCMV-VSV-G) and transfer plasmid (pBOB-EF1-FastFUCCI-Puro). As a reference concentration, the amount to receive a final concentration of 1.3 pmol psPAX.2 per 10 cm dish was determined [9]. To convert 1.3 pmol to DNA mass, the following formula was used [10]:

$$\text{mass of dsDNA (g)} = \text{moles of dsDNA (mol)}$$
$$\left(\left(\text{length of dsDNA (bp)} \cdot 617.96\ \text{g}\frac{\text{g}}{\text{mol} \cdot \text{bp}}\right) + 36.04\ \text{g/mol}\right)$$

As a result, 1.3 pmol psPAX.2 corresponds to 8.6 µg of DNA per 10 cm dish. Based on this, the required DNA mass of transfer and envelope plasmid for a final 1:1:1 molar ratio were calculated according to the following formulas [11]:

$$\text{required mass of transfer plasmid (g)} = \text{desired molar ratio} \cdot \text{mass of packaging plasmid (g)} \cdot \frac{\text{transfer plasmid length (bp)}}{\text{packaging plasmid length (bp)}}$$

$$\text{required mass of envelope plasmid (g)} = \text{desired molar ratio} \cdot \text{mass of packaging plasmid (g)} \cdot \frac{\text{envelope plasmid length (bp)}}{\text{packaging plasmid length (bp)}}$$

4. For a successful infection of suspension cells, cells should be centrifuged first. The resulting cell pellet can then directly be resuspended in the thawed lentivirus. For the transduction of primary cells, the lowest possible amount of virus that still allows for a high transduction efficiency should be chosen, since the subsequent selection step is omitted.
5. Alternatively, the MOI (multiplicity of infection) can be determined, to optimize transduction efficiency. The MOI is the desired number of viral particles per cells. To determine this number, the lentiviral titer has to be measured first, which can be achieved by commercial assays, e.g., ELISA-based determination of the HIV-1 p24 antigen. Subsequently, the number of required viral particles can be determined by

$$\text{viral particles } (n) = \text{MOI} \cdot \text{number of seeded cells}$$

6. The FUCCI system can be used with different types of fluorescence microscope systems, such as epifluorescence or confocal systems, upright or inverted systems, depending on the type of sample, specific question and microscope availability. A confocal system will reduce out-of-focus emission light, which will be particularly useful when capturing image stacks in 3D samples (Fig. 4) [12], since maximum-projection of image stacks increases background fluorescence. In general, the microscope needs to be equipped with GFP/RFP filter types. While the majority of fluorescence microscopes is equipped with these or similar "standard" filter sets, compatibility of the filters with kusabira-orange and azami-green fluorescent protein excitation/emission spectra should be verified prior to imaging.
7. While FUCCI reporter cells can certainly be fixed and imaged without further immunostaining, one key advantage of the system is the possibility to image a living sample multiple times for an efficient observation of cell proliferation over time (Fig. 5) [13]. Further culture of the samples between imaging cycles might benefit from proper environmental conditions during imaging. While short-term imaging of most cells may not require $CO_2$ atmosphere, long-term imaging should be performed in the presence of $CO_2$ to prevent alkaline pH within the sample. In addition, as most microscope systems do not allow for imaging under sterile cell culture conditions, imaging is mostly performed under semi-sterile conditions.

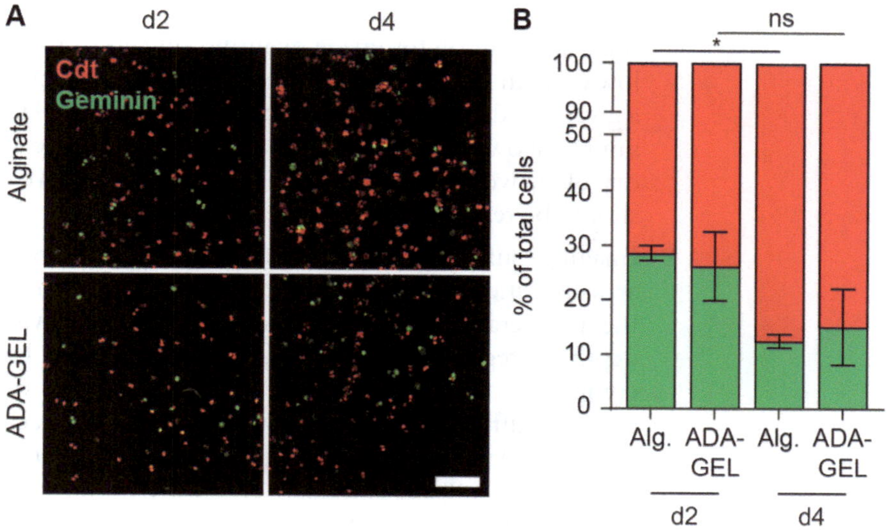


**Fig. 4** Proliferation of NIH/3T3 cells in different 3D matrices using the FastFUCCI reporter system. (**a**) Confocal image stacks of FastFUCCI expressing NIH/3T3 cells seeded into 2% alginate (upper panel) or 2.5% ADA-GEL (lower panel) after 48 h (d2) or 96 h (d4). (**b**) Quantification of mean fractions of red and green nuclei, as described in (**a**). Note significant drop of cell proliferation over time in alginate. Error bars: SD. * $p < 0.05$. (Figure was adapted from Ref. [12])

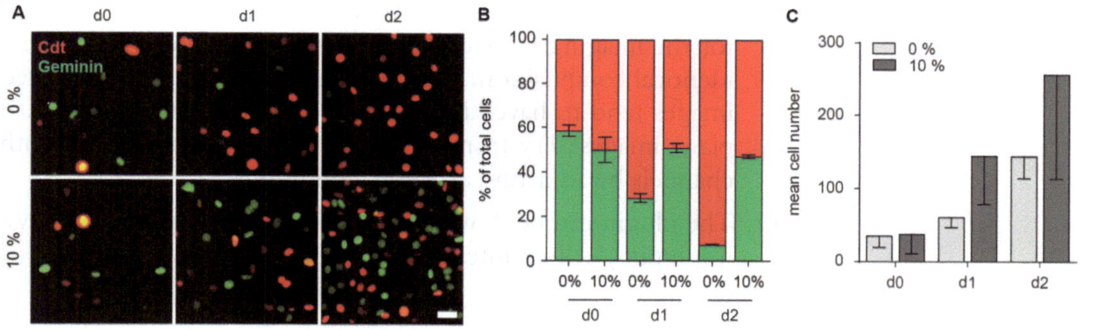


**Fig. 5** Serum deprivation as a positive control for the FUCCI reporter system. (**a**) Epifluorescence images of FastFUCCI expressing NIH/3T3 fibroblasts. Cells were seeded at 20% confluency (d0) on 2D cell culture dishes and cultured for 24 h (d1) and 48 h (d2) in the presence of 0% or 10% serum. Scale bar: 50 μm. (**b**) Mean fractions of green or red nuclei. Error bars: SD. (**c**) Mean cell counts per field of view of cells. Note constant proliferation rate over time and fast increase in cell number in the presence of serum, and a decreasing proliferation rate over time and a slow increase in cell number in the absence of serum. (Figure was adapted from Ref. [13])

Even though contamination risks are generally low due to the presence of antibiotics in most cell culture media, sample exposure to room atmosphere should be reduced to a minimum, imaging should be performed in a chamber when possible, and Dip-In objective lenses should be ethanol-treated prior to imaging.

8. The objective lens used for imaging should provide sufficient optical resolution for image quantification and a magnification low enough to capture a sufficient number of cells per image at a given cell density such that a reasonable number of images per sample allows for solid statistical evaluation. While a 10–20× dry objective is usually a good choice, the objective of choice should be tested prior to final experimentation.
9. Imaging sufficient fields-of-view covering different sample areas is critical, particularly when imaging 3D samples, where the proliferative activity of the cells may vary with sample depth, as regions deep within a 3D matrix might be hypoxic compared to peripheral sample areas. Usually, 10–15 fields of view are sufficient to cover most sample regions. Each individual field of view should contain sufficient cells for solid statistical analyses, but not too many cells either, since excessive cell density might impair image analysis. Depending on the cell density in the sample, either individual images might be captured and analysed (high cell density) or stacks of increasing thickness (low cell densities) which are subsequently maximum-projected for the image analysis.
10. If the background at 488 nm excitation (geminin-azami-green) is very high, consider replacing phenol-containing cell culture medium (phenol-autofluorescence at blue excitation) for the imaging by phenol-free imaging medium of otherwise identical composition. Keep in mind that maximum-projected images tend to have higher background levels than individual plane images. Try to match nuclear signal intensities in both channels by adjusting exposure conditions.
11. The plugin can be downloaded from https://imagej.nih.gov/ij/plugins/cell-counter.html.

## References

1. Sakaue-Sawano A, Kurokawa H, Morimura T et al (2008) Visualizing spatiotemporal dynamics of multicellular cell-cycle progression. Cell 132:487–498. https://doi.org/10.1016/j.cell.2007.12.033
2. Li X, Zhao Q, Liao R et al (2003) The SCFSkp2 ubiquitin ligase complex interacts with the human replication licensing factor Cdt1 and regulates Cdt1 degradation. J Biol Chem 278:30854–30858. https://doi.org/10.1074/jbc.C300251200
3. McGarry TJ, Kirschner MW (1998) Geminin, an inhibitor of DNA replication, is degraded during mitosis. Cell 93:1043–1053. https://doi.org/10.1016/S0092-8674(00)81209-X
4. Nishitani H, Lygerou Z, Nishimoto T (2004) Proteolysis of DNA replication licensing factor Cdt1 in S-phase is performed independently of Geminin through its N-terminal region. J Biol Chem 279:30807–30816. https://doi.org/10.1074/jbc.M312644200
5. Vodermaier HC (2004) APC/C and SCF: controlling each other and the cell cycle. Curr Biol 14:787–796. https://doi.org/10.1016/j.cub.2004.09.020
6. Zielke N, Edgar BA (2015) FUCCI sensors: powerful new tools for analysis of cell proliferation. Wiley Interdiscip Rev Dev Biol 4:469–487. https://doi.org/10.1002/wdev.189

7. Koh S-B, Mascalchi P, Rodriguez E et al (2017) A quantitative FastFUCCI assay defines cell cycle dynamics at a single-cell level. J Cell Sci 130:512–520. https://doi.org/10.1242/jcs.195164
8. Schindelin J, Arganda-Carrera I, Frise E et al (2009) Fiji – an open platform for biological image analysis. Nat Methods 9. https://doi.org/10.1038/nmeth.2019.Fiji
9. Addgene (2016) Lentivirus production. https://www.addgene.org/protocols/lentivirus-production/
10. New England Biolabs NEBioCalculator: dsDNA: Mass to/from Moles Convertor. In: 2021. https://nebiocalculator.neb.com/#!/dsdnaamt
11. New England Biolabs NEBioCalculator: ligation Calculator. In: 2021. https://nebiocalculator.neb.com/#!/ligation
12. Karakaya E, Fischer L, Hazur J et al (2020) Strategies to evaluate alginate based bioinks applying extrusion printing for biofabrication. Trans Addit Manuf Meets Med 2:1–2. https://doi.org/10.18416/AMMM.2020.2009022
13. Fischer L, Nosratlo M, Hast K et al (2022) Calcium supplementation of bioinks reduces shear stress-induced cell damage during bioprinting. Biofabrication 14:045005

# Chapter 25

# Cell Viability Assays for 3D Cellular Constructs

Zachary Congress, Matthew Brovold, and Shay Soker

## Abstract

In vitro models fall short of replicating the complex in vivo processes including cell growth and differentiation. For many years, molecular biology research and drug development have relied on the use of cells grown within tissue culture dishes. These traditional in vitro two-dimensional (2D) cultures fail to recapitulate the 3D microenvironment of in vivo tissues. Due to inadequate surface topography, surface stiffness, cell-to-cell, and cell-to-ECM matrices, 2D cell culture systems are incapable of mimicking cell physiology seen in living healthy tissues. These factors can also place selective pressure on cells that substantially alter their molecular and phenotypic properties. With these disadvantages in mind, new and adaptive cell culture systems are necessary to recapitulate the cellular microenvironment in a more accurate manner for drug development, toxicity studies, drug delivery, and much more. Newly developed biofabrication technologies capable of creating 3D tissue constructs can open new opportunities for cell growth and developmental modeling. These constructs show great promise in representing an environment that allows cells to interact with other cells and their microenvironment in a much more physiologically accurate manner. When transitioning from 2D to 3D systems, there is the need to translate common cell viability analysis techniques from that of 2D cell culture to these 3D tissue constructs. Cell viability assays are critical in evaluating the health of cells in response to drug treatment or other stimuli to better understand how these factors effect the tissue constructs. As 3D cellular systems become the new standard in biomedical engineering, this chapter provides different assays used to assess cell viability qualitatively and quantitatively in 3D environments.

**Key words** Bioprinting, Biofabrication, ATP, ECM, Cell viability, 3D assays, Proliferation, Hydrogels, Organoids, Cell culture models

## 1 Introduction

Two-dimensional cell culture systems have been the gold standard since the early 1900s; however, when cultured in three dimensions, these culture systems are able to show vast improvement in terms of cell morphology, proliferation, drug metabolism, and differentiation by creating a 3D niche more representative of natural ECM [1–4]. When studying cells in 3D tissue culture, the overarching goal is to accurately recapitulate the in vivo microenvironment such that cells replicate their natural and innate cellular dynamics with

Oliver Friedrich and Daniel F. Gilbert (eds.), *Cell Viability Assays: Methods and Protocols*, Methods in Molecular Biology, vol. 2644, https://doi.org/10.1007/978-1-0716-3052-5_25,

high fidelity. Cell-cell and cell-ECM interactions are imperative in allowing cells to proliferate properly, express natural genes, respond to stimuli, and differentiate as they would in the body, so the lack of these interactions in 2D systems presents major difficulties in recapitulating the natural cellular environment [5]. Furthermore, drug diffusion kinetics vary dramatically, and drug doses effective in 2D systems are often ineffective when scaled to the patient level [6, 7]. Such drastic differences between 2D laboratory cell culture and cells within native tissue have deemed many drugs that were initially effective in the laboratory ineffective and/or toxic when translated to patients.

Cell behavior is highly dependent on physical and chemical environmental cues, which is why bioprinting 3D cell-laden constructs allows for a more controlled and sustainable environment that can support prolonged cellular viability [8] in comparison to 2D systems. These constructs take advantage of synthetic and naturally derived biomaterials as well as advanced biofabrication techniques such as bioprinting to maintain a microenvironment similar to that of the native tissues and can be specifically tuned to mimic in vivo conditions. In addition, 3D tissue constructs provide the cells with a tunable physicomechanical environment allowing for both cell-cell and cell-ECM interactions in multiple dimensions, which is crucial for mimicking natural cellular function [9]. As 2D applications have increasingly been found to become less accurate due to their overly simplified physiological conditions, three-dimensional cellular viability assays [9] have become paramount in the field of biomedical engineering to study how these systems help support cell interaction, growth, and survival. Cell viability assays are critical in evaluating the health of cells in response to drug treatment or other stimuli. As 3D cellular systems become the new standard in biomedical engineering, it is necessary to translate the standard techniques used to assess cell viability qualitatively and quantitatively in 2D to that of a 3D environment.

There is a critical need for reliable methods to monitor cell viability, cytotoxicity, and proliferation in vitro, as the measurement of healthy cell behavior is critical in understanding how various stimuli or factors impact the tissue system. When attempting to recapitulate an in vivo tissue system, it is imperative to understand whether specific cell types are proliferating at the proper rate, displaying healthy phenotypes, or reacting to stimuli in an in vivo like manner. Whether one is studying how their experiment eradicates cancer cells or how specific cell types proliferate in regenerating tissue, cell viability techniques are crucial in determining the health of cells as well as the efficacy of the system as a whole. In 2D systems, assays that measure mitochondrial activities are typically used to determine cellular viability, such as MTS or alamarBlue™ assays [10–12]. MTS assays are capable of quantifying cell growth based on the reduction of tetrazolium reagents that only occurs in metabolically active cells, carried out by dehydrogenase enzymes dependent on NAD(P)H, to form a formazan product. The reagent is soluble in the cell culture medium and creates a stable formazan dye that can be measured with spectrophotometry plate reader [10]. AlamarBlue™ assays utilize a cell membrane permeable oxidation-reduction indicator that fluoresces in response to the reducing environment found only within viable cells [11]. In this assay, the active resazurin agent in alamarBlue™ (AB) permeates cells and is reduced to resorufin only in metabolically active cells, which can be studied using an absorbance or fluorescence-based reader [12]. Despite their accuracy in 2D cell culture systems, these colorimetric assays may not be as useful in the study of 3D systems due to the increased surface area and tighter cell-cell junctions, limiting the respective uptake of formazan or resazurin [13, 14]. Due to the depth of the 3D system, these dyes are not always able to fully diffuse through the substrate and permeate cells within the entire construct and therefore may display inaccurate results in terms of cell counts.

Recently, a new technique was developed to measure cellular viability by measuring the amount of ATP being produced by the cells, which represents the number of metabolically active cells. As cells lose their viability or their function weakens, their cell membrane integrity also weakens as they lose the ability to produce ATP. As this occurs, native ATPases eliminate any remaining ATP within the cytoplasm [15]. Detecting ATP via CellTiter-Glo utilizes a detergent to lyse cells, an ATPase inhibitor to protect ATP released from cells, and an enzyme luciferase which acts on luciferin when $Mg^{2+}$ and ATP are present to produce energy in the form of light [16]. A CellTiter-Glo 3D® (CT3D, e.g., Promega) assay has been developed, making this quick and simple measurement of cell viability easily translatable to a 3D cellular environment by using a bioluminescence meter that quantitatively measures light intensity following the reaction of the reagent and ATP released from cells within a hydrogel [17]. Unlike the colorimetric assays used in 2D systems, CT3D allows for the penetration of larger 3D constructs due to its increased lytic capability, which provides a much more accurate measure of viability in 3D systems [18].

Another approach involves the use of fluorometric assays to determine cellular viability visually and quantitatively in 3D constructs. The LIVE/DEAD™ Viability/Cytotoxicity Kit (e.g., Thermo Fisher Scientific) works by fluorescently staining living cells with a calcein AM dye that detects intracellular esterase activity and labeling dead cells with a non-permeable ethidium bromide dye that binds to nucleic acids [19]. Since ethidium bromide is excluded by intact plasma membranes of living cells, it only binds to nucleic acids within nonliving cells [19]. This method can be translated to 3D cellular systems by using confocal or two-photon microscopy to measure the fluorescence of cells through the entire depth of LIVE/DEAD™ stained constructs [20].

Reliable viability assays, such as those described above, are critical for a number of reasons. They are extremely useful for determining outcomes of drug response and for more reliable drug efficacy testing, especially in a 3D environment. Such tests provide quantifiable outcomes in response to growth factors, media composition, or differentiation induction. Particularly in the realm of tissue engineering, these assays provide us with an insight into the overall biocompatibility of hydrogels, bioinks, or other biomaterials. Overall, within this chapter, we describe the successfully translated cell viability techniques commonly used to measure cell viability in 2D systems and, more recently, 3D cultured cells: luminescent and fluorescent assays.

## 2 Materials

Prepare all solutions at room temperature (RT), unless otherwise stated, using aseptic technique to avoid potential contamination, preferably in a cell culture hood.

### 2.1 Solutions, Reagents, and Tissue Constructs

1. Cell culture media. Store at 4 °C (*see* **Note 1**).
2. Sterile PBS.
3. Isopropyl alcohol: 91%.
4. CT3D reagent. Store at −20 °C.
5. LIVE/DEAD™ Viability/Cytotoxicity Kit for mammalian cells. Store at −20 °C.
6. CellTiter 96® $AQ_{ueous}$ One Solution Cell Proliferation Assay. Store at −20 °C.
7. alamarBlue™ HS Cell Viability Reagent. Store at 4 °C.
8. Prepared 3D constructs containing cells. Store in media of choice in an incubator at 37 °C, 5% $CO_2$ (*see* **Notes 1** and **2**).

### 2.2 General Equipment

1. Manual or automated pipette 1: A pipette that can dispense volumes from 2 to 20 μL.
2. Manual or automated pipette 2: A pipette that can dispense volumes of 100 and 200 μL.
3. Manual or automated pipette 3: A pipette that can dispense volumes from 0 to 20 mL.
4. Centrifuge tubes: 15 mL conical.
5. Plastic spatulas (e.g., LevGo).
6. Sterile petri dish: 100 $mm^2$.
7. 48-well tissue culture plates.

8. 96-well opaque, white-walled microplates.
9. 96-well black, clear-bottom microplates.
10. Plate reader capable of measuring absorption, fluorescence, and luminescence from a 96-well microplate.

### 2.3 Software

1. Program for statistical data analysis and graphing (*see* **Note 3**).
2. Program for image analysis (*see* **Note 3**).

### 2.4 CT3D Assay

1. Plate shaker that can reach 500 RPM.
2. CT3D working solution: 5 mL CT3D Reagent, 5 mL cell culture media (*see* **Notes 1** and **4**).

### 2.5 Live/Dead Assay

1. 8-well chamber slide.
2. Kimwipes.
3. Confocal microscope capable of 488 and 561/594 nm laser lines.
4. Live/Dead working solution: 1 mL sterile PBS, 1 μL Calcein AM, 1.8 μL ethidium homodimer-1 (*see* **Note 5**).

### 2.6 AB Assay

1. AB working solution: 200 μL alamarBlue™ HS Cell Viability Reagent, 1.8 mL cell culture media (*see* **Notes 1** and **6**).

## 3 Methods

### 3.1 CellTiter-Glo 3D (CT3D) Viability Assay

The aim of the following steps is to prepare the CT3D working solution that will be used to lyse cells within 3D constructs to quantitatively determine the amount of ATP within each construct. After completing this step, you will have a complete working solution that will be used with 3D culture systems for further analysis. Conduct all procedures at RT using aseptic technique unless stated otherwise.

1. Allow the frozen CT3D stock solution to thaw slowly by placing an aliquot or an entire bottle of CT3D in a 4 °C refrigerator overnight (do not attempt to thaw the solution rapidly by placing it in a water or bead bath as the bottle may rupture).
2. After the solution has thawed, equilibrate to RT by removing the bottle from the 4 °C fridge and placing it in a 22 °C water or bead bath. Let the solution remain for at least 30 min to equilibrate properly.
3. Place the cell culture media of choice into a 22 °C water or bead bath, and allow it to warm to RT for at least 30 min (*see* **Note 1**).

4. Gently agitate the reagents to ensure homogeneity.
5. Determine the amount of CT3D working solution to be prepared (one tissue construct will be placed in each well of a 48-well plate). Multiply the number of samples that will be assayed for the experiment by 200 (μL). Add an additional 1 mL to account for no cell control wells (or blanks) as well as transfer losses.

$$200\ \mu\text{L} \times \text{total number of samples} + 1000\ \mu\text{L} = \text{total volume } (\mu\text{L})$$

The working solution consists of 50% CT3D and 50% cell culture media. To obtain the volume required for each component, divide the product of the above formula by 2.
6. Mix equal volumes of the media and the CT3D into a 15 mL centrifuge tube.
7. Gently mix the reagents, and let equilibrate (*see* **Note 7**).
8. Prepare the TC plate and the opaque, white-walled plate by equilibrating to RT.

   The aim of the following steps is to use the aforementioned CT3D working solution along with 3D constructs stored in an incubator to run an ATP assay. Following this protocol, the viability of cells within 3D constructs can be quantified and plotted using a luminescence plate reader and software such as GraphPad.
9. Remove the constructs from the incubator and place into a sterile hood. Using a spatula, transfer each 3D tissue construct into the 48-well TC plate, placing one construct into each well (*see* **Note 8**).
10. Dispense 200 μL of the CT3D working solution into each well containing a tissue construct (*see* **Note 9**).
11. Additionally, dispense 200 μL of prepared CT3D working solution into three empty wells to serve as no-cell blank controls.
12. Place the plate on a rotational shaker at RT, set the agitation to 250 RPM, and let agitate for 5 min (*see* **Note 10**).
13. Let the plate sit for 25 min (*see* **Note 10**).
14. For each well containing the working solution, gently mix up and down with a pipettor several times before transferring 100 μL to the opaque, 96-well opaque, white-walled plate (*see* **Note 11**).
15. Read the 96-well opaque, white-walled plate using a luminescence plate reader (*see* **Note 12**).

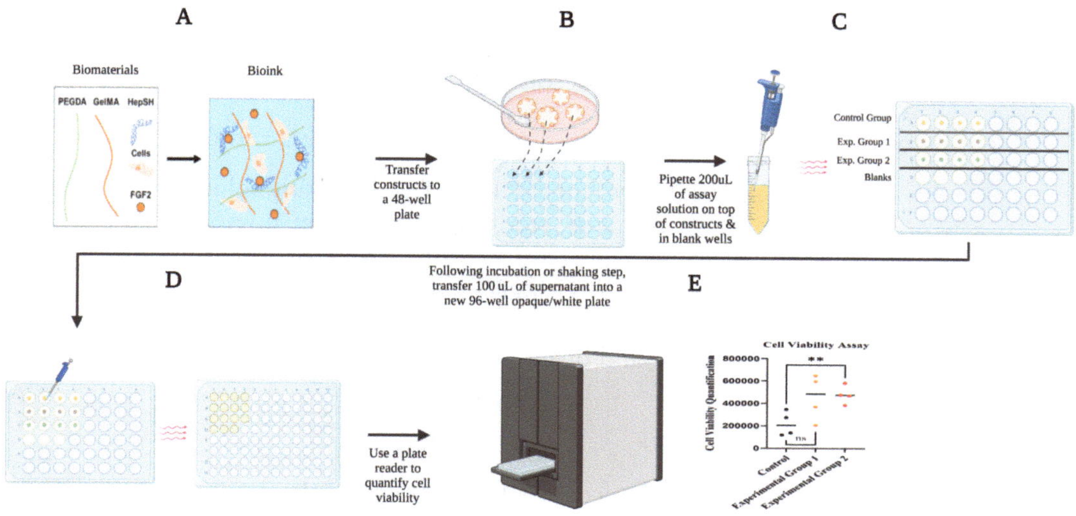


**Fig. 1** A fluorescence/luminescence cell viability assay for 3D bioprinted constructs. (**a**) Hydrogel biomaterials and growth factor mixture are combined with the desired cell types and used as a bioink to bioprint 3D constructs. Constructs are cultured in a cell culture incubator until the desired times of analysis. (**b**) Using a spatula, individual constructs are transferred into a 48-well plate, with each row representing a different experimental group. (**c**) Assay solution is added into each construct-containing well. Additional empty wells receive the same volume of assay solution to serve as "blanks" (controls). (**d**) After the recommended incubation time, some of the solution in each well is transferred to a clean well of a 96-well opaque/white plate that can be read on a standard fluorescent or luminescent plate reader. (**e**) Using a plate reader, fluorescent or luminescent signals are quantified and results from the different experimental groups are graphed against each other. Created with BioRender

16. Using collected data, you may plot the data along with desired statistics on your software of choice such as GraphPad, MATLAB, etc.
17. For a visual overview of the steps performed in a luminescent assay, please refer to Fig. 1.

### 3.2 Live/Dead Assay

The aim of the following steps is to develop a Live/Dead working solution, which is volume dependent, based on the number of 3D constructs being studied. Following these steps, a complete Live/Dead solution will be available to conduct the remainder of the protocol.

1. Remove both ethidium homodimer-1 and calcein AM solutions from −20 °C, and bring to RT by leaving the solutions out on the benchtop. This will take approximately 10 min (*see* **Note 13**).
2. Transfer the ethidium homodimer-1, calcein AM, and RT sterile PBS into a sterile hood.
3. Determine the volume of Live/Dead working solution to be prepared (each well within an 8-well chamber slide will contain

one construct) by multiplying the number of samples that will be assayed for the experiment by 500 (μL). Add an additional 500 (μL) to account for any transfer losses.

$$500\ \mu\text{L} \times \text{number of samples} + 500\ \mu\text{L} = \text{total volume}\ (\mu\text{L})$$

4. The total volume of Live/Dead working solution will consist of three reagents: sterile PBS, ethidium homodimer-1, and calcein AM. Per 1 mL of PBS, 1.8 μL of ethidium homodimer-1 and 1 μL of calcein AM will be used. To determine the amount of ethidium homodimer-1 to add, multiply the total volume by 1.8 μL, and divide by 1000.

$$\frac{\text{Total volume}\ (\mu\text{L}) * 1.8\ \mu\text{L}}{1000(\mu\text{L})} = x$$

For calcein AM simply divide the total volume in μL by 1000.

$$\frac{\text{Total volume}\ (\mu\text{L}) * 1.0\ (\mu\text{L})}{1000(\mu\text{L})} = x$$

5. Mix appropriate volumes of sterile PBS, ethidium homodimer-1, and calcein AM in a 15 mL centrifuge tube. Gently mix the reagents together by pipetting up and down several times using a 1 mL pipette to ensure a homogeneous solution. Cover the tube with aluminum foil, and avoid light for optimal results.

   The aim of the following steps is to wash 3D constructs with PBS to remove any media residue and then use the prepared Live/Dead working solution to quantitatively assess cell viability. Following this protocol, a raw image displaying live and dead cells will be developed that allows for visual and quantitative assessment of cellular health in 3D constructs.

6. Using a spatula, remove each tissue construct from the incubator, and add one construct per well in an 8-well chamber slide (*see* **Note 14**).
7. Wash each sample with PBS by pipetting 500 μL of PBS into each well. Pipette the PBS up and down at least five times to ensure all media is washed from the constructs and then dispose of the used PBS. Repeat this step a second time by adding another 500 μL of PBS per well and washing the cells once more. Remove all PBS from the well, leaving only the washed samples in the wells (*see* **Note 15**).
8. After each sample is washed, use a sterile pipette tip or sterile spatula to gently make sure the sample is flat on the bottom of the 8-well coverslip plate prior to adding the Live/Dead solution. This ensures that the samples are fully submerged in the Live/Dead working solution.

9. Pipette 500 μL of the Live/Dead working solution per well on top of each construct, ensuring that the constructs are fully submerged in the liquid and still adherent to the bottom of the well. If a construct has moved during this process, use a sterile pipette tip or sterile spatula to fully submerge the construct.
10. Place the 8-well chamber slide (with the cover on) in a sterile 100 $mm^2$ petri dish, spray the petri dish with 70% ethanol, and place the petri dish in a cell culture incubator for 60 min.
11. After 60 min, remove the petri dish from the incubator, and suction out all of the Live/Dead working solution so that only the constructs remain in each well. For proper imaging, make sure that the constructs are at the bottom of the well and not stuck to the side. If necessary, move the constructs to the center of the wells using a sterile pipette tip or sterile spatula.
12. Using a confocal microscope equipped with lasers capable of laser lines 488 and 561/594 nm laser excitation, place the 8-well chamber slide on the microscope stage and image at an appropriate focal plane (*see* **Note 16**).
13. Quantify Live/Dead results using image analysis software such as ImageJ. Refer to Fig. 2 to see example results including the Live/Dead images and cell viability quantification.

### 3.3 MTS Assay

The aim of the following steps is to immerse 3D constructs in a working solution containing 20 μL of MTS assay per 100 μL culture medium prior to incubating the constructs. Following incubation, a colored formazan product will be solubilized only in metabolically active cells. After transferring the solubilized supernatant to a 96-well plate, the absorbance can be read which will be indicative of the degree of viable cells.

1. Remove MTS solution from the −20 °C freezer, and allow it to thaw and equilibrate to RT by leaving the reagent out on the benchtop. This will take approximately 90 min to equilibrate to RT.
2. Using a plastic spatula, place one construct into each well of a 96-well TC plate.
3. In each well containing a construct, add 200 μL of culture medium used during the experiment (*see* **Note 1**). Using a sterile pipette tip or sterile spatula, ensure that the construct is completely submerged in media.
4. After the addition of the media, add 40 μL of MTS aqueous solution to each well.
5. Create a series of three no-cell control wells by adding 200 μL media and 40 μL of MTS aqueous solution to any unused wells.

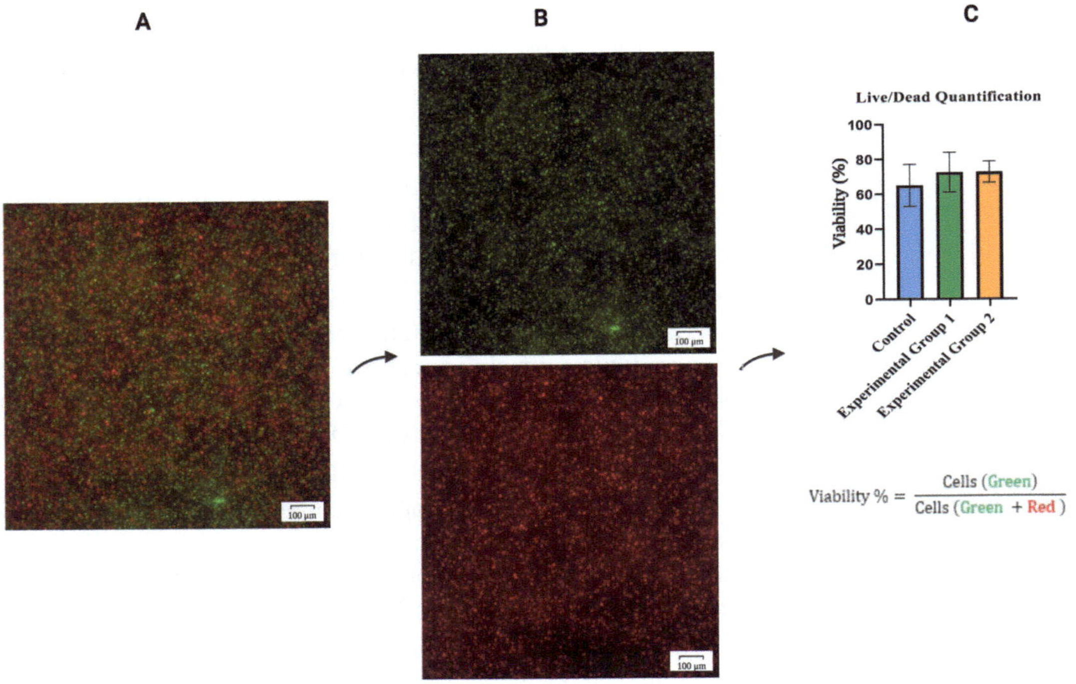


**Fig. 2** Cell viability assessment using the Live/Dead viability/cytotoxicity stains. (**a**) Raw image of Live/Dead cells within a 3D construct using confocal microscopy. (**b**) Separate images for green and red fluorescence for quantification using ImageJ™ software. (**c**) The percent of viable cells is calculated using the equation provided and the results of different experimental groups are graphed against each other. Created with BioRender

6. Place the TC plate in a tissue culture incubator 1.5 h at 37C, 5% $CO_2$ to allow the uptake and reduction of the dye (*see* **Note 17**).
7. Remove the plate from the incubator, and gently mix each well by pipetting several times to ensure a homogenous solution.
8. After the solution is mixed, gently mix up and down with a pipettor several times before transferring 100 μL to a 96-well black, clear-bottom microplate (*see* **Note 18**).
9. Place the 96-well plate into a plate reader, and set it to read fluorescence at a wavelength of 490 nm and analyze the plate.
10. Quantify and analyze using appropriate statistical software.

### 3.4 AlamarBlue (AB) Assay

The aim of the following steps is to prepare the AB working solution that will be used to determine the amount of metabolically active cells within 3D constructs. After completing this step, you will have a complete AB working solution that will be mixed with incubated 3D culture systems for further analysis.

1. Remove the AB reagent from the refrigerator, and allow it to warm to RT.
2. A working solution of AB consists of 10% reagent and 90% cell culture media. First, calculate the volume of AB working solution to be prepared by determining the amount of samples that will be assayed for the experiment and multiplying that number by 200 (μL). Add an additional 1000 μL to allow for no cell control wells (or blanks) to be assayed as well as to account for any transfer losses.

   $$200\ \mu L \times \text{number of samples} + 1000\ \mu L = \text{total volume}$$

   Multiply the total volume by 0.1 to obtain the amount of AB reagent required,

   $$\text{total volume} \times 0.1 = \text{alamarBlue reagent volume (in } \mu L)$$

   then make up the remainder of the total volume by adding the appropriate cell culture media (*see* **Note 1**).
3. Mix the determined volumes of the media and AB in a 15 mL centrifuge tube.
4. Gently mix the reagents, pipetting up and down to fully homogenize the solution.

   The aim of the following steps is to incubate 3D constructs submerged in the AB working solution to quantify the fluorescence of living cells. Only living cells will have resazurin reduced to resorufin which fluoresces bright red under a plate reader, allowing viability to be determined. Following this protocol, the viability of cells within 3D constructs can be quantified and plotted using a fluorescent or absorption-based plate reader and statistical analysis software.
5. Using a plastic spatula, transfer each 3D tissue construct into the 96-well TC plate, placing one construct into each well.
6. Dispense 200 μL of the AB working solution into each well containing a tissue construct.
7. Additionally, dispense 200 μL of solution into three empty wells to serve as blank controls.
8. Incubate the 96-well plate containing 3D constructs and AB working solution at 37 °C, 5% $CO_2$ for 1.5 h (*see* **Note 17**).
9. After removal from the incubator, mix each well gently with a pipettor three to four times before transferring 100 μL from each well to a 96-well black, clear-bottom microplate (*see* **Note 18**).

10. Read the 96-well clear bottom plate using a fluorescence or absorption plate reader. Fluorescence can be read at a fluorescent excitation wavelength of 560 nm and fluorescence emission wavelength of 590 nm. Absorption may be measured at a wavelength of 570 nm.
11. Quantify and analyze using appropriate statistical software.

## 4 Notes

1. Since 3D culture systems can comprise any cell type, we are unable to provide any specific details on the type of cell culture media required. Using a reliable source (e.g., ATCC.org), please look up the cell type used during your experiment to determine the proper media and media components to be used during these protocols. If you already have media that you are using to culture cells incorporated into your 3D constructs, continue using that same media formulation throughout these protocols.
2. 3D tissue constructs consist of cells embedded within various biomaterials (natural or synthetic), which create systems more representative of the native tissue environment. There are various methods used to create these hydrogels; however, the 3D constructs we use involve biofabrication to crosslink appropriate biomaterials that provide the cells with the ability to display more physiologically accurate cell-cell and cell-ECM interactions. For example, our constructs use the biomaterials PEG-DA & Gel-MA crosslinked by UV light. Cells are added (ten million per 1 mL bioink), and 50 μL 3D constructs are printed using DLP printing techniques. However, there are many different materials and cells for which these techniques will work. The maximum volume of tissue construct we have successfully used is 100 μL, but generally, a lower volume of 50 μL or less will have the best results.
3. Prior to starting your experiments, it is important to recognize that certain software will be needed to analyze and display results. For the protocols that retrieve quantitative analysis of relative viability between experimental groups (CT3D, MTS, & AB), you will need a program capable of data entry, statistical analysis, and simple graphing to understand the significance of your results. Our recommendation is to use GraphPad due to its user-friendly nature, Microsoft Excel, or MATLAB for a more advanced programmer; however, any program capable of data entry and statistical analysis may be used. For the Live/Dead protocol, a couple of additional software programs are needed. Following the imaging of Live/Dead using a confocal microscope, images are to be exported as TIFs for further

analysis. An image analysis software will be required to threshold levels of live and dead cells to further compare viability between experimental groups. In our case, the imaging software CellSens paired with our confocal microscope, which will vary depending on the microscope used, works as a good method for image analysis. Alternative programs such as ImageJ that are capable of thresholding images based on the degree of red or green cells can also be used to quantify relative viability.

4. This is just an example regarding the amount of solution to be made. CT3D working solution will consist of 50% CT3D reagent and 50% cell culture media. The amount of total working solution needed will rely upon how many 3D constructs are being analyzed. This is described in further detail in the Subheading 3; however, in brief this will require 200 uL per sample of CT3D working solution plus an additional 1 mL of CT3D working solution to account for any transfer losses and blank control wells. Therefore, if you are analyzing five samples, you will need to create 1 mL ($5 \times 200$ μL) + 1 mL = 2 mL of total CT3D working solution consisting of 50% CT3D reagent (1 mL) and 50% cell culture media (1 mL).

5. This represents one unit volume of Live/Dead working solution. For every 1 mL of PBS used, 1 μL of calcein AM and 1.8 uL of ethidium homodimer-1 will be added to form the Live/Dead working solution. The total amount of working volume will be dependent on the amount of 3D constructs being analyzed. This is described in further detail in the Subheading 3; however, in brief this will require 500 uL of Live/Dead working solution per sample plus an additional 500 μL to account for any transfer losses. For example, if you are analyzing six samples, you will need to create 3 mL of Live/Dead working solution ($6 \times 500$ μL) + 500 μL Live/Dead working solution = 3.5 mL of total working solution. This will be created by combining 3.5 mL of sterile PBS, 3.5 μL (1 μL $\times$ 3.5) of calcein AM, and 6.3 μL (1.8 μL $\times$ 3.5) of ethidium homodimer-1.

6. These concentrations of AB reagent and cell culture media are just examples of the proportion of the reagents (10% AB reagent, 90% cell culture media). The total amount of working volume will be dependent on the amount of 3D constructs being analyzed. This is described in further detail in the Subheading 3; however, in brief the total AB working solution can be determined by multiplying the amount of samples by 200 μL and adding an additional 1 mL to account for transfer losses and blank control wells. For example, if you are analyzing six samples, you will need to create a total of 1.2 mL (200 μL $\times$ 6) + 1 mL = 2.2 mL total AB working solution.

10% of this solution (0.22 mL) will consist of the AB reagent, and the remaining 90% (0.98 mL) will consist of cell culture media.

7. Adequately mix the CT3D and cell culture media working solution by gently mixing to fully homogenize the solution. Then wait 30 min prior to using solution for complete mixing.
8. Use a separate spatula for each group when transferring samples into the 96-well plate to reduce contamination between groups.
9. Following the addition of CT3D working solution into the 96-well plate, use a spatula to ensure that samples are fully submerged at the bottom of the plate and are adherent to the bottom prior to shaking. We have experienced that if not all 3D constructs are not fully submerged, the solution does not penetrate each construct equally, which will increase variability between groups.
10. For some formulations of hydrogels, it was found that increasing the plate shaker to 500 rpm for exactly 30 min allows the solution to better penetrate samples. Do not exceed this time limit or use a lower magnitude of shaking for optimal results. There is a slight increase in background noise, but this is negligible when high cell concentrations are used.
11. When transferring 100 μL from each well in the transfer process to the 96-well opaque, white-walled plate, use a new pipette tip after each group to ensure no mixing of solution between groups.
12. Subtract the average of the "blank" control groups from the results following ATP measurement to determine the real measurement of ATP from each sample. The blank groups are representative of the raw luminescence produced from the working solution by itself, which is why it must be subtracted to attain true ATP values produced by the actual cells.
13. Prior to thawing calcein AM and ethidium homodimer-1, sterilize the hood, and prepare all necessary plates/items for the protocol. This ensures that the reagents are not sitting out too long. Due to the small size of the reagent vials (200 μL), they thaw quickly and should not be left out for long prior to use. We have found that leaving these solutions out too long prior to use negatively affects the Live/Dead image quality.
14. When transferring constructs into the slide, use a different spatula for each different experimental/control group to avoid cross-group contamination.
15. When washing each sample with sterile PBS, use a new pipette tip for each group and each wash session to avoid any potential mixing of PBS between groups.

16. Using a Kimwipe and a small amount of 91% isopropyl alcohol, gently clean the bottom of the slide prior to placing it on the confocal microscope to ensure visibility of the samples.
17. Samples can be incubated between 1 and 4 h, which will vary between different compositions of hydrogel. A preliminary test of your material should be conducted beforehand.
18. Occasionally we found that large constructs or some compositions of hydrogel can interfere with the plate reader analysis. However, if your sample is sufficiently small, or you have a low cell density, skip this step and place it directly into the plate reader.

## References


1. Kunz-Schughart LA, Freyer JP, Hofstaedter F, Ebner R (2004) The use of 3-D cultures for high-throughput screening: the multicellular spheroid model. J Biomol Screen 9(4):273–285. https://doi.org/10.1177/1087057104265040
2. Bhattacharya S, Zhang Q, Carmichael PL, Boekelheide K, Andersen ME (2011) Toxicity testing in the 21 century: defining new risk assessment approaches based on perturbation of intracellular toxicity pathways. PLoS One 6(6):e20887. https://doi.org/10.1371/journal.pone.0020887
3. Arnold M, Karim-Kos HE, Coebergh JW, Byrnes G, Antilla A, Ferlay J, Renehan AG, Forman D, Soerjomataram I (2015) Recent trends in incidence of five common cancers in 26 European countries since 1988: analysis of the European Cancer Observatory. Eur J Cancer 51(9):1164–1187. https://doi.org/10.1016/j.ejca.2013.09.002
4. Jensen C, Teng Y (2020) Is it time to start transitioning from 2D to 3D cell culture? Front Mol Biosci 7. https://doi.org/10.3389/fmolb.2020.00033
5. Kapalczynska M, Kolenda T, Przybyla W, Zajaczkowska M, Teresiak A, Filas V, Ibbs M, Blizniak R, Luczewski L, Lamperska K (2018) 2D and 3D cell cultures – a comparison of different types of cancer cell cultures. Arch Med Sci 14(4):910–919. https://doi.org/10.5114/aoms.2016.63743
6. Ho WJ, Pham EA, Kim JW, Ng CW, Kim JH, Kamei DT, Wu BM (2010) Incorporation of multicellular spheroids into 3-D polymeric scaffolds provides an improved tumor model for screening anticancer drugs. Cancer Sci 101(12):2637–2643. https://doi.org/10.1111/j.1349-7006.2010.01723.x
7. Drewitz M, Helbling M, Fried N, Bieri M, Moritz W, Lichtenberg J, Kelm JM (2011) Towards automated production and drug sensitivity testing using scaffold-free spherical tumor microtissues. Biotechnol J 6(12):1488–1496. https://doi.org/10.1002/biot.201100290
8. Fetah K, Tebon P, Goudie M, Eichenbaum J, Ren L, Barros N, Nasiri R, Ahadian S, Ashammakhi N, Dokmeci M, Khademhosseini A (2019) The emergence of 3D bioprinting in organ-on-chip systems. Prog Biomed Eng 1:012001. https://doi.org/10.1088/2516-1091/ab23df
9. Dominijanni AJ, Devarasetty M, Forsythe SD, Votanopoulos KI, Soker S (2021) Cell viability assays in three-dimensional hydrogels: a comparative study of accuracy. Tissue Eng Part C Methods 27(7):401–410. https://doi.org/10.1089/ten.TEC.2021.0060
10. Kamiloglu S, Sari G, Ozdal T, Capanoglu E (2020) Guidelines for cell viability assays. Food Front 1(3):332–349. https://doi.org/10.1002/fft2.44
11. G-Biosciences (2016) Alamar blue cell viability assay reagent. https://cdn.gbiosciences.com/pdfs/protocol/786-921_protocol.pdf. Accessed 21 May 2022
12. ThermoFisher Scientific alamarBlue™ Cell Viability Reagent. https://www.thermofisher.com/order/catalog/product/DAL1025. Accessed 21 May 2022
13. Bonnier F, Keating ME, Wrobel TP, Majzner K, Baranska M, Garcia-Munoz A, Blanco A, Byrne HJ (2015) Cell viability assessment using the Alamar blue assay: a comparison of 2D and 3D cell culture models. Toxicol In Vitro 29(1):124–131. https://doi.org/10.1016/j.tiv.2014.09.014

14. Walzl A, Unger C, Kramer N, Unterleuthner D, Scherzer M, Hengstschlager M, Schwanzer-Pfeiffer D, Dolznig H (2014) The resazurin reduction assay can distinguish cytotoxic from cytostatic compounds in spheroid screening assays. J Biomol Screen 19(7):1047–1059. https://doi.org/10.1177/1087057114532352
15. Riss TL, Moravec RA, Niles AL, Duellman S, Benink HA, Worzella TJ, Minor L (2004) Cell viability assays. In: Markossian S, Grossman A, Brimacombe K et al (eds) Assay guidance manual, Bethesda, MD
16. Hannah R, Beck M, Moravec R, Riss T (2002) Celltiter-Glo™ luminescent cell viability assay: a sensitive and rapid method for determining cell viability. https://www.promega.com/~/media/Files/Resources/Cell%20Notes/cn002/CellTiter-Glo%20Luminescent%20Cell%20Viability%20Assay-%20A%20Sensitive.ashx. Accessed 21 May 2022
17. Sanna T, Dallolio L, Raggi A, Mazzetti M, Lorusso G, Zanni A, Farruggia P, Leoni E (2018) ATP bioluminescence assay for evaluating cleaning practices in operating theatres: applicability and limitations. BMC Infect Dis 18(1):583. https://doi.org/10.1186/s12879-018-3505-y
18. Lee J, Mohns M (2019) How to choose a cell viability or cytotoxicity assay. https://www.promega.com/resources/guides/cell-biology/cell-viability. Accessed 21 May 2022
19. Molecular Probes Invitrogen detection technologies (2005) Live/dead viability/cytotoxicity kit for mammalian cells. https://tools.thermofisher.com/content/sfs/manuals/mp03224.pdf. Accessed 18 May 2022
20. Dittmar R, Potier E, van Zandvoort M, Ito K (2012) Assessment of cell viability in three-dimensional scaffolds using cellular auto-fluorescence. Tissue Eng Part C Methods 18(3):198–204. https://doi.org/10.1089/ten.TEC.2011.0334

# Chapter 26

# Toxicological Analysis by Assessment of Vascularization and Cell Viability Using the Chicken's Chorioallantoic Membrane (CAM Assay)

**Nadine Wiesmann, Jürgen Brieger, and Jonas Eckrich**


## Abstract

The chorioallantoic membrane (CAM) assay is an increasingly popular method using a hen's egg as an experimental organism. Animal models have been established in scientific research for centuries. Yet, awareness of animal welfare in society rises, and the transferability of findings obtained in rodent models to human physiology is challenged. Thus, using fertilized eggs as an alternative platform for animal experimentation might be a promising alternative. The CAM assay is utilized for toxicological analysis by determination of CAM irritation as well as analysis of organ damage and ultimately death of the embryo. Furthermore the CAM provides a micromilieu suited for the implantation of xenografts. Xenogene tissues and tumors grow on the CAM due to a lack of rejection by the immune system and a dense vascular network providing oxygen and nutrients. Multiple analytical methods including in vivo microscopy and various imaging techniques are applicable to this model. Additionally, ethical aspects, a comparatively low financial burden, and low bureaucratic hurdles legitimize the CAM assay.

We here describe an in ovo model utilized for xenotransplantation of a human tumor. The model can be used to evaluate the efficacy as well as the toxicity of different therapeutic agents after intravascular injection. Additionally, we present the evaluation of vascularization and viability by intravital microscopy, ultrasonography, and immunohistochemistry.


**Key words** Animal models, In vivo models, CAM assay, CAM model, Chorioallantoic membrane assay, In vivo imaging, Toxicology

## 1 Introduction

Using chicken eggs as an experimental organism has been established since the nineteenth century. Exposure of the chorioallantoic membrane (CAM) by (partial) removal of the eggshell enables the direct sight onto a densely vascularized membrane which is easily accessible for various experimentation, for example for microvascular research or implantation of xenografts. The CAM is developed by the partial fusion of the mesodermal layer of the allantois with the adjacent mesodermal layer of the chick's chorion in the early

Oliver Friedrich and Daniel F. Gilbert (eds.), *Cell Viability Assays: Methods and Protocols*, Methods in Molecular Biology, vol. 2644, https://doi.org/10.1007/978-1-0716-3052-5_26,

stages of embryonal development. The exchange of respiratory gases, the calcium- and acid-base homeostasis, as well as the reabsorption of ions and $H_2O$ from the allantoic fluid, are the main organ functions of the CAM [1]. Since a dense vascularization is essential for these metabolic processes, the CAM is permeated by a complex vascular network that is connected to the embryonic bloodstream by two allantoic arteries and one allantoic vein. The CAM's surface with the inherent vascular network rapidly expands during the early stages of embryonal development [2–5]. This expansion stagnates on day 11 of egg development eventually leading to a fully developed and differentiated CAM on day 13 [2, 4].

The dense vascularization in combination with good accessibility and a lack of innervation are what makes the CAM so appealing as an experimental platform. The dense vascularization facilitates a ubiquitous supply of oxygen and nutrients [4]. This milieu allows the inoculation of cells both from cell culture [6, 7] as well as from heterotransplantation [8]. Additionally, in the hen's egg, a specific immune system is absent in the early stage of development [9]. Hence, the implantation of xenografts does not require artificial induction of immunodeficiency [6, 10].

Today, multiple (tumor) cell lines including papilloma [11], head and neck, liver, bladder, renal, ovarian, pancreatic, and prostate cancers as well as sarcomas, and neuroblastoma have been established on the CAM [12–21]. Solid tumors have been shown to grow into the CAM, attracting vessel ingrowth into the tumor from days 2 to 5 after inoculation [6, 12–14, 22]. In contrast, a functioning immune system in rodent models prevents the growth of xenografts. Consequently, severely immunodeficient rodent strains or rodents with a humanized immune system are needed to successfully transplant xenografts into a foreign organism [23]. Frequently used rodent-xenograft models are mice and rats with an absence of B cells (nude mice/rats) or prevalence of severe combined immunodeficiency (SCID) [24, 25].

Besides tumor research, the CAM assay is an established model for the analysis of angiogenesis [26], biomaterial research [27, 28], toxicology [21, 29–33], wound healing [34], tissue engineering [35], or bone regeneration [9, 36, 37]. As the CAM is fully accessible for experimentation and observation, local application of therapeutic agents [38, 39], as well as intravascular administration [21, 31, 40], are possible. A complete shell removal and transfer of the egg content into an artificial, transparent surrounding, like a petri dish, is possible and allows a 360° insight into the developing organism.

Various optical analytical techniques ranging from reflected light microscopy to fluorescence microscopy and luminescence assays have been established for this model. Furthermore, imaging techniques, like magnetic resonance imaging (MRI) [39], computed tomography (CT) [41], ultrafast ultrasound microvessel imaging (UMI) with power Doppler imaging [42] or ultrasonography [6, 40], can be used to visualize both the embryo as well as xenografts implanted onto the CAM.

Analytical methods further comprise methods of toxicological evaluation. Luepke initially established the CAM assay to analyse the irritative potential of substances as an alternative method replacing the Draize rabbit eye test in 1985 [1, 43]. Since then, multiple substances have been tested for their irritative potential on the CAM [43, 44]. Apart from effects on the CAM itself, organ damage of specific substances can be determined by dissection of the organism and histological analysis of organ and tissue-specific effects in the embryo as well as in xenografts implanted onto the CAM [21]. Finally, effects on the embryonic organism are ultimately reflected by dropout rates after administration of a specific substance. Moreover, determination of the dropout rate allows the calculation of toxicological parameters, like the median lethal dose ($LD_{50}$) [21, 31].

Using the CAM assay requires a minimum of specialized equipment and, from a technical standpoint, is a comparatively simple methodology with moderate space and maintenance requirements. The methodology has very low bureaucratic hurdles, especially when compared to animal experiments with mammallians, birds, reptiles, fish, amphibians and *Cephalopoda*, and it allows a high quantitative output and a good reproducibility without the need for excessive investments in personnel or equipment [6, 45]. The CAM has no nociception, and the chick embryo is unable to experience pain until day 14 of incubation due to the ongoing development of the nervous system [2]. Currently, until day 14, experimentation with the CAM assay does not require approval by a governmental organization or ethics committee for animal experimentation [45]. However, for the regulation of experimentation with the CAM assay, regional differences apply, and before any experiments with chicken embryos, the current legal situation should be carefully reconsidered.

Today, animal research causes controversy in society due to increasing awareness of animal welfare [46]. The "3R principle" (reduce, refine, replace) has been established as a benchmark in scientific research today [47]. According to Russell's and Burch's "Principles of Humane Experimental Technique," the CAM assay, when compared to rodent models, is a refinement, in the regard that the method causes significantly less pain and suffering.

Overall, the CAM assay is a highly versatile method in the field of in vivo experimentation. It is cost-efficient, is relatively easy to perform, and allows high quantitative output.

## 2 Materials

### 2.1 General Equipment

1. Surgical scissors.
2. Spatula.
3. Surgical forceps.
4. Single-use scalpels.
5. Cell culture flasks with a growth area of 175 $cm^2$.
6. 6-well plates.
7. 10 mL sterile tubes.
8. Sterile 10 mL syringes.
9. 21G 1 1/2 needles (0.8 × 40 mm).
10. 30G 1/2 needles (0.3 × 13 mm).
11. Adhesive tape.
12. Leukopor® tape 1.25 and 2.5 cm width.
13. PARAFILM®.
14. Silver nitrate applicator.
15. Filter paper strips for embedding.
16. Embedding plastic cassettes.
17. Sponges for embedding.
18. Staining cuvettes.
19. Superfrost adhesion microscope slides (e.g., Epredia™ SuperFrost Plus™ Adhesion, PHC Holdings Corporation, Tokyo, Japan).
20. Neubauer counting chamber.

### 2.2 Solutions and Reagents

1. Autoclaved, distilled water.
2. 5% (w/v) fluorescein isothiocyanate (FITC)-labeled dextran (average molecular weight 500 kDa) dissolved in an isotonic solution of sodium chloride (0.9 g NaCl per 100 mL $H_2O$).
3. 0.05% (w/v) tetramethylrhodamine isothiocyanate (TRITC)-labeled dextran (70 kDa) dissolved in an isotonic solution of sodium chloride (0.9 g NaCl per 100 mL $H_2O$).
4. Trypsin-EDTA 1× solution.
5. 4% (v/v) formaldehyde stabilized, buffered (pH 7.0 ± 0.2).
6. Phosphate-buffered saline (PBS).
7. Xylene.
8. Isopropanol solutions in distilled water with 70%, 80%, and 90% isopropanol (v/v) as well as pure isopropanol.
9. Eukitt®.

10. 10 mM citrate buffer (pH 6.0).
11. Methanol/$H_2O_2$: 80 mL methanol, 2 mL 30% $H_2O_2$.
12. PBS-BSA: phosphate-buffered saline (PBS), 1% (w/v) bovine serum albumin (BSA).
13. Goat serum (normal).
14. TBST20 buffer I: 400 mM NaCl, 50 mM Tris, 0.05% (v/v) Tween 20, pH adjusted to 7.6 with HCl.
15. TBST20 buffer II: 147 mM NaCl, 50 mM Tris, 0.05% (v/v) Tween 20, pH adjusted to 7.6 with HCl.
16. Full cell culture medium: 500 mL Dulbecco's Modified Eagle's Medium, 2% (v/v) penicillin/streptomycin, 5% (v/v) fetal calf serum.
17. 0.9% (w/v) NaCl.
18. Paraffin.
19. Mayer's hemalum.
20. Eosin.
21. Horseradish peroxidase-conjugated streptavidin (e.g., Dako Denmark A/S, Glostrup, Denmark).
22. DAB substrate (e.g., Dako Denmark A/S, Glostrup, Denmark).

### 2.3 Laboratory Equipment and Software

1. Intravital microscope (e.g., Olympus BXFM, Olympus GmbH, Hamburg, Germany) with a green filter (excitation [Ex]:470 nm/emission [Em]: 525 nm) and/or orange filter (Ex: 545 nm; Em: 605 nm) (e.g., Chroma Technology Corp, Bellows Falls, VT, USA).
2. Automated vessel analysis software (e.g., IKOSA CAM; KML Vision GmbH, Graz, Austria).
3. Manual vessel analysis software (e.g., Fiji; open-source software; https://imagej.net/software/fiji/downloads).
4. Standard ultrasound scanner (e.g., 15 MHz linear transducer GE Healthcare Ultrasound LOGIQ E9 (GE Healthcare Little Chalfont, UK)).
5. Microtome (e.g., Leica CM1900, Leica Biosystems Nussloch GmbH, Nussloch, Germany).
6. Embedding machine.
7. Water bath.
8. Drying cabinet.
9. Incubator for eggs.
10. Electrical pipette.

#### *2.4 Cells and Biological Materials*

1. Human hepatocellular carcinoma cell line HuH-7 (e.g., from RIKEN BioResource Center, Kyoto, Japan) or any other suitable tumor cell line.
2. Fertilized hen's eggs.
3. Corning® Matrigel® Basement Membrane Matrix.

#### *2.5 Antibodies for Immunohistochemistry*

1. Monoclonal alpha-SMA antibody (A2547, Sigma-Aldrich, St. Louis, Missouri, USA).
2. Monoclonal cleaved caspase-3 antibodies (Asp175, Cell Signaling Technology, Cambridge, UK).
3. Monoclonal mouse anti-proliferating cell nuclear antigen, clone PC10 (Dako Denmark A/S, Glostrup, Denmark).
4. Secondary antibody biotinylated polyclonal goat anti-mouse immunoglobulin (P0447, Dako Denmark A/S, Glostrup, Denmark).

## 3 Methods

#### *3.1 Chicken Egg Development and Incubation*

1. On **day 0**, place the eggs horizontally on a plastic egg tray in an incubator at 38 °C and an approximate relative humidity of 55%. Storing the eggs horizontally ensures that the chorioallantoic membrane (CAM) detaches from the eggshell.
2. On **day 3** of incubation, the eggs are opened. Remove one egg from the incubator (*see* **Note 1**).
3. First, remove 5–6 mL of albumen from the egg depending on the size of the egg (*see* Fig. 1a, b). This serves the purpose to lower the liquid level in the egg and allows the opening of the egg on the upward-facing side without damaging the CAM. To do this, pierce a small hole with a cannula at the blunt end of the egg, lower the cannula downwards, and remove 6 mL by aspiration. Carefully seal the hole with adhesive tape.
4. Now open the egg on the side facing upwards. To do this, first stabilize the eggshell with a strip of Leucopor® tape. This ensures that the eggshell does not crumble so easily and that fewer particles of the shell fall onto the CAM.
5. Finally open the shell of the egg carefully by drilling a small hole in the shell using sterile scissors (*see* Fig. 1c, d). Starting from this hole, a bigger circular hole can be cut into the top of the egg. Discard the removed shell.
6. After opening the eggs, unfertilized eggs can be removed from the experiment (*see* **Note 2**).
7. After opening the eggshell, cover the aperture with PARAFILM® to avoid evaporation, and put the egg back into the incubator (*see* Fig. 1e).

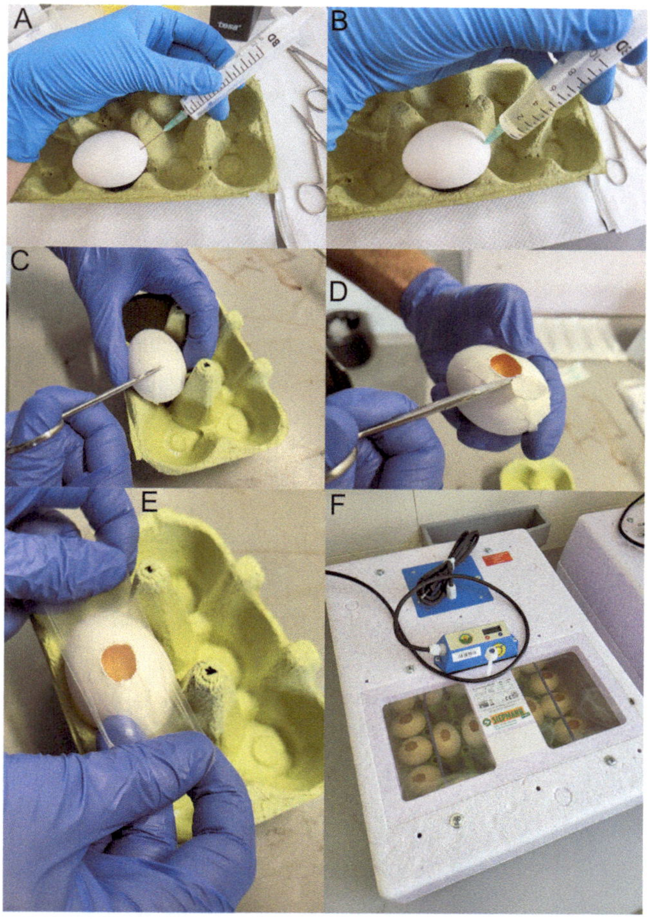


**Fig. 1** Cutting of a window into the eggshell. (**a**, **b**) Removal of 5–6 mL albumen on day 3 of egg incubation. (**c**, **d**) Opening of the eggs on the upward-facing side, as described in Subheading 3.1. (**e**) Coverage of the egg with parafilm. (**f**) Incubation of the egg in a commercially available incubator

### 3.2 Tumor Cell Culture

Perform the cell culture of the tumor cells as required by the cell line (*see* **Note 3**).

For HuH-7 cells:

1. Culture the HuH-7 cells in full cell culture medium in an incubator at 37 °C and 5% $CO_2$.
2. Passage cells at 1:3 to 1:5 dilution every 3–5 days. The doubling time of HuH-7 cells is about 30 h.
3. Propagate the tumor cells to achieve a cell number sufficient to use five million cells per egg in the experiment. Depending on the tumor cell line, the number of eggs on which a solid tumor is formed might vary. Thus, the number of eggs required might vary, and the propagation of the cells must be started at an early stage.

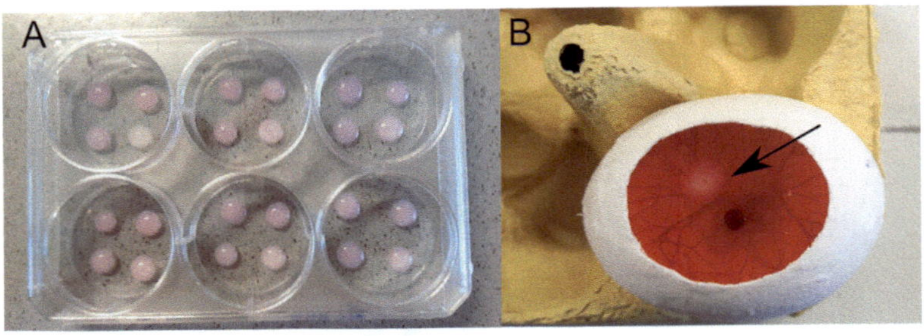


**Fig. 2** Tumor cell inoculation onto the CAM. (**a**) Incorporation of tumor cells into Matrigel™ to form a 3D culture. (**b**) Transfer of 3D cultures (arrow) onto the CAM (**b**)

4. Harvest the tumor cells on **day 6** of the incubation of the eggs, i.e., 1 day before transferring the cells onto the egg, by tryptic digestion. The conditions for harvesting the cells depend on the cell line used and its adherence to the cell culture plate.
5. Count the cells using a Neubauer counting chamber or any other tool you preferably use.
6. To transfer the cells onto the CAM, they need to be assembled using the basement membrane matrix Matrigel™. To do so, distribute the cells at five million per 1.5 mL tube.
7. Centrifuge at 260 × *g* for 10 min. Carefully pipette off the supernatant and discard it. Now mix the cell pellet carefully and quickly with the 25 μL ice-cold Matrigel™, as far as possible without trapping air bubbles.
8. Pipette the cells in Matrigel™ onto the bottom of a 6-well plate so that the mass can harden, and a small round 3D culture is created (*see* Fig. 2a).
9. Incubate the 6-well plate for 30 min at 37 °C in an incubator. After incubation, the Matrigel™ should have been set to a firm consistency.
10. Cover the 3D cultures with cell culture medium, and further incubate them overnight.

### 3.3 Tumor Cell Inoculation onto the CAM

1. On **day 7** of incubation of the eggs, the 3D cultures with tumor cells that are prepared the day before can be transferred onto the CAM. Remove one egg from the incubator (*see* **Note 1**).
2. Select a site on the ventral side of the embryo which is well vascularized. Carefully incise the CAM using a single-use scalpel over a distance of approximately 0.5 cm opening the upper cellular layer.
3. Place the 3D culture onto the incision, and cover it with 20 μL ice-cold Matrigel™ by pipetting it onto the culture (*see*

Fig. 2b). This protects against cell desiccation and further immobilizes the tumor on the CAM. Put the egg back into the incubator.

### *3.4 Administration of Therapeutic Agents to the Chicken Embryo*

Approximately from **day 9** of egg development, intravascular administration can be performed as the vascular diameter has to be wide enough for a needle to be inserted into the lumen without excessive damage, although earlier time points of intravascular injection have been published (*see* table in [21]).

1. Perform injections using a 30G1/2 needle (0.3 × 13 mm). Volumes between 10 and 50 μL are feasible as injection volume.
2. Select a vessel with a diameter of at least 0.3 mm for administration.
3. Carefully insert the needle into the vessel lumen (*see* **Note 4**). Avoid piercing through the vessel as this will result in excessive bleeding into the chorion (*see* Fig. 3c).
4. Slowly inject the fluid into the vascular network.
5. To prevent bleeding after injection, the punctuation site should be immediately coagulated by etching with silver nitrate (*see* Fig. 3d).

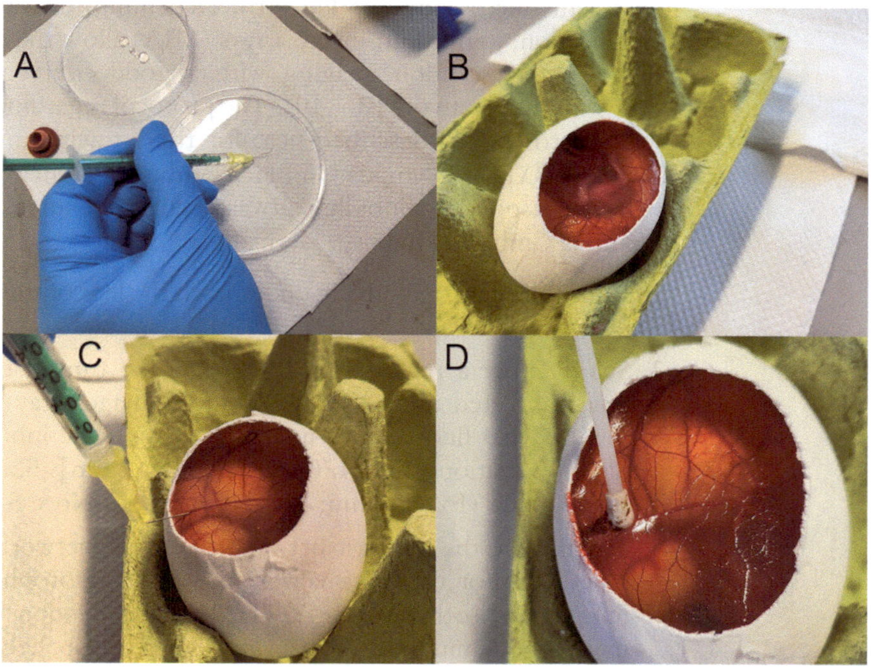


**Fig. 3** Intravascular application of substances to the CAM. From day 9 of egg incubation onwards, the intravascular injection of therapeutic agents is possible as described in Subheading 3.4. (**a**) Sterile tilting of the needle by approximately 30°. (**b**) Slight tilting of the egg to increase tension of the CAM. (**c**) Punctuation of the vessel lumen. (**d**) Coagulation of the punctuation side by etching with silver nitrate

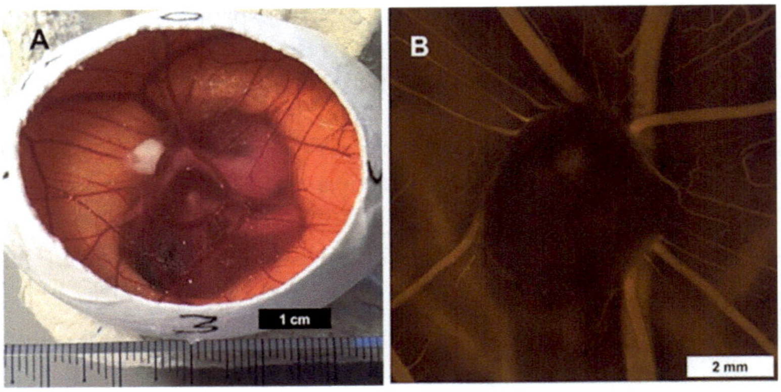


**Fig. 4** Photographic (**a**) and fluorescence-microscopy (**b**) imaging of the CAM. (**a**) Photographic documentation of tumor growth and embryonic development. (**b**) Fluorescence microscopy of the tumor ingrown into the CAM after intravascular application of TRITC dextrane

For local administration, the specific substance is either administered onto the CAM or the tumor growing on the CAM, respectively.

### *3.5 In Ovo Imaging with (Fluorescence) Microscopy*

Visualization of the tumor and the CAM's vascular network can be performed using photography (*see* Fig. 4a), microscopy as well as fluorescent microscopy (*see* Fig. 4b). Depending on the specific magnification, joint multi-image arrays (MIA) allow the combination of high-resolution imaging with a good overview of the CAM's vascular network [36, 40]. The protocol regarding fluorescence microscopy is described here for the intravital microscope (Olympus BXFM, Olympus GmbH, Hamburg, Germany) as well as the corresponding inbuilt software cellSens Dimension 1.14 (Olympus GmbH, Hamburg, Germany).

1. Place the eggs horizontally underneath the lens of the intravital microscope.
2. Select either a region of interest, like the tumor, or a joint image (created using the inbuilt multi-image array function) for intravital fluorescence microscopy using the inbuilt green filter (excitation [Ex]: 470 nm/emission [Em]: 525 nm) or orange filter (Ex: 545 nm; Em: 605 nm) (*see* **Note 5**).
3. To enhance the contrast between the vessel network and the membrane for better visualization, various fluorophores, like fluorescein isothiocyanate (FITC)-labeled dextran or tetramethylrhodamine isothiocyanate (TRITC)-labeled dextran, can be administered into the vascular network (*see* Fig. 4b) [40, 48].

4. To determine the functional vessel density and the number of junctions and branches, either manually mark the vessels in an overlay technique [36] or automatically with specialized software, like IKOSA® CAM.
5. The software automatically calculates the length of all vessels marked and the vessel density (vd) per $\mu m^2$ is determined by dividing the overall vessel length by the ROI area.
6. Perform a vessel junction analysis using the acquired data saved in JPEG image format. Subsequently import the images into the open-source software Fiji.
7. Convert the images then into black and white using Fiji's "Make binary" function.
8. Afterwards, skeletonize the binary file using the "Skeletonize" function in Fiji.
9. Analyze the skeleton of the vascular network utilizing the "Analyze Skeleton" function which automatically determines the number of branches and junctions [49].

### 3.6 Ultrasonography in the CAM Assay

Ultrasonography is suited for the visualization of tumor volume and tumor vascularization. It can be repetitively performed without a significant increase in dropout rates [6]. Any modern, high-resolution ultrasound should be able to allow sufficient imaging of the abovementioned parameters. In this specific protocol, ultrasonography is performed with the GE Healthcare Ultrasound LOGIQ E9 15 MHz linear transducer in the B-Mode (Gain 35).

1. Place one egg on a plain surface on a mount allowing very little movement.
2. Fill the space between the CAM and shell opening with 3–5 mL of 0.9% NaCl until the liquid level is in line with the apical opening in the eggshell.
3. Position the transducer on the liquid surface right above the tumor until imaging of the tumor is possible (*see* Fig. 5a and *see* **Note 6**).
4. Visualize the maximum longitudinal, as well as the maximum transversal tumor diameter, to allow measuring of tumor size in three plains.
5. Freeze the respective pictures using the inbuilt "Freeze" function. Use the split-screen function, installed in most modern ultrasound devices, to visualize the tumor's maximum diameters in one image (*see* Fig. 5b, c).
6. Measure the tumor length, width, and thickness using the "measure distance" function.

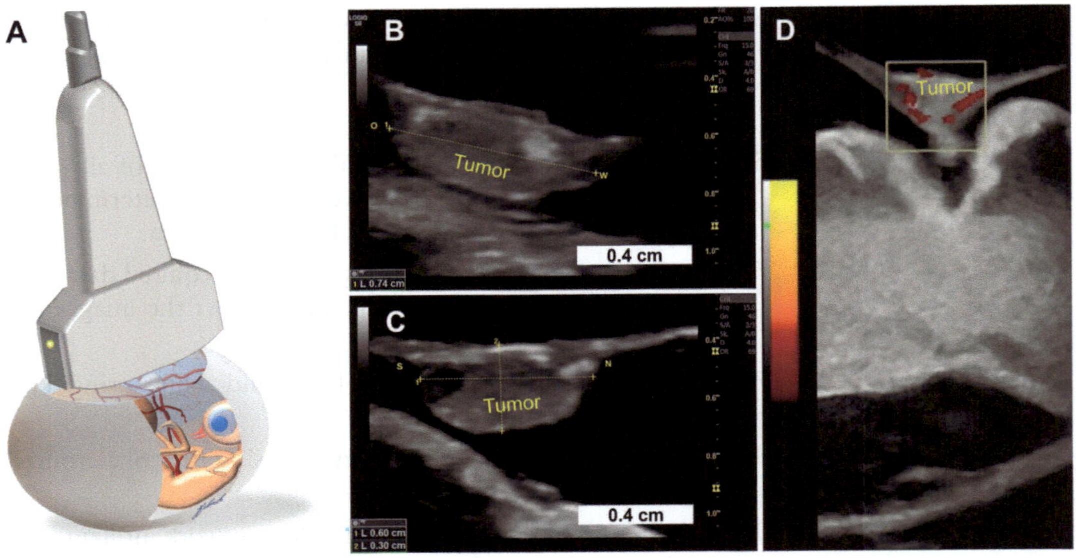


**Fig. 5** Ultrasonography of tumor in the CAM and chicken embryo. (**a**) Schematic setup of ultrasonography to visualize tumor, CAM, and the embryo. (**b**, **c**) Measurements of tumor size using ultrasonography of the inoculated tumor in ovo in the longitudinal (**b**) and coronar (**c**) plane. (**d**) Visualization of intratumor blood flow with fine flow color-duplex ultrasonography

7. Visualize the vascular network inside the tumor using the built-in duplex mode. Video sequences of duplex imaging allow an offline analysis regarding the intensity of perfusion of the tumor stroma (*see* Fig. 5d).
8. For repetitive measurements, remove the 0.9% NaCl solution using an electrical pipette with a 10 mL sterile tube (*see* **Note 7**).

As tumors grow in a rounded shape, calculation of an estimated tumor volume is performed by inserting the specific length of the three diameters in the tri-axial ellipsoid equation

$$V = \frac{4}{3} \times \pi \times [0.5 \times d1(\text{sagittal})] \times [0.5 \times d2(\text{transversal})] \times [0.5 \times d2(\text{coronar})].$$

### *3.7 Preparation of Tumor, CAM, and Organs for Pathological Analysis*

1. After completion of the study protocol, usually on **day 14** of incubation, kill the embryo by decapitation using surgical scissors.
2. To harvest the samples for immunohistochemical analysis, break the egg over a petri dish (*see* **Note 8**).
3. If the CAM does not detach from the eggshell on its own, carefully help with a spatula.

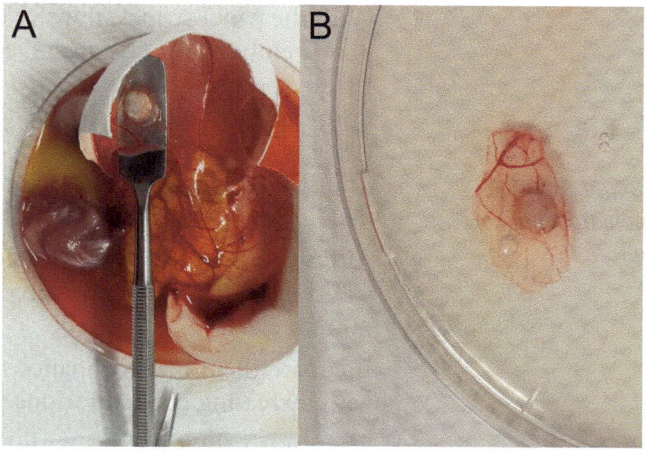


**Fig. 6** Harvesting of tissue samples for immunohistochemistry. (**a**) After completion of the study protocol, usually on day 14 of incubation, the embryo is killed. (**b**) Tissue samples from the CAM, the tumor, and the organs of the embryo can be isolated

4. In the next step, cut out a piece of the CAM, and transfer it with a spatula into a petri dish with autoclaved and distilled water. The water helps the CAM to unfold, and you can transfer the fragile tissue to a strip of paper without folds (*see* Fig. 6).
5. Proceed with tumors in the CAM in the same way as described for the CAM itself (*see* **Note 9**).
6. Release the head and body of the embryo from the amniotic sac. Fix the embryo's extremities with pins on a polystyrene board to allow clean scalpel guidance. Open the abdomen of the embryo with a superficial skin incision. Remove the sternum and the ribs.
7. Cut the vessels leading to the heart and remove the heart. Transfer it to paper stripes (*see* **Note 10**).
8. Carefully expose the liver which can be identified by its typical color. Transfer it to paper stripes (*see* **Note 10**).
9. To retrieve to the kidneys, it is necessary to remove the entire gastrointestinal tract of the embryo with forceps and a scalpel. Separate the kidneys from their vascular supply through a careful incision at the renal hilus. Detach the organ carefully from the surrounding structures, and carefully lift it with a spatula on paper (*see* **Note 10**).
10. For the preparation of the brain, fix the head at the beak, neck, and eye with pins on a polystyrene board. With a carefully guided incision remove the scalp. Remove the brain from the eyes with the scalpel. A deep incision below the brain separates the organ from other structures. Transfer the brain onto paper strips.

11. With the paper stripes, transfer the harvested tissue probes into plastic embedding cassettes. Depending on the thickness of the tissue, sponges can be used to fixate the tissue in the embedding cassette.

### 3.8 Preparation for Immunohistochemical Staining

1. Transfer the immobilized tissue samples into a 4% formaldehyde solution for at least 24 h, not more than 72 h.
2. Afterwards, remove the samples from the 4% formaldehyde solution, and wash them three times with distilled water for 20 min each.
3. Incubate them in isopropanol solution in distilled water (v/v) with increasing concentrations (80%/90%/100%) for 1 h each.
4. Then wash the cassettes with the samples again with distilled water and incubate them in pure xylene for 2 h.
5. Transfer the samples into a fresh bath with xylene, and incubate them overnight so that the samples are covered completely.
6. Embed the samples in paraffin at 58 °C. Tissues can be embedded into paraffin using specialized embedding machines. To make cross-sections through the tumor and the CAM, the correct orientation of the tissue in the paraffin must be ensured.
7. Cut 4 μm-thick tissue sections using a microtome.
8. Float the sections in a 56 °C water bath.
9. Mount the sections onto histological super-frost slides, and dry them overnight in a drying cabinet.

### 3.9 Hematoxylin and Eosin (HE) Staining

The HE staining provides an overview of the morphology and the structure of the examined tissues (*see* Fig. 7).

1. To perform HE stains, the tissue samples must be rehydrated. To do this, start by incubating the samples two times for 5 min in xylene.

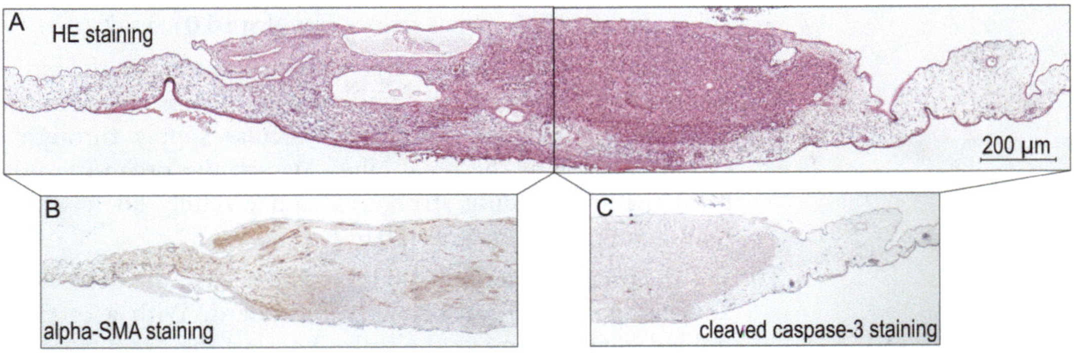


**Fig. 7** Example images of immunohistochemical staining. (**a**) Exemplary HE staining of a tumor growing in the CAM. (**b**) Section of the alpha-SMA staining of this tumor. (**c**) Section of the cleaved caspase-3 staining of this tumor

2. Continue by incubating them in isopropanol solution in distilled water (v/v) with decreasing concentrations (100%/90%/80%/70%) for 5 min each.
3. Incubate slides in distilled water for 5 min.
4. Incubate slides in Mayer's hemalum for 5 min.
5. Rinse the slides under tap water until the blue discoloration of the water is no longer visible.
6. Incubate slides for another 5 min in tap water.
7. Incubate slides in distilled water for 2 min.
8. Subsequently, incubate slides in eosin for 2 min for the counterstaining.
9. Rinse slides with tap water.
10. Incubate slides in isopropanol solution with increasing concentrations (80%/90%/100%) for 2 min each and xylene for 10 min.
11. Finally, prepare slides for microscopy by embedding the specimen in Eukitt®.

### 3.10 Immunohistochemical Staining for Alpha-smooth Muscle Actin (a-SMA) and Cleaved Caspase-3

1. First, the tissue samples must be rehydrated. To do this, start by incubating the samples two times for 5 min in xylene.
2. Continue by incubating them in isopropanol solution in distilled water (v/v) with decreasing concentrations (100%/90%/80%/70%) for 5 min each.
3. Incubate slides in distilled water for 5 min.
4. Cook slides in citrate buffer (pH 6.0) for 30 min for antigenic demasking.
5. Incubate slides for 20 min in methanol/$H_2O_2$.
6. Block with goat normal serum for 45 min.
7. After preparation of the slides, incubate slides with monoclonal alpha-SMA antibody dissolved (1/1500) in PBS-BSA overnight at 4 °C.
8. The next day, incubate slides in $TBST_{20}$ buffer I for 5 min.
9. Rinse slides with PBS.
10. Incubate slides for 5 min in $TBST_{20}$ buffer II.
11. Use as secondary antibody biotinylated polyclonal goat anti-mouse immunoglobulin dissolved (1:250) in PBS-BSA for 30 min.
12. Wash slides two times with TBST20 buffer II.
13. For visualization, use horseradish peroxidase-conjugated streptavidin (1/200) dissolved in PBS.
14. Wash two times with TBST20 buffer II.

15. Incubate slides with DAB substrate.
16. For counterstaining staining perform staining with hemalum as described above.
17. Finally, slides can be prepared for microscopy by embedding the specimen in Eukitt®.
18. Carry out the staining of cleaved caspase-3 analogously to the above-described protocol with the first antibody in a mixing ratio of 1/50.

## 4 Notes

1. Whenever eggs are removed from the incubator, only remove one egg at a time from the incubator. The loss of heat can disrupt the development cycle of the eggs, thus minimize it.
2. A fertilized egg can be identified by a small white circle on the yolk which is called the germinal disc. Depending on the time point of opening, you may already be able to recognize the developing chicken embryo.
3. Perform the experiments at a passage number below 50, and test the cell culture regularly for mycoplasma to rule out contamination.
4. Before the punctuation of the vessel, tilting the needle by approximately 30° allows easier access to the vessel lumen (*see* Fig. 3a). Injection close to a vessel junction prevents excessive movement of the vessel. Tilt the egg slightly sideways to increase the CAM's tension and to limit the movement of the vessel plexus (*see* Fig. 3b).
5. The curvature of the CAM may result in a partial blur of the image when a MIA is stacked together. To avoid image blur, use the manual MIA function, and perform repetitive refocusing before moving to another area.
6. Most transducers exceed the diameter of the opening within the eggshell in size. Nevertheless, partial immersion of the ultrasound transducer still allows sufficient visualization of the CAM, and specific areas of interest can be magnified using the inbuilt "zoom" function.
7. It is of high importance to avoid direct mechanical trauma caused by the pipette when removing the NaCl. Trauma will result in bleeding or rupture of the CAM and at worst, death of the embryo.
8. To open the egg for histological examination, cut into the eggshell with surgical scissors about 1 cm deep on both sides of the egg. Then open the egg with a clean crack.

9. It is important that not only the tumor is excised, but that the surrounding CAM tissue is also cut out to allow complete tumor assessment and the evaluation of ingrowth of the tumor into the CAM later in the immunohistochemical staining.
10. It is very important that the histological preparation of the embryo is carried out very carefully. As the embryonic liver tissue is very fragile, lift it with a spatula on paper stripes for embedding. The same applies to the kidneys which are usually even more fragile in texture than the liver. Whenever possible, avoid using forceps which would destroy the fragile tissue, and instead use a spatula to move the tissue samples.

## References


1. Luepke NP (1985) Hen's egg chorioallantoic membrane test for irritation potential. Food Chem Toxicol 23(2):287–291
2. Ribatti D (2016) The chick embryo chorioallantoic membrane (CAM). A multifaceted experimental model. Mech Dev 141:70–77
3. DeFouw DO, Rizzo VJ, Steinfeld R, Feinberg RN (1989) Mapping of microcirculation in the chick chorioallantoic membrane during normal angiogenesis. Microvasc Res 38:136–147
4. Nowak-Sliwinska P, Segura T, Iruela-Arispe ML (2014) The chicken chorioallantoic membrane model in biology, medicine and bioengineering. Angiogenesis 17(4):779–804
5. Ribatti D, Tamma R, Annese T (2021) Chorioallantoic membrane vascularization. A meta-analysis. Exp Cell Res 405(2):112716
6. Eckrich J, Kugler P, Buhr CR, Ernst BP, Mendler S, Baumgart J, Brieger J, Wiesmann N (2020) Monitoring of tumor growth and vascularization with repetitive ultrasonography in the chicken chorioallantoic-membrane-assay. Sci Rep 10(1):18585
7. Chu P-Y, Koh AP-F, Antony J, Huang RY-J (2022) Applications of the chick chorioallantoic membrane as an alternative model for cancer studies. Cells Tissues Organs 211(2): 222–237
8. Komatsu A, Matsumoto K, Saito T, Muto M, Tamanoi F (2019) Patient derived chicken egg tumor model (PDcE model): current status and critical issues. Cell 8(5)
9. Moreno-Jiménez I, Hulsart-Billstrom G, Lanham SA, Janeczek AA, Kontouli N, Kanczler JM, Evans ND, Oreffo RO (2016) The chorioallantoic membrane (CAM) assay for the study of human bone regeneration: a refinement animal model for tissue engineering. Sci Rep 6: 32168
10. Marga Janse E, Jeurissen SH (1991) Ontogeny and function of two non-lymphoid cell populations in the chicken embryo. Immunobiology 182(5):472–481
11. Uloza V, Kuzminienė A, Palubinskienė J, Balnytė I, Ulozienė I, Valančiūtė A (2017) Model of human recurrent respiratory papilloma on chicken embryo chorioallantoic membrane for tumor angiogenesis research. Histol Histopathol 32(7):699–710
12. Dohle DS, Pasa SD, Gustmann S, Laub M, Wissler JH, Jennissen HP, Dünker N (2009) Chick ex ovo culture and ex ovo CAM assay: how it really works. J Vis Exp 33
13. Kunz P, Schenker A, Sähr H, Lehner B, Fellenberg J (2019) Optimization of the chicken chorioallantoic membrane assay as reliable in vivo model for the analysis of osteosarcoma. PLoS One 14(4):e0215312
14. Kunzi-Rapp K, Genze F, Küfer R, Reich E, Hauptmann RE, Gschwend JE (2001) Chorioallantoic membrane assay: vascularized 3-dimensional cell culture system for human prostate cancer cells as an animal substitute model. J Orol 166:152–157
15. Lokman NA, Elder ASF, Ricciardelli C, Oehler MK (2012) Chick chorioallantoic membrane (CAM) assay as an in vivo model to study the effect of newly identified molecules on ovarian cancer invasion and metastasis. Int J Mol Sci 13(8):9959–9970
16. Ferician O, Cimpean AM, Avram S, Raica M (2015) Endostatin effects on tumor cells and vascular network of human renal cell carcinoma implanted on p chick embryo chorioallantoic membrane. Anticancer Res 35(12):6521–6528
17. Marzullo A, Vacca A, Roncali L, Pollice L, Ribatti D (1998) Angiogenesis in hepatocellular carcinoma: an experimental study in the

chick embryo chorioallantoic membrane. Int J Oncol 13(1):17–21

18. Liu M, Scanlon CS, Banerjee R, Russo N, Inglehart RC, Willis AL, Weiss SJ, D'Silva NJ (2013) The histone methyltransferase EZH2 mediates tumor progression on the chick chorioallantoic membrane assay, a novel model of head and neck squamous cell carcinoma. Transl Oncol 6(3):273–281
19. Vu BT, Shahin SA, Croissant J, Fatieiev Y, Matsumoto K, Le-Hoang Doan T, Yik T, Simargi S, Conteras A, Ratliff L, Jimenez CM, Raehm L, Khashab N, Durand J-O, Glackin C, Tamanoi F (2018) Chick chorioallantoic membrane assay as an in vivo model to study the effect of nanoparticle-based anticancer drugs in ovarian cancer. Sci Rep 8(1):8524
20. Ribatti D, Nico B, Pezzolo A, Vacca A, Meazza R, Cinti R, Carlini B, Parodi F, Pistoia V, Corrias MV (2006) Angiogenesis in a human neuroblastoma xenograft model: mechanisms and inhibition by tumour-derived interferon-gamma. Br J Cancer 94(12): 1845–1852
21. Buhr CR, Wiesmann N, Tanner RC, Brieger J, Eckrich J (2020) The chorioallantoic membrane assay in nanotoxicological research-an alternative for in vivo experimentation. Nano 10(12)
22. Knighton D, Ausprunk D, Tapper D, Folkman J (1977) Avascular and vascular phases of tumour groath in the chick embryo. Br J Cancer 35:347–356
23. Richmond A, Su Y (2008) Mouse xenograft models vs GEM models for human cancer therapeutics. Dis Model Mech 1(2–3):78–82
24. Bosma GC, Custer RP, Bosma MJ (1983) A severe combined immunodeficiency mutation in the mouse. Nature 10(301):527–530
25. Maruo K, Ueyama Y, Kuwahara Y, Hioki K, Saito M, Nomura T, Tamaoki N (1982) Human tumour xenografts in athymic rats and their age dependence. Br J Cancer 786-789(45)
26. Ribatti D (2008) The chick embryo chorioallantoic membrane in the study of tumor angiogenesis. Romanian J Morphol Embryol 49(2): 131–135
27. Baiguera S, Gonfiotti A, Jaus M, Comin CE, Paglierani M, Del Gaudio C, Bianco A, Ribatti D, Macchiarini P (2011) Development of bioengineered human larynx. Biomaterials 32(19):4433–4442
28. Wiesmann N, Mendler S, Buhr CR, Ritz U, Kämmerer PW, Brieger J (2021) Zinc oxide nanoparticles exhibit favorable properties to promote tissue integration of biomaterials. Biomedicine 9(10)
29. Vinardell M, Mitjans M (2015) Antitumor activities of metal oxide nanoparticles. Nano 5(2):1004–1021
30. Buhr CR, Eckrich J, Kluenker M, Bruns K, Wiesmann N, Tremel W, Brieger J (2021) Determination of the LD50 with the chick embryo chorioallantoic membrane (CAM) assay as a promising alternative in nanotoxicological evaluation. Nanotoxicology 15(5): 690–705
31. Kue CS, Tan KY, Lam ML, Lee HB (2015) Chick embryo chorioallantoic membrane (CAM): an alternative predictive model in acute toxicological studies for anti-cancer drugs. Exp Anim 64(2):129–138
32. Grodzik M, Sawosz E (2006) The influence of silver nanoparticles on chicken embryo development and bursa of Fabricius morphology. J Anim Feed Sci 15(Suppl. 1):111–114
33. Blasi P, Schoubben A, Traina G, Manfroni G, Barberini L, Alberti PF, Cirotto C, Ricci M (2013) Lipid nanoparticles for brain targeting III. Long-term stability and in vivo toxicity. Int J Pharm 454(1):316–323
34. Ribatti D, Vacca A, Ranieri G, Sorino S, Roncali L (1996) The chick embryo chorioallantoic membrane as an in vivo wound healing model. Pathol Res Pract 192(10):1068–1076
35. Moreno-Jiménez I, Kanczler JM, Hulsart-Billstrom G, Inglis S, Oreffo ROC (2017) * The chorioallantoic membrane assay for biomaterial testing in tissue engineering: a short-term in vivo preclinical model. Tissue Eng Part C Methods 23(12):938–952
36. Heimes D, Wiesmann N, Eckrich J, Brieger J, Mattyasovszky S, Proff P, Weber M, Deschner J, Al-Nawas B, Kämmerer PW (2020) In vivo modulation of angiogenesis and immune response on a collagen matrix via extracorporeal shockwaves. Int J Mol Sci 21(20)
37. Peters K, Wiesmann N, Heimes D, Schwab R, Kämmerer PW, Al-Nawas B, Unger RE, Hasenburg A, Brenner W (2022) Extracorporeal shock wave therapy improves in vitro formation of multilayered epithelium of oral mucosa equivalents. Biomedicine 10(3)
38. Maia ALC, de Aguiar Ferreira C, de Barros ALB, Silva ATME, Ramaldes GA, da Silva Cunha Júnior A, de Pádua Oliveira DC, Fernandes C, Soares DCF (2018) Vincristine-loaded hydroxyapatite nanoparticles as a potential delivery system for bone cancer therapy. J Drug Target 26(7):592–603

39. Irvin MW, Zijlstra A, Wikswo JP, Pozzi A (2014) Techniques and assays for the study of angiogenesis. Exp Biol Med (Maywood) 239(11):1476–1488
40. Kugler P, Becker S, Welz C, Wiesmann N, Sax J, Buhr CR, Thoma MH, Brieger J, Eckrich J (2022) Cold atmospheric plasma reduces vessel density and increases vascular permeability and apoptotic cell death in solid tumors. Cancers (Basel) 14(10):2432
41. Henning AL, Jiang MX, Yalcin HC, Butcher JT (2011) Quantitative three-dimensional imaging of live avian embryonic morphogenesis via micro-computed tomography. Dev Dyn 240(8):1949–1957
42. Huang C, Lowerison MR, Lucien F, Gong P, Wang D, Song P, Chen S (2019) Noninvasive contrast-free 3D evaluation of tumor angiogenesis with ultrasensitive ultrasound microvessel imaging. Sci Rep 9(1):4907
43. Kishore AS, Surekha PA, Sekhar PVR, Srinivas A, Murthy PB (2008) Hen egg chorioallantoic membrane bioassay: an in vitro alternative to draize eye irritation test for pesticide screening. Int J Toxicol 27(6):449–453
44. de Silva O, Rougier A, Dossou KG (1992) The HET-CAM test: a study of the irritation potential of chemicals and formulations. Altern Lab Anim 20(3):432–437
45. Vargas A, Zeisser-Labouèbe M, Lange N, Gurny R, Delie F (2007) The chick embryo and its chorioallantoic membrane (CAM) for the in vivo evaluation of drug delivery systems. Adv Drug Deliv Rev 59(11):1162–1176
46. Knudsen LE, Smith A, Törnqvist E, Forsby A, Tähti H (2019) Nordic symposium on "toxicology and pharmacology without animal experiments-will it be possible in the next 10 years?". Basic Clin Pharmacol Toxicol 124(5):560–567
47. Russell WMS, Burch RL (1960) The principles of humane experimental technique. Med J Aust 1(13):500
48. Weiss A, van Beijnum JR, Bonvin D, Jichlinski P, Dyson PJ, Griffioen AW, Nowak-Sliwinska P (2014) Low-dose angiostatic tyrosine kinase inhibitors improve photodynamic therapy for cancer: lack of vascular normalization. J Cell Mol Med 18(3):480–491
49. Eckrich J, Hoormann N, Kersten E, Piradashvili K, Wurm FR, Heller M, Becker S, Anusic T, Brieger J, Strieth S (2021) Surface modification of porous polyethylene implants with an albumin-based nanocarrier-release system. Biomedicine 9(10)

# Chapter 27

# Assessment of Tissue Viability by Functional Imaging of Membrane Potential

**Peter Kohl and Callum M. Zgierski-Johnston**

## Abstract

Electrical activity plays a key role in physiology, in particular for signaling and coordination. Cellular electrophysiology is often studied with micropipette-based techniques such as patch clamp and sharp electrodes, but for measurements at the tissue or organ scale, more integrated approaches are needed. Epifluorescence imaging of voltage-sensitive dyes ("optical mapping") is a tissue non-destructive approach to obtain insight into electrophysiology with high spatiotemporal resolution. Optical mapping has primarily been applied to excitable organs, especially the heart and brain. Action potential durations, conduction patterns, and conduction velocities can be determined from the recordings, providing information about electrophysiological mechanisms, including factors such as effects of pharmacological interventions, ion channel mutations, or tissue remodeling. Here, we describe the process for optical mapping of Langendorff-perfused mouse hearts, highlighting potential issues and key considerations.

**Key words** Optical mapping, Electrophysiology, Cardiac, Langendorff, Voltage imaging

## 1 Introduction

Electrical activity plays an important role in physiology, not only in excitable cells such as of the brain and heart but also in non-excitable cells, where changes in membrane voltage are involved in development and molecular signal transduction [1]. Cellular electrophysiology is most commonly studied using the patch-clamp technique, which – depending on the configuration used – can allow one to explore electrical activity from single channels to whole cells. Patch clamp can be used on single cells and cells in 2D monolayers, but larger multicellular samples including tissues and organs require alternative techniques such as sharp electrode recordings, multi-electrode arrays, or epifluorescence

**Supplementary Information** The online version contains supplementary material available at https://doi.org/10.1007/978-1-0716-3052-5_27.

Oliver Friedrich and Daniel F. Gilbert (eds.), *Cell Viability Assays: Methods and Protocols*, Methods in Molecular Biology, vol. 2644, https://doi.org/10.1007/978-1-0716-3052-5_27,

measurements of voltage-sensitive dyes (optical mapping). The sharp electrode technique permits assessment of single-cell voltage dynamics within a tissue but does not support tissue-level electrophysiology assessment. Multi-electrode arrays allow one to assess electrophysiology across tissue or organs but are limited in their spatial resolution and only record extracellular electrical potentials. While extracellular electrical potentials give a summation of the underlying activity, they miss transmembrane voltage dynamics that reflect underlying cell health. Optical mapping provides high spatiotemporal resolution recordings of changes in transmembrane voltage and can be used from single cells to whole organs in a contactless, tissue non-destructive manner, reducing the likelihood of altering the underlying physiology and facilitating the ability to combine measurements with other techniques.

First demonstrated for cardiac research in the 1970s, optical mapping is a key method in basic science for the investigation of cardiac electrophysiology in cells, tissue, and organs [2]. Optical mapping usually refers to the epifluorescence-based measurement of changes in transmembrane voltage but may include other parameters such as calcium or NADH transients. Measuring changes in transmembrane voltage allows one to discern local action potential (AP) shapes and spatial patterns of excitation and repolarization. As a result, key properties such as AP duration or conduction velocity can be determined with high spatial and temporal resolution, thereby providing insight into the electrophysiological state of the underlying tissue. This allows one to assess changes as a result of disease and treatment, providing insight into electrophysiological mechanisms which are often investigated in combination with genetic and optogenetic models [3–7].

In the past decade, decreasing camera costs, development of dyes with improved properties, 3D printing of setup components, and open-source approaches for optical mapping data acquisition and analysis have reduced the accessibility threshold for what used to be a highly specialized method requiring skills that were not readily available in many basic science research laboratories [8–10]. Therefore, it is increasingly important to understand how to interpret the signals obtained in terms of the underlying tissue physiology. Here, we describe a method for optical mapping of transmembrane voltage in isolated Langendorff-perfused mouse hearts, including data analysis and interpretation of the obtained results.

## 2 Materials

### 2.1 Solutions and Reagents

1. Physiological saline solution: 140 mM NaCl, 5.4 mM KCl, 10 mM HEPES, 1 mM $MgCl_2$, 10 mM glucose, 1.8 mM $CaCl_2$, pH 7.4, osmolarity 300 mOsm/L (*see* **Note 1**).

2. Heparinized physiological saline solution: as in 1, but with 10,000 IU/L heparin.
3. Blebbistatin: 10 mM stock dissolved in DMSO.
4. Voltage-sensitive dye (Di-4-ANBDQPQ; *see* **Note 2**): 1.4 mM stock dissolved in ethanol.

### 2.2 Heart Extraction and Cannulation

1. Mouse (*see* **Note 3**).
2. Short-term animal holding facility (*see* **Note 4**).
3. Stereomicroscope including light source.
4. Tools: aortic cannula (*see* **Note 5**), system for holding cannula (*see* **Note 6**), forceps and scissors.
5. 100 mL of heparinized physiological saline solution (*see* Subheading 2.1, **item 2**).
6. Glassware: 60 mL containers with screw caps, 50 mm petri dish.
7. Consumables: suture material, syringe.

### 2.3 Langendorff Perfusion

1. Circulating water bath.
2. Glassware: water-jacketed solution reservoir, water-jacketed heat exchanger with an additional top port to act as a bubble trap, and water-jacketed measurement bath.
3. Pump with dual head or two pumps.
4. Oxygenation system: e.g. sintered glass pieces or a membrane oxygenator.
5. 0.5 L of physiological saline solution (*see* Subheading 2.1, **item 1**).
6. Consumables: tubing, injection port for dye loading, tubing connectors, inline filters (*see* **Note 7**).
7. Data acquisition system including sensors: depending on physiological parameters of interest, e.g., pressure sensor, flow sensor, temperature sensor, and electrodes for pseudo-electrocardiogram (ECG) recordings.

### 2.4 Optical Mapping

1. Recording: camera, optical emission filter (collecting light above approximately 700 nm, e.g., Semrock FF01–775/140, *see* **Note 2**), light shaping approach: microscope, macroscope, or lens (*see* **Note 8**).
2. Illumination: LED selection based on dye excitation parameters, optical excitation filters (centered around 640 nm, e.g., Chroma ZET642/20X, *see* **Note 2**), LED driver.
3. 500 μL of blebbistatin (*see* Subheading 2.1, **item 3**).
4. 15 μL of voltage-sensitive dye (*see* Subheading 2.1, **item 4**).
5. Tools: microsyringe and repeating syringe dispenser.

### 2.5 Data Analysis

1. Workstation (*see* **Note 9**).
2. Analysis software (*see* **Note 10**).

## 3 Methods

### 3.1 Langendorff System Preparation (see Note 11)

1. Turn on thermocirculator and rinse Langendorff system with distilled water.
2. Fill reservoir with solution and bubble with oxygen for at least half an hour prior to perfusion.
3. Fill Langendorff system, bubble-free, with oxygenated solution just prior to commencing experiment.

### 3.2 Heart Extraction and Cannulation

1. Fill containers and syringe with saline + heparin solution. Place one bottle in a 37 °C water bath; place the other bottle and the syringe in a 4 °C refrigerator. Allow time for equilibration of fluids to target temperature.
2. Prepare two sutures by forming a loop and wrapping one end twice around the other (*see* Fig. 1b).
3. Remove the syringe from fridge and prefill aortic cannula; place it in the holder under the stereomicroscope. Place both sutures over the cannula.
4. Sacrifice the mouse according to local guidelines.

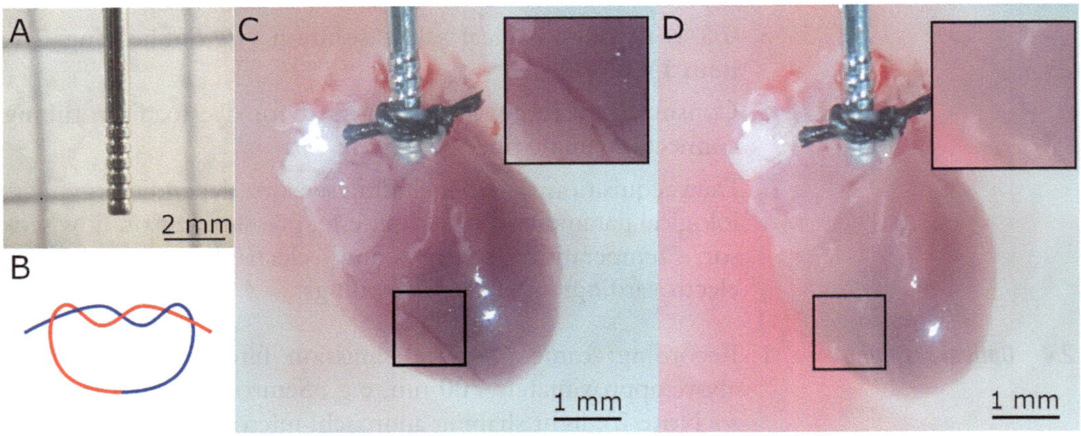


**Fig. 1** Key considerations for aortic cannulation. (**a**) Magnified image of our aortic cannula showing the flat tip and grooves that help the aorta and sutures stay on the cannula. (**b**) Schematic showing the configuration of the sutures once pre-tied and ready to be used. (**c**) Cannulated mouse heart prior to perfusion with physiological saline solution. Importantly, the tip of the cannula is still visible in the semi-transparent aorta, if the cannula is inserted too far it prevents proper closing of the aortic valve. (**d**) Same heart as in (c) but following perfusion, leading to blood clearing from the coronary vasculature. Insets in (c) and (d) show magnification of visible coronary vasculature

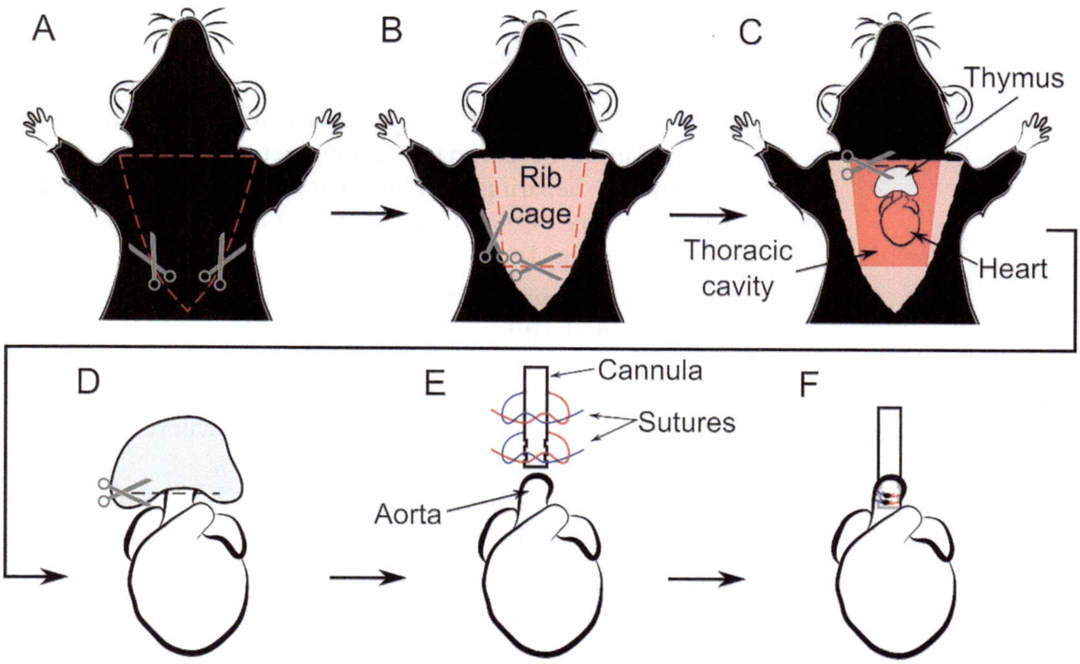


**Fig. 2** Heart extraction and cannulation. (**a**) Schematic of a mouse indicating the wide v that needs to be cut to remove the skin from the thorax. (**b**) Subsequently the rib cage is opened by cutting through the diaphragm and up the rib cage on either side. (**c**) The heart is then extracted by grasping the thymus with fine forceps and cutting above and behind the thymus and heart. (**d**) The thymus and other extracardiac tissue is then removed from the heart using fine forceps. (**e**) The aorta is cleaned and inserted into the cannula and sutures tied to hold the aorta in place leading to the cannulated mouse heart (**f**)

5. Holding the skin above the sternum with forceps, use scissors to cut the skin in a wide V from the tip of the sternum towards the shoulders (*see* Fig. 2).
6. Cut open the thorax below the sternum then through the diaphragm and up the rib cage on either side in a U-shaped incision. Carefully open the rib cage, cutting the tissue between the inside of the chest wall and the organs to avoid placing excess mechanical stress on the heart (*see* Fig. 2).
7. Using fine forceps, grasp the thymus and aorta and cut above the thymus and around the back of the heart (*see* Fig. 2). Immediately place the heart into the container with 37 °C solution for a few heartbeats to remove blood from the cardiac cavities prior to transferring to the container with 4 °C solution to stop contractions and slow metabolism. Transfer heart and cooled solution to a petri dish at the stereomicroscope.
8. Gently remove extracardiac tissue using forceps, taking care to avoid touching the myocardium. Clean the aorta, cut across the distal part of the ascending aorta just below the first cephalic branch, and slide it on bubble-free to the aortic cannula (*see*

**Note 12** and Fig. 2). Tie off the first suture around the aorta, then tie off the second using either your hands or blunt forceps to ensure the suture is tight and will not slip. Cut open the pulmonary artery to avoid right-ventricular fluid build-up from accumulation of coronary venous return. Perfuse a small amount of heparinised solution to check that the coronary arteries empty of blood as visible by a change in color of the arteries (*see* Fig. 1c/d).

9. Transfer to Langendorff system and start perfusion; the heart should begin beating within a few seconds.

### *3.3 Langendorff Perfusion and Optical Mapping*

1. Once connected to the Langendorff system adjust flow rate (between 2 mL/min and 4 mL/min) to achieve a perfusion pressure of 8 kPa (60 mmHg; *see* **Note 13**).
2. Place ECG electrodes onto the right atrium and the apex to provide a surface ECG.
3. Add blebbistatin to the reservoir to a final concentration of 10 μM, contractions should cease within 10–20 min (*see* **Note 14**).
4. Inject 15 μL of 1.4 mM Di-4ANBDQPQ (dissolved in ethanol) in boluses of 0.5 μL using the syringe dispenser into the aortic cannula (*see* **Note 15**).
5. Turn room lights off to limit photobleaching and photodegradation of Di-4-ANBDQPQ and blebbistatin.
6. Position LED and focus camera to get bright signals with homogeneous illumination (*see* **Note 16**).
7. Optical mapping can then be performed by turning LED on, starting to save any physiological signals (e.g., ECG), and recording using the camera. The optical recordings are typically limited to a few seconds to avoid generating files that are too large for easy analysis. LED should be turned off between recordings to limit photobleaching of dye.

### *3.4 Data Analysis*

1. The data analysis workflow is shown in Fig. 3.
2. Remove background either manually or using an automated thresholding approach.
3. Invert signals (*see* **Note 17**) and normalize on a pixel-by-pixel basis (*see* **Note 18**).
4. Play signals frame by frame to gain first impression of conduction (*see* **Note 19** and Video 1).
5. Average signals and perform spatial and temporal filtering to increase the signal-to-noise ratio (*see* **Note 20**).
6. Perform spatial binning to further reduce noise, and then use tools within the software to obtain activation time maps and AP duration maps (*see* **Note 21**).

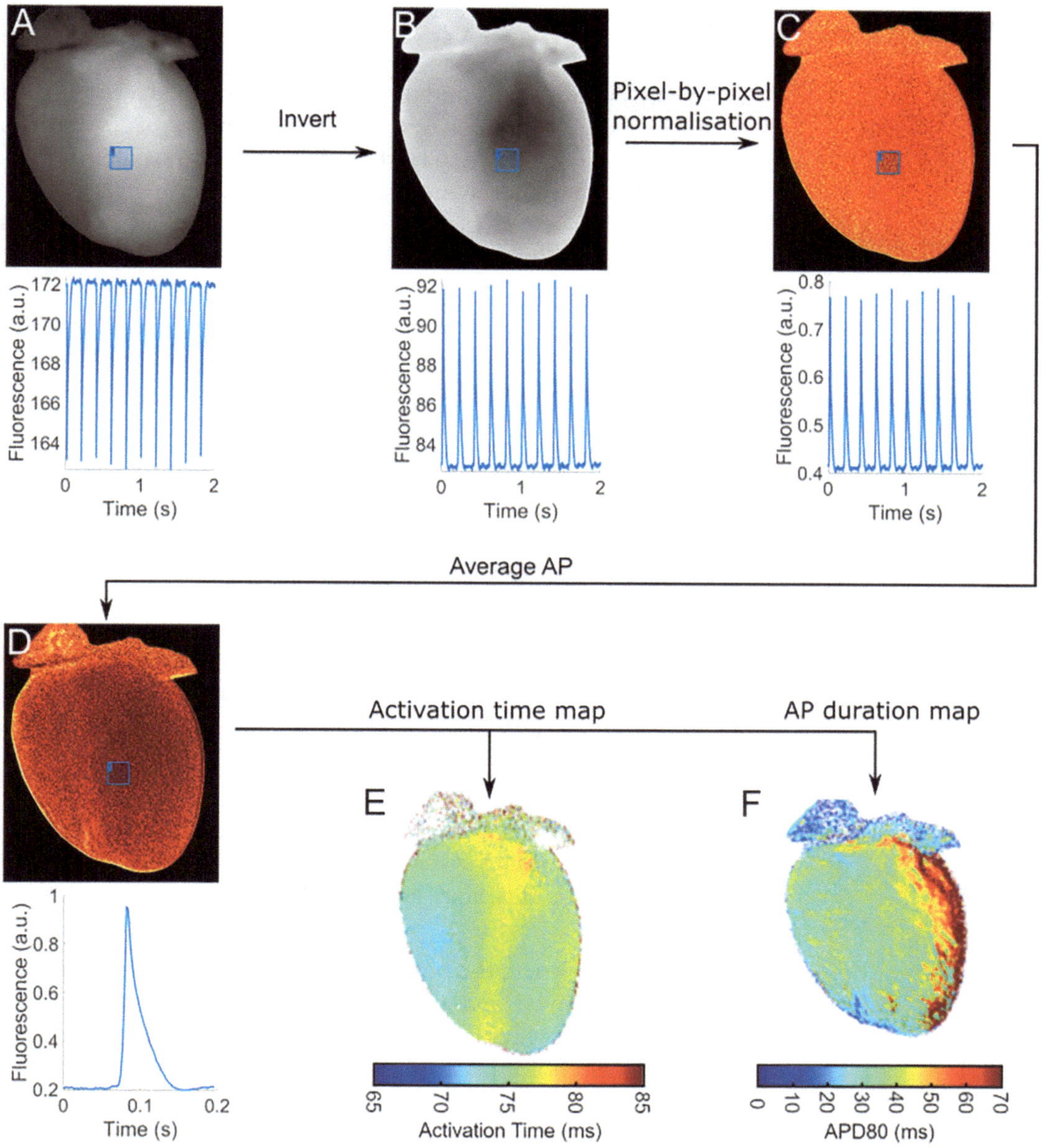


**Fig. 3** Optical mapping data processing. (**a**) Top image shows the first frame of an optical mapping video with background removed and a region of interest (ROI) on the left ventricle. The bottom trace shows the mean signal intensity from the ROI over time. (**b**) The data are then inverted by subtracting all values from the maximum value to yield upward-swinging AP more closely resembling typical AP recordings. (**c**) Subsequently, data are normalized in a pixel-by-pixel fashion to account for inhomogeneities in dye loading and illumination. (**d**) The data can then be averaged across the multiple consecutive AP, greatly reducing the noise while preserving signal amplitude. The resulting data can then be used to generate an activation time map (**e**), here showing the activation across the ventricles with 0 ms corresponding to the start of atrial activation, or an AP duration map showing the time from AP start to 80% repolarization (**f**), here showing the shorter AP durations in atria than in ventricles (note that the long AP duration on the right hand size of the image is likely an artefact resulting from residual motion). Interpretation of data at the edge of the visible area of the curved cardiac surface should be performed with care due to possible residual movement of the heart

## 4 Notes

1. Ideally, solutions should be prepared on the day of use. If solutions must be prepared in advance or are made up from stock, glucose and $CaCl_2$ should be added just before use. Osmolarity should be measured using an osmometer, as compounds do not dissociate equally well so that calculated values can be misleading.
2. Di-4-ANBDQPQ is used as an example in this chapter, but the techniques described are applicable to other voltage-sensitive dyes. When using a different dye, the light source and optical filters must be chosen to take into account dye-dependent excitation and emission spectra [11, 12].
3. The use of animals is necessary to study electrophysiology on the whole organ scale. Insights can be obtained from cell culture and computational modelling, but these still cannot adequately capture the complex electrophysiology of the heart, particularly when studying novel drugs or disease targets. Animal sacrifice must be performed in accordance with local guidelines and ethical approvals; handling of animals should only be performed by trained and approved personnel.
4. If mice need to be transferred to the laboratory from a central facility, they should have time to recover from transport. The exact time will depend on the local animal authority but should be at least an hour. During this time, their cage should be in an appropriate holding system ensuring species-appropriate temperature, humidity, and light control, while avoiding further disturbance.
5. In our experience, the best aortic cannula is a 21G needle with the tip removed and circular grooves near the outflow (*see* Fig. 1a). The circular grooves provide an additional surface grip to help hold the sutures in place.
6. Cannula holder could be a micro-manipulator or something as simple as a spring from a ballpoint pen fixed to a dish.
7. Mouse hearts are particularly sensitive to blockage of the coronary arteries by circulating particulates leading to ischemia. Therefore, inline meshes or filters with a fairly large pore size (we typically use 45 μm pore size mesh) should be used. The use of a microbial syringe filter is not recommended as this can remove blebbistatin which will be visible by a loss of color from the solution and accumulation in the filter.
8. The choice of camera and light shaping approach depend on the question of interest and the interplay of imaged area and spatial resolution required to address it. A macroscope or lens are typically the best approach for imaging the whole mouse

heart. If specific areas are of interest, opting for the higher magnification of a microscope may be useful. The choice of camera depends on the spatial and temporal resolution desired. A higher spatial resolution reduces temporal resolution, although a more expensive camera will typically shift this balance to higher spatial and temporal resolutions.

9. The workstation should be chosen with the technical specifications of the camera in mind (interface, data transfer rates, required write speeds). Approximately 0.2 GB of data are generated per second of recording with high bit-depth (12-bit), and spatial (~600 x 400 pixels) and temporal (~2 ms) resolution. Therefore, having high RAM (64 GB or more) to allow rapid post-processing of data is an advantage.

10. Most groups use custom-developed software, based on their specific hardware and physiological questions, but a range of open-source solutions exist and most groups are happy to share their software upon request [13–15].

11. We assume the reader is somewhat familiar with Langendorff perfusion and so we do not describe the setup of the perfusion system in-depth. More details regarding possible issues and considerations for Langendorff perfusion can be found in the review by Bell et al. [16].

12. It is very important to ensure that no air bubbles enter the coronary circulation as these can block cardiac vessels, leading to regional underperfusion. Before attaching the aorta, ensure the cannula tip is submerged in the perfusate, and then perfuse a small amount of solution through the cannula to check for bubbles prior to insertion of the cannula into the aorta.

13. Osmotic swelling increases myocardial tissue volume throughout perfusion, especially during the first 30 min [17]. This compresses coronary blood vessels effectively reducing their diameter. Therefore, ideally, a feedback system to control pump speed to maintain the aortic pressure should be used. If not, the perfusion pressure should be checked and the flow rate readjusted regularly to avoid damaging the blood vessels and heart due to excessive perfusion pressure.

14. Blebbistatin was first thought to be highly specific and to not alter the underlying electrophysiology. More recent studies indicate that AP duration and frequency of fibrillation are altered with blebbistatin [18, 19]. The lowest concentration of blebbistatin possible is best; however, this may lead to artefacts resulting from residual motion. Furthermore, blebbistatin is sensitive to blue light, leading to its inactivation and toxicity, so illumination of the blebbistatin-containing solution should be limited.

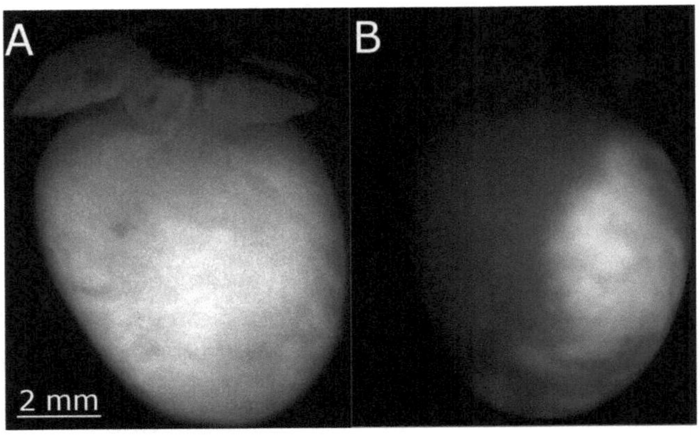


**Fig. 4** Hearts following dye loading. (**a**) Heart with good dye loading where atria and ventricles can be clearly seen in the fluorescence image. (**b**) Heart where partial blockage of the coronary vasculature has occurred meaning that only part of the left ventricle shows good dye loading. When examining signals there is still some dye present in other regions, but these regions are likely to have been rendered ischemic due to limited coronary flow

15. Allow 10 s between each injection to limit damage that may occur from the ethanol, in which the dye is dissolved (overall dye injection takes ~5 min). Ethanol still may cause some slowing of the heart rate, atrioventricular block, and arrhythmias. If the heart has not recovered to a stable heart rate of at least 300 bpm within 10 min following the last injection, the heart should be excluded (unless using a model where an alteration in heart rate is expected).
16. Another exclusion criterion is the absence of even dye loading across the heart. If areas of the heart appear especially dim, this often indicates that a coronary artery has been blocked during the isolation and perfusion process (*see* Fig. 4). Scarred areas will also appear dim due to the potentially limited coronary supply of the area—this is to be expected and so should not form an exclusion criterion.
17. Di-4-ANBDQPQ undergoes a reduction in fluorescence with increased transmembrane voltage when illuminated with red light, so emitted signals are typically inverted to resemble standard AP shapes. To reduce artefacts due to non-homogenous dye loading and motion, ratiometric imaging may be used by alternating blue and red illumination. This comes at the cost of halving the temporal resolution [11].
18. Inhomogeneities in dye loading and illumination as well as the relatively small percentage change in fluorescence upon changes in transmembrane voltage can partially obscure signals. By normalizing each pixel, based on its maximum and minimum values, the signal is homogenized, which simplifies

visualization and further processing. One issue here is that this can amplify noise in regions that have no or very low change in fluorescence with transmembrane voltage—non-excitable or "dead" tissue, scars, or areas with poor dye loading (*see* **Note 16**). If the non-excitable tissue presents acutely as a result of an intervention, normalization can be done using a previous recording when excitation still occurred. In all the other cases, care must be taken when interpreting results—one option is to define a minimum amplitude for the normalization.

19. A video of the processed fluorescence signals can give a first impression of viability—e.g., illustrate whether conduction spreads across atria and the ventricles in an orderly manner (*see* Video 1).
20. Averaging beats should only be performed if the heart is in a stable rhythm. If heart rate is changing or there are arrhythmias, averaging will instead reduce signal quality and result in a loss of information. Activation or AP duration maps can be generated on a single beat without averaging but spatial and temporal filtering may be necessary to reduce the noise.
21. *See* Fig. 3 for an example of normal activation timing across the ventricle and AP duration mapping. The activation timing allows one to identify areas of slowed down conduction, such as occurs in fibrosis or otherwise damaged tissue. AP duration maps allow one to identify alterations in repolarization that again may indicate damaged tissue.

## Acknowledgments

CM Zgierski-Johnston and P Kohl are members of SFB1425, funded by the Deutsche Forschungs-gemeinschaft (DFG, German Research Foundation)—Project #422681845.

## References

1. Levin M (2014) Molecular bioelectricity: how endogenous voltage potentials control cell behavior and instruct pattern regulation in vivo. Mol Biol Cell 25:3835–3850. https://doi.org/10.1091/mbc.E13-12-0708
2. Salama G, Morad M (1976) Merocyanine 540 as an optical probe of transmembrane electrical activity in the heart. Science 191:485–487
3. Han B, Trew ML, Zgierski-Johnston CM (2021) Cardiac conduction velocity, remodeling and arrhythmogenesis. Cells 10:2923. https://doi.org/10.3390/cells10112923
4. Zgierski-Johnston CM, Ayub S, Fernández MC et al (2020) Cardiac pacing using transmural multi-LED probes in channelrhodopsin-expressing mouse hearts. Prog Biophys Mol Biol 154:51–61. https://doi.org/10.1016/j.pbiomolbio.2019.11.004
5. Hansen BJ, Zhao J, Csepe T a., et al. (2015) Atrial fibrillation driven by micro-anatomic intramural re-entry revealed by simultaneous sub-epicardial and sub-endocardial optical mapping in explanted human hearts. Eur Heart J 1–12. https://doi.org/10.1093/eurheartj/ehv233

6. Hanna P, Dacey MJ, Brennan J et al (2021) Innervation and neuronal control of the mammalian sinoatrial node a comprehensive atlas. Circ Res 128:1279–1296. https://doi.org/10.1161/CIRCRESAHA.120.318458
7. Wang Z, Tapa S, Francis Stuart SD et al (2020) Aging disrupts normal time-of-day variation in cardiac electrophysiology. Circ Arrhythmia Electrophysiol:1019–1032. https://doi.org/10.1161/CIRCEP.119.008093
8. Huang YL, Walker AS, Miller EW (2015) A photostable silicon rhodamine platform for optical voltage sensing. J Am Chem Soc 137(10767–10):776. https://doi.org/10.1021/jacs.5b06644
9. Cathey B, Obaid S, Zolotarev AM et al (2019) Open-source multiparametric optocardiography. Sci Rep 9:721. https://doi.org/10.1038/s41598-018-36809-y
10. Marina-Breysse M, García-Escolano A, Vila-García J et al (2021) A complete and low-cost cardiac optical mapping system in translational animal models. Front Physiol 12:696270. https://doi.org/10.3389/fphys.2021.696270
11. Lee P, Bollensdorff C, Quinn TA et al (2011) Single-sensor system for spatially resolved, continuous, and multiparametric optical mapping of cardiac tissue. Hear Rhythm 8:1482–1491. https://doi.org/10.1016/j.hrthm.2011.03.061
12. Uzelac I, Crowley CJ, Iravanian S et al (2022) Methodology for cross-talk elimination in simultaneous voltage and calcium optical mapping measurements with semasbestic wavelengths. Front Physiol 13:1–13. https://doi.org/10.3389/fphys.2022.812968
13. Gloschat C, Aras K, Gupta S et al (2018) RHYTHM: an open source imaging toolkit for cardiac panoramic optical mapping. Sci Rep 8:1–12. https://doi.org/10.1038/s41598-018-21333-w
14. O'Shea C, Holmes AP, Yu TY et al (2019) ElectroMap: high-throughput open-source software for analysis and mapping of cardiac electrophysiology. Sci Rep 9:1–13. https://doi.org/10.1038/s41598-018-38263-2
15. Cooper BL, Gloschat C, Swift LM et al (2021) KairoSight: open-source software for the analysis of cardiac optical data collected from multiple species. Front Physiol 12:1–13. https://doi.org/10.3389/fphys.2021.752940
16. Bell RM, Mocanu MM, Yellon DM (2011) Retrograde heart perfusion: the Langendorff technique of isolated heart perfusion. J Mol Cell Cardiol 50:940–950. https://doi.org/10.1016/j.yjmcc.2011.02.018
17. Bub G, Camelliti P, Bollensdorff C et al (2010) Measurement and analysis of sarcomere length in rat cardiomyocytes in situ and in vitro. AJP Hear Circ Physiol 298:H1616–H1625. https://doi.org/10.1152/ajpheart.00481.2009
18. Kappadan V, Telele S, Uzelac I et al (2020) High-resolution optical measurement of cardiac restitution, contraction, and fibrillation dynamics in beating vs. Blebbistatin-uncoupled isolated rabbit hearts. Front Physiol 11:1–18. https://doi.org/10.3389/fphys.2020.00464
19. Brack KE, Narang R, Winter J, Ng GA (2013) The mechanical uncoupler blebbistatin is associated with significant electrophysiological effects in the isolated rabbit heart. Exp Physiol 98:1009–1027. https://doi.org/10.1113/expphysiol.2012.069369

# Chapter 28

# Multiplexed Viability Assays for High-Throughput Screening of Spheroids of Multiple Sizes

**Mohana Marimuthu and Thomas Gervais**

## Abstract

High-throughput (HT) drug screening is in high demand for successful drug discovery and personalized medicine. Spheroids act as a promising preclinical model for HT drug screening, which may decrease drug failures in clinical trials. Numerous spheroid-forming technological platforms are currently under development, which include synchronous, jumbo-sized, hanging drop, rotary, and nonadherent surface spheroid growth. Initial cell seeding concentration and time of culture play a vital role for spheroids to mimic the extracellular microenvironment of natural tissue, especially for HT preclinical evaluation. Hence microfluidic platforms become a potential technology to provide a confined space for the oxygen and nutrient gradients within the tissues while controlling the cell count and spheroid size in an HT manner. We describe here a microfluidic platform capable of generating spheroids of multiple sizes in a controlled manner with a predefined cell concentration for HT drug screening. Ovarian cancer spheroids grown on this microfluidic platform were evaluated for viability using a confocal microscope and flow cytometer. In addition, screening of the HT chemotherapeutic drug carboplatin was carried out on-chip to evaluate the impact of spheroid size on drug toxicity. This chapter summarizes a detailed protocol on microfluidic platform fabrication for spheroid growth, on-chip multi-sized spheroid analysis, and chemotherapeutic drug screening.

**Key words** Multi-sized spheroids, Microfluidics, On-chip imaging, Perfusion, High-throughput, Onco screening

## 1 Introduction

It is well established that two-dimensional (2D) cell cultures in static conditions or animal models do not replicate human physiology in a meaningful way, leading to a number of promising therapeutics being rejected [1] before clinical trials such as NCT04225299, NCT04105374. Cancer spheroids as tumor model for the study of pharmaceutical drug candidates have gained enormous interest through their physiological relevancy in terms of cell-cell/cell-extracellular matrix (ECM) interactions and signaling with gas/nutrient gradients to recreate tumor microenvironment (TME) [2]. Since 2020, there have been >30 clinical trials using

Oliver Friedrich and Daniel F. Gilbert (eds.), *Cell Viability Assays: Methods and Protocols*, Methods in Molecular Biology, vol. 2644, https://doi.org/10.1007/978-1-0716-3052-5_28,

organoid models for drug response and chemosensitivity predictions against various cancers. Examples of which include lung (NCT03655015), ovarian (NCT04768270), pancreatic (NCT03544255), and colon cancer (NCT04906733). Despite this increase in organoid models, difficulties in the formation of standardized and synchronized organoids still persist due to large disparities in organoid morphologies, cellular concentration, and culture times. Recent reports highlight the influence of spheroid (as a preclinical organoid model) age and size on the drug delivery and chemotherapeutic response [3, 4]. Nevertheless, most academic and commercial spheroid formation platforms still perform static culture of uniformly sized spheroids on nonadherent surfaces (e.g., Aggrewell plates of StemCell™ Technologies) or in hanging droplets (HD) (GravityPLUS™ plates of InSphero). Further, HD and spinner flasks approaches are limited in providing uniformity over size and shape, an important parameter to control and identify the IC50 of a therapeutic [3].

A promising approach to circumvent these drawbacks is the development of microfluidics-based "tumor spheroids or organoids on chip" [5]. While microfluidics-based chips can confer flow-mediated applications of physiologically relevant biochemical and mechanical stimuli (including shear stress and pressure), multicellular spheroids or organoids allow tumor cells to interact with other nonmalignant stromal and immune cells. As such, their combination as "organoids-on-a chip platform" can serve as excellent models to replicate the complex physicochemical and multicellular properties of the TME [6].

In the last decade, microfluidics platforms have emerged as an effective alternative for the generation of multi-sized spheroids as they present strategies able to control spheroid size and oxygen gradients through perfusion culture [6–10]. Efforts on studying multi-sized spheroids on microfluidics platforms are only restricted to nonadherent culture (lack of air-liquid interface) leading to inconsistent spheroid circularity and integrity. The combination of a conventional HD method within a microfluidics platform with a control over cellular concentration gradient could enable the size regulation of spheroids. We have demonstrated that spheroids of tunable size can be formed in a high-throughput (HT) microfluidics format that is both easy to use and reproducible [11]. In this chapter, we will detail the fabrication and operation of such devices to improve their dissemination. All designs introduced are patent free and can be used without intellectual property constraints.

The device relies on the HD method that incorporates a microfluidic trick—the use of geometrically tailored funnels with varying apex angles—to precisely define the number of cells accumulated in a microdrop, allowing precise control of spheroid diameter. The design is adapted to a HT format and enables perfusion without perturbing the spheroids in the HD, a very significant advantage for drug testing and toxicology studies. Further, HT multi-sized spheroid formation platform enables on-chip spheroid imaging and also an easy spheroid retrieval for downstream analysis. As a proof of concept, ovarian cancer spheroids of tunable sizes were synthesized on-chip. A chemotherapeutic drug, carboplatin, was used to test the impact of spheroid size against the chemo toxicity. On-chip imaging including size distribution measurements and successful retrieval of spheroids for flow cytometry analysis before and after drug treatment were also demonstrated. Key innovation is that the end user can generate multi-size or fixed-size spheroids of a predetermined size for drug evaluation under biomimetic environments.

## 2 Materials

### 2.1 Microchip

1. Poly(methyl methacrylate) (PMMA) sheets of different thicknesses such as 1, 2, 3, 6, and 8 mm.
2. Polydimethylsiloxane (PDMS) with curing agent (Sylgard 184).
3. Isopropanol.
4. Weighing boats.
5. Disposable plastic Pasteur pipettes.
6. Syringes.
7. Hollow nylon cylindrical fittings.
8. Silicone tubing.
9. Commercial plastic polishing kit.
10. Paper cups.

### 2.2 Cell Lines

1. Human ovarian cancer-derived ascites cell line, OV90.
2. Human epithelial ovarian cancer cell line, TOV112D.

### 2.3 Fluorescent Stains and Markers

1. CellTracker™ Green (CTG) CMFDA (5-chloromethylfluorescein diacetate).
2. Annexin V PE (phycoerythrin)-labeled antibodies.
3. 7-AAD (7-aminoactinomycin D) in binding buffer.
4. 5 μM CTG in PBS.
5. 1.5 μM propidium iodide (PI) in PBS.

### 2.4 Solutions and Reagents

1. Complete OSE medium: Mix 10 mL of fetal bovine serum in 90 mL of OSE medium.
2. 10% fetal bovine serum.

3. 55 mg/L gentamycin.
4. 600 μg/L amphotericin B.
5. 350 μM carboplatin.
6. 10 mg/mL Pluronic F-108.
7. Phosphate-buffered saline.
8. Trypsin.
9. Dichloromethane.

### 2.5 General Equipment and Others

1. Computer numerical control (CNC) milling machine.
2. 3D designing software – CATIA.
3. FeatureCAM.
4. Fluorescence microscope.
5. Inverted microscope.
6. Confocal microscope.
7. Flow cytometry (FACS).
8. $CO_2$ incubator.
9. $CO_2$ laser cutter.
10. Atmospheric plasma torch.
11. 35-μm cell strainer.
12. ImageJ – an open-source freeware for image processing.

## 3 Methods

The HT spheroid chip comprises three layers. Bottom layer: A channel with pinholes for droplet formation. Middle layer: conical cell funnels. Top layer: chimney or vent layer as shown in Fig. 1 to prevent spillover. Draw the 3D design of each layer using CATIA software and generate G-code using FeatureCAM software. Fabricate each layer of PMMA mold using a computerized numerical control machine.

### 3.1 Development of PMMA Molds for Spheroid Chip

1. Use the 3D design dimensions of the bottom, middle, and top layers of PMMA molds for designing in the 3D CATIA software (*see* **Notes 1–3**).
2. Use FeatureCAM software to generate 3D designs into G-code for micromachining in the CNC machine.
3. For milling the funnel layer, use 3-mm flat end mill, for milling HD layer (pinhole and seeding channel), use 1-mm flat end mill and for milling vent layer use a combination of 3- and 1-mm flat end mills.

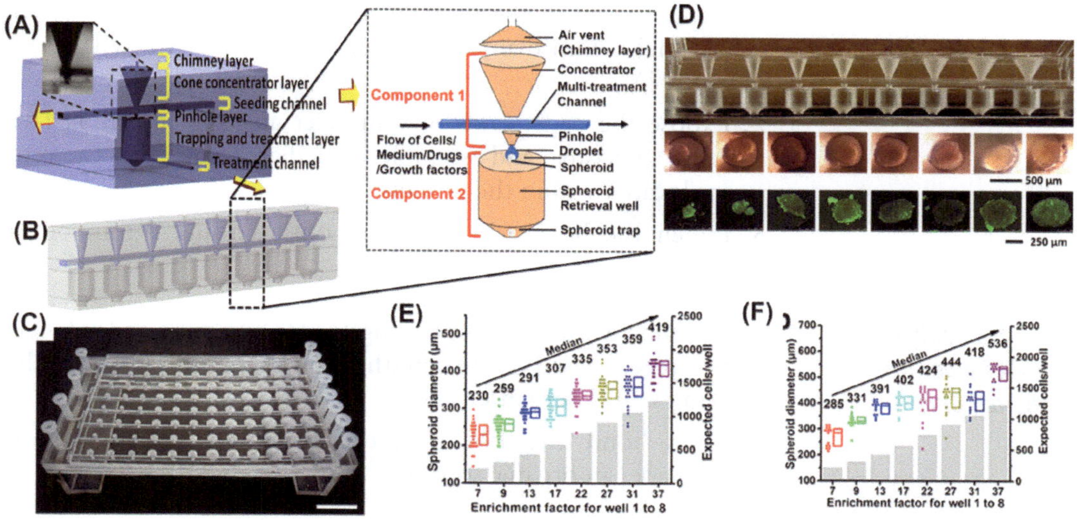


**Fig. 1** Schematic representation of (**a**) single-well assembly, (**b**) 8-well assembly. Inset: Layer-by-layer assembly of component 1 and component 2, (**c**) 96-well assembly; (**d**) top panel: Picture of 8-well chip; middle panel: microscopic images; and lower panel: 3D confocal microscopic images of multi-sized spheroids. (**e**) And (**f**) Box plot representation of diameters of day 5 spheroids of OV90 and TOV112D cells, respectively. The number of expected cells in each pinhole as per enrichment factor is represented as grey bars. (Reproduced from Marimuthu et al. 2018)

4. *Development of mold for spheroid retrieval*: Use the following assembly pattern and dimension for the synthesis of mold for spheroid retrieval chip.

   *Assembly*: Cylindrical well with 6-mm diameter and 5-mm height is assembled with inverse design of the vent layer.

5. Micromachine the G-codes of the design patterns to obtain the PMMA molds. Clean the molds with isopropanol. Polish the sides of the molds using a commercial plastic polishing kit.

### 3.2 Fabrication of Spheroid Chip (See Note 4)

1. Mix PDMS and curing agent in the ratio of 10:1, and degas under vacuum desiccator for 10–15 min.
2. Pour PDMS mixture into the PMMA molds, and degas again under vacuum desiccator for 10–15 min or until the air bubbles are removed (*see* **Notes 5** and **6**).
3. Then keep the molds in hot air oven at 80 °C for 1.5–2 h.
4. Peel off the PDMS layers of the spheroid chip from the molds carefully, and then clean the surface using adhesive tape.
5. Bond the three PDMS layers upon exposure to an atmospheric plasma torch.
6. Expose upper side of HD layer having the interior of the channel/pinholes, and also expose bottom side of cell funneling layer having 1-mm diameter to an atmospheric plasma for 30 seconds. Assemble the layers immediately such that 1-mm

diameter end of the funnels of cell funneling layer is located within 1.5-mm diameter end of the pinhole of HD layer (*see* **Note 7**).

7. Expose the wider ends of vent layer and cell funneling layer again to plasma, and assemble immediately such that the vents are located within the funnels.
8. Punch the holes for inlets and outlets using 3-mm biopsy punches.

### 3.3 Fabrication of PMMA Holder for Spheroid Chip

1. Cut the PMMA sheets into length-wise one piece and side-wise two pieces based on the length and width of the spheroid chip using $CO_2$ laser cutter.
2. Then bond the PMMA pieces together using dichloromethane, an organic solvent.

### 3.4 Synthesis of Spheroids

1. Thoroughly clean the spheroid chips with adhesive tapes for the removal of dust particles.
2. Sterilize the hollow nylon cylindrical fittings, silicone tubings, and spheroid chip with 70% ethanol prior to cell seeding.
3. Inject 10 mg/mL solution of the surfactant Pluronic F-108 into the chip and incubate for 30 min (*see* **Note 8**).
4. Then remove the Pluronic solution and wash the chip again with 70% ethanol (*see* **Note 9**).
5. Vortex the cancer cell suspension of 12,000 cells/mL concentration taken in a 3-mL syringe immediately before seeding in the chip.
6. Inject the cell suspension using syringe into the chip through silicone tubing attached to the inlet of the chip (*see* **Note 10**).
7. Once all the funnels are filled, incubate the chip for 1.5 h to allow the cells to sediment into the pinholes.
8. After incubation, remove the cell suspension from the chip using the syringe leaving only the sedimented cells in the pinholes.
9. Flush back fresh complete OSE medium into the chip.
10. Then remove the silicone tubing from the inlet, and place the hollow nylon cylindrical fittings in the inlet and outlets to act as reservoirs to create a positive hydrostatic pressure within the device (*see* **Note 11**).
11. Fill the inlet and outlet reservoirs with fresh medium to create a HD in each pinhole.
12. Then keep the chip inside a pipette tip box on top of the tip rack to provide humidified environment. Keep a sterile water-soaked wipes at the bottom of the tip rack to counteract the evaporation during the incubation at 37 °C with 5% $CO_2$ and 95% ambient air.

13. Top up the medium in the reservoir every 3 days, and allow the spheroids to grow in the spheroid culture chip for 7–10 days.
14. On day 3, 5, 7, and 10, before topping up of chip with fresh culture medium, image the spheroids using an inverted microscope.
15. For imaging purpose, keep the chips in 150-mm petri plate for transport purpose to prevent contamination, and imaging of spheroids in HD can be done directly on-chip.
16. Analyze the spheroid images using ImageJ software for diameter analysis.

### 3.5 Chemotherapeutic Drug Response Analysis

1. On Day 5 of the spheroid culture on-chip, withdraw the old medium using a syringe, leaving spheroids within the pinholes.
2. Flush 350-μM carboplatin in culture medium into the chip, and incubate in humidified cell incubator (37 °C, 5% $CO_2$, 95% ambient air) for 24 h.
3. Image the spheroids after 24-h chemo treatment and before collection for FACS analysis under inverted microscope for morphological and size evaluations.

### 3.6 Spheroid Staining for Confocal Imaging

1. On day 7, remove the culture medium from the chip, and wash multiple times with PBS solution (*see* **Note 12**).
2. Flush 5-μM concentration of CTG in PBS into the chip, and incubate for 1 h in a humidified cell incubator.
3. Then remove CTG from the chip, and wash the spheroids several times with PBS solution.
4. Then treat the spheroids with 1.5-μM concentration of PI in PBS by flushing the solution into the chip and incubate for 20–30 min.
5. Finally, remove the dye and wash thoroughly using PBS.
6. Then contact transfer the spheroids into the spheroid capture chip (*see* **Note 13**).

### 3.7 Spheroid Capture and Staining for FACS Analysis

1. Contact transfer the control and carboplatin-treated spheroids into the spheroid capture platform.
2. Collect the spheroids of individual sizes in Eppendorf tubes containing fresh medium and centrifuge at 300 × g for 5 min (*see* **Note 14**).
3. Remove the supernatant, and add freshly prepared PBS into the tubes and mix gently to wash the spheroids.
4. Then centrifuge the tubes again at 300 × g for 5 min. Remove the supernatant carefully.
5. Digest the spheroids by adding 100 μL of trypsin and incubation for 2 min.

6. After 2 min, arrest the action of trypsin by adding FBS. Disintegrate spheroids into single cells by gentle pipetting.
7. Filter the suspension using a 35-μm cell strainer to obtain cell aggregate-free single-cell suspension.
8. Centrifuge the suspension at 300 × g for 5 min.
9. Resuspend the cell pellets and stain with Annexin V PE-labeled antibodies and 7-AAD in binding buffer as per the manufacturer's protocol (559763, BD Pharmingen, San Diego, USA) to measure cell viability.

## 4 Notes

1. *Dimensions of Bottom Layer (Hanging Drop Layer): (Refer* Fig. 2)

   For channel, width is 2 mm and height is 1 mm. For pinhole, height is 1 mm, diameter of top end of the conical pinhole is 1.5 mm, and the diameter of tip of conical pinhole is 0.75 mm.
2. *Dimensions of Middle Layer (Cell Funneling Layer)*

   Height of each funnel of cell funneling layer is 5 mm, and the diameter of the tip of funnel is 1 mm but the apex angles vary in the range of 12, 19, 26, 33, 40, 47, 54, and 61 degrees as shown in Fig. 2c.

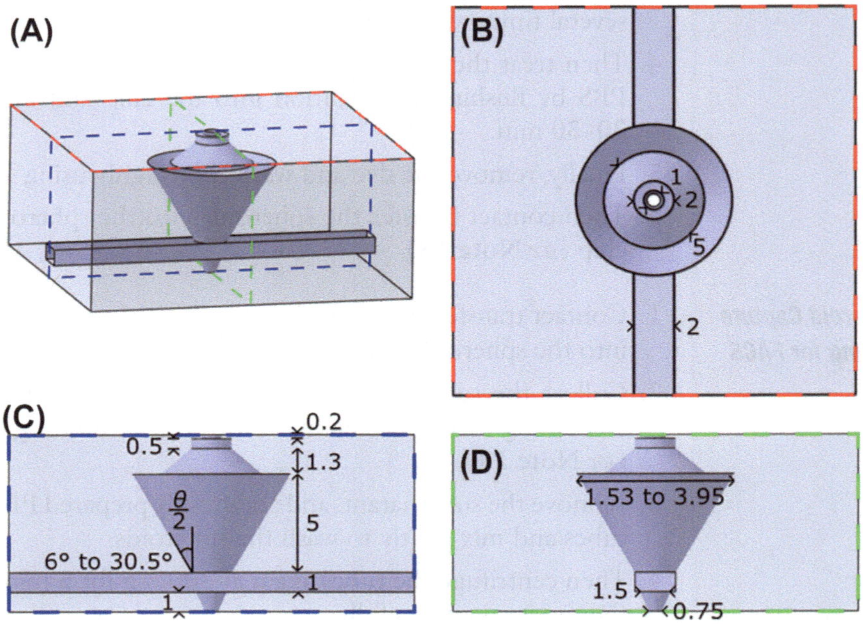


**Fig. 2** Schematic representation of dimensions of microfluidic chip. (**a**) Subunit consists of a pinhole, cell concentrator, and a vent. (**b**) Top view. (**c**) Side view demonstrating the heights of the parts of the device. (**d**) Side view demonstrating the width and diameters of the parts of the device. Unit in mm

3. *Dimensions of Top Layer (Chimney or Vent)*

   Vent assembly is done as double cylinder on top of a cone as shown in Fig. 2c.

   *Cylinder 1 has a d*iameter of 1 mm with a height of 0.2 mm. *Cylinder 2 has a d*iameter of 2 mm with a height of 0.5 mm. *Conical vent has a h*eight of 1.3 mm with a diameter of 2 mm at the top end of the cone and 5-mm diameter at the bottom end of the cone. Total height is 2 mm (1.3-mm height of cone +0.5-mm height of the first cylinder +0.2-mm height of the second cylinder).

4. The vent layer and pinhole layer may be quite tricky during the molding process. One should make sure the vent and pinhole are open to air as purpose may not be served if the vent or pinhole is closed. It is possible to miss if the vent/pinhole is closed as the PDMS is transparent and the vent/pinhole opening dimension is small.

5. Pinhole dimension is the key component to determine the HD volume, shape, curvature on which spheroid diameter and circularity depended on. Therefore, various dimensions and shapes of pinholes can be tested. Various geometries of pinholes such as pyramid, conical, cone cylindrical, step cone and cylindrical are shown in Fig. 3. Among which conical-shaped pinhole is identified to be more appropriate for spheroid culture application. However, the pinhole shape and dimension can be explored for independent application perspective.

6. It is known that inverse pattern for funnels and pinholes of the chip in a mold are pillars. While pouring PDMS mixture in the mold, PDMS level should be carefully checked not to overflow the pillars of the mold, because, for both funnels and pinholes, the design is supposed to be open. If PDMS mixture is over the pillar, after polymerization, it is required to use specific diameter biopsy punches to make an opening of the funnels and pinholes.

7. For the quick and proper alignment of the layers of the chip after plasma exposure, it is suggested to use any organic solvent such as ethanol or isopropanol on the surface immediately after plasma treatment. This will prevent the contact of the plasma-treated surface with the atmospheric air and provides extra time for the alignment. A microscope may also be used to align the layers such that the tip of the funnels falls within the wider end of the pinhole. Therefore, the sedimentation of the cells inside the pinhole is facilitated.

8. Pluronic F-108 treatment will prevent the adhesion of cells to the surface of the channel, pinholes, and funnels. Thereby, unwanted cell aggregation at unwanted spot of the chip may be avoided.

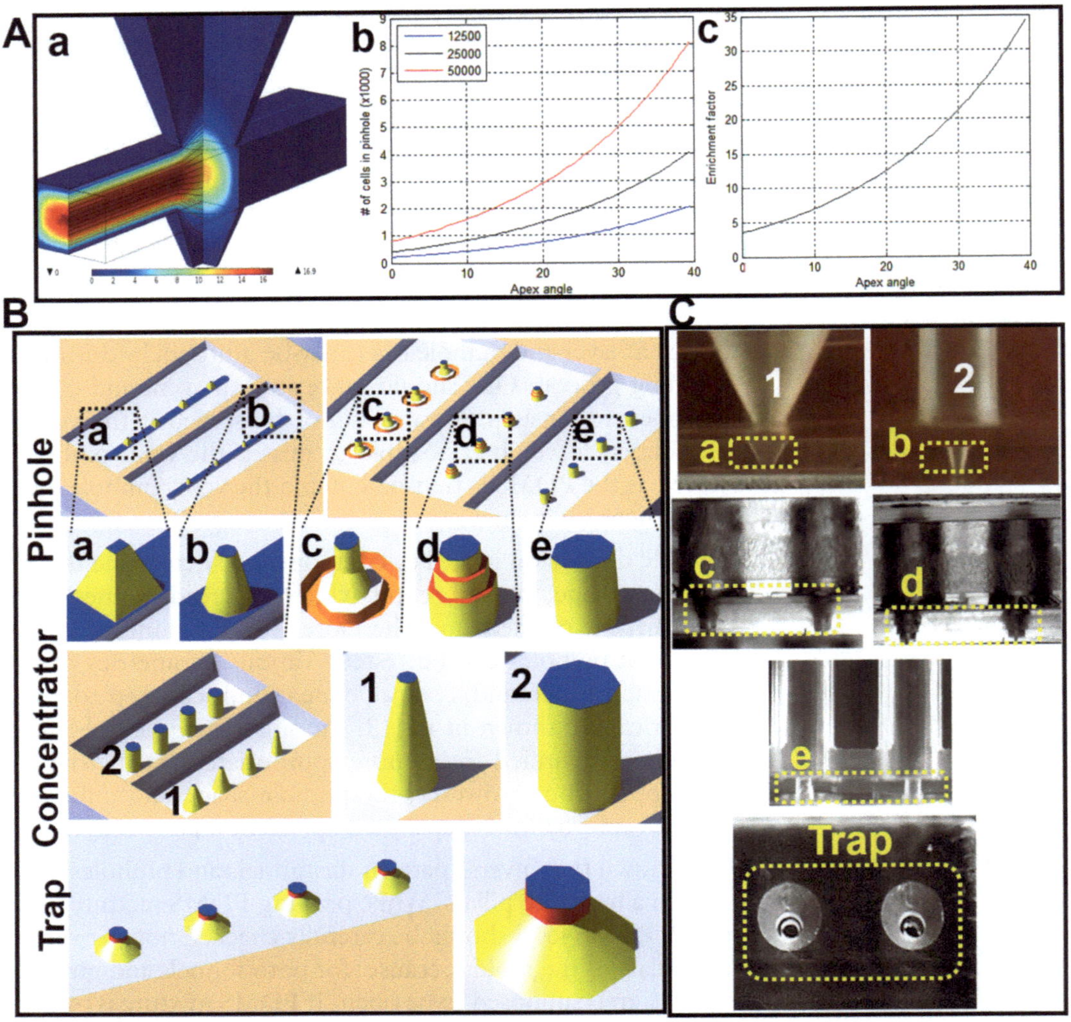


**Fig. 3** (**a**) Theoretical evaluation of flow of liquid and expected number of cells vs apex angle. Hydrodynamic simulation model of the flow of medium inside the channel (a). Graphs to show the relation between apex angle and number of cells in each pinhole (b) and also between apex angle and enrichment factor (c). (**b**) Various geometries of pinholes and cell concentrator, i.e., pinhole shapes such as pyramid (a), conical (b), cone cylindrical (c), step cone (d), and cylindrical (e); cell concentrator shapes such as conical (1) and cylindrical (2) have been studied for spheroid generation experiments. Similarly, a shallow cone-cylindrical pattern was utilized to capture the spheroid for downstream analysis (Fig. 1b, c trap). (**c**) Microscopic images of each geometry of pinholes, cell concentrators, and trap of spheroid chip (unpublished data)

9. Before cell seeding into the chip, it is required to wait for complete ethanol drying, especially the tip of pinholes. Hydrophobicity along the opening of the tip of the pinhole is the key factor for keeping the HD in place with a defined curvature. If the bottom of the pinhole is wet with ethanol, this will make the spot hydrophilic and results in dripping or modified curvature of the HD.
10. The flow rate of the cell suspension injection into the chip needs to be kept steady and constant. The injection of cell suspension into the spheroid chip does not require a syringe pump which is an important advantage. However, it is possible for the user to push hard to exceed the threshold flow rate which may result in dripping of the HD from the first pinhole due to the exceeding pressure drop along the channel. Hydrodynamics of the flow of cell suspension without dripping during injection should be taken care of by the user through a controlled flow rate.
11. The number of inlets and outlets varies for each spheroid-forming chip formats. There are four inlets with eight outlets for 96-well format. For 24-well chip, there is one inlet with three outlets. For other chip formats, there is an inlet and an outlet. So, the device is user customizable.
12. As discussed in **Note 10**, removal of waste/media out of the chip must also be done with a controlled flow rate to avoid the lifting of formed spheroids from the pinholes.
13. For the transfer of spheroids, medium should be filled up to the surface of the spheroid capture chip. The contact between bottom of spheroid chip and surface of the spheroid capture chip containing medium will pull down the spheroids into the conical trap of the spheroid capture chip as demonstrated by Cavnar et al. [12].
14. After carboplatin treatment, at higher drug concentrations it is observed under optical microscope that the spheroids started to disintegrate (Fig. 4). So, during the media exchange, there is a possibility that dead cells (floating above the spheroids and in the channels) from the spheroids may be lost. This may result in a variation in viability percentage in the final FACS analysis. Therefore, it is suggested to collect the media during media exchange and centrifuge down and include the dead cells count in the final viability analysis.

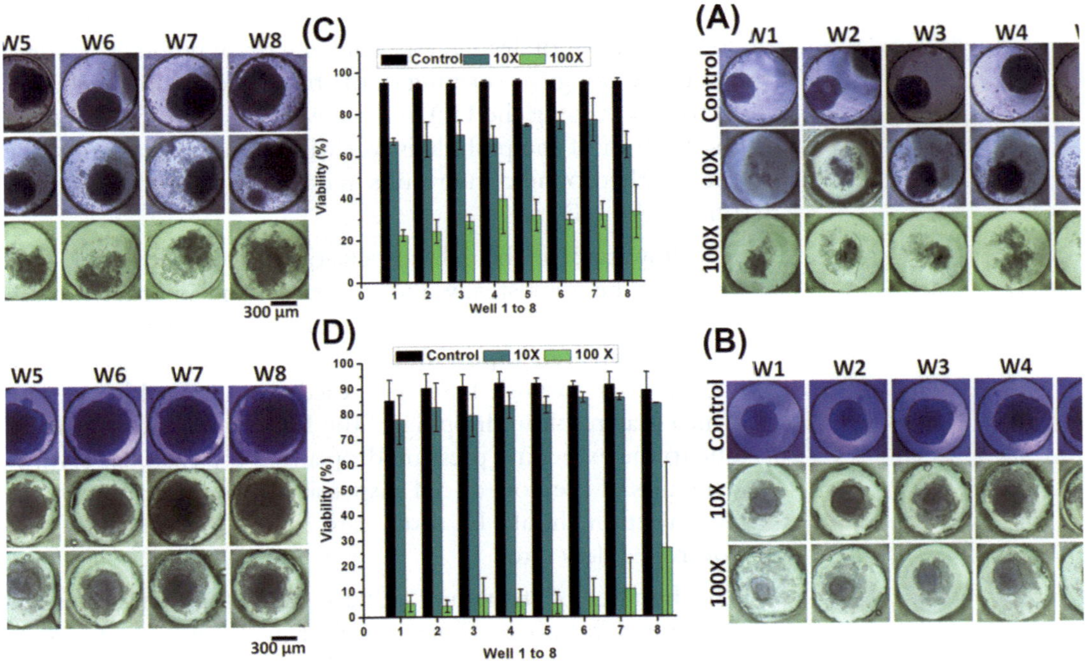


**Fig. 4** Carboplatin drug response analysis. Microscopic images after drug treatment to determine the morphological changes of (**a**) OV90 and (**b**) TOV112D spheroids. Viability analysis of (**c**) OV90 and (**d**) TOV112D spheroids by FACS. Percentage of live cells with 10X (purple)/100X (blue) and without (black) carboplatin treatment

## Acknowledgments

This work was supported by NSERC (RGPIN-06409), the Cancer Research Society, and CMC Microsystems (cmc.ca), the Fond de Recherche du Québec—Santé (FRQS) and FRQ-NT Quebec-India post-doctoral fellowship. M Marimuthu acknowledges the support of Dhanalakshmi Srinivasan University.

### References

1. van Norman GA (2020) Limitations of animal studies for predicting toxicity in clinical trials: part 2: potential alternatives to the use of animals in preclinical trials. JACC Basic Transl Sci 5:387–397
2. Ando Y, Ta HP, Yen DP, Lee SS, Raola S, Shen K (2017) A microdevice platform recapitulating hypoxic tumor microenvironments. Sci Rep 7:1–12
3. Eilenberger C, Rothbauer M, Selinger F, Gerhartl A, Jordan C, Harasek M, Schädl B, Grillari J, Weghuber J, Neuhaus W, Küpcü S, Ertl P (2021) A microfluidic multisize spheroid array for multiparametric screening of anticancer drugs and blood–brain barrier transport properties. Adv Sci (Weinh) 8:1–20
4. Refet-Mollof E, Najyb O, Chermat R, Glory A, Lafontaine J, Wong P, Gervais T (2021) Hypoxic jumbo spheroids on-A-chip (HOnAChip): insights into treatment efficacy. Cancers (Basel) 13:1–20
5. Moshksayan K, Kashaninejad N, Warkiani ME, Lock JG, Moghadas H, Firoozabadi B, Saidi MS, Nguyen NT (2018) Spheroids-on-a-chip:

recent advances and design considerations in microfluidic platforms for spheroid formation and culture. Sensors Actuators B Chem 263: 151–176

6. Bhattacharya S, Calar K, de La Puente P (2020) Mimicking tumor hypoxia and tumor-immune interactions employing three-dimensional in vitro models. J Exp Clin Cancer Res 39:1–16
7. Liu X, Fang J, Huang S, Wu X, Xie X, Wang J, Liu F, Zhang M, Peng Z, Hu N (2021) Tumor-on-a-chip: from bioinspired design to biomedical application. Microsyst Nanoeng 7: 50
8. Wu Q, Liu J, Wang X, Feng L, Wu J, Zhu X, Wen W, Gong X (2020) Organ-on-a-chip: recent breakthroughs and future prospects. Biomed Eng Online 19:1–19
9. Marimuthu M, Kim S (2013) Pumpless steady-flow microfluidic chip for cell culture. Anal Biochem 437:161–163
10. Marimuthu M, Kim S (2015) Continuous oxygen supply in pump-less micro-bioreactor based on microfluidics. Biochip J 9:1–9
11. Marimuthu M, Rousset N, St-Georges-Robillard A, Lateef MA, Ferland M, Mes-Masson A-M, Gervais T (2018) Multi-size spheroid formation using microfluidic funnels. Lab Chip 18:304–314
12. Cavnar SP, Salomonsson E, Luker KE, Luker GD, Takayama S (2014) Transfer, imaging, and analysis plate for facile handling of 384 hanging drop 3D tissue spheroids. J Lab Autom 19:208–214

# Chapter 29

# Tracking Gut Motility in Organ and Cultures

**Peng Du, Vikram Joshi, and Arthur Beyder**

## Abstract

Gastrointestinal (GI) motility is a key component of digestive health, and it is complex, involving a multitude of cell types and mechanisms to drive both rhythmic and arrhythmic activity. Tracking GI motility in organ and tissue cultures across multiple temporal (seconds, minutes, hours, days) scales can provide valuable information regarding dysmotility and to evaluate treatment options. Here, the chapter describes a simple method to monitor GI motility in organotypic cultures, using a single video camera is placed perpendicularly to the surface of the tissue. A cross-correlational analysis is used to track the relative movements of tissues between subsequent frames and subsequent fitting procedures to fit finite element functions to the deformed tissue to calculate the strain fields. Additional motility index measures from the displacement information are used to further quantify the behaviors of the tissues that are maintained in organotypic culture over days. The protocols presented in this chapter can be adapted to study organotypic cultures from other organs.

**Key words** Gastrointestinal, Organotypic, Cross-correlation, Motility, Contraction, Smooth muscle, Stomach, Intestine

## 1 Introduction

The gastrointestinal (GI) tract displays a number of complex motility patterns, which are governed by an interplay between neuronal, myoelectrical, and hormonal factors [1]. Abnormal GI motility, also known as dysmotility, is associated with and can lead to both frequent and rare digestive disorders, such as irritable bowel syndrome (IBS) and functional dyspepsia, pseudo-obstruction, and slow transit constipation (STC) [2]. STC is a colonic motility disorder in which colonic transit is markedly delayed without a concomitant pelvic floor dysfunction [3]. However, the pathophysiology of STC appears to be multifactorial, involving smooth muscle cells (SMCs) [2], interstitial cells of Cajal (ICCs) [4, 5], enteric neurons [6], and aging [7, 8], but overall, the mechanisms are yet to be elucidated [9]. Investigations have suggested that damage to enteric nerves and the GI pacemaker cells, the ICC, may be factors

Oliver Friedrich and Daniel F. Gilbert (eds.), *Cell Viability Assays: Methods and Protocols*, Methods in Molecular Biology, vol. 2644, https://doi.org/10.1007/978-1-0716-3052-5_29,

that underpin loss of motility in STC [4, 5]. However, diagnostics based on histology is challenging in a majority of cases, and therefore, a functional pathology at the molecular level will likely provide a useful biomarker of STC [10]. Given the lack of mechanistic understanding in STC, current treatment approaches are often not effective, and eventually patients may require a colectomy [11].

SMC dysfunction is classically associated with GI motility disruptions, including STC. The individual GI SMCs are spindle-shaped cells that are usually 2–5 μm in diameter and 20–500 μm in length. Like all cells, ion channels are critical players in managing ionic concentrations, and they are required for the electrical excitability and $Ca^{2+}$ handling which is critical for actin-myosin contractility, and therefore, they are indispensable for electromechanical coupling of SMCs throughout the GI tract [12, 13]. SMC ions permeate the SMC membranes using many types of ion channels, each forming their own electrochemical gradient. Ion channels and most notably potassium ($K^+$) channels contribute to the transmembrane potential. In the SMC at rest, the resting membrane potential is approximately −70 mV in the colonic circular muscle [14] and −68 mV in the stomach [15], and there is also a transmural gradient in SMC resting potential [16]. From this resting potential, SMCs undergo action potentials which involve membrane depolarization, plateau, and repolarization. Preceding a contraction, ion channels that carry ions which depolarize the membrane (e.g., $Na^+$, $Ca^{2+}$) allow a flux of ions across the cell membrane, which results in a depolarization that in turn allows for both $Ca^{2+}$ influxes from outside the cell and release from intracellular stores and that is critical for initiating contractions. It has been postulated that the force of contraction may be dependent on the amplitude of depolarization, with additional factors such as neural, hormonal, and paracrine influences all having a major bearing on the contractile response [17, 18]. There is a range of ion channels involved in maintaining the depolarized state (plateau) and repolarizing the cell membrane, which is required to halt SMC contraction [12].

Recent investigations have focused on the role of voltage-gated $Na^+$ channels in GI smooth muscle excitability, even where $Ca^{2+}$ signaling is fundamental, $Na^+$ channels contribute in a species-dependent fashion to the regulation of electrical and contractile function [19, 20]. In particular, a voltage-gated $Na^+$ channel, $Na_V1.5$, encoded by the *SCN5A* gene, is expressed in human GI smooth muscle and ICC in the small and large intestine [21], where it exhibits a significant degree of influence on the electromechanical coupling [20]. $Na_V1.5$ inhibition by lidocaine and ranolazine alters human colonic smooth muscle contractility [22, 23], leading to a strong association with STC [24]. Mutations in ion channels lead to a group of diseases called channelopathies [25], which manifest as functional disruptions in electrically excitable organs, such as heart and gut [26]. The *SCN5A* channelopathies in the heart include long QT syndrome and Brugada syndrome [27], and *SCN5A* is the only nonstructural gene implicated in dilated cardiomyopathy, a cardiac contractility disorder [28]. Interestingly, patients with cardiac *SCN5A* channelopathies have an increased prevalence of functional GI diseases [29], and conversely patients with irritable bowel syndrome are more likely to have *SCN5A* channelopathies, with a greater percentage of patients with *SCN5A* channelopathies having constipation-predominant IBS [30, 31]. Despite the clear evidence of the role of channelopathies in a number of GI diseases [26], more common diseases are caused by altered ion channel expression and regulation, where even relatively small change in the total ion current could lead to fatality [32]. Multiple mechanisms contribute to the tight regulation of ion channel expression. MicroRNAs (miRNAs) are a class of small noncoding RNAs that modulate the expression of many ion channels, such as the density of $Na_V$1.5 which is associated with cardiac conduction disorders [33]. Similar level of regulations has also been reported in the GI tract, with miRNA regulation of ion channel densities contributing to functional GI disorders [34]. For example, in the case of STC, inhibition of $Na_V$1.5 by miRNAs leads to reduction in the colonic smooth muscle motility [2].

Since ion channel disruptions frequently manifest as abnormalities in GI motility, it is critical to closely assay GI motility using methods that span in vivo to in vitro. Assessment of in vivo GI motility can be done using several techniques, each with varying degree of advantages and limitations. Clinically, GI motility has been used as a key diagnostic biomarker through quantification of GI transit, such as gastric emptying, small bowel, and colonic transit, using scintigraphy or radiopaque markers, which provide a general guide of the amount of content being transported in each region of the GI tract. Other measurements of GI motility such as MRI, antroduodenal manometry, smart pill, and force transducers are often heterogeneous in response and consistency, which suggest the need to investigate GI smooth muscle motility at a fundamental level before any clinical biomarker can be adopted as a true reflection of GI motility in health and disease. Thus, while in vivo gut transit measurements establish diagnoses for diseases, these measurements, however, are lacking information on the degree of movements exhibited by the GI smooth muscles due to the inability to tracking beyond the boundary of the GI tract. This information is critical for better understanding of mechanisms underlying diseases.

Most ex vivo and in vitro studies of GI motility rely on direct capture of the movements of the GI tract, usually from the serosal surface, via videographic capture or cinematography. Since GI motility takes place over large distances and time scales, it is critical to examine motility at a range of spatial and temporal scales. Due to the limited means of image processing techniques and computational power, early cinematographic analyses were largely qualitative rather than quantitative. In more recent times, analyses focused on tracking the propagation of contractility and periodic GI wall motions [35], with later improvements in the capability of detecting contractions in a specific direction based on captures of markers placed on the GI tract [36]. Unlike the heart, however, GI motility requires measurements that span seconds to days. Tracking of markers in two dimensions also allowed quantification of longitudinal and circumferential contractions [37]. The standard is to quantify the degree of movements either as diameter maps (Dmaps) or longitudinal (Lmaps) [38, 39]. However, the constructions of Dmaps and Lmaps assumed an a priori knowledge of the orientation of the organ relative to the field of view (FOV) and potentially limit the resolution of analysis, especially when marker-based measurements are required [40]. On the other hand, marker-free tracking and computation of strain fields have been proposed to overcome those limitations. The advantage of a marker-free system is the elimination of the need for physical markers from the preparation which potentially eases the preparation and shortens time of exposure [41], with the ability to track the detailed spatiotemporal changes in motility in a two-dimensional setting.

The main objective of this chapter is to present the techniques for ex vivo motility tracking of GI tissues, in a standard large animal model setup and in a specialized preparation that allows tracking of tissues over multiple days from tissues gathered in a small animal model. These techniques can be extended to analyze different regions of the GI tract and other muscular organ systems, as well as assaying GI motility impact of therapies.

## 2 Materials

### 2.1 General Material

1. Isoflurane (2.5–5% with an oxygen flow of 400 mL).
2. Zoletil (tiletamine HCl 50 mg $mL^{-1}$).
3. Zolazepam (HCl 50 mg $mL^{-1}$).
4. Magnesium sulfate (50 mL).
5. 0.9% saline (38 °C, 1 L).
6. Tegaderm (3 M).
7. 60 mL petri dish covered by a thin layer of Sylgard©.

### 2.2 Solutions and Reagents

1. Krebs solution composed of (in mM): NaCl 120.3, KCl 5.9, $MgCl_2$ 1.2, $NaHCO_3$ 15.5, $NaH_2PO_4$ 1.2, glucose 11.5, $CaCl_2$ 2.5, pH 7.35–7.45 when bubbled with 97% $O_2$ + 3% $CO_2$ at 37 ± 0.5 °C.

2. Culture medium: M199 without phenol red, supplemented with 4.5 g/l of glucose, 10% FBS, 2 mM L-glutamine and 2% antibiotic-antimycotic, pH of 7.35–7.45 when incubated at 37 °C in a 5% $CO_2$ incubator.

### 2.3 Hardware and Software

1. Video camera (e.g., Sony NEX-VG10, Tokyo, Japan) for organ recording.
2. CCD camera connected to a stereomicroscope (e.g., SZ61, Olympus) for tissue recording.
3. Two linear polarizers.
4. A closed-circuit anesthetic system for administration of isoflurane.
5. ECG machine.
6. Rectal temperature probe.
7. MATLAB R2021b.
8. ImageJ v1.53k.

### 2.4 Animal Models

1. Cross-breed, weaner pig of either sex, 3–4 months old, 38–43 kg.
2. BALB/c mice of either sex, embryonic day 15 (E15) to post-natal day 10–21 [42–44] or rats (Sprague Dawley and Dark Agouti) of either sex, 6–8 weeks old [2, 19].

## 3 Methods

The following protocols provide instructions on intestinal motility in tracking animal maintained under general anesthesia as well as tissues from organotypic cultures. A general workflow of the protocol is illustrated in Fig. 1. Additional manipulations can be imposed on the organ and/or organotypic cultures by pharmacological means or modified gene expression in targeted cell types [2, 19, 42, 43, 45].

### 3.1 Large Animal Surgery and Preparation

1. Allow up to 12 h of acclimation followed by 8 h of overnight fast to each study (*see* **Note 1**).
2. Administer general anesthesia (*see* **Note 2**).
3. Place animal in a stable supine position with legs secured by rope to the operating table.
4. Cannulate the right femoral artery for monitoring of blood pressure.
5. Apply needle ECG leads/dots to the fore and hind legs to monitor cardiac electrical activity.

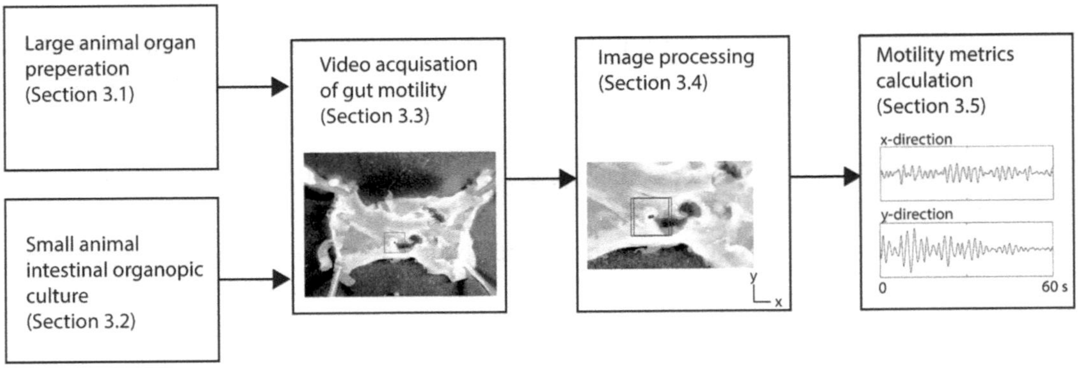


**Fig. 1** General overview of the protocols for measuring and tracking gastric and intestinal motility. Experimental preparation protocols of large and small animal models are described in Subheadings 3.1 and 3.2. Video acquisition using a single camera setup is described in Subheading 3.3. Imaging processing and motility tracking are described in Subheading 3.4, with the MATLAB scripts required for implementing the algorithms included in **Notes 8** and **9**. Motility metrics calculation protocol is described in Subheading 3.5, with the MATLAB scripts required for implementing the algorithms included in **Note 14**

6. Insert a rectal temperature probe to measure body temperature.
7. Clear the abdominal wall of debris and other loose tissues.
8. Perform a 10–15 cm laparotomy using a surgical scalpel along the *linea alba* (*see* **Note 3**).
9. Place a Tegaderm on the abdominal surface below the incision.
10. Exteriorize 20–30 cm of the intestine (or organ of interest) through the incision, and then place it on top of the Tegaderm (*see* **Note 4**).
11. Spray 10 mL of warm saline on the exteriorized segment at 5 min intervals to keep it moist and warm throughout recording.

### 3.2 Small Animal Surgery and Organotypic Cultures

1. Perform cervical dislocation (mouse).
2. Quickly dissect out *tunica muscularis* from jejunum.
3. Place the tissue sample in oxygenated Krebs solution, and loosely pin (either circular or longitudinal muscle facing upward) onto a sterile dish covered by Sylgard©.
4. Perform five rapid washes in culture medium.
5. Incubate for 3 h at 37 °C for equilibration (*see* **Note 5**).
6. Repeat **step 5** every 24 h with fresh medium.

### 3.3 Video Acquisition

1. Place and secure a single video camera above the exteriorized organ/culture, in the perpendicular direction to the tissue (*see* **Note 6**).
2. Place a 1 cm scale in the field of view.

3. Place a linear polarizer in front of an illumination source at angle to the tissue and another polarizer between the tissue and camera to minimize refection [46]. The two polarizers should be orientated at 90° to each other.
4. Mark the position of the organotypic culture relative to the camera by marking the corners of the dish.
5. Record the motility of the organ/tissue at 14 frames per second (1920×1440 pixels per frame) in the recording software, for 300 s per recording, and save in .avi format [47] (*see* **Note 7**).
6. Align the dish with the positional markings from **step 3** for recordings of tissues over multiple days.

### 3.4 Image Processing

1. Execute the "avi2png.m" script in MATLAB to export each frame of the .avi video into a grayscale .png image (*see* **Note 8**). Input, [example].avi; output, [example].png images. The video file should be in the same directory as the script.
2. Execute the "tracking.m" script in MATLAB script to track the movements of individual squares of 100 by 100 pixels (*see* **Note 9**). Input, [example].png images; output, [track_example].mat containing the movements data. Determine the following parameters on the specific image set (*see* **Notes 10–13**):
   - fps: frame rate.
   - ts: start time of analysis in s.
   - dn: increment of analysis, 1 means every frame is analyzed.
   - n: half of the size of each square region of interest for tracking.
   - sh: maximum allowable shift between successive frames for analysis.

### 3.5 Contractility Frequency Measurements

1. Open ImageJ, click on File then select Open, open a frame with the calibration marker from **step 2** of Subheading 3.3 in the field of view.
2. Click on the "Straight Line" tool, and then mark the calibration marker in the field of view. Record the length measurement in pixels.
3. Execute the "measurements.m" script in MATLAB to quantify the motility metrics (*see* **Note 14**). Input, [track_example].mat; output, c.mat array, mode of frequency (*see* **Note 15**), standard deviation of frequency, mean amplitude of contraction (*see* **Note 16**) in *x*-direction, standard deviation of amplitude of contraction in *x*-direction, maximum amplitude of contractions in *x*-direction, mean amplitude of contraction in y-direction, standard deviation of amplitude of contraction in *y*-direction, maximum amplitude of contractions in *y*-direction,

mean area under curve in $x$-direction (*see* **Note 17**), standard deviation of area under curve in $x$-direction, mean area under curve in $y$-direction, standard deviation of area under curve in $y$-direction, mean coordination index (*see* **Note 18**), standard deviation of coordination index. Determine the following parameters on the specific image set:

- calib: calibration of pixels to μm based on the measurement from **step 2**/10000.
- F_threshold: the expected lower and upper frequency threshold in the dataset.
- t_threshold: the ratio of the absolute "refractory period" to expected interval.

## 4 Notes

1. Ethics should consider the 3R (reduce, refine, reuse) principle and follow the guide for the care and use of laboratory animals [48].
2. After induction with Zoletil, the animal can be maintained under isoflurane or an appropriate anesthetic protocol. For example, general anesthesia with Zoletil (tiletamine HCl 50 mg $mL^{-1}$ and zolazepam HCl 50 mg $mL^{-1}$). After induction, the subject shall be maintained with isoflurane (2.5–5%) with an oxygen flow of 400 mL within a closed-circuit anesthetic system, which can be adjusted based on the state of anesthesia.
3. It is advisable to use a scalpel to gently pierce a hole in the omentum so the underlying organs are not damaged by the surgical instruments. The exposure of the abdominal cavity should be minimized in order to maintain the normal homeostatic conditions.
4. In cases where further drug interventions are to be performed on the segment, the superior mesenteric arcade supplying that segment can be isolated and cannulated with a 23 gauge angiocatheter. A 0.9% saline flush can be performed to confirm the supply of the segment by the temporary blanching of the vessels from the flush.
5. For additional lentiviral-mediated transduction with short hairpin RNA (shRNA) to knockout (KO) gene of interest, the tissues are treated with either non-targeting (NT) or targeting shRNA lentiviral transduction particles against gene of interest (T-KD) with appropriate multiplicity of infection and appropriate incubation time [2]. Like lentiviral treatment, drugs can also be applied to the tissues in the organotypic culture before contractility measurements [43].

6. Mount the camera using a secure tripod or frame above the organ/tissue. Care must be taken to reduce any movements (walking, talking, and traffic) while recording.
7. The rate of capture can vary depending on the system setup, but generally between 14 and 20 frames per second (FPS) would be acceptable to capture most types of movements in the intestine [2]. The FPS and size of each frame may also depend on the bandwidth of the capture system and the speed of saving to the hard drive, but generally it is recommended to save the frames in a format without compression so all of the data can be retained.
8. Choose New from the File menu and select Script, copy and paste the following commands into the text editor, and then save as "avi2png.m." To execute, Click on Run in Editor.

```
clearvars;
close all;
filelist = {'example.avi'};
for j = 1:length(filelist)
filename = filelist{j};
v = VideoReader([filename]);
subfolder = filename(1:length(filename)-4);
if ~exist(['images/',subfolder],'dir')
mkdir(['images/',subfolder])
else
end

i = 0;
while hasFrame(v)
vidFrame = readFrame(v);
frame = rgb2gray(vidFrame);
 imwrite(frame,['images/',subfolder,'/',strcat('frame-',
num2str(i),'.png')]);
i = i + 1;
%disp(i);
end
end
```

9. Choose New from the File menu and select Script, copy and paste the following commands into the text editor, and then save as "tracking.m." To execute, Click on Run in Editor.

```
clearvars;
close all;
filelist = {'exampe.avi'};
```

```
fps = 14;
ts = 1;
fs = ts*fps;
dn = 1;

for k = 1:length(filelist)
subject = filelist{k};
subject = subject(1:length(subject)-4);
disp(subject);
listing = dir(['images/',subject]);
tmax = round((size(listing,1)/fps-ts*4));
I1 = histeq(imread(['images/',subject,'/frame-',int2str
(fs),'.png']));
ni = 1;
ti = fs:dn:(fs+fps*tmax);
if max(ti) > size(listing,1)
disp('max time exceeds video time');
else
n = 100;
sh = 200;
img_width = size(I1,2);
img_height = size(I1,1);
n_width = floor((img_width-sh*2)/(2*n+1));
n_height = floor((img_height-sh*2)/(2*n+1));
Xi = zeros(n_height,n_width,length(ti)+1);
Yi = Xi;
for i = 1:n_height
for j = 1:n_width
y_c = sh+n+1+2*n*(i-1);
x_c = sh+n+1+2*n*(j-1);
Xi(i,j,ni) = x_c;
Yi(i,j,ni) = y_c;
end
end
for m = ti
I2 = histeq(imread(['images/',subject,'/','frame-',int2str
(m),'.png']));
for i = 1:n_height
for j = 1:n_width
y_c = Yi(i,j,1);
x_c = Xi(i,j,1);
P = I1((y_c-n):(y_c+n),(x_c-n):(x_c+n));
Q = I2((y_c-n-sh):(y_c+n+sh),(x_c-n-sh):(x_c+n+sh));
C = normxcorr2(P, Q);
v = max(max(C));
[row,col] = find(C==v);
ypeak = round(sum(row)/length(row));
xpeak = round(sum(col)/length(col));
```

```
yoffSet = ypeak-(size(C,2)-1)/2+y_c;
xoffSet = xpeak-(size(C,1)-1)/2+x_c;
Xi(i,j,ni+1) = xoffSet;
Yi(i,j,ni+1) = yoffSet;
end
end
disp(ni)
ni = ni + 1;
end
filename = ['track_',subject];
save(['outputs/',filename],'Xi','Yi','ti')
end
end
```

10. The "histeq" in **Note 9** equalizes the histogram of the gray scale. The use of contrast enhancement is not always necessary if the source video and the tissue itself offer sufficient contrast for tracking.
11. The "normxcorr2" in **Note 9** is a normalized 2D cross-correlational technique used to track motility of gastric and intestinal tissues [39, 49, 50]. It effectively identifies the most similar frame within a vicinity defined by the "sh" parameter between frames.
12. The "n" parameter controls the number of regions of interest (ROIs) tracked in the frame. The number and size of the ROIs is dependent on the specific setup, with the trade-off being that the bigger the ROI, the more accurate its location can be detected in the subsequent frames but with spatial resolution of the movements. In the organotypic setup as specified in Subheading 3.3, 7×5 equally spaced ROIs of 201×201 pixels (i.e., $n = 100$) between an initial reference frame and every subsequent frame are recommended [2].
13. Subheadings 3.3, and 3.4 are the most computationally demanding steps. For a typical desktop computer, it is recommended to test the initial setup on a shorter segment (i.e., <60 s) before doing the full analysis.
14. Choose New from the File menu and select Script, copy and paste the following commands into the text editor, and then save as "measurements." To execute, Click on Run in Editor. Type "c" into the Command Window to view the calculated motility metrics.

```
clearvars;
close all;

calib = 0.0022;
t_threshold = 0.6;
```

```
F_threshold = [17 45];

filelist = {'example.avi'};
filename = filelist{1};
filename = filename(1:length(filename)-4);
load(['outputs/track_',filename]);
[Xi, Yi, eFs, mask] = array_allocate(filename, Xi, Yi);
for i = 1:size(Xi,3)
Xi(:,:,i) = Xi(:,:,i).*mask;
Yi(:,:,i) = Yi(:,:,i).*mask;
end
Fs = eFs/1;
t = linspace(min(ti)/(eFs),max(ti)/(eFs),length(ti));
x_diff = zeros(size(Xi,1), size(Xi,2), size(Xi,3)-1);
y_diff = x_diff;
d_diff = x_diff;
for i = 1:size(x_diff,3)
for j = 1:size(x_diff,2)
for k = 1:size(x_diff,1)
x_diff(k,j,i) = Xi(k,j,i+1)-Xi(k,j,i);
y_diff(k,j,i) = Yi(k,j,i+1)-Yi(k,j,i);
end
end
end
dis = zeros(size(Xi,1)*size(Xi,2),size(x_diff,3));
dis_x = dis;
dis_y = dis;
for i = 1:size(dis,2)
temp = x_diff(:,:,i)';
temp = reshape(temp,1,size(Xi,1)*size(Xi,2));
dis_x(:,i) = temp;

temp = y_diff(:,:,i)';
temp = reshape(temp,1,size(Xi,1)*size(Xi,2));
dis_y(:,i) = temp;
end
fc = 1;
[b,a] = butter(6,fc/(Fs/2));
order = 4;
framelen = 21;
opol = 4;
N = 20;
for i = 1:size(dis,1)
dis_x(i,:) = filter_signal(dis_x(i,:),a,b, N).*calib;
dis_y(i,:) = filter_signal(dis_y(i,:),a,b, N).*calib;
dis(i,:) = sqrt(dis_x(i,:).^2 + dis_y(i,:).^2);
end
Fx = zeros(size(Xi,1)*size(Xi,2),1);
```

```
Fy = Fx;
for i = 1:(size(Xi,1)*size(Xi,2))
Fx(i) = fft_signal(dis_x(i,:),Fs);
Fy(i) = fft_signal(dis_x(i,:),Fs);
end

Fx = Fx.*60;
Fy = Fy.*60;
F = [Fx, Fy];
freq = F(F>F_threshold(1)&F<F_threshold(2));
Fc = mode(freq);
interval = 60/Fc;
ni = 1;
mi = 1;
t_threshold = interval*t_threshold;
coord = zeros(size(dis_x,2),1);
for i = 1:size(dis_x,2)
Norm = sqrt(dis_x(:,i).^2 + dis_y(:,i).^2);
ind = find(Norm>0);
dis_x_norm = dis_x(ind,i)./Norm(ind);
dis_y_norm = dis_y(ind,i)./Norm(ind);
 coord(i)  =  sqrt(mean(dis_x_norm).^2+mean(dis_y_norm).^2)./
mean(sqrt(dis_x_norm.^2+dis_y_norm.^2));
end
for i = 1:(size(Xi,1)*size(Xi,2))
if Fx(i) > Fc-5 && Fx(i) < Fc+5 && Fy(i) > Fc-5 && Fy(i) < Fc+5
 [pks1,locs1]  =  findpeaks(dis_x(i,:),t,'MinPeakDistance',in-
terval);
 amp1  =  amplitude_calc_v2(t,  dis_x(i,:),  pks1,  locs1,  round
(t_threshold*Fs));
 [pks2,locs2]  =  findpeaks(dis_y(i,:),t,'MinPeakDistance',in-
terval);
 amp2  =  amplitude_calc_v2(t,  dis_y(i,:),  pks2,  locs2,  round
(t_threshold*Fs));
Ax(ni,1) = mean(amp1);
Ax(ni,2) = std(amp1);
Amax_x(ni,1) = max(amp1);
Ay(ni,1) = mean(amp2);
Ay(ni,2) = std(amp2);
Amax_y(ni,1) = max(amp2);
ind1 = [];
ind1 = find(dis_x>10);
if ~isempty(ind1)
dis_x(ind1) = 0;
end
ind1 = find(dis_y>10);
if ~isempty(ind1)
dis_y(ind1) = 0;
```

```
end
AUCx(ni,1) = trapz(t,abs(dis_x(i,:)));
AUCy(ni,1) = trapz(t,abs(dis_y(i,:)));
ind(ni) = i;
ni = ni + 1;
end
end
c(1) = Fc;
c(2) = std(freq);
c(3) = mean(Ax(:,1));
c(4) = sqrt(mean(Ax(:,2).^2));
c(5) = max(Amax_x);
c(6) = mean(Ay(:,1));
c(7) = sqrt(mean(Ay(:,2).^2));
c(8) = max(Amax_y);
n = (max(t)-min(t))./60;
c(9) = mean(AUCx./n);
c(10) = std(AUCx./n);
c(11) = mean(AUCy./n);
c(12) = std(AUCy./n);
c(13) = mean(coord);
c(14) = std(coord);
return

function filter_sig = filter_signal(signal,a,b, N)
sig_a = signal(1:N);
sig_b = signal(end-N+1:end);
signal = [sig_a, signal, sig_b];
filter_sig = filtfilt(b,a,signal);
filter_sig = filter_sig(N+1:end-N);
end

function f = fft_signal(y,Fs)
L = length(y);
Y = fft(y-mean(y));
P2 = abs(Y/L);
P1 = P2(round(1:L/2)+1);
P1(2:end-1) = 2*P1(2:end-1);
f = Fs*(0:(L/2))/L;
ind = find(P1==max(P1));
f = f(ind);
end

 function amp = amplitude_calc(pks1, locs1, pks2, locs2,
t_threshold)
amp = [];
j = 1;
for i = 1:length(pks1)
```

```
%if i <= length(pk2)
temp = locs2 - locs1(i)-t_threshold;
ind = find(temp<0);
if ~isempty(ind)
temp(ind) = inf;
end
[m, mi] = min(temp);
if m < t_threshold
amp(j) = pks1(i) - pks2(mi);
j = j + 1;
end
end
end
```

15. The frequency of contractions is characterized by the mode of the frequencies of the tracked ROIs, calculated using Fast Fourier Transform (FFT), under "fft_signal" in **Note 14** in displacements both x- and y-directions. The mode of the frequencies is then taken as the reported frequency metrics in cycles per minute (cpm).
16. Amplitude is identified by identifying the peaks in the signal at an interval specified by mode of frequency ("fc" in **Note 14**) and the "t_threshold" variable, i.e., 1/fc*t_threshold.
17. Area under the curve is calculated using trapezoidal numerical integration ("trapz" in **Note 14**) of the displacement signals in x- and y-directions separately.
18. The coordination index (the "coord" variable in **Note 14**) is calculated as a measure of the ratio of vectors pointing in the same direction by using the following equation:

$$CI = \frac{\sqrt{\overline{x}^2 + \overline{y}^2}}{\overline{\sqrt{x^2 + y^2}}}$$

where $x$ and $y$ are the normalized displacements relative to the frame in $x$- and $y$-directions, respectively.

## References


1. Huizinga JD, Lammers WJ (2009) Gut peristalsis is governed by a multitude of cooperating mechanisms. Am J Physiol Gastrointest Liver Physiol 296:G1–G8. https://doi.org/10.1152/ajpgi.90380.2008
2. Mazzone A, Strege PR, Gibbons SJ, Alcaino C, Joshi V, Haak AJ, Tschumperlin DJ, Bernard CE, Cima RR, Larson DW, Chua HK, Graham RP, El Refaey M, Mohler PJ, Hayashi Y, Ordog T, Calder S, Du P, Farrugia G, Beyder A (2020) microRNA overexpression in slow transit constipation leads to reduced Na V 1.5 current and altered smooth muscle contractility. Gut 69:868–876. https://doi.org/10.1136/gutjnl-2019-318747
3. Knowles CH, Farrugia G (2011) Gastrointestinal neuromuscular pathology in chronic constipation. Best Pract Res Clin Gastroenterol 25: 43–57. https://doi.org/10.1016/j.bpg.2010.12.001

4. He CL, Burgart L, Wang L, Pemberton J, Young-Fadok T, Szurszewski J, Farrugia G (2000) Decreased interstitial cell of cajal volume in patients with slow-transit constipation. Gastroenterology 118:14–21
5. Lyford GL, He C-L, Soffer E, Hull TL, Strong SA, Senagore AJ, Burgart LJ, Young-Fadok T, Szurszewski JH, Farrugia G (2002) Pan-colonic decrease in interstitial cells of Cajal in patients with slow transit constipation. Gut 51:496–501. https://doi.org/10.1136/gut.51.4.496
6. Porter AJ, Wattchow DA, Hunter A, Costa M (1998) Abnormalities of nerve fibers in the circular muscle of patients with slow transit constipation. Int J Color Dis 13:208–216. https://doi.org/10.1007/s003840050163
7. Izbeki F, Asuzu DT, Lorincz A, Bardsley MR, Popko LN, Choi KM, Young DL, Hayashi Y, Linden DR, Kuro-o M, Farrugia G, Ordog T (2010) Loss of Kitlow progenitors, reduced stem cell factor and high oxidative stress underlie gastric dysfunction in progeric mice. J Physiol 588:3101–3117. https://doi.org/10.1113/jphysiol.2010.191023
8. Hayashi Y, Toyomasu Y, Saravanaperumal SA, Bardsley MR, Smestad JA, Lorincz A, Eisenman ST, Cipriani G, Nelson Holte MH, Al Khazal FJ, Syed SA, Gajdos GB, Choi KM, Stoltz GJ, Miller KE, Kendrick ML, Rubin BP, Gibbons SJ, Bharucha AE, Linden DR, Maher LJ, Farrugia G, Ordog T (2017) Hyperglycemia increases interstitial cells of Cajal via MAPK1 and MAPK3 signaling to ETV1 and KIT, leading to rapid gastric emptying. Gastroenterology 153:521–535.e20. https://doi.org/10.1053/j.gastro.2017.04.020
9. Bharucha AE, Lacy BE (2020) Mechanisms, evaluation, and management of chronic constipation. Gastroenterology 158:1232–1249.e3. https://doi.org/10.1053/j.gastro.2019.12.034
10. Tillou J, Poylin V (2017) Functional disorders: slow-transit constipation. Clin Colon Rectal Surg 30:76–86. https://doi.org/10.1055/s-0036-1593436
11. Knowles CH, Grossi U, Horrocks EJ, Pares D, Vollebregt PF, Chapman M, Brown SR, Mercer-Jones M, Williams AB, Hooper RJ, Stevens N, Mason J, NIHR CapaCiTY working group, Pelvic floor Society (2017) Surgery for constipation: systematic review and clinical guidance: paper 1: Introduction & Methods. Color Dis 19(Suppl 3):5–16. https://doi.org/10.1111/codi.13774
12. Beyder A, Farrugia G (2012) Targeting ion channels for the treatment of gastrointestinal motility disorders. Ther Adv Gastroenterol 5:5–21. https://doi.org/10.1177/1756283X11415892
13. Joshi V, Strege PR, Farrugia G, Beyder A (2021) Mechanotransduction in gastrointestinal smooth muscle cells: role of mechanosensitive ion channels. Am J Physiol Gastrointest Liver Physiol 320:G897–G906. https://doi.org/10.1152/ajpgi.00481.2020
14. Serio R, Barajas-Lopez C, Daniel EE, Berezin I, Huizinga JD (1991) Slow-wave activity in colon: role of network of submucosal interstitial cells of Cajal. Am J Phys 260:G636–G645. https://doi.org/10.1152/ajpgi.1991.260.4.G636
15. Hirst GDS, Edwards FR (2006) Electrical events underlying organized myogenic contractions of the Guinea pig stomach. J Physiol 576:659–665. https://doi.org/10.1113/jphysiol.2006.116491
16. Sha L, Farrugia G, Linden DR, Szurszewski JH (2010) The transwall gradient across the mouse colonic circular muscle layer is carbon monoxide dependent. FASEB J 24:3840–3849. https://doi.org/10.1096/fj.10-156232
17. Ozaki H, Gerthoffer WT, Publicover NG, Fusetani N, Sanders KM (1991) Time-dependent changes in Ca2+ sensitivity during phasic contraction of canine antral smooth muscle. J Physiol 440:207–224. https://doi.org/10.1113/jphysiol.1991.sp018704
18. Choi KM, Gibbons SJ, Roeder JL, Lurken MS, Zhu J, Wouters MM, Miller SM, Szurszewski JH, Farrugia G (2007) Regulation of interstitial cells of Cajal in the mouse gastric body by neuronal nitric oxide. Neurogastroenterol Motil 19:585–595. https://doi.org/10.1111/j.1365-2982.2007.00936.x
19. Beyder A, Gibbons SJ, Mazzone A, Strege PR, Saravanaperumal SA, Sha L, Higgins S, Eisenman ST, Bernard CE, Geurts A, Kline CF, Mohler PJ, Farrugia G (2016) Expression and function of the Scn5a-encoded voltage-gated sodium channel NaV 1.5 in the rat jejunum. Neurogastroenterol Motil 28:64–73. https://doi.org/10.1111/nmo.12697
20. Strege PR, Mazzone A, Kraichely RE, Sha L, Holm AN, Ou Y, Lim I, Gibbons SJ, Sarr MG, Farrugia G (2007) Species dependent expression of intestinal smooth muscle mechanosensitive sodium channels. Neurogastroenterol Motil 19:135–143. https://doi.org/10.1111/j.1365-2982.2006.00844.x
21. Holm AN, Rich A, Miller SM, Strege P, Ou Y, Gibbons S, Sarr MG, Szurszewski JH, Rae JL, Farrugia G (2002) Sodium current in human jejunal circular smooth muscle cells.

Gastroenterology 122:178–187. https://doi.org/10.1053/gast.2002.30346

22. Neshatian L, Strege PR, Rhee P-L, Kraichely RE, Mazzone A, Bernard CE, Cima RR, Larson DW, Dozois EJ, Kline CF, Mohler PJ, Beyder A, Farrugia G (2015) Ranolazine inhibits voltage-gated mechanosensitive sodium channels in human colon circular smooth muscle cells. Am J Physiol Gastrointest Liver Physiol 309:G506–G512. https://doi.org/10.1152/ajpgi.00051.2015
23. Strege PR, Ou Y, Sha L, Rich A, Gibbons SJ, Szurszewski JH, Sarr MG, Farrugia G (2003) Sodium current in human intestinal interstitial cells of Cajal. Am J Physiol Gastrointest Liver Physiol 285:G1111–G1121. https://doi.org/10.1152/ajpgi.00152.2003
24. Salazar CA, Basilio Flores JE, Veramendi Espinoza LE, Mejia Dolores JW, Rey Rodriguez DE, Loza Munárriz C (2017) Ranolazine for stable angina pectoris. Cochrane Database Syst Rev 2:CD011747. https://doi.org/10.1002/14651858.CD011747.pub2
25. Kass RS (2005) The channelopathies: novel insights into molecular and genetic mechanisms of human disease. J Clin Invest 115:1986–1989. https://doi.org/10.1172/JCI26011
26. Beyder A, Farrugia G (2016) Ion channelopathies in functional GI disorders. Am J Physiol Gastrointest Liver Physiol 311:G581–G586. https://doi.org/10.1152/ajpgi.00237.2016
27. Ackerman MJ (1998) The long QT syndrome: ion channel diseases of the heart. Mayo Clin Proc 73:250–269. https://doi.org/10.4065/73.3.250
28. McNair WP, Sinagra G, MRG T, Di Lenarda A, Ferguson DA, Salcedo EE, Slavov D, Zhu X, Caldwell JH, Mestroni L, Familial Cardiomyopathy Registry Research Group (2011) SCN5A mutations associate with arrhythmic dilated cardiomyopathy and commonly localize to the voltage-sensing mechanism. J Am Coll Cardiol 57:2160–2168. https://doi.org/10.1016/j.jacc.2010.09.084
29. Locke GR, Ackerman MJ, Zinsmeister AR, Thapa P, Farrugia G (2006) Gastrointestinal symptoms in families of patients with an SCN5A-encoded cardiac channelopathy: evidence of an intestinal channelopathy. Am J Gastroenterol 101:1299–1304. https://doi.org/10.1111/j.1572-0241.2006.00507.x
30. Beyder A, Mazzone A, Strege PR, Tester DJ, Saito YA, Bernard CE, Enders FT, Ek WE, Schmidt PT, Dlugosz A, Lindberg G, Karling P, Ohlsson B, Gazouli M, Nardone G, Cuomo R, Usai-Satta P, Galeazzi F, Neri M, Portincasa P, Bellini M, Barbara G, Camilleri M, Locke GR, Talley NJ, D'Amato M, Ackerman MJ, Farrugia G (2014) Loss-of-function of the voltage-gated sodium channel NaV1.5 (channelopathies) in patients with irritable bowel syndrome. Gastroenterology 146:1659–1668. https://doi.org/10.1053/j.gastro.2014.02.054
31. Strege PR, Mazzone A, Bernard CE, Neshatian L, Gibbons SJ, Saito YA, Tester DJ, Calvert ML, Mayer EA, Chang L, Ackerman MJ, Beyder A, Farrugia G (2018) Irritable bowel syndrome patients have SCN5A channelopathies that lead to decreased NaV1.5 current and mechanosensitivity. Am J Physiol Gastrointest Liver Physiol 314:G494–G503. https://doi.org/10.1152/ajpgi.00016.2017
32. Bennett PB, Yazawa K, Makita N, George AL (1995) Molecular mechanism for an inherited cardiac arrhythmia. Nature 376:683–685. https://doi.org/10.1038/376683a0
33. Kim GH (2013) MicroRNA regulation of cardiac conduction and arrhythmias. Transl Res 161:381–392. https://doi.org/10.1016/j.trsl.2012.12.004
34. Zhou Q, Costinean S, Croce CM, Brasier AR, Merwat S, Larson SA, Basra S, Verne GN (2015) MicroRNA 29 targets nuclear factor-κ B-repressing factor and Claudin 1 to increase intestinal permeability. Gastroenterology 148:158–169.e8. https://doi.org/10.1053/j.gastro.2014.09.037
35. Benard T, Bouchoucha M, Dupres M, Cugnenc PH (1997) In vitro analysis of rat intestinal wall movements at rest and during propagated contraction: a new method. Am J Phys 273:G776–G784. https://doi.org/10.1152/ajpgi.1997.273.4.G776
36. Hennig GW, Costa M, Chen BN, Brookes SJ (1999) Quantitative analysis of peristalsis in the Guinea-pig small intestine using spatio-temporal maps. J Physiol 517(Pt 2):575–590. https://doi.org/10.1111/j.1469-7793.1999.0575t.x
37. Lammers WJ, Dhanasekaran S, Slack JR, Stephen B (2001) Two-dimensional high-resolution motility mapping in the isolated feline duodenum: methodology and initial results. Neurogastroenterol Motil 13:309–323. https://doi.org/10.1046/j.1365-2982.2001.00270.x
38. D'Antona G, Hennig GW, Costa M, Humphreys CM, Brookes SJ (2001) Analysis of motor patterns in the isolated Guinea-pig large intestine by spatio-temporal maps. Neurogastroenterol Motil 13:483–492. https://doi.org/10.1046/j.1365-2982.2001.00282.x

39. Janssen PWM, Lentle RG, Chambers P, Reynolds GW, De Loubens C, Hulls CM (2014) Spatiotemporal organization of standing postprandial contractions in the distal ileum of the anesthetized pig. Neurogastroenterol Motil 26:1651–1662. https://doi.org/10.1111/nmo.12447
40. Lentle RG, Hulls CM (2018) Quantifying patterns of smooth muscle motility in the gut and other organs with new techniques of video spatiotemporal mapping. Front Physiol 9:338. https://doi.org/10.3389/fphys.2018.00338
41. Du P, Mazzone A, Calder S, Qian A, Gibbons SJ, Farrugia G, Beyder A (2019) A method for multi-day tracking of gastrointestinal smooth muscle contractile patterns in Organotypic culture. Annu Int Conf IEEE Eng Med Biol Soc 2019:4791–4794. https://doi.org/10.1109/EMBC.2019.8857365
42. Ward SM, Harney SC, Bayguinov JR, McLaren GJ, Sanders KM (1997) Development of electrical rhythmicity in the murine gastrointestinal tract is specifically encoded in the tunica muscularis. J Physiol 505(Pt 1):241–258. https://doi.org/10.1111/j.1469-7793.1997.241bc.x
43. Mazzone A, Eisenman ST, Strege PR, Yao Z, Ordog T, Gibbons SJ, Farrugia G (2012) Inhibition of cell proliferation by a selective inhibitor of the Ca(2+)-activated Cl(−) channel, Ano1. Biochem Biophys Res Commun 427: 248–253. https://doi.org/10.1016/j.bbrc.2012.09.022
44. Singh RD, Gibbons SJ, Saravanaperumal SA, Du P, Hennig GW, Eisenman ST, Mazzone A, Hayashi Y, Cao C, Stoltz GJ, Ordog T, Rock JR, Harfe BD, Szurszewski JH, Farrugia G (2014) Ano1, a Ca2+−activated Cl- channel, coordinates contractility in mouse intestine by Ca2+ transient coordination between interstitial cells of Cajal. J Physiol 592:4051–4068. https://doi.org/10.1113/jphysiol.2014.277152
45. Angeli TR, Du P, Paskaranandavadivel N, Janssen PWM, Beyder A, Lentle RG, Bissett IP, Cheng LK, O'Grady G (2013) The bioelectrical basis and validity of gastrointestinal extracellular slow wave recordings. J Physiol 591: 4567–4579. https://doi.org/10.1113/jphysiol.2013.254292
46. Kuruppu S, Cheng LK, Nielsen PMF, Gamage TPB, Avci R, Angeli TR, Paskaranandavadivel N (2022) High-resolution spatiotemporal quantification of intestinal motility with free-form deformation. IEEE Trans Biomed Eng 69:2077–2086. https://doi.org/10.1109/TBME.2021.3135855
47. Du P, Mazzone A, Calder S, Qian A, Gibbons SJ, Farrugia G, Beyder A (2019) A method for multi-day tracking of gastrointestinal smooth muscle contractile patterns in Organotypic culture. In: 41st international engineering in medicine and biology conference. Berlin
48. National Research Council (2011) Guide for the care and use of laboratory animals. National Academies Press, Washington, D.C.
49. Lentle RG, De Loubens C, Hulls C, Janssen PWM, Golding MD, Chambers JP (2012) A comparison of the organization of longitudinal and circular contractions during pendular and segmental activity in the duodenum of the rat and Guinea pig. Neurogastroenterol Motil 24: 686–95, e298. https://doi.org/10.1111/j.1365-2982.2012.01923.x
50. Lentle RG, Janssen PWM, Asvarujanon P, Chambers P, Stafford KJ, Hemar Y (2007) High definition mapping of circular and longitudinal motility in the terminal ileum of the brushtail possum Trichosurus vulpecula with watery and viscous perfusates. J Comp Physiol B 177:543–556. https://doi.org/10.1007/s00360-007-0153-8

# Chapter 30

# Optimized Method of 3D Scaffold Seeding, Cell Cultivation, and Monitoring Cell Status for Bone Tissue Engineering

**Adrian Krolinski, Kai Sommer, Johanna Wiesner, Oliver Friedrich, and Martin Vielreicher**

## Abstract

The cultivation of cells in 3D systems is commonly regarded to be more physiological than in 2D as it comes much closer to the natural situation in tissues in many different aspects. However, 3D cell culture is much more complex. Cells within the pores of a printed 3D scaffold face a special situation concerning cell-material interaction and cell adhesion, cell proliferation, and supply of medium and oxygen into the core of the scaffolds. Biological assays (for cell proliferation, viability, and activity) have been validated primarily for 2D cell cultures and need to be adapted for 3D cultures. Likewise, in imaging, a number of points need to be taken into account in order to get a clear picture of the cells in 3D scaffolds, preferably with the method of multiphoton microscopy. Here, we describe a method for pretreatment and cell seeding of porous inorganic composite scaffolds (α-TCP/HA) for bone tissue engineering and for cultivation of the cell-scaffold constructs. The analytical methods described are the cell proliferation assay and the ALP activity assay. A step-by-step protocol is provided here that safely tackles typical difficulties that arise with this 3D cell-scaffold setting. In addition, MPM imaging of cells is described both with and without labeling. The combination of biochemical assays and imaging provides valuable insights into the possibilities of analysis with this 3D cell-scaffold system.

**Key words** 3D scaffolds, Cell seeding, Cell adhesion, Cell proliferation, Scaffold population, Osteogenic differentiation, Collagen-I formation, Quantitative assays, Multiphoton imaging, Second harmonic generation (SHG)

## 1 Introduction

In this chapter, a method for the effective population of porous 3D scaffolds [1] with osteogenic cells and subsequent cell culture with monitoring of cell behavior in this 3D setting is described.

The behavior of cells in 3D constructs is complex in several ways. The 3D cell arrangement represents a more physiologic situation compared to 2D, as more cell-cell contacts and a more complex spatial architecture forms in 3D. In addition, much more compact cell bodies and a more complex ECM, compared to a

Oliver Friedrich and Daniel F. Gilbert (eds.), *Cell Viability Assays: Methods and Protocols*, Methods in Molecular Biology, vol. 2644, https://doi.org/10.1007/978-1-0716-3052-5_30,

planar surface, are formed [2–4]. In addition, cell-cell communication differs from that in 2D cell cultures, where cells are anchored in a layer next to each other on a substrate [5].

However, unlike in the more established 2D cell culture, a number of difficulties arise in 3D with porous scaffolds. Already during cell seeding, it is vital that the inner surfaces of the scaffold (pores) are effectively and stably colonized. The cells within the interior are usually not visible or, in cases where they are visible, far less easily delineated compared to a 2D culture in the absence of external staining. With increasing cell density, diffusion-limited undersupply of nutrients and oxygen as well as waste removal can quickly set in, resulting in hypoxic scaffold cores. To effectively counteract such deficiencies, sufficient amount of medium (relative to amount of seeded cells) and media exchanges are required as well as precautions on the scaffold level (size and interconnectivity of pores, open porosity to all sides of the scaffold). Many biochemical assays for measuring cell proliferation, viability, and activity have been established and validated primarily for 2D cell cultures, but they need to be adapted for 3D cultures.

Biochemical assays provide bulk measurements that average over all cells in a population. Individual cell behavior is primarily accessible via microscopy, but usually with the limitation of the need for invasive staining and limited depth resolution. These issues can be partially resolved by multiphoton microscopy (MPM). However, for global statements, biochemical assays have an indispensable value. To render tissue formation inside a porous scaffold structure accessible for analysis, sectioning, e.g., by microtome sectioning or thin sectioning after embedding, is often inevitable.

Here, we describe a method for pretreatment and colonization of porous inorganic composite scaffolds for bone tissue engineering (α-TCP/HA) and cultivation with osteogenic differentiation [6] of the cells. The scaffold material mimics the behavior of natural bone, can be 3D-printed accurately, and exhibits suitable and well-defined degradation behavior. The protocol works well with mouse MC3T3-E1 pre-osteoblasts, a cell line established in bone tissue engineering [7, 8] that reliably differentiates into osteoblasts, but other osteogenic cell lines could be used with appropriate modifications. The analytical methods described are the cell proliferation assay (proliferation inside the scaffold) and the ALP activity assay (quantitative information on osteogenic differentiation inside the scaffold). However, other assays (e.g., ELISA assays) could be used as well, for which the procedure hardly needs to be modified. A step-by-step protocol is provided here that safely tackles typical difficulties that arise with this 3D cell-scaffold setting. In addition, MPM imaging of cells is described both with and without labeling (label-free MPM). The combination of biochemical assays and imaging provides valuable insights into the possibilities of analysis in 3D tissue engineering with this cell-scaffold system.

## 2 Materials

Prepare all solutions using ultrapure water (prepared by purifying deionized water), and analytical grade reagents. Prepare and store all reagents at room temperature unless indicated otherwise. Diligently follow all waste disposal regulations when disposing of waste materials.

### 2.1 General Equipment

1. Cell culture: biological safety cabinet, $CO_2$ incubator, water bath, centrifuges, pipettes with tips, multi-well plates (12-well, 6-well, 96-well, nonadhesive culture plates for cell seeding of scaffolds), cell counting chamber.
2. Scaffold handling: tweezers, UV lamp, filter paper.
3. Multimode plate reader (absorbance and luminescence).
4. Liquid nitrogen.
5. Glass-bottom dishes (diameter, 35 mm; glass thickness, ≤170 μm) suitable for high-resolution microscopy (low autofluorescence, high spectral transmission, flatness).
6. Multiphoton microscope: inverted microscope, femtosecond-pulsed laser with dispersion pre-compensation and tunable wavelengths in the near-infrared (NIR) range with high repetition rate (e.g., 80 MHz), high-speed XY-scanner, XYZ-stage with micromechanical movement, high-sensitivity PMT detectors mounted in non-descanned configuration close to the back aperture of the objective [9–12].
7. Objective: 40× water immersion LD C-apochromat (e.g., 40× with high numerical aperture of 1.1 and working distance of 0.62 mm) with transparency in the UV-vis and NIR range.
8. Optical filters: for label-free imaging (SHG/AF): longpass dichroic mirror (T495lpxr) with bandpass filters 405/20 (for SHG) and 525/50 (for AF); for fluorescence imaging, longpass dichroic mirror (DC560) with bandpass filters 620/60 (for PF546-phalloidin/DAPI) and 525/50 (for labeled hydroxyapatite).

### 2.2 Solutions and Reagents

1. Growth medium: MEM-alpha basal medium (without nucleosides, with phenol red) supplemented with 10% fetal bovine serum (FBS), 1% penicillin/streptomycin (10,000 U/mL), and 1% L-glutamine (200 mM).
2. Osteogenic medium: Normal growth medium supplemented with 50 μg/mL ascorbic acid, 10 mM β-glycerophosphate, and 10 nM dexamethasone [13].
3. Cell preparation solutions: PBS (phosphate-buffered saline, pH 7.4) without $Ca^{2+}$ and $Mg^{2+}$, 0.25% trypsin-EDTA solution.

4. Sterilization solution: Isopropanol (70%).
5. Cell proliferation assay: Cell counting kit-8 (CCK-8, colorimetric detection).
6. Cell lysis buffer: PBS with 0.5% Triton® X-100, 25 mM TRIS, 1 mM $MgCl_2$, and 0.1 mM $ZnCl_2$.
7. Alkaline phosphatase (ALP) activity assay with chemiluminometric detection.
8. Protein concentration measurement: Bradford assay, albumin fraction V (protein standards).
9. Cell fixation solution: Histofix (PBS with 4% paraformaldehyde).
10. Cell permeabilization buffer: PBS with 0.1% Triton X-100.

### 2.3 Labels

1. Mineralization assay. Kit used for fluorescence-labeling of hydroxyapatite (HA) particles.
2. Nuclear dye: DAPI (4′, 6-diamidino-2-phenylindole).
3. Actin cytoskeleton dye: PromoFluor546 (PF546)-phalloidin.

### 2.4 Software

1. Analysis of plate reader data: Analysis software.
2. Image acquisition software (instrument specific).
3. Image analysis and processing: Fiji software (NIH, USA).

### 2.5 Biological Material

Mouse pre-osteoblast cell line MC3T3-E1 (parental cell line; [14], (*see* **Note 1**).

### 2.6 Scaffolds

3D-printed cylindrical, porous α-TCP/HA scaffolds (diameter, 9.4 mm; height, 3.0 mm) consisting of synthetic calcium phosphates, mainly α-tricalcium phosphate (α-TCP), and microcrystalline, calcium-deficient hydroxyapatite (Fig. 1a) [15–17] (*see* **Notes 2–5**).

## 3 Methods

Carry out all procedures at room temperature unless otherwise specified. Figure 2 illustrates the major steps described in this protocol. Take into account a suitable number of scaffolds and cells required for optimizations of scaffold pretreatment, cell seeding, and later systematic studies during planning of experiments. Often, the number of variables to be tested is larger than expected.

### 3.1 Pretreatment of Scaffolds

1. Sterilize the scaffolds prior to cell seeding for 15 min in 70% isopropanol. An UV treatment (UV lamp in sterile hood for 5 min) can be performed in addition. Wash the scaffolds 3× with PBS (5 min each) before equilibration in growth medium overnight.

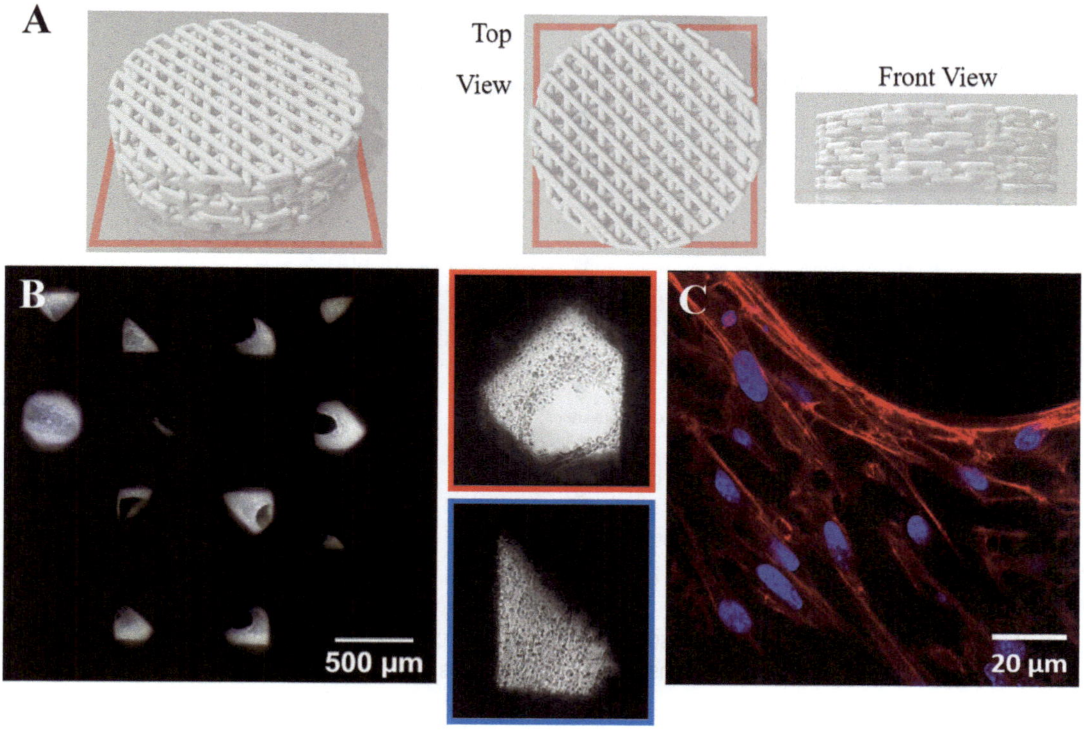


**Fig. 1** Scaffold structure and cell population. (**a**) 3D-printed α-TCP/HA scaffold illustrated as 3D model. (**b**) Transmission microscopy images taken after 14 days of cultivation in normal growth medium. Overview of visible pore structures filled with adhered MC3T3-E1 cells (bright regions). Zoomed images: pores partially (red frame) and fully populated with cells (blue frame). Microscope: Nikon Eclipse TS100 (camera, Canon EOS M10; magnification, 4×). (**c**) Cell distribution and morphology close to pore center visualized after fluorescence-labeling with DAPI (nuclei, blue) and PF546-phalloidin (cytoskeleton, red)

2. Air-dry the scaffolds completely by soaking them with pre-sterilized (treated with 70% isopropanol) and dried filter paper. Keep scaffolds in air stream of the sterile hood for complete drying (at least 45 min) (*see* **Note 6**).
3. Depending on the scaffold material used, coating may be required in case cell adhesion is not sufficient. For this, the scaffold should be incubated in a solution of suitable extracellular matrix (ECM) components (*see* **Note** 7).

### *3.2 Cell Seeding and Cultivation*

1. Culture MC3T3-E1 cells in a $CO_2$ incubator at 37 °C, 5% $CO_2$, and 95% relative humidity (cells have a doubling time of ~15 h). Replace growth medium every 2–3 days. Verify the cells as mycoplasma-free, e.g., by PCR testing.
2. Prepare a cell suspension with a concentration of 6,250 cells/μL from cells in proliferative phase, and use 80 μL (~500,000 cells) for scaffold seeding (*see* **Note 8**).

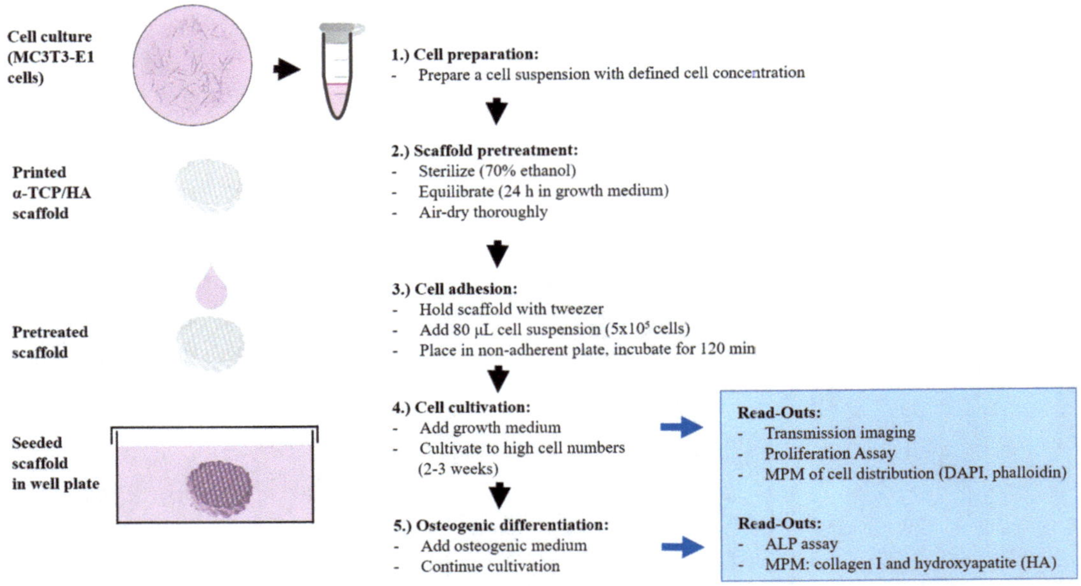


**Fig. 2** Flow chart illustrating the steps of scaffold pretreatment, cell seeding, and cultivation of the constructs. Blue box: read-outs to monitor cell expansion and osteogenesis

3. Drop 40 μL on one side of the scaffold, followed by the other half onto the opposite side. Hold the scaffold with tweezers to avoid contact of the cell suspension to the well bottom (*see* **Note 9**).
4. Gently place the scaffold in a 6-well plate (*see* **Note 10**), and incubate for an adhesion time of 120 min in the $CO_2$ incubator (*see* **Note 11**).
5. At the end of the adhesion time, immerse the scaffolds in 4 mL growth medium and incubate further with media exchanges every 2–3 days (*see* **Note 12**). Use seeding efficiency analysis to quantify and optimize the cell seeding process (*see* **Note 13**).
6. Analyze scaffold population with cells at desired time points with transmission microscopy (Fig. 1b) and/or other suitable biological assays (*see* **Note 14**).
7. At a suitable cell density (e.g., after 2–3 weeks), switch from growth medium to osteogenic medium. Keep some scaffolds in growth medium as references. Replace medium every 2–3 days.

### 3.3 Cell Proliferation Assay

Test cell proliferation using the cell counting kit-8 (CCK-8) assay (*see* **Notes 15–17**). Prolong the duration of incubation and washing steps (compared to normal 2D cell cultures) for them to work effectively (*see* **Note 18**). Test for interference of the scaffold material with the assay using non-seeded scaffolds as negative controls (*see* **Note 19**).

1. Perform assay according to manufacturer's instructions (*see* **Note 20**).
2. Fully submerge the α-TCP/HA scaffolds in 2 mL growth medium in a 12-well plate.
3. Add 10% (v/v) of CCK-8 reagent to the wells, and incubate for 2 h ($CO_2$ incubator). Manually mix the medium every 30 min with a pipette to ensure homogenous distribution of the reagent to all cells of the scaffold.
4. Transfer 110 μL of CCK-8-containing medium to a 96-well plate, and measure the absorbance (e.g., at 450 nm) using a plate reader. Time points for detection are every 3–4 days for a total of 2–4 weeks.
5. Normalize absorbance values to the initial measurement and plot the results against days after seeding (*see* **Note 21**).

### 3.4 Alkaline Phosphatase Activity Assay

Prolong the duration of incubation and washing steps (compared to 2D cell cultures) for them to work effectively. Test for interference of the scaffold material with the assay using unseeded scaffolds as negative controls (*see* **Notes 18** and **19**). Measure ALP activity according to manufacturer's recommendations (*see* **Notes 22** and **23**).

1. Collect cell-seeded scaffolds after switch to osteogenic medium in regular intervals to track the time course of osteogenesis (e.g., on days 0, 5, 10, 15, and 21). Use scaffolds cultivated in normal growth medium as reference.
2. Incubate scaffolds for 1 h in serum-free medium to remove serum, wash an additional three times with PBS (5 min).
3. Incubate in 2 mL of lysis buffer for 20 min on ice. To ensure full lysis, shock-freeze (liquid nitrogen) and thaw samples (three cycles).
4. Centrifuge at high speed, collect supernatant (lysate), and keep on ice or freeze for later use.
5. Mix 20 μL of the lysate (triplicates) with 100 μL of the ALP kit reagent (CSPD) in a nontransparent 96-well, and store for 20 min at room temperature.
6. Measure chemiluminescence with the plate reader according to manufacturer's recommendations. Use lysis buffer mixed with ALP reagent as a negative control.
7. In order to normalize ALP activity to total protein content, determine protein concentration in lysates in a 96-well plate using the Bradford assay according to manufacturer's recommendations. Measure the absorbance of the lysate samples (e.g., 3 s at 595 nm).

8. For reference, generate a standard curve using serial dilutions of albumin fraction V in lysis buffer (0 to 200 μg/mL in 20 μg/mL steps).
9. Relate luminometry counts to protein concentration values.
10. For data presentation, relate ALP activities in the constructs (osteogenic medium) to the initial measurements (at day 0) and plot together with reference samples (kept in normal growth medium).

### 3.5 Cell Preparation for Microscopy

For imaging of fluorescence-labeled samples, collect the cell-scaffold constructs at selected time points after switch to osteogenic medium.

1. Wash the constructs with PBS 3× for 5 min in a 6-well plate, and fix the cells for 10 min (Histofix).
2. After three washing steps (PBS, 5 min), permeabilize the cells in cell permeabilization buffer for 5 min.
3. DAPI/PF546-phalloidin staining: Wash the scaffolds (PBS, 3× 5 min), submerge in staining solution (DAPI and PF546-phalloidin in PBS), and incubate light protected for 25 min. Following this, again wash samples with PBS (3× 5 min).
4. In the case of HA staining, incubate the samples (after fixation, permeabilization, and washing steps) in a solution containing the HA-specific dye for 30 min. Wash samples with PBS (3× for 5 min).
5. For imaging of non-labeled samples (SHG/AF imaging) at desired time points, wash the scaffolds once with sterile PBS, transfer to glass-bottom dish, add PBS, and place the lid on the plate to keep sterile conditions during microscopy.

### 3.6 Multiphoton Microscopy

1. Place constructs on glass-bottom dish, position on the microscope, and bring cells into focus (*see* **Notes 24** and **25**).
2. Image cell-scaffold interfaces and cell assemblies within the scaffold pores using the image acquisition software (*see* **Note 26**).
3. For DAPI/PF546-phalloidin-labeled constructs, set excitation wavelength to 775 nm. Separate emission signals using a dichroic mirror at 560 nm. Block fluorescence from PF546 using a 620/60 bandpass filter (Fig. 1c) (*see* **Note 27**).
4. For imaging of collagen-I formation together with cells, set laser excitation to 810 nm. Ensure separation of emission signals using a 495 nm dichroic mirror. Isolate the SHG signal from collagen-I (reflected) using a 405/20 bandpass filter and the autofluorescence (AF) signal from cells using a 525/50 bandpass filter (*see* **Note 28**).

5. For HA-labeled constructs set excitation wavelength to 775 nm. Separate emission signals using a dichroic mirror at 560 nm. Co-localize collagen-I fiber networks (SHG imaging) with fluorescence-labeled HA particles imaging to analyze the state of ECM formation and mineralization.
6. Imaging with 4× line averaging and 200 Hz line frequency will yield good results (*see* **Note 29**).

## 4 Notes

1. We use the parental cell line of MC3T3-E1 [20] (#ACC-210, DSMZ, Germany). The cells were established from the calvaria of an embryo/fetus C57BL/6 mouse. These are adherent cells with fibroblast morphology that were described to differentiate to osteoblasts. MC3T3-E1 are popular in studies of osteogenesis, because of their clear developmental sequence of proliferation and differentiation [21]. Exposed to osteogenic medium, proliferation rate continuously decreases, whereas collagen-I accumulates in the ECM. ALP activity is low. From around day 9 to 16, collagen-I increases as well as ALP activity. The final stage of differentiation is characterized by matrix mineralization with formation of HA [22].
2. This material composition mimics natural bone. The α-TCP/HA has a clear and relatively fast degradation profile. The printed scaffolds exhibit an open porosity to all sides, which is required to warrant optimum supply with nutrients and oxygen in static culture that supports cells to proliferate to high densities.
3. The scaffolds are manufactured by 3D extrusion printing by Innotere GmbH (Radebeul, Germany, #211CC4) using a paste-like calcium phosphate cement which is well printable. The printing is followed by various washing and air-drying steps, without need for heating/sintering [15–18].
4. The α-TCP/HA scaffolds have a cylindrical shape (volume: ~208 $mm^3$). The formation occurs layer-by-layer (with a total of twelve layers). The strands are ~330 μm in diameter (strand gaps: ~390 μm). Strand orientation changes from layer to layer in a 45° pattern. This pattern was selected as it provides more mechanical stability for certain applications compared to a 90° pattern [19]. The scaffolds exhibit a high degree of interconnected porosity (50%) and a high specific surface area favorable for cell attachment. The average pore size is ~300 μm. The pore structure is rhombic and more complex than cubic pores.

5. The exact internal geometry has a significant influence on the stability of the constructs and the diffusion of media components, waste products, and oxygen. Furthermore, the internal geometry impacts the surface-to-volume ratio of the pores (and thus cell spreading) as well as growth direction and observability of cells in microscopy.
6. The drying step is very important and must be completed in order to keep the added cell suspension within the interior of the scaffold by capillary forces.
7. Coating with ECM components could enhance cell-material interaction and, thus, cell adhesion. The solution composition and concentration and the duration of incubation need to be determined empirically for a specific ECM component (e.g., gelatin, collagens, GelMA, fibronectin) or, alternatively, poly-L-lysine. An alternative could be to subject scaffolds to plasma treatment (e.g., if scaffolds initially are too hydrophobic at the surface).
8. For the determination of the required cell number and suspension volume, a CAD model of the printed scaffold is very helpful, but at least some basic information on scaffold volume and porosity is necessary. The empty scaffold volume results from the overall scaffold volume (208 $mm^3$) and porosity (50%). Using 80% of that volume was helpful in our experiments to determine the cell suspension volume (80 μL) that fills the empty scaffold volume. The required total cell number needs to be determined experimentally. Transmission imaging may provide information on cultivation duration, after which visible cell complexes can be observed in the scaffold pores. In our work, we were interested to start with a maximum cell adhesion density for fast population of the pores.
9. This step is important to avoid overfilling of the scaffold and contact of the added cell suspension to the bottom of the plate that may lead to leaking out of suspension from the scaffold. In our hands, the described procedure helped to keep the cells inside the scaffold and thereby to increase seeding efficiency.
10. It is also recommended to use nonadhesive culture plates as these are hydrophobic and minimize fluid leakage, thus supporting the adhesion process. Such plates will also decrease the tendency of cells to leave the scaffold and adhere to the plate bottom during incubation in growth medium, which could lead to increased duration of scaffold population.
11. The adhesion time is very important, because it allows cells to stabilize their initial contacts to the scaffold surface, so that cells do not detach and get rinsed out again by subsequent addition of medium. We had good experience with an extended adhesion time of 100–180 min.

12. We recommend to flip scaffold orientation regularly as the bottom side is less well supplied with nutrients compared to the top side, which is oriented towards the free medium.
13. Determination of seeding efficiency: After 24 h post-seeding, cells that were not adherent to the scaffold were counted. The seeding efficiency was calculated as the number of cells adherent to the scaffolds (total minus counted free number of cells) divided by the total number of cells (in %).
14. It is important to regularly check with transmission light microscopy, if scaffold pores are still open. Significant diffusion limitations will occur if pores are overpopulated with too many cells, which will lead to hypoxic conditions and cell death in the scaffold core.
15. An example for tests of intracellular enzyme activity as measure of cell proliferation activity is the CCK-8 assay. This assay is based on the observation that the amount of produced WST-8 formazan (reduced from WST-8 by cellular dehydrogenase activity) is directly proportional to the number of proliferative cells and thus, is a direct measure of cell proliferation. The assay does not require cell lysis. According to the manufacturer, the assay can be used repeatedly with the same sample to measure the development of proliferative cell numbers (kit components noncytotoxic). We, however, have observed that repeatedly using the assay has a negative effect on the proliferation of the cells.
16. Possible additional assays: The supernatant from a cell-seeded scaffold can be used as well and is suitable for repeated monitoring of osteogenesis-specific soluble markers, e.g., with ELISA testing [23]. Biological assays are bulk methods and provide a statement about the sum of all cells.
17. For a rough estimate of cell quantity, transmission microscopy can be performed. To enhance contrast and to enable differentiation and counting of cells, we recommend staining of the cell nuclei with DAPI. As this requires fixation and staining, a construct cultured in parallel must be used. In our experiments with various scaffold types, the cells typically arranged themselves in a concentric manner within the pores over time.
18. For assays with 3D scaffolds, it is important to understand that a high number of cells is present in a very small space, and the diffusion of substances to them may be limited; therefore, increasing the duration of incubations steps is necessary.
19. It should be considered that the scaffold material may interact with the biological assay, so that appropriate control experiments are necessary to separate out such influences.

20. Manufacturer's manuals recommend establishing a standard curve with defined cell numbers (e.g., ranging from 0 to 40,000 cells/well of a 96-well plate), cultured as 2D layer. In our view, such a reference has only limited informative value for cells cultured in 3D scaffolds. In 2D cell cultures, at low cell density, all cells proliferate until a denser cell layer is formed. In this situation, the proliferation assay provides information about the total cell number. In a porous 3D scaffold after initial seeding, however, cells begin to line the pores, where they arrange densely in 3D formations with a more limited nutrient supply compared to a 2D cell layer exposed to a large volume of medium. Our microscopic observations showed that the pores filled only relatively slowly, which was in accordance with the absorbance values in the assay being constantly lower than expected according to a standard curve. Our conclusion is that only a small proportion of the cells, likely those close to the pore centers, will actively proliferate.
21. In our experiments, cell proliferation initially increases until a maximum is reached after ~14 days (1.8-fold increase compared to initial value at day 4). At later time points, cell proliferation decreased again (to finally 0.8-fold at day 25). The essential information can be derived from the measured values and the characteristic curve progression for the purpose of comparing equally cell-seeded scaffolds cultured under variable conditions. More precise statements on the number of proliferating cells are not possible due to the lack of comparability with the standard measurements as proposed by the manufacturer.
22. Alkaline phosphatase (ALP) has the physiological role of dephosphorylating compounds. In bone, ALP is found on the surface of osteoblasts and reflects the biosynthetic activity of these bone-forming cells. Bone ALP activity has been shown to be a sensitive and reliable indicator of bone metabolism and is commonly used to monitor the osteogenic differentiation of cells like pre-osteoblasts and MSCs.
23. CSPD is a chemiluminescent substrate for ALP that enables detection with very high sensitivity and signal-to-noise ratio by producing visible light.
24. Finding the focus plane with cells here is a bit more difficult compared to a cell layer on cover slips. Here, knowledge of scaffold geometry, strut size, and surface properties are helpful for orientation. The cell-scaffold interface can be used as a helpful landmark.
25. The fluorescence properties of the scaffold material are a critical parameter and should be tested ahead of imaging, as a significant AF signal may derive from it.

26. MPM enables increased depth penetration (a few hundred μm) and allows imaging deeper inside the scaffold pore system. MPM is, furthermore, a less invasive technique, which significantly reduces photo bleaching and phototoxicity effects.
27. Imbalances between intensities of both dyes can be counteracted by adapting the staining protocol (concentration of dyes and incubation time) or the imaging parameters (laser excitation intensity, detector settings). Co-imaging of DAPI and PF546-phalloidin provides spatial information on cell numbers (DAPI signal form nuclei) and cell morphology (PF546-phalloidin signal from cytoskeleton).
28. Fine-tuning of imaging parameters will be required. SHG may appear as clear signals only at medium to high abundance of the protein. Label-free collagen-I fiber (SHG) and cell detection (AF) in combination can be used to monitor the maturation grade of the artificial bone tissue during culture.
29. Image stacking followed by 3D image reconstructions provides a comprehensive view of the localization of cells and ECM within the scaffold pores. Image stitching to large area mosaics helps to gather an overview of the localization of cell and ECM structures in the respective imaging plane and therefore, it helps to identify typical cell and ECM distributions.

## Acknowledgments

We would like to thank Sascha Heinemann (Innotere GmbH, Radebeul, Germany) for discussions on α-TCP/HA scaffolds.
This work was funded by the Deutsche Forschungsgemeinschaft (grant numbers VI 770/3-1 (M.V.), FR 2993/25-1 (O.F.), and TRR 225 "Biofabrication" (project 326998133, subproject Z02) (O.F.)). We acknowledge ongoing support through the Erlangen Graduate School in Advanced Optical Technologies (SAOT) within the German Excellence Initiative.

## References

1. Bose S, Vahabzadeh S, Bandyopadhyay A (2013) Bone tissue engineering using 3D printing. Mater Today 16:496–504. https://doi.org/10.1016/j.mattod.2013.11.017
2. Pontes Soares C, Midlej V, Weschollek de Oliveira ME, Benchimol M, Costa ML, Mermelstein C (2012) 2D and 3D-organized cardiac cells shows differences in cellular morphology, adhesion junctions, presence of myofibrils and protein expression. PLoS One 7:e38147. https://doi.org/10.1371/journal.pone.0038147
3. Haycock JW (2011) 3D cell culture: a review of current approaches and techniques. In: Haycock JW (ed) Methods in molecular biology. Springer Science+Business Media, pp 1–15. https://doi.org/10.1007/978-1-60761-984-0_1
4. Duval K, Grover H, Han L, Mou Y, Pegoraro AF, Fredberg J, Chen Z (2017) Modeling physiological events in 2D vs 3D cell culture. Physiology 32:266–277. https://doi.org/10.1152/physiol.00036.2016

5. Wang YK, Chen CS (2013) Cell adhesion and mechanical stimulation in the regulation of mesenchymal stem cell differentiation. J Cell Mol Med 17:823–832. https://doi.org/10.1111/jcmm.12061
6. Infante A, Rodríguez CI (2018) Osteogenesis and aging: lessons from mesenchymal stem cells. Stem Cell Res Ther 9:244. https://doi.org/10.1186/s13287-018-0995-x
7. Amini AR, Laurencin CT, Nukavarapu SP (2012) Bone tissue Engineering: recent advances and challenges. Crit Rev Biomed Eng 40:363–408
8. Reichert J, Hutmacher D (2011) Bone tissue engineering. In: Pallua N, Suscheck CV (eds) Tissue engineering: from Lab to Clinic. Springer Berlin Heidelberg, Berlin, Heidelberg, pp 431–456. https://doi.org/10.1007/978-3-642-02824-3_21
9. Helmchen F, Denk W (2005) Deep tissue two-photon microscopy. Nat Methods 2:932–940. https://doi.org/10.1038/NMETH818
10. Ustione A, Piston DW (2011) A simple introduction to multiphoton microscopy. J Microsc 243:221–226. https://doi.org/10.1111/j.1365-2818.2011.03532.x
11. Yeh AT, Gibbs H, Hu J-J, Larson AM (2008) Advances in nonlinear optical microscopy for visualizing dynamic tissue properties in culture. Tissue Eng Part B Rev 14:119–131. https://doi.org/10.1089/teb.2007.0284
12. Zoumi A, Yeh A, Tromberg BJ (2002) Imaging cells and extracellular matrix in vivo by using second-harmonic generation and two-photon excited fluorescence. PNAS 99:11014–11019. https://doi.org/10.1073?pnas.172368799
13. Langenbach F, Handschel J (2013) Effects of dexamethasone, ascorbic acid and β -glycerophosphate on the osteogenic differentiation of stem cells in vitro. Stem Cell Res Ther 4:117. https://doi.org/10.1186/scrt328
14. Kodama S, Amagai H, Sudo Y, Yamamoto H (1981) Establishment of a clonal osteogenic cell line from newborn mouse calvaria. J Oral Biosci 23:899–901
15. Hettich G, Schierjott RA, Epple M, Gbureck U, Heinemann S, Mozaffari-Jovein H, Grupp TM (2019) Calcium phosphate bone graft substitutes with high mechanical load capacity and high degree of interconnecting porosity. Materials (Basel) 12:3471. https://doi.org/10.3390/ma12213471
16. Klein A, Baranowski A, Ritz U, Götz H, Heinemann S, Mattyasovszky S, Rommens PM, Hofmann A (2018) Effect of bone sialoprotein coated three-dimensional printed calcium phosphate scaffolds on primary human osteoblasts. J Biomed Mater Res B Appl Biomater 106:1–11. https://doi.org/10.1002/jbm.b.34073
17. Lode A, Meissner K, Luo Y, Sonntag F, Glorius S, Nies B, Vater C, Despang F, Hanke T, Gelinsky M (2014) Fabrication of porous scaffolds by three-dimensional plotting of a pasty calcium phosphate bone cement under mild conditions. J Tissue Eng Regen Med 9:682–693. https://doi.org/10.1002/term.1563
18. Ahlfeld T, Lode A, Richter RF, Pradel W, Franke A, Rauner M, Stadlinger B, Lauer G, Gelinsky M, Korn P (2021) Toward biofabrication of resorbable implants consisting of a calcium phosphate cement and fibrin—a characterization in vitro and in vivo. Int J Mol Sci 22:1218. https://doi.org/10.3390/ijms22031218
19. Zadpoor AA (2015) Bone tissue regeneration: the role of scaffold geometry. Biomater Sci 3:231–245. https://doi.org/10.1039/c4bm00291a
20. Sudo H, Kodama H, Amagai Y, Yamamoto S, Kasai S (1983) In vitro differentiation and calcification in a new clonal osteogenic cell line derived from newborn mouse Calvaria. J Cell Biol 96:191–198. https://doi.org/10.1083/jcb.96.1.191
21. Peterson WJ, Tachiki KH, Yamaguchi DT (2004) Serial passage of MC3T3-E1 cells down-regulates proliferation during osteogenesis in vitro. Cell Prolif 37:325–336. https://doi.org/10.1111/j.1365-2184.2004.00316.x
22. Chung CY, Iida-klein A, Wyatt LE, Rudkin GH, Ishida K, Yamaguchi DT, Miller TA (1999) Serial passage of MC3T3-E1 cells alters osteoblastic function and responsiveness to transforming growth factor-β1 and bone morphogenetic protein-2. Biochem Biophys Res Commun 265:246–251
23. Choi J, Lee B-H, Song K, Park R, Kim I, Sohn K, Jo J-S, Ryoo H-M (1996) Expression patterns of bone-related proteins during osteoblastic differentiation in MC3T3-E1 cells. J Cell Biochem 61:609–618. https://doi.org/10.1002/(SICI)1097-4644(19960616)61:4%3C609::AID-JCB15%3E3.0.CO;2-A

# INDEX




Oliver Friedrich and Daniel F. Gilbert (eds.), *Cell Viability Assays: Methods and Protocols*, Methods in Molecular Biology, vol. 2644, https://doi.org/10.1007/978-1-0716-3052-5,

## C

## D

## E

## F

## G

## J



## K

## N

## Q



## R

## S

## U



## V

MIX
Papier aus verantwortungsvollen Quellen
Paper from responsible sources
FSC® C105338

If you have any concerns about our products,
you can contact us on
**ProductSafety@springernature.com**

In case Publisher is established outside the EU,
the EU authorized representative is:
**Springer Nature Customer Service Center GmbH**
**Europaplatz 3, 69115 Heidelberg, Germany**

Printed by Libri Plureos GmbH
in Hamburg, Germany